Lecture Notes in Computer Science   6733

Commenced Publication in 1973
Founding and Former Series Editors:
Gerhard Goos, Juris Hartmanis, and Jan van Leeuwen

Antoine Joux (Ed.)

Fast
Software Encryption

18th International Workshop, FSE 2011
Lyngby, Denmark, February 13-16, 2011
Revised Selected Papers

 Springer

Volume Editor

Antoine Joux
DGA and Université de Versailles Saint-Quentin-en-Yvelines
45 avenue des Etats-Unis, 78035 Versailles Cedex, France
E-mail: antoine.joux@m4x.org

ISSN 0302-9743   e-ISSN 1611-3349
ISBN 978-3-642-21701-2   e-ISBN 978-3-642-21702-9
DOI 10.1007/978-3-642-21702-9
Springer Heidelberg Dordrecht London New York

Library of Congress Control Number: 2011929384

CR Subject Classification (1998): E.3, K.6.5, D.4.6, C.2, J.1, G.2.1

LNCS Sublibrary: SL 4 – Security and Cryptology

Typesetting: Camera-ready by author, data conversion by Scientific Publishing Services, Chennai, India

Printed on acid-free paper

Springer is part of Springer Science+Business Media (www.springer.com)

Preface

You are holding the proceedings of FSE 2011, the 18th International Workshop on Fast Software Encryption. This workshop was organized in cooperation with the International Association for Cryptologic Research. It was held in Lyngby, Denmark, during February 13–16, 2011.

The FSE 2011 Program Committee (PC) consisted of 21 members, listed on the next page. There were 106 submissions and 22 were selected to appear in this volume. Each submission was assigned to at least three PC members and reviewed anonymously. During the review process, the PC members were assisted by 66 external reviewers. Once the reviews were available, the committee discussed the papers in depth using the EasyChair conference management system. At the workshop, papers were made available to the audience in electronic form. After the conference, the authors of accepted papers were given six weeks to prepare the final versions included in these proceedings. The revised papers were not reviewed again and their authors bear the responsibility for their content.

In addition to the papers included in this volume, the conference also featured a Rump Session. Vincent Rijmen served as the chair of the Rump Session. The conference also had the pleasure of hearing invited talks by Willy Meier and Ivan Damgård. An invited paper corresponding to Willy Meier's talk is included in the proceedings. Ivan Damgård considered that the material presented in his talk was already published and thus did not wish to send an invited paper.

The PC decided to give the Best Paper Award to Takanori Isobe for his paper titled "A Single-Key Attack on the Full GOST Block Cipher." In addition, the committee selected another paper for invitation to the *Journal of Cryptology*: "Cryptanalysis of PRESENT-Like Ciphers with Secret S-Boxes" by Julia Borghoff, Lars Ramkilde Knudsen, Gregor Leander and Søren Thomsen.

I wish to thank all the people who contributed to this conference. First, all the authors who submitted their work. Second the PC members and their external reviewers for the thorough job they did while reading and commenting the submissions. Without them, selecting the papers for this conference would have been an impossible task. I thank Andrei Voronkov for his review system EasyChair. Once again, I was very glad to have access to his magic tools that helped me assemble this volume. I would also like to thank the General Chairs, Lars Knudsen and Gregor Leander for making this conference possible.

Being the Program Chair for FSE 2011 was a great honor and I may only hope that the readers of these proceedings find them as interesting as I found the task of selecting their content.

April 2011   Antoine Joux

Conference Organization

General Chairs

Lars R. Knudsen and Gregor Leander, Technical University of Denmark

Program Chair

Antoine Joux, DGA and Université de Versailles Saint-Quentin-en-Yvelines

Program Committee

John Black	University of Colorado, USA
Anne Canteaut	INRIA, France
Pierre-Alain Fouque	ENS, Paris, France
Helena Handschuh	Intrinsic-ID, USA and KU Leuven, Belgium
Tetsu Iwata	Nagoya University, Japan
Pascal Junod	HEIG-VD, Switzerland
Stefan Lucks	Bauhaus University Weimar, Germany
David M'Raihi	Apple Inc., USA
Marine Minier	CITI INSA-Lyon - ARES INRIA Project, France
Shiho Moriai	Sony, Japan
Maria Naya-Plasencia	FHNW, Switzerland
Kaisa Nyberg	Helsinki University of Technology, Finland
Elisabeth Oswald	University of Bristol, UK
Thomas Peyrin	NTU, Singapore
Christian Rechberger	IAIK, Graz University of Technology, Austria
Greg Rose	Qualcomm, USA
Martijn Stam	EPFL, Switzerland
Søren Thomsen	Technical University of Denmark, Denmark
Marion Videau	LORIA and Université Henri Poincare, Nancy 1, France
Ralf-Philipp Weinmann	Université du Luxembourg, Luxembourg

External Reviewers

Aumasson, Jean-Philippe
Avoine, Gildas
Benadjila, Ryad
Blazy, Olivier
Bos, Joppe
Bouillaguet, Charles

Brumley, Billy
Ciet, Mathieu
De Canniere, Christophe
Degabriele, Jean Paul
Finiasz, Matthieu
Fleischmann, Ewan

Forler, Christian
Fuhr, Thomas
Gauravaram, Praveen
Gorski, Michael
Guo, Jian
Hakala, Risto
Halevi, Shai
Hirose, Shoichi
Icart, Thomas
Isobe, Takanori
Jacobson, David
Jean, Jeremy
Järvinen, Kimmo
Khovratovich, Dmitry
Kindarji, Bruno
Knellwolf, Simon
Knudsen, Lars
Krause, Matthias
Leander, Gregor
Lee, Jooyoung
Lehmann, Anja
Leurent, Gaëtan
Lomne, Victor
Lyubashevsky, Vadim
Macchetti, Marco
Mandal, Avradip
McDonald, Cameron

Meier, Willi
Mennink, Bart
Minematsu, Kazuhiko
Motre, Stephanie
Nguyen, Phuong Ha
Nikolic, Ivica
Ozen, Onur
Pietrzak, Krzysztof
Poschmann, Axel
Preneel, Bart
Priemuth-Schmid, Deike
Reinhard, Jean-René
Röck, Andrea
Saarinen, Markku-Juhani
Sarinay, Juraj
Sarkar, Palash
Sasaki, Yu
Schläffer, Martin
Sekar, Gautham
Sendrier, Nicolas
Seurin, Yannick
Shirai, Taizo
Steinberger, John
Tessaro, Stefano
Weinmann, Ralf-Philipp
Yasuda, Kan
Zenner, Erik

Table of Contents

Stream Ciphers

Hash Functions II

Block Ciphers and Modes

Linear and Differential Cryptanalysis

Hash Functions III

Differential Cryptanalysis of Round-Reduced PRINTCIPHER: Computing Roots of Permutations

Mohamed Ahmed Abdelraheem, Gregor Leander, and Erik Zenner

Technical University of Denmark, DK-2800 Kgs. Lyngby, Denmark
{M.A.Abdelraheem,G.Leander,E.Zenner}@mat.dtu.dk

Abstract. At CHES 2010, the new block cipher PRINTCIPHER was presented. In addition to using an xor round key as is common practice for round-based block ciphers, PRINTCIPHER also uses key-dependent permutations. While this seems to make differential cryptanalysis difficult due to the unknown bit permutations, we show in this paper that this is not the case. We present two differential attacks that successfully break about half of the rounds of PRINTCIPHER, thereby giving the first cryptanalytic result on the cipher.

In addition, one of the attacks is of independent interest, since it uses a mechanism to compute roots of permutations. If an attacker knows the many-round permutation π^r, the algorithm can be used to compute the underlying single-round permutation π. This technique is thus relevant for all iterative ciphers that deploy key-dependent permutations. In the case of PRINTCIPHER, it can be used to show that the linear layer adds little to the security against differential attacks.

Keywords: symmetric cryptography, block cipher, differential cryptanalysis, permutations.

1 Introduction

After the establishment of Rijndael as AES, the need for new block ciphers has greatly diminished. However, given that the future IT-landscape is supposed to be dominated by tiny computing devices such as RFID tags or sensor networks, the need for low cost security has grown substantially. This need opened up the research field of light-weight cryptography. Quite a number of light-weight block ciphers have been proposed in the last couple of years, examples among others are PRESENT [3], HIGHT [7] and KATAN/KTANTAN [4].

PRINTcipher. One recent proposal in this direction is the block cipher PRINT-CIPHER presented at CHES 2010. PRINTCIPHER is an SP-network and comes in two versions, PRINTCIPHER-48 and PRINTCIPHER-96 with block sizes of 48 and 96 bits. PRINTCIPHER is targeted at IC-printing and makes use of the fact that this technology allows to make the circuit implementing the cipher key-dependent. This allows PRINTCIPHER to be implemented with a considerably

A. Joux (Ed.): FSE 2011, LNCS 6733, pp. 1–17, 2011.

smaller circuit compared to other light-weight ciphers. In order to maximize the profit from a key-dependent circuit all round keys in PRINTCIPHER are identical. To increase the size of the key space beyond the block size, the key in PRINTCIPHER consists not only of a (constant) round key xored to the state, but also parts of the linear layer are made key-dependent.

Differential Cryptanalysis. This attack, invented by Biham and Shamir [2], is one of the most powerful and most general attacks on block ciphers known. The main idea is to encrypt pairs of plaintexts and trace the evolution of their difference through the encryption process. As most modern block ciphers are round based, an attacker usually starts by analyzing one round of the cipher with respect to difference propagation and extends to multiple rounds afterwards. Under well established independence assumptions the probability that a plaintext pair with a given difference α leads to a ciphertext pair with difference β can easily be computed by studying single rounds. Thus, differential attacks are most often based on so-called differential characteristics, that is a sequence of intermediate differences for all rounds together with their associated probabilities.

Our Results. In this paper we mount a differential attack on round-reduced versions of PRINTCIPHER. The main technical problem while doing so is that the differential characteristics are key-dependent, more precisely they depend on the (key-dependent) choice of the linear layer. That is to say, without knowing the key, we do not know the best differential characteristics. At first glance, this seems to complicate a differential attack on PRINTCIPHER. There is another way to look at this, though. If the differential characteristics are key-dependent, then conversely, knowing the best differential one might be able to deduce information about the key. In general this dependency might be very complex. However, in the case of PRINTCIPHER we show that given the best differentials, computing the key-dependent linear layer can be reduced to computing roots of permutations in S_{48} or S_{96}. The remaining key bits, that is the constant round key xored to the state, can then be recovered using a standard differential attack, or, at a higher cost, simply by brute force.

Now, computing roots of permutations is a well-studied problem and our attack will profit from known algorithms. However, note that a permutation can have a huge number of roots and this causes two problems. First, this makes algorithms computing all possible roots eventually slow and second, in the case of PRINTCIPHER this means that many possible linear layers are proposed. We explain how both problems can be overcome.

In particular, our results show that making the linear layer of PRINTCIPHER key-dependent adds little to no additional security against differential attacks.

Related Work. PRINTCIPHER is not the first block cipher with key-dependent components. Other well known examples are Khufu [10], the Khufu variation Blowfish [12] and Twofish [13]. Along with those proposals, several attempts to cryptanalyze those block ciphers, see for example [5] for a differential attack on Khufu or Vaudenay's attack on round-reduced Blowfish [14] have been published.

2   A Short Description of PRINTCIPHER

This section holds a short description of PRINTCIPHER, focusing only on the parts that are of interest for our analysis. For more details we refer to [8]. PRINTCIPHER-48 (resp -96) is an SP-network with a block size of $b = 48$ (resp $b = 96$) bits and 48 (resp 96) rounds. The key size is 80 bits for PRINTCIPHER-48 and 160 bits for PRINTCIPHER-96. It is closely related to the block cipher PRESENT in the sense that both ciphers use small s-boxes and a simple bit permutation as the linear layer. PRINTCIPHER uses a single 3 bit s-box shown in the following table.

x	0	1	2	3	4	5	6	7
$S[x]$	0	1	3	6	7	4	5	2

In the non-linear layer the current state is split into 16 words of 3 bits for PRINTCIPHER-48 and into 32 words of 3 bits for PRINTCIPHER-96 and each word is processed by the s-box in parallel. The linear layer consists of a bit permutation, where bit i of the current state is moved to bit position $P(i)$ where

$$P(i) = \begin{cases} 3i - 2 \bmod b - 1 & \text{for } 1 \leq i \leq b - 1, \\ b & \text{for } i = b, \end{cases}$$

where $b \in \{48, 96\}$ is the block size.

The peculiar part of PRINTCIPHER is to have all rounds identical up to adding a round constant on a small number of bits. Here identical has to be understood as including the round key, in other words, all round keys are identical. As a simple round key xored to the state in each round limits the key size to 48 resp 96 bits, an additional key-dependent permutation layer was introduced. This permutation layer permutes the input bits of each s-box individually. Out of 6 possible permutations on 3 bits, only four are valid permutations for PRINTCIPHER.

For PRINTCIPHER-48 the 80-bit user-supplied key k is split into two subkeys $k = \text{sk}_1 \| \text{sk}_2$ where sk_1 is 48 bits long and sk_2 is 32 bits long. The first subkey sk_1 is xored to the state at the beginning of each round. The second subkey sk_2 is used to generate the key-dependent permutations in the following way. The 32-bits are divided into 16 sets of two bits and each two-bit quantity $a_1 \| a_0$ is used to pick one of four of the six available permutations of the three input bits. Specifically, the three input bits $c_2 \| c_1 \| c_0$ are permuted to give the following output bits according to two key bits $a_0 \| a_1$.

$a_1 \| a_0$	
00	$c_2 \| c_1 \| c_0$
01	$c_1 \| c_2 \| c_0$
10	$c_2 \| c_0 \| c_1$
11	$c_0 \| c_1 \| c_2$

One round of PRINTCIPHER-48 is shown in Figure 1.

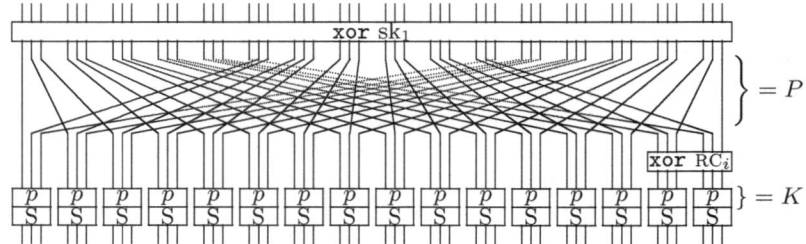

Fig. 1. One round of PRINTCIPHER-48 illustrating the bit-mapping between the 16 3-bit S-boxes from one round to the next. The first subkey is used in the first **xor**, the round counter is denoted RC_i, while key-dependent permutations are used at the input to each S-box.

3   Using Differential Cryptanalysis to Recover the Permutation Key

A classical differential attack against an SP-network finds an input difference α that produces a certain output difference β with high probability (a so-called *differential*). The attacker then analyses a large number of input pairs (x, x') with $x \oplus x' = \alpha$ and their corresponding output pairs (y, y'), hoping to find the expected difference $\beta = y \oplus y'$. Once this difference actually occurs, the attacker learns something about the internal behaviour of cipher. In particular, he can often use this knowledge to recover parts of the key.

For PRINTCIPHER, this attack can not be directly applied in a straightforward fashion, since finding good differentials requires the knowledge of the linear layer, which for PRINTCIPHER is key-dependent and thus unknown. As already pointed out, however, this disadvantage can also be turned into an advantage for the attacker: It can be used to learn something about the part of the key that defines the linear layer.

3.1   Optimal Differential Characteristic

We start our analysis by proving the following fact about the optimal PRINTCIPHER characteristic.

Theorem 1. *Given an input difference α of weight one, the unique most probable r-round differential characteristic is*

$$\alpha \to (PK)(\alpha) \to (PK)^2(\alpha) \to (PK)^r(\alpha),$$

which will occur with probability $(1/4)^r$.

Proof. The difference distribution table for the PRINTCIPHER S-box (see Table 1) shows that all occuring differences are equally probable (prob. 1/4) and that for every 1-bit input difference, there exists exactly one 1-bit output difference.

Table 1. Difference distribution table for PRINTCIPHER S-box. Note that the difference table is symmetric. 1-bit to 1-bit differences are marked with boxes.

		Δy							
		000	001	010	011	100	101	110	111
	000	8	-	-	-	-	-	-	-
	001	-	2	-	2	-	2	-	2
	010	-	-	2	2	-	-	2	2
Δx	011	-	2	2	-	-	2	2	-
	100	-	-	-	-	2	2	2	2
	101	-	2	-	2	2	-	2	-
	110	-	-	2	2	2	2	-	-
	111	-	2	2	-	2	-	-	2

From this, it follows that starting with a 1-bit input difference, a 1-bit differential trail through r rounds of PRINTCIPHER occurs with probability $(1/4)^r$. Note also that this trail has the minimum possible number of r active S-boxes and that no other S-box difference is more probable, meaning that this trail is the most probable one.

Also note that the 1-bit output difference always occurs in the same bit position as the 1-bit input difference. This means that if the 1-bit differential occurs, the S-box does not permute the active bit - its position on the differential trail is only influenced by the fixed permutation P and the key-dependent permutation K. Thus, the difference α is indeed mapped to $(PK)^r(\alpha)$, which proves the theorem.   □

The probability of the differential characteristic is based on assumptions, in particular the assumption of independent round-keys. This assumption is in particulary questionable for PRINTCIPHER as all round-keys are identical. Therefore, we ran (limited) tests to see if the theoretical probability of $(1/4)^r$ is actually met. Our experimental data depicted in Figure 2 suggest that indeed the probability is slightly higher than expected.

3.2 Targeting the xor Key

In the following, we assume that the attacker has the full code book at his disposal (i.e. 2^{48} plaintext/ciphertext pairs for r rounds of PRINTCIPHER-48). For every 1-bit input difference $\alpha_1 = (100...0), \alpha_2 = (010...0), \ldots, \alpha_{48} = (000...1)$, the attacker now forms all 2^{47} input pairs with $x \oplus x' = \alpha_i$ and checks whether the output difference also has weight one. If yes, he assumes that he has found the above optimal characteristic. It turns out that as long as $r \leq 22$, this is very likely to happen[1].

---

[1] We have 2^{47} pairs and a success probability of $(1/4)^{22} = 2^{-44}$, yielding a success probability close to 1 for any single index i and of ≈ 0.984 for all 48 indices. When increasing the number of rounds to $r = 23$, the success probability drops to 0.865 for any single index and to 0.001 for all 48 indices.

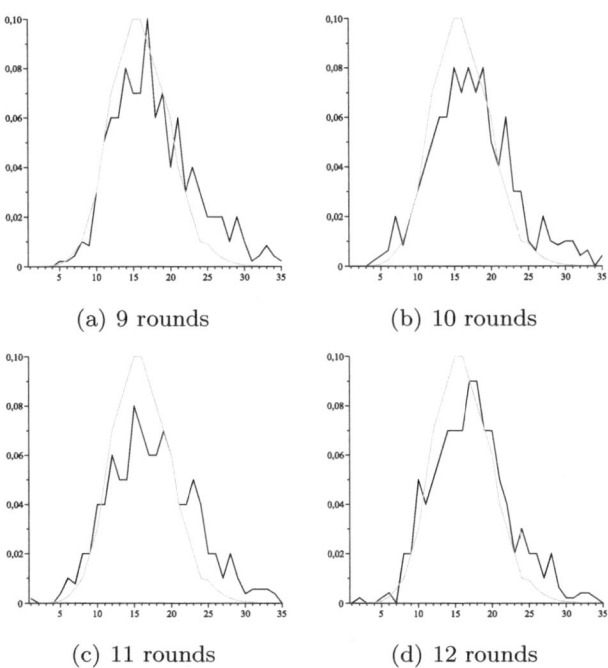

Fig. 2. Experimental vs. theoretical estimates for the optimal differentials. The x-axis shows the number of pairs yielding the correct output difference within 2^{2r+4} tries. The y-axis shows the relative frequency.

Every successful 1-bit differential gives the attacker information about the internal behaviour of the cipher which can be used to reconstruct part of the xor key. Consider the first round of the cipher and note that according to [8], the order of S-box and key-dependent permutation can be inversed by adding two constants c and d that do not affect the differential. Thus, we can alternatively consider one PRINTCIPHER round to consist of key addition, fixed permutation, round constant, adding c, S-box, key-dependent permutation, and adding d. In particular, for the purposes of differential cryptanalysis, we can assume the S-box to follow directly after the key addition.

Now consider a successful differential with input difference $\alpha_1 = (100...$
$0)$. Three key bits (with indices 1, 17 and 33) will affect the bits that go into the first S-box. There are *a priori* 8 possible choices for these bits, generating all possible 3-bit S-box input pairs with difference α_1. However, as shown in Table 1, only 2 of them will lead to a 1-bit output difference after running through the S-box. Thus, only 1/4 of all keys meet the condition for the first S-box, reducing the key entropy by 2 bit. Thus, finding 16 successful 1-bit to 1-bit differentials (one for each S-box) will reduce the key entropy by 32 bit, leaving a brute-force

effort of 2^{48} steps. This work factor could be reduced further, but without greatly affecting the overall running time, which is dominated by the 2^{48} steps of computing the full code book anyway.

The false positive problem: The above description is a simplification since it does not take false positives into account. For every 1-bit differential, trying out 2^{47} plaintext pairs will yield $2^{47} \cdot \frac{48}{2^{48}} = 24$ false positives on average, i.e. 1-bit output differences that occur accidentally and not as a result of the correct differential. The question remains how they can be distinguished from the cases where the 1-bit output differences really result from the desired differential. It turns out that for 22 rounds, the probability that all 48 differentials are met at least three times is 0.514, meaning that in more than half of the cases, the correct 1-bit difference should be recognizable by occuring more often than the false positives, which very rarely occur more than twice.

3.3   Targeting the Linear Layer

As it turns out, there is also a different way of using the above differential to cryptanalyse PRINTCIPHER. Remembering that according to Theorem 1, *every* 1-bit to 1-bit characteristic is optimal and describes the mapping $\alpha \rightarrow (PK)^r(\alpha)$, the following corollary immediately follows:

Corollary 1. *Learning all optimal characteristics is the same as learning* $(PK)^r$.

If the attacker has the full code book available, he can form 2^{47} plaintext pairs for every 1-bit input difference. The probability that at least one example of all 48 1-bit differentials is found is 0.984, and as stated above, the probability that they all can be distinguished successfully from false positives is 0.514. Thus, for up to $r = 22$ rounds of PRINTCIPHER-48, the attacker can learn the permutation $(PK)^r$.

   If he can find the r-th root of this permutation, then he has derived PK and thus the linear layer key K. Once this has been done, the xor key can be retrieved bitwise, using a simple divide-and-conquer attack similar to the one described in Subsection 3.2. It turns out that here too, the overall running time is dominated by computing the code book, i.e. the attack requires about 2^{48} computational steps.

   This type of differential attack is the dual to the one targeting the xor key and is relevant for all SPN-like ciphers that use key-dependent permutations. For this reason, it is not only interesting for the analysis of PRINTCIPHER, but also for the understanding of key-dependent permutations in general. In the rest of this paper, we will thus discuss the computation of permutation roots in more detail.

4   Finding (PRINTCIPHER)-Roots of a Permutation

From the previous section, we see that our problem of finding the permutation key can be reduced to the problem of finding the r-th roots of a given permutation in the symmetric groups, S_{48} and S_{96}, where r is the number of rounds.

Any permutation can be expressed as a product of disjoint cycles, and it is this representation that is most useful when computing roots. In particular, the permutation found through differential cryptanalysis can be expressed as a product of disjoint cycles in S_{48} and S_{96}.

Before describing how to find a root for a permutation in general, we outline the basic ideas. For this, let us first see what happens when we raise a single cycle to the power r.

Let $c = (c_0, c_1, \ldots, c_{l-1})$ be a cycle of length l in S_n. Then c^2 will remain a single cycle when l is odd, namely, $c^2 = (c_0, c_2, \ldots, c_{l-1}, c_1, c_3, \ldots, c_{l-2})$, and will be decomposed into 2 cycles when l is even, namely, $c^2 = (c_0, c_2, \ldots, c_{l-2})(c_1, c_3, \ldots, c_{l-1})$. In general, depending on l, c^r will either remain a single cycle or be decomposed into a number of cycles having the same length (see Lemma 1). Each element c_i will be in a cycle, say, $(c_i, c_{i+r}, c_{i+2r}, \ldots, c_{i+(k-1)r})$, where $i + kr \equiv i$ (mod l) and $i + jr$ is reduced modulo l for each j. So in order to find the r-th root we have two cases, the first one is when c^r is a single cycle, and here c^r equals exactly $(c_0, c_r, c_{2r}, \ldots, c_{(l-1)r})$. The second case is when c^r consists of a number of disjoint cycles, and here we combine these disjoint cycles into a single cycle in a certain way in order to get c (see the proof of Theorem 2). To illustrate this, let us find the square root of the permutation $\sigma^2 = (1, 3, 2)(4, 6, 7)(5)(8)$ in S_8. According to the above explanation, we know that cycles of the same length are either a decomposition of a single cycle in the root σ or a reordering of a single cycle in the root σ. Considering cycles of length 1, (5) and (8), it is obvious that they arise from either $(5)(8)$ or $(5, 8)$.

Thus, there are two possibilities for cycles of length 1 in σ. Cycles of length 3, $(1, 3, 2)$ and $(4, 6, 7)$, are either a decomposition of a single cycle in σ, this could be $(1, 4, 3, 6, 2, 7)$, $(1, 6, 3, 7, 2, 4)$ or $(1, 7, 3, 4, 2, 6)$; or a reordering of disjoint cycles in σ and this could only be $(1, 2, 3)(4, 7, 6)$. Summarizing, there are four possibilities for cycles of length 3 in σ. So the total number of square roots for the permutation σ^2 is 8.

4.1 The General Case

The procedure for constructing an r-th root for a permutation, described in [15], is based on the following basic fact in the theory of symmetric groups which can be easily deduced from the previous explanation.

Lemma 1. *Let $C \in S_n$ be a cycle of length l and let r be a positive integer. Then C^r consists of $\gcd(l, r)$ disjoint cycles, each of length $\frac{l}{\gcd(l,r)}$.*

The following theorem is due to A. Knopfmacher and R. Warlimont [15, p. 148]. We recall its proof, as the proof describes how to construct an r-th root. Throughout the rest of this paper, we use the notation l-cycle to mean a cycle of length l.

Theorem 2. *[15,1] Let $r = p_1^{i_1} p_2^{i_2} \ldots p_n^{i_n}$, where $p_1, p_2, \ldots, p_n$ are the prime factors of r. A permutation $Q \in S_n$ has an r-th root, iff for every integer $l \geq 1$, the number of l-cycles in Q is divisible by $((l, r)) := \prod_{\{j : p_j \mid l\}} p_j^{i_j}$.*

Proof. ($\Leftarrow$): to prove this, we construct an r-th root, R, of Q. Let a_l be the number of l-cycles in Q. Let $g = ((l, r))$. Then $a_l = gm$, where m is an integer, so we can divide the l-cycles of Q into m groups where each group consists of g l-cycles. Assume that we have the cycles, $c_{ij} = (c_{ij}^{(0)}, c_{ij}^{(1)}, \ldots, c_{ij}^{(l-1)})$ where $1 \le i \le g$ and $1 \le j \le m$. For each j, we construct a cycle of length gl, say $R_j = (c_{1j}^{(0)}, c_{2j}^{(0)}, \ldots, c_{gj}^{(0)}, c_{1j}^{(d)}, c_{2j}^{(d)}, \ldots, c_{gj}^{(d)}, \ldots, c_{1j}^{((l-1)d)}, c_{2j}^{((l-1)d)}, \ldots, c_{gj}^{((l-1)d)})$, where $d = \frac{r}{g}$ and sd is reduced modulo l for each $1 \le s \le l - 1$. Now R_j is a cycle of length gl, so according to the previous lemma, R_j^r consists of $gcd(gl, r)$ cycles of length $\frac{gl}{gcd(gl,r)}$. Now, since $g = ((l, r))$, then $gcd(l, \frac{r}{g}) = 1$ and so $gcd(gl, r) = g$, which means that R_j^r consists of g cycles of length l, namely, $c_{1j}, c_{2j}, \ldots, c_{gj}$. So $\prod_{j:1 \le j \le m} R_j$ is an r-th root for the l-cycles of Q. Repeating the same procedure for all l will yield an r-th root of Q. For the proof of ($\Rightarrow$), see [1]. □

In [6,9], a procedure to find all the roots of Q is described. Going back to the previous theorem, we see that the main property that enables us to construct an r-th root for the l-cycles of Q is having $gcd(gl, r) = g$. Repeating the same procedure for all the g's that satisfy $gcd(gl, r) = g$ will allow us to find all the possible roots that can come from the l-cycles. Note that g is bounded by a_l (the number of l-cycles). To find all the roots, for each group consisting of l-cycles in Q, we proceed as follows.

First we construct the set $G_r(l, a_l) = \{g_i : gcd(g_i l, r) = g_i \text{ and } 1 \le g_i \le a_l\}$. Now, this tells us that the roots have cycles of length $g_i l$, but we do not know how many of them. For this, we solve the following Frobenius equation for $x_i \ge 0$:

$$g_1 x_1 + g_2 x_2 + \cdots + g_k x_k = a_l \qquad \text{where } k = |G| \qquad (1)$$

This equation will usually have more than one solution. Each solution corresponds to a possible cycle structure of the roots. For instance, the solution $x = (x_1, x_2, \ldots, x_k)$, tells us that each corresponding root for the l-cycles of Q consists of x_i cycles of length $g_i l$ for $1 \le i \le k$.

The efficiency of computing all roots is of course bounded by the total number of roots. If a permutation has a huge number of roots, computing all of them is very time consuming. It is therefore of interest to know the number of roots in advance.

In [9], using the above information about the cycle structure of permutations that have an r-th root, the following explicit formula[2] for calculating the number of all the possible roots is provided.

Theorem 3. *[9] Let r be a positive integer and $Q \in S_n$. Let a_l be the number of l-cycles in Q, where $1 \le l \le n$. Let $X(l, a_l)$ be the set of all the possible solutions of equation (1). Then the number of r-th roots of Q is*

$$\prod_{a_l \ne 0} a_l! \left(\sum_{x \in X(l,a_l)} \prod_{i=1}^{k} \frac{l^{(g_i-1)x_i}}{g_i^{x_i} x_i!} \right) \qquad (2)$$

where $x = (x_1, x_2, \ldots, x_k)$ and $\{g_i : 1 \le i \le k\}$ are the elements of $G_r(l, a_l)$.

---

[2] A more complicated formula was previously found by Pavlov in [11].

To get a feeling of how many roots of a permutation can be expected for the case of PRINTCIPHER-48, let us take the following permutation in S_{48}, suppose we have

$$
\begin{aligned}
\tau^{24} =& (1,7,47)(2,19,45)(3,48,17)(4,9,38)(5,16)(6,33,32)(8,28)(10,35) \\
& (11,27,18)(12,20)(14,19,41)(15,46)(21,26,30)(22,34)(23,36)(25, \\
& 42,39)(40,44)(13)(24)(31)(37)(43)
\end{aligned} \tag{3}
$$

So we have $a_1 = 5, a_2 = 8, a_3 = 9$ and $a_l = 0$ for $4 \leq l \leq 48$. $G(1,a_1) = \{1,2,3,4\}$, $X(1,a_1) = \{(0,1,1,0),(1,0,0,1),(1,2,0,0),(2,0,1,0),(3,1,0,0), (5,0,0,0)\}$, $G(2,a_2) = \{8\}$, $X(2,a_2) = \{(1)\}$ and $G(3,a_3) = \{3,6\}$, $X(3,a_3) = \{(3,0),(1,1)\}$. Plugging these values into equation (2), we find that the number of roots is $\simeq 2^{51.3}$. Moreover, the case where τ^{22} is the *Identity* has $\simeq 2^{192}$ roots in S_{48}.

Note that out of all 48!(96!) permutations only a tiny fraction of $2^{32}(2^{64})$ permutations actually correspond to a valid key in PRINTCIPHER-48(96). We can therefore expect that in the above example out of the $\simeq 2^{51.3}$ only a very small number will actually correspond to a PRINTCIPHER-permutation. In particular there is only one root for equation (3) that corresponds to a PRINTCIPHER permutation.

The main purpose of the next section is to describe a method that filters out wrong candidates as soon as possible, allowing to considerably speed up the computation of all valid PRINTCIPHER-roots.

4.2   PRINTCIPHER-**Roots**

As discussed in the last section, computing all the roots of $(PK)^r$ in order to find the right permutation key is inefficient. In this section we describe a method that finds the permutation roots PK belonging to the $2^{32}(2^{64})$ possible permutations in PRINTCIPHER-48(96). Throughout the rest of this paper, we only discuss PRINTCIPHER-48 and unless mentioned explicitly, the assumption is that everything about PRINTCIPHER-48 follows for PRINTCIPHER-96 with a slight modification.

Our method uses the fact that when we apply the fixed permutation, P, for all $1 \leq i \leq 16$, the 3 bits $i, i+16$ and $i+32$ go to the ith Sbox, where depending on the permutation key, they are permuted to only four out of the six possible permutations. So the result of applying the fixed permutation, P, and then applying the keyed permutation, K, on a 48 bits plain text, is a permutation PK that satisfies the following two properties:

1. *Property* 1: For all $1 \leq i \leq 48$, $PK(i)$ equals one of the following three possible values depending on K,

$$
PK(i) = \begin{cases} 3i-2 \pmod{48} & \text{if } 3i-2 \neq 48) \\ 3i-1 \pmod{48} & \text{if } 3i-1 \neq 48) \\ 3i \pmod{48} & \text{if } 3i \neq 48) \end{cases}
$$

2. *Property* 2: Only 4 out of the 6 possible 3-bit permutations are valid, namely, $PK(i)$, $PK(i+16)$ and $PK(i+32)$ are permuted to one of the four possible permutations, i.e., for all $1 \leq i \leq 48$, the following two permutations are not allowed:
   (a) $PK(i) = 3i - 1$, $PK(i + 16) = 3i$ and $PK(i + 32) = 3i - 2$.
   (b) $PK(i) = 3i$, $PK(i + 16) = 3i - 2$ and $PK(i + 32) = 3i - 1$.

Definition 1. *A* PRINTCIPHER *permutation root is any permutation on 48 elements satisfying both Property 1 and Property 2.*

Definition 2. *A* PRINTCIPHER *permutation(cycle) is any permutation(cycle) on less than 48 elements satisfying both Property 1 and Property 2.*

To explain these definitions, consider the following two cycles $(14, 41, 25, 26, 30, 42, 29, 39, 21)$ and $(14, 42, 30, 41, 25, 26, 29, 39, 21)$. We want to investigate whether these cycles are PRINTCIPHER cycles or not. The latter cycle satisfies the two properties and so it is a PRINTCIPHER cycle, in other words it can be part of a PRINTCIPHER permutation root, PK. The former cycle satisfies only *Property* 1 but not *Property* 2 since we have $PK(14) = 41$ and $PK(30) = 42$ and therefore $PK(42) = 40$, and this is one of the two disallowed permutations (see item (a) in *Property* 2) and so it cannot be part of a valid PRINTCIPHER permutation root, PK. Sometimes we can have a permutation consisting of two or more cycles having same or different lengths, that satisfies *Property* 1 but not *Property* 2. For example, the following permutation, $(1, 2, 6, 17)(5, 15, 44, 34)$, satisfies *Property* 1 but not *Property* 2 as we have $PK(2) = 6$ and $PK(34) = 5$ and therefore $PK(18) = 4$, which is an invalid PRINTCIPHER permutation (see item (b) in *Property* 2).

   Since the same cycles and permutations can be written in different ways, our method adopts the notion that starts writing each cycle by its smallest element and lexicographically order the disjoint cycles of the same length of a permutation in order to avoid repetitions in the permutation roots of $(PK)^r$. Our method consists of two algorithms: the first one constructs a PRINTCIPHER cycle of length gl and the second one uses the first algorithm to construct k combined disjoint cycles, each of length gl. In what follows, we shall give a detailed description of the two algorithms and end this section by showing how to use Algorithm 2 to find the whole PRINTCIPHER permutation roots of $(PK)^r$.

Finding single PRINTcipher cycles. Given a_l cycles of length l, the following algorithm constructs all the possible PRINTCIPHER cycles of length gl beginning with an element called *first* specified in the input (must be one the the first elements in one of these a_l cycles). The algorithm performs a depth first search to find all the other possible $g - 1$ cycles with minimal elements larger than *first* and can be combined with the cycle containing *first* as described in Theorem 2 in order to form a PRINTCIPHER cycle (or just reorder the given cycle in the case $g = 1$ as described previously).

Algorithm 1. finds a PRINTCIPHER cycle of length gl

find-cycle(cycle, current, g, l-cycles)

Require: l-cycles numbered from 1 to a_l where $a_l \geq g$
Require: current $= first$, cycle $= first$
Ensure: *cycle* is a PRINTCIPHER cycle of length gl
 1: **for** count=0 to 2 **do**
 2:     next $= 3 \times$ current - count
 3:     **if** next $\in l$-cycles **then**
 4:         **if** next $> first$ **and** next.cycleno $\neq first$.cycleno **and** *cycle*.length $< g$ **then**
 5:             **if** next.cycleno $\neq$ the cycleno of all the elements of *cycle* **then**
 6:                 Add next to *cycle*
 7:                 current = next
 8:                 Perform again this algorithm on *cycle*, *find-cycle(cycle, current, g, l-cycles)*
 9:             **end if**
10:         **else if** *cycle*.length $= g$ **and** next.cycleno $= first$.cycleno **then**
11:             Complete the construction of *cycle* by combining the g different cycles to get a single cycle of length gl as shown in the proof of Theorem 1 (when $g = 1$, reorder the cycle containing *first* as described previously and assign it to *cycle*)
12:             **if** *cycle* satisfies *Property* 2 **then**
13:                 *cycle* is a PRINTCIPHER cycle of length gl
14:             **end if**
15:         **end if**
16:     **end if**
17: **end for**

Plugging all the 2-cycles of equation (3) and setting $first = 5$ and $g = 8$ will produce the following PRINTCIPHER cycle of length 16

$$(5, 15, 44, 36, 12, 35, 8, 22, 16, 46, 40, 23, 20, 10, 28, 34).$$

Algorithm 1 enables us to find a PRINTCIPHER permutation consisting of only one cycle of length gl but note that some of the x_i's in equation (1) can be more than 1. So we need another algorithm which can find a PRINTCIPHER permutation consisting of k disjoint cycles where $k \geq 1$.

Finding k combined PRINTcipher cycles. Given a_l cycles of length l, the following algorithm constructs a permutation beginning with an element called *first* specified in the input (must be the first element in one of these a_l cycles) and consisting of k combined and disjoint cycles ordered lexicographically. It basically performs a recursive depth first search. The recursive algorithm begins by invoking Algorithm 1 which outputs single cycles of length gl beginning with *first*. It then proceeds from each cycle found by Algorithm 1 and concatenates it with the previously $i - 1$ concatenated disjoint cycles found after the ith recursive call and if the concatenation satisfies *Property* 2, it recursively calls itself a number of times, each time with a different *first* element to begin the required permutation with as this will enable us to find all the possible $i + 1$ disjoint cycles, on a reduced number of l-cycles (exactly $a_l - gi$ cycles) consisting of all the l-cycles except the gi cycles involved on the i concatenated disjoint cycles (in each invocation *first* is set to the smallest element on one of the currently available $a_l - gi$ cycles). Each recursive call stops when $i = k$, or when Algorithm 1 returns nothing, or when each concatenation of i cycles does not satisfy *Property* 2.

Algorithm 2. finds a PRINTCIPHER permutation that has k disjoint gl-cycles *find-k-cycles(C, current, k, g, l-cycles)*

Require: l-cycles numbered from 1 to a_l where $a_l \geq g$
Require: current = *first*
Require: $C = \{\}$
Ensure: k disjoint PRINTCIPHER gl-cycles, or return $\{\}$ if there is no k disjoint
    PRINTCIPHER cycles
1: Invoke Alg. 1 on the current l-cycles
2: **if** number of *cycles* found by Alg. 1 > 0 **then**
3:    **if** the number of disjoint cycles in C consists of $k - 1$ disjoint cycles **then**
4:       **for** each permutation *cycle* found by Alg. 1 **do**
5:          $C = C \cup cycle$
6:          **if** C satisfies *Property* 2 **then**
7:             **return** C
8:          **else**
9:             **return** $\{\}$
10:         **end if**
11:      **end for**
12:    **else**
13:       **for** each *cycle* found by Alg. 1 **do**
14:          $C = C \cup cycle$
15:          **if** C satisfies *Property* 2 **then**
16:             Delete all the l-cycles involved in C from the a_l cycles of length l
17:             **for** each $cycle \in$ currently available l-cycles **do**
18:                {Perform again this algorithm on the current l-cycles to find the other $k - 1$ cycles}
19:                current = first element in *cycle*
20:                *find-k-cycles(C, current, k, g, l-cycles)*
21:             **end for**
22:          **end if**
23:       **end for**
24:    **end if**
25: **else**
26:    **return**
27: **end if**

Using Algorithm 1 for all the possible g's along with all the possible *first* values and setting $a_1 = 48$, we can find all the possible PRINTCIPHER cycles. For instance, when $g = 1$ and $a_1 = 48$, Algorithm 1 returns four 1-cycles when trying all the possible values for *first*, namely, (1), (24), (25) and (48). When $g = 2$ and $a_2 = 48$, we found that there are six possible 2-cycles, namely, $(6, 18)$, $(7, 19)$, $(12, 36)$, $(13, 37)$, $(30, 42)$ and $(31, 43)$. When $g = 3$, we found there are eight possible 3-cycles. This information enables us to reduce the size of the cycle structure of the roots by removing any structure containing more than four 1-cycles, six 2-cycles and eight 3-cycles. It also enables us to easily find some roots, for example, knowing all PRINTCIPHER cycles of length 1 and 2, we can easily find that $(24)(13, 37)(30, 42)$ is a PRINTCIPHER permutation that is a root for the 1-cycles of equation (3).

Moreover, using Algorithm 2 we find that we cannot have a permutation consisting of more than 6 disjoint cycles of length 5 in PRINTCIPHER-48 and not more than 9 cycles of length 4, 12 cycles of length 5, 13 cycles of length 6 and 12 cycles of length 7 in PRINTCIPHER-96. This will generally reduce the number of solutions of equation (1) and therefore the size of the cycle structure which will speed up the process of finding PRINTCIPHER permutations roots.

Finding PRINTcipher permutations. Now, when given a_l cycles of length l, Algorithm 2 enables us to find PRINTCIPHER permutations beginning with a specified element and consisting of k cycles, each of length gl. But in order to find the rth permutation roots for all the l-cycles we use Algorithm 2 together with the elements of the sets $G(l, a_l)$ and $X(l, a_l)$. Each entry $x_j = (x_{j1}, x_{j2}, \ldots, x_{jk}) \in X(l, a_l)$ where $k = |G(l, a_l)|$, represents the cycle structure of many rth roots for the l-cycles and it might correspond to few or none PRINTCIPHER permutations, so for each $x_j \in X(l, a_l)$, we try to find all the possible PRINTCIPHER permutations beginning with a specific element called *first* (must be the first element in one of these a_l cycles) and that can be roots for the l-cycles by applying Algorithm 2 through all the nonzero entries of x_j. Trying all the possible values for *first* gives us all PRINTCIPHER permutations that are roots for all the l-cycles.

Now, assume that we find all the possible PRINTCIPHER permutations for each l, say σ_{l_i}, for $1 \le i \le \eta_l$ where η_l is the number of permutation roots of the l-cycles of $(PK)^r$, so all the possible products $\prod_{a_l > 0} \sigma_{l_i}$ where $1 \le l \le 48$ and $1 \le i \le \eta_l$, represent the PRINTCIPHER permutation roots which are the possible values for PK and by brute forcing these PK values we can recover the permutation key, K.

Let us try to find PRINTCIPHER permutations that are roots for the nine 3-cycles in equation (3). We have $G(3, a_3) = \{3, 6\}$ and $X(3, a_3) = \{(3, 0), (1, 1)\}$. We start with, $x_1 = (3, 0)$, here we only need to apply Algorithm 2 using any possible *first* because the 3 disjoint cycles of length 9 would come from all the 9 cycles. Setting *first* $= 1$ and applying Algorithm 2 doesn't give us 3 disjoint cycles of length 9, so we conclude that there is no root having the cycle structure x_1. So we go to the next cycle structure, $x_2 = (1, 1)$, we start with $x_{21} = 1$ and use Algorithm 2 on all the possible *first* values. Setting *first* $= 1, 2, 3, 4, 6$ and 11 doesn't yield a single cycle of length 9, while *first* $= 14$ yields the cycle $(14, 42, 30, 41, 25, 26, 29, 39, 21)$, we save it and continue to the next element $x_{22} = 1$ where we use Algorithm 2 on the 6 cycles that are not involved in the previous found cycle. Now we want to construct a cycle of length 18, so all the 6 cycles would be involved in it, setting $first = 1$, yields $(1, 2, 4, 11, 33, 3, 7, 19, 9, 27, 32, 48, 47, 45, 38, 18, 6, 17)$. Concatenating this cycle with the previous found cycle, we get $(14, 42, 30, 41, 25, 26, 29, 39, 21)(1, 2, 4, 11,$ $33, 3, 7, 19, 9, 27, 32, 48, 47, 45, 38, 18, 6, 17)$ which satisfies *Property* 2. This means that it is a PRINTCIPHER permutation that is a root for all the 3-cycles in equation (3). Now, we have found the roots for all the l-cycles in equation (3). Concatenating them together gives us the following PRINTCIPHER permutation root: $(1, 2, 4, 11, 33, 3, 7, 19, 9, 27, 32, 48, 47, 45, 38, 18, 6, 17)(5, 15, 44, 36, 12, 35, 8,$ $22, 16, 46, 40, 23, 20, 10, 28, 34)(14, 42, 30, 41, 25, 26, 29, 39, 21)(13, 37)(30, 42)(24)$.

5   Experimental Verifications

To demonstrate the efficiency of our attack we implemented the above algorithms. Experiments show that $(PK)^r$ could yield more than one PRINTCIPHER root when $(PK)^r$ contains several 1-cycles, but in most cases there was exactly one PRINTCIPHER root.

To derive bounds for the number of PRINTCIPHER permutations roots, we computed the number of all PRINTCIPHER permutation roots for $(PK)^r = Identity$ where $2 \leq r \leq 22$. This seems the worst case that could happen for any r since $a_1 = 48$, which is a_1's largest value, and as shown in Table 2, the number of PRINTCIPHER roots when $r = 22$ is $2^{22.04}$. These roots are found within less than 3 hours on a standard PC.

Furthermore, we tried 10^4 random PRINTCIPHER-48 permutation keys excluding the ones that yield $(PK)^r = Identity$. Note that, for a random key, the probability for the worst case is $\frac{2^{22.04}}{2^{32}} = 0.001$ for 22 rounds and less than that for $r < 22$. These experiments took a few seconds on average on a standard PC and they show that most of the time there is a unique PRINTCIPHER permutation root. Table 2 shows the number of keys (n_k), out of the 10^4 random keys, that yield more than one PRINTCIPHER permutation root. It also shows the number of PRINTCIPHER permutation roots in the worst case (n_w) for each number of rounds.

Table 2. Results of the 10^4 trials and the worst case for $2 \leq r \leq 22$, $n_k \equiv$ the number of keys that yield more than one PRINTCIPHER permutation root, $n_w \equiv$ the number of PRINTCIPHER permutation roots in the worst case.

r	$\log_2 n_k$	$\log_2 n_w$	r	$\log_2 n_k$	$\log_2 n_w$	r	$\log_2 n_k$	$\log_2 n_w$
2	-	-	9	7.66	8.58	16	10.80	18.95
3	-	-	10	8.33	11.90	17	8.71	-
4	6.11	2	11	7.94	9.31	18	11.16	20.67
5	2	-	12	11.46	17.39	19	8.77	-
6	9.30	4.17	13	8.47	-	20	10.68	21.54
7	3.70	-	14	9.10	16.27	21	9.18	18.73
8	9.59	10.07	15	9.77	16.63	22	9.59	22.04

6   Conclusions

We have described two differential attacks against 22 rounds of PRINTCIPHER-48, requiring the full code book and about 2^{48} computational steps. While this is far from breaking the full 48 rounds of the cipher, it is the best currently known result against the cipher. Similar results can be obtained for the 96-bit version of the cipher.

One of the attacks is a new technique targeting the key-dependent permutations used in PRINTCIPHER. Since such key-dependent permutations are currently not well-studied, the attack is of importance to past and future designs that use them. We introduced a novel technique for computing permutation roots, making it possible to retrieve the key-dependent single-round permutation π given nothing but the r-round permutation π^r and the cipher description. While our technique so far applies only to the case where the linear layer is a (key-depended) bit permutation, future designers of cryptographic primitives using key-dependent permutations should be aware of this technique when choosing parameters like round numbers or S-box layout for their algorithms.

References

1. Annin, S., Jansen, T.: On kth roots in the symmetric and alternating groups. Pi Mu Epsilon Journal 12(10), 581–589 (2009)
2. Biham, E., Shamir, A.: Differential Cryptanalysis of DES-like Cryptosystems. In: Menezes, A., Vanstone, S.A. (eds.) CRYPTO 1990. LNCS, vol. 537, pp. 2–21. Springer, Heidelberg (1991)
3. Bogdanov, A.A., Knudsen, L.R., Leander, G., Paar, C., Poschmann, A., Robshaw, M.J.B., Seurin, Y., Vikkelsø, C.: PRESENT: An Ultra-Lightweight Block Cipher. In: Paillier, P., Verbauwhede, I. (eds.) CHES 2007. LNCS, vol. 4727, pp. 450–466. Springer, Heidelberg (2007)
4. De Cannière, C., Dunkelman, O., Knežević, M.: KATAN and KTANTAN — A Family of Small and Efficient Hardware-Oriented Block Ciphers. In: Clavier, C., Gaj, K. (eds.) CHES 2009. LNCS, vol. 5747, pp. 272–288. Springer, Heidelberg (2009)
5. Gilbert, H., Chauvaud, P.: A Chosen Plaintext Attack of the 16-Round Khufu Cryptosystem. In: Desmedt, Y. (ed.) CRYPTO 1994. LNCS, vol. 839, pp. 359–368. Springer, Heidelberg (1994)
6. Groch, A., Hofheinz, D., Steinwandt, R.: A practical attack on the root problem in braid groups. In: Algebraic Methods in Cryptography, vol. 418, pp. 121–132. American Mathematical Society, Providence (2006)
7. Hong, D., Sung, J., Hong, S.H., Lim, J.-I., Lee, S.-J., Koo, B.-S., Lee, C.-H., Chang, D., Lee, J., Jeong, K., Kim, H., Kim, J.-S., Chee, S.: HIGHT: A New Block Cipher Suitable for Low-Resource Device. In: Goubin, L., Matsui, M. (eds.) CHES 2006. LNCS, vol. 4249, pp. 46–59. Springer, Heidelberg (2006)
8. Knudsen, L.R., Leander, G., Poschmann, A., Robshaw, M.J.B.: PRINTCIPHER: A Block Cipher for IC-Printing. In: Mangard, S., Standaert, F.-X. (eds.) CHES 2010. LNCS, vol. 6225, pp. 16–32. Springer, Heidelberg (2010)
9. Leaños, J., Moreno, R., Rivera-Martínez, L.M.: A note on the number of m-th roots of permutations. Arxiv preprint arXiv:1005.1531 (2010)

10. Merkle, R.C.: Fast software encryption functions. In: Menezes, A., Vanstone, S.A. (eds.) CRYPTO 1990. LNCS, vol. 537, pp. 476–501. Springer, Heidelberg (1991)
11. Pavlov, A.I.: On the number of solutions of the equation $x^k = a$ in the symmetric group S_n. Mathematics of the USSR-Sbornik 40(3), 349–362 (1981)
12. Schneier, B.: Description of a new variable-length key, 64-bit block cipher (Blowfish). In: Anderson, R.J. (ed.) FSE 1993. LNCS, vol. 809, pp. 191–204. Springer, Heidelberg (1994)
13. Schneier, B., Kelsey, J., Whiting, D., Wagner, D., Hall, C., Ferguson, N.: Twofish: A 128-bit block cipher. Submitted as candidate for AES (February 5, 2010), http://www.schneier.com/paper-twofish-paper.pdf
14. Vaudenay, S.: On the weak keys of Blowfish. In: Gollmann, D. (ed.) FSE 1996. LNCS, vol. 1039, pp. 27–32. Springer, Heidelberg (1996)
15. Wilf, H.S.: Generatingfunctionology. Academic Press, London (1993)

Search for Related-Key Differential Characteristics in DES-Like Ciphers

Alex Biryukov and Ivica Nikolić*

University of Luxembourg
{alex.biryukov,ivica.nikolic}uni.lu

Abstract. We present the first automatic search algorithms for the best related-key differential characteristics in DES-like ciphers. We show that instead of brute-forcing the space of all possible differences in the master key and the plaintext, it is computationally more efficient to try only a reduced set of input-output differences of three consecutive S-box layers. Based on this observation, we propose two search algorithms – the first explores Matsui's approach, while the second is divide-and-conquer technique. Using our algorithms, we find the probabilities (or the upper bounds on the probabilities) of the best related-key characteristics in DES, DESL, and s^2DES.

Keywords: Cryptanalysis tool, automatic search, differential characteristic, related-key attack, DES.

1 Introduction

The Data Encryption Standard (DES) [8], adopted by the U.S. National Bureau of Standards in 1977, was a block cipher standard for several decades. Some of the design principles of DES were fully understood by the public only after the first cryptanalysis presented by Biham and Shamir [2]. They introduced the idea of differential analysis and differential characteristics, and showed that if one encrypts with DES a pair of plaintexts with a specific XOR difference, then the pair of corresponding ciphertexts will have some predictable difference with a probability higher than expected.

In [7] Matsui showed that the differential characteristics found by Biham and Shamir were indeed the best, i.e. they have the highest probability among all characteristics. He was able to prove this fact by running a full search on the space of all possible characteristics, using a special algorithm that speeds up the search. Matsui's algorithm was adopted and applied for search of the best characteristics in LOKI and s^2DES [10], Twofish [9], FEAL [1], and others. In all of these cases, the search was targeting only single-key characteristics, i.e. the characteristics that have a difference in the plaintext, but not in the master key. Biryukov and Nikolić in [4] showed that Matsui's idea indeed can be used

---

* This author is supported by the Fonds National de la Recherche Luxembourg grant TR-PHD-BFR07-031.

A. Joux (Ed.): FSE 2011, LNCS 6733, pp. 18–34, 2011.

to build a search algorithm that finds the best related-key characteristics (the difference can be in the key as well as in the ciphertext) for some classes of byte-oriented ciphers. To the best of our knowledge, there are no published results on a search for related-key characteristics in any bit-oriented cipher.

Our contribution. We present algorithms for finding the best (with the highest probability) round-reduced related-key differential characteristics in DES and DES-like ciphers. We show that instead of trying all differences in the key and in the plaintext, which would result in a search space of size 2^{120}, it is computationally more efficient to try only a reduced set of input-output differences of three consecutive S-boxes layers. Based on this observation, we are able to propose two algorithms for automatic search of related-key differential characteristics in DES-like ciphers – the first is based on Matsui's approach, while the second is in line with the technique of divide-and-conquer. We apply our algorithms to DES, DESL [6], and s^2DES [5] and find either the probabilities of the best round-reduced related-key differential characteristics, or the upper bounds on these probabilities. Interestingly, although for lower number of rounds these probabilities are much higher than in the case of single-key characteristics, for higher number of rounds, the best characteristics are single-key characteristics. We obtain an interesting result regarding DES. By providing the probability of the best related-key characteristic on 13 rounds, we show that Biham-Shamir attack cannot be improved if one uses related-key characteristic (instead of single-key). Moreover, the low probabilities of the best related-key characteristics on higher rounds indicate that NSA did not introduce any weakness (or trapdoor) in the key schedule of DES with regard to differential attacks. Although in this paper we apply our algorithms only to the DES-like ciphers, we believe that our approaches can be used as well to search for high probability related-key differential characteristics in any bit-oriented ciphers with linear key schedule.

2   Description of DES-Like Block Ciphers

DES [8] is 64-bit block cipher with 56-bit key[1]. It is 16-round Feistel cipher with additional permutations IP, IP^{-1} at the beginning and at the end. The 64-bit plaintext, after the application of the initial permutation IP is divided into two halves L_0 and R_0 - each half has 32 bits. Then, the halves are updated 16 times with the round function:

$$L_i = R_{i-1}$$
$$R_i = L_{i-1} \oplus f(R_{i-1}, K_i),$$

where $i = 1, \ldots, 16$ and K_i are 48-bit round keys, obtained from the initial key K with some linear transforms (rotations that depend on the round number and bit selection function PC-2). The ciphertext is defined as $IP^{-1}(R_{16} \| L_{16})$.

---

[1] Officially, the key has 64 bits, but 8 bits are only used to check the parity, and then discarded.

The round function $f(R, K_i)$ takes 32-bit state R and 48-bit round key K_i and produces 32-bit output. First it expands the 32-bit value of R to 48 bits with the linear function E and then it XORs the values of $E(R)$ and K_i to produce some intermediate result, which we further denote as f_i. This 48-bit value is divided into 8 six-bit values, and each of these values goes through a separate 6x4 S-box. Finally, the 32-bit output of the S-boxes goes through a bit permutation P and the output $\tilde{f}_i$ of the round function is produced.

The DES-like block ciphers DESL [6] and s^2DES [5], differ from DES only in the definition of the S-boxes and the initial and final permutations. Since these permutations have no cryptographic values, we can assume that the only difference among the ciphers of the DES-family is in the S-boxes.

3   Automatic Search for Related-Key Differential Characteristics in DES-Like Ciphers

The best characteristic on r rounds, i.e. the best r round-reduced characteristic, is the one that has the highest probability among all characteristics on r rounds of the cipher. In this section we propose two methods for building efficient automatic search algorithms for finding the best round-reduced related-key differential characteristics in DES-like ciphers. When constructing these algorithms, the main problem that has to be tackled is how to deal with the enormous search space. There are 64 bits in the state and 56 bits in the key, hence in total there are 2^{120} starting values for differential characteristics. However, in general, this number can be reduced significantly. Our first method is based on Matsui's search tool applied for finding the best single-key round-reduced characteristics in DES. The second method, which we call *the split* approach, can be used when Matsui's approach fails – to find characteristics on high number of rounds when not all the characteristics on lower number of rounds are known.

Due to the complementation property of DES, there are related-key characteristics (including round-reduced) that hold with probability 1. Further, we do not consider these characteristics.

Considering the different rotation amounts in the key-schedule, the probability of the best round-reduced related-key characteristic depends on the rounds covered by the characteristic. For example, the best 5-round related-key characteristic covering rounds 0-4, can have different probability from the best characteristic that covers rounds 1-5. The best related-key characteristics in our paper, always cover the last rounds, e.g. the characteristic on 7 rounds, covers the rounds 9-15.

3.1   Matsui's Approach for Single-Key Characteristics

The search for the best single-key differential characteristics in DES was successfully performed by Matsui in [7]. Note that even in this case, when there is no difference in the key, the search space is rather large – 2^{64} starting differences. However, Matsui presented several useful approaches how to deal with a large number of starting differences and how to significantly reduce the search space.

A naive approach to search for the best n-round characteristic would be to try all possible starting differences in the plaintext and try to extend each of them to n rounds. The non-linearity of the S-boxes will introduce branching, and a k-round characteristic ($k < n$) is extended for an additional round only if its probability is higher than the probability P_n^* of some known characteristic on n rounds.

Matsui's approach on the other hand, cuts out a large number of round-reduced characteristics in the early stage. Given the probabilities $\overline{P_1}, \ldots, \overline{P_{n-1}}$ of the best characteristics on the first $n - 1$ rounds, and some estimate[2] P_n^* for the probability of the characteristic on n rounds, the algorithm produces the best characteristic on n rounds. Hence, the attacker can sequentially produce, starting from 1 or 2-round reduced, characteristics on all rounds of DES. In short, the attacker, as in the naive approach, tries all possible starting differences[3]. For each of them he produces 1-round characteristic (there can be many one-round characteristics, and the following procedure is repeated for each of them) that holds with probability P_1. Then, he tries to extend it to two rounds only if $P_1 \cdot \overline{P_{n-1}} > P_n^*$. This is because in order to extend 1-round characteristic to n rounds, one should use an additional $(n - 1)$-round characteristic. Since the best one has probability $\overline{P_{n-1}}$, the total probability of the n-round characteristic will be at most $P_1 \cdot \overline{P_{n-1}}$ and this value should be better than the probability P_n^* of the best known characteristic on n-rounds. Similarly, if the attacker has built k-round characteristic with probability P_k than he tries to extended for an additional round only if $P_k \cdot \overline{P_{n-k}} > P_n^*$. Note that in the naive approach, the attacker only checks if $P_k > P_n^*$. Therefore, Matsui's approach stops the extension of many round-reduced characteristics and that way speeds up the search.

Now let us take a closer look how to reduce the number of possible starting differences. Interestingly, the same approach as above can be used. First note that a characteristic on the first two rounds (assuming this 2-round characteristic is part of the best n-round characteristic) has a probability P_2 such that $P_2 < P_n^*/\overline{P_{n-2}}$. The following observation is used to explore this property of 2-round single-key characteristics.

Observation 1. *Given the input and the output differences $(\Delta f_1, \Delta \tilde{f}_1), (\Delta f_2, \Delta \tilde{f}_2)$ of the S-boxes layers in the first two rounds, one can find the difference in the plaintext ΔP and the difference $(\Delta L_2, \Delta R_2)$ in state at the beginning of the third round.*

Proof. From the Feistel construction it leads that $\Delta R_0 = E^{-1}(\Delta f_1)$ and $\Delta L_0 = \Delta \tilde{f}_1 \oplus E^{-1}(\Delta f_2)$. Then the difference ΔP in the plaintext is $\Delta P = IP^{-1}(\Delta R_0 || \Delta L_0)$. Similarly, for the difference at the beginning of the third round we get $\Delta R_2 = \Delta R_0 \oplus \Delta \tilde{f}_2$ and $\Delta L_2 = \Delta L_0 \oplus E^{-1}(\Delta f_2)$.   □

---

[2] For example, the attacker can use the probability of the already known characteristic on n rounds as an estimate.

[3] We will see later, that this requirement can be omitted.

Therefore, instead of fixing all possible differences ΔP in the plaintext one can fix only the input and the output differences to the S-boxes in rounds 1,2. But, since the active S-boxes of the first round have to hold with a probability of at least $P_n^*/\overline{P_{n-1}}$, and in the first and the second round with at least $P_n^*/\overline{P_{n-2}}$, the number of 2-round characteristics is significantly reduced. For each such characteristic, one can proceed with Matsui's technique, and try to extend it to n-rounds (since the difference at the beginning of round 3 is fixed).

3.2   Applying Matsui's Approach for Related-Key Characteristics

One can easily reconstruct Matsui's algorithm to search for related-key characteristics. Note that for a fixed difference in the key, the algorithm still works and it finds the best characteristic with this specific difference. However, since the key has 56 bits, this search has to be repeated 2^{56} times and hence this naive approach is not feasible. We can still run a so-called *limited search* for related-key characteristics, by allowing low Hamming difference in the key. For example, to find the best characteristic that has at most 2-bit difference in the key, we have to rerun Matsui's algorithm $1 + C_{56}^1 + C_{56}^2 = 1597$ times.

Indeed, finding the best related-key characteristic using Matsui's approach can be done efficiently. We only have to find a way to efficiently limit the number of possible differences in the key and in the plaintext. We want to reduce the search space, yet to perform a full search of all possible related-key differential characteristics. The following observation can be used for that purpose.

Observation 2. *Given the input and the output differences* $(\Delta f_1, \Delta \tilde{f}_1)$, $(\Delta f_2, \Delta \tilde{f}_2)$, $(\Delta f_3, \Delta \tilde{f}_3)$ *of the S-boxes layers in the first three rounds, one can find the difference in the plaintext* ΔP, *the difference* $(\Delta L_3, \Delta R_3)$ *in state at the beginning of the fourth round, and all* 2^8 *values for the difference* ΔK *in the master key.*

Proof. Again we use the property of the Feistel construction and the linearity of the key schedule. From the definition of DES we get:

$$\Delta f_1 = E(\Delta R_0) \oplus \Delta K_1 \tag{1}$$
$$\Delta f_3 = E(\Delta R_0 \oplus \Delta \tilde{f}_2) \oplus \Delta K_3 \tag{2}$$

Since E is linear, we get:

$$\Delta K_1 \oplus \Delta K_3 = \Delta f_1 \oplus \Delta f_3 \oplus E(\Delta \tilde{f}_2)$$

The key schedule is linear, and both K_1 and K_3 are obtained from the master K with some linear transformation. Therefore $\Delta K_1 \oplus \Delta K_3$ can be expressed as $\mathcal{L}(\Delta K)$, where $\mathcal{L}$ is a linear transformation. On the other hand, the input-output differences of the S-boxes are given, and therefore, the value $V = \Delta f_1 \oplus \Delta f_3 \oplus E(\Delta \tilde{f}_2)$ is known. Hence, the master key difference ΔK can be found as $\Delta K = \mathcal{L}^{-1}(V)$. However, the key is 56 bits, while V only 48 bits. Therefore we get an underdefined system of linear equations with 2^8 solutions. If we fix a particular solution for the system, and thereby the difference in the key K, we can easily find $\Delta K_1, \Delta K_3$ (and ΔK_2). Then $\Delta R_0 = E^{-1}(\Delta f_1 \oplus \Delta K_1)$ and $\Delta L_0 = E^{-1}(\Delta f_2 \oplus K_2) \oplus \Delta \tilde{f}_1$. Similarly can be found the differences $\Delta L_3, \Delta R_3$.   □

The above observation clearly indicates how to reduce the search space. Instead of trying all possible differences in the key K and running Matsui's algorithm for each of them, one should only fix the input and the output differences to the S-box layers in the first three rounds. Due to restrictions on the probability, all the active S-boxes in first, in the first and second, and in the first, second and third round, should have a combined probability of at least $P_n^*/\overline{P_{n-1}}, P_n^*/\overline{P_{n-2}}, P_n^*/\overline{P_{n-3}}$, respectively. Once the active S-boxes for the first three rounds are fixed, one can easily find all 2^8 candidates for the difference in the master key and the difference in the state after the third round and hence produce 3-round differential characteristic with a fixed difference in the master key. Further, Matsui's approach can be used, and this characteristic can be extended to any number of rounds. The pseudo-code of the whole algorithm is given at Alg. 1.

On the complexity and optimization of the search. Calculating the exact time complexity of the whole search is complex and probably impossible. However, some estimate can be given, under a certain assumption. Our experiments indicate that once the difference in the state (after the third round) and in the key is fixed, extending the characteristic to n rounds becomes fairly easy and computationally cheap task. The main complexity lies in generating all 3-round related-key characteristics that have a certain probability. More precisely, from observation 2 it follows that one should generate all active S-boxes in the first round that hold with a combined probability P_1 of not less than $P_n^*/\overline{P_{n-1}}$, then all active S-boxes in the second round with a combined probability not less than $P_n^*/(\overline{P_{n-2}} \cdot P_1)$ and all active in the third round with probability of not less than $P_n^*/(\overline{P_{n-3}} \cdot P_2)$ (where P_2 is the probability of the active S-boxes in the first two rounds). Therefore, the number of all 3-round related-key characteristics depends only on the values $P_n^*/\overline{P_{n-1}}, P_n^*/\overline{P_{n-2}}$ and $P_n^*/\overline{P_{n-3}}$ – higher the values, less characteristics exist, and the search is faster.

The complexity of creating all these 3-round characteristics is not the same (or proportional) as the number of such characteristics. This comes from the fact that the linear transform E is not a surjective, since it has 32-bit input and 48-bit output. For example, after ΔK_1 is found (see the proof of the observation 2), the value $\Delta R_0 = E^{-1}(\Delta f_1 \oplus \Delta K_1)$ exists only with a probability 2^{-16}. Similar holds for ΔL_0. Hence, the optimal strategy for creating the 3-round characteristics would be to:

1. Fix the probabilities of the first four active S-boxes in the first and the third round and all the active S-boxes of the second round (that have the above limitations), without fixing the exact input-output differences. This can be done by fixing only the possible values from the difference distribution tables of the S-boxes.

Algorithm 1. Search for RK differential characteristic

FullSearch()
{
// The first three rounds
for all $\Delta f_1 \to \Delta \tilde{f}_1 | P(\Delta f_1 \to \Delta \tilde{f}_1)\overline{P_{n-1}} > P_n^*$ **do**
   for all $\Delta f_2 \to \Delta \tilde{f}_2 | P(\Delta f_1 \to \Delta \tilde{f}_1)P(\Delta f_2 \to \Delta \tilde{f}_2)\overline{P_{n-2}} > P_n^*$ **do**
      for all $\Delta f_3 \to \Delta \tilde{f}_3 | P(\Delta f_1 \to \Delta \tilde{f}_1)P(\Delta f_2 \to \Delta \tilde{f}_2)P(\Delta f_3 \to \Delta \tilde{f}_3)\overline{P_{n-3}} > P_n^*$
      do
         $V = \Delta f_1 \oplus \Delta f_3 \oplus E(\Delta \tilde{f}_2)$
         for all $\Delta K | \mathcal{L}(\Delta K) = V$ **do**
         $\Delta K_1 = PC2(rot(\Delta K, 1))$
         $\Delta K_2 = PC2(rot(\Delta K, 2))$
         if $E^{-1}(\Delta K_1 \oplus \Delta f_1)$ and $E^{-1}(\Delta K_2 \oplus \Delta f_2)$ **then**
           $\Delta R_0 = E^{-1}(\Delta K_1 \oplus \Delta f_1)$
           $\Delta L_0 = E^{-1}(\Delta K_2 \oplus \Delta f_2) \oplus \Delta \tilde{f}_1$
           $\Delta R_3 = \Delta L_0 \oplus \Delta \tilde{f}_1 \Delta \tilde{f}_3$
           $\Delta L_3 = \Delta R_0 \oplus \Delta \tilde{f}_2$
           Call NextRound(ΔL_3, ΔR_3, ΔK, $P(\Delta f_1 \to \Delta \tilde{f}_1)P(\Delta f_2 \to \Delta \tilde{f}_2)P(\Delta f_3 \to \Delta \tilde{f}_3)$, 4)
         end if
         end for
      end for
   end for
end for
}

NextRound($\Delta L, \Delta R, \Delta K, p, round$)
{
$\Delta K_r = PC2(rot(\Delta K, round))$
$\Delta f = \Delta K_r \oplus E(\Delta R)$
for all $\Delta f \to \Delta \tilde{f} | P(\Delta f \to \Delta \tilde{f}) \cdot p \cdot \overline{P_{n-round}} > P_n^*$ **do**
   $\Delta L_{new} = \Delta R$
   $\Delta R_{new} = \Delta L \oplus \Delta \tilde{f}$
   if $round == n$ **then**
      if $P(\Delta f \to \Delta \tilde{f}) \cdot p > P_n^*$ **then**
         $P_n^* = P(\Delta f \to \Delta \tilde{f}) \cdot p$
      end if
   else
      Call NextRound($\Delta L_{new}, \Delta R_{new}, \Delta K, P(\Delta f \to \Delta \tilde{f}) \cdot p, round + 1$)
   end if
end for
}

---

2. Fix the input differences to the four S-boxes of round 1,3, and the output differences of the S-boxes of round 2 (that correspond to the previously fixed distribution values).
3. Find 28 bits of ΔK, then find 28 bits of ΔK_1 and check if there exist preimage of 24 bits of $\Delta f_1 \oplus \Delta K_1$ for E. This can be done, since the left and the right 28-bit halves of the key are independent.
4. If exists, fix the probabilities of the last four active S-boxes in the first and the third round.
5. Fix the input differences to these 8 S-boxes.
6. Find the rest 28 bits of ΔK, then of ΔK_1 and check if there exist preimage of last 24 bits of $\Delta f_1 \oplus \Delta K_1$ for E.
7. If exists, find ΔK_2, fix the input difference to the S-boxes in the second round and check if there exist a preimage of $\Delta f_2 \oplus K_2$ for E.
8. If exists, fix the output differences for the S-boxes of round 3 (it is not necessary to fix the outputs of S-boxes of round 1).

Although we cannot give a precise estimate for the complexity of creating all 3-round characteristics, we can give such estimates for some particular fixed values of $P_n^*, \overline{P_{n-1}}, \overline{P_{n-2}}$, and $\overline{P_{n-2}}$. For example, when $P_n^*/\overline{P_{n-1}} = 2^{-3}, P_n^*/\overline{P_{n-2}} = 2^{-6}, P_n^*/\overline{P_{n-3}} = 2^{-9}$, then steps 1-8 are repeated $2^{16.7}, 2^{28.9}, 2^{32.9}, 2^{27.3}, 2^{30.8}, 2^{34.9}, 2^{27.6}, 2^{20.3}$ times, respectively, leading to a total complexity of around 2^{35}. On the other hand, when $P_n^*/\overline{P_{n-1}} = 2^{-3}, P_n^*/\overline{P_{n-2}} = 2^{-7}, P_n^*/\overline{P_{n-3}} = 2^{-10}$, then steps 1-8 are repeated $2^{18.7}, 2^{32.4}, 2^{36.4}, 2^{30.8}, 2^{34.3}, 2^{38.4}, 2^{30.9}, 2^{22.6}$ times, respectively, and hence the complexity is around 2^{39}, while there exist around $2^{22.6}$ (step 8) good 3-round related-key characteristics.

3.3   The Split Approach

To build the best n-round characteristic Matsui's approach requires first to build the best characteristics on $1, 2, \ldots, n-1$ rounds because it uses the probabilities of these characteristics. One may be able to skip building the characteristics on some rounds and to assume that they have the same probability as the characteristic on lower number of rounds. Under this assumption, the algorithm still works and finds the best characteristic on n rounds, however the time complexity usually suffers significantly.

Avoiding building all round-reduced characteristics can be done with a different approach. Let us assume we search for characteristic on n rounds that has a probability of at least P_n^*. This n-round characteristic can be seen as a concatenation of two $n/2$-round characteristics, with a combined probability of at least P_n^*. Therefore, one of these two characteristics has a probability of at least $\sqrt{P_n^*}$. Indeed we can split the n-round characteristic on any (reasonable) number of k characteristics, each on n/k rounds, and claim that at least one of them has a probability of $\sqrt[k]{P_n^*}$.

Now, let us assume that $n = 3k$, and the n-round characteristic has been split into k three-round characteristics. One of these characteristics (we do not know exactly which), has to have a probability of at least $\sqrt[k]{P_n^*}$. Since it is on three rounds, and it has a bound on its probability, we can use our previous method (observation 2), to build all such characteristics. However, unlike in Matsui's approach, where each of the three rounds has some bound on probability, now we build 3-round characteristics that only have the bound on the combined probability (of all three rounds). Once we have built all such the 3-round characteristics we try to extend them to n rounds (recall that if the difference in the state and in the key is fixed, then it is easy to extend it to more rounds – the difficulty lies in creating all such 3-round characteristics). Interestingly, when extending the three round characteristics, we can use the bounds from Matsui's approach.

For example, let us assume we want to build a characteristic on 9 rounds with a probability at least 2^{-24}. Then we know that one of the three 3-round characteristics has a probability of at least 2^{-8}. First we assume that this is the characteristic on the first three rounds. We build all first 3-round characteristics with probability at least 2^{-8}, i.e. $P_3 \geq 2^{-8}$, and then try to extend them 6 rounds forward, thus obtaining a characteristic on 9 rounds. If we have the probabilities $\overline{P_1}, \ldots, \overline{P_6}$ for the best characteristics on the last 6 rounds, then for rounds 4-9, we can use Matsui's approach, e.g. for 4 rounds we take only those with P_4 such that $P_4 \cdot \overline{P_5} \geq 2^{-24}$, for 5 rounds $P_5 \cdot \overline{P_4} \geq 2^{-24}$, etc. If we do not have the best probabilities than for each round i $(i \geq 4)$ we only check if $P_i \geq 2^{-24}$. Then we assume the characteristic on rounds 4-6 has a probability of at least 2^{-8}. Again, we build all 3-round characteristics with at most 2^{-8} and extend them three rounds forward and three backwards (by using Matsui's bounds). Finally, we assume this is the 3-round characteristic on the last three rounds (7-9). We build all such characteristics and extend them 6 rounds backwards (again we can use Matsui's bounds if we have the best probabilities for the first 6 rounds). Among all 9-round characteristics we have produced in these three iterations, we take the one with the highest probability. If such characteristic exist than it is the best characteristic on 9 rounds and it has a probability at least 2^{-24}. If it does not exist then it means all the characteristics on 9 rounds have probability lower than 2^{-24}.

What is the real advantage of this approach compared to related-key Matsui's approach? To find this out, we have to compare the number of possible 3-round related-key characteristic built in the two approaches. In Matsui's algorithm, this number depends on the values $P_n^*/\overline{P_{n-1}}$, $P_n^*/\overline{P_{n-2}}$ and $P_n^*/\overline{P_{n-3}}$, while in the split approach, the number depends only on P_n^*. Hence, when the probabilities $\overline{P_{n-1}}$, $\overline{P_{n-2}}$, $\overline{P_{n-3}}$ are really high, then it is computationally cheaper to build the n-round characteristic with the split approach.

4   The Case of DES

The notion of (single-key) differentials and differential characteristics was introduced in the seminal paper of Biham and Shamir [2] on cryptanalysis of DES, where the authors presented characteristic on 15 rounds of DES with a probability higher than 2^{-56}. Later in [3], the authors used 13-round characteristic to give the first attack on all 16 rounds of DES. By performing a full search, Matsui [7] has shown that the characteristics found by Biham and Shamir were actually the best round-reduced single-key characteristics for DES. It is well known that S-boxes and the permutation used in the round function of DES are very carefully chosen to avoid single-key differential cryptanalysis and even subtle changes in them can weaken the cipher [3]. Our study of related-key attacks on DES is motivated by the fact that differences in the subkeys could violate some of the design principles and this could lead to new attacks on DES.

We would like to run a full search of the space of all related-key differential characteristics in DES by using the approaches of the previous section. We start with the related-key version of Matsui's algorithm and try to find the best related-key characteristics on as many rounds as possible. Although our search will always find the best characteristics, we should keep in mind that we have a limited computational power. For example, if we try to find the best n-round related-key characteristic that holds with a probability at least P_n^*, then the time complexity of the search mostly depends on the probability $\overline{P_{n-3}}$ of the best characteristics on $(n-3)$ rounds (but also depends on $\overline{P_{n-1}}, \overline{P_{n-2}}$). Our experimental results show that when $P_n^*/\overline{P_{n-3}} < 2^{-12} \sim 2^{-14}$ we do not have the resources to perform the search, hence if for some n this holds, then we will switch to the split approach and continue further with this approach. Note that even in the case of single-key characteristics a similar limitation holds when for some n the ratio $P_n^*/\overline{P_{n-2}}$ is too low.

We start the search by finding the best related-key characteristic on 3 rounds (we assume that $\overline{P_0} = \overline{P_1} = \overline{P_2} = 1$). We fix P_3^* (the probability of the best related-key 3-round characteristic) to 2^{-1} and then gradually decrease by a factor of 2^{-1} if we do not find a characteristic that holds with this probability. There is always a lower bound on this probability – the case of the single-key characteristic (our tool does not make distinction between these two cases, and searches for both). Hence, we can be sure that P_3^* cannot be lower than 2^{-4} (this is the probability of the best single-key characteristic on 3 rounds). Having found the highest P_3^*, we fix $\overline{P_3} = P_3^*$, and then search for $\overline{P_4}$. We fix P_4^* to $\overline{P_3}$, i.e. we assume that the characteristic on 4-rounds has the same probability as the best characteristic on 3 rounds, and then gradually decrease this probability by a factor 2^{-1} each time when we cannot find 4-round characteristic with such probability. Up to $\overline{P_6}$ we could easily perform the search. However, when searching for $\overline{P_7}$ we could not find anything even when P_7^* was set up to 2^{-18}. We knew that $\overline{P_7}$ could not be lower than $2^{-23.6}$ (the probability of the single-key characteristic on 7 rounds), however if we set $P_7^* = 2^{-23.6}$, then $P_7^*/\overline{P_4} = 2^{-19}$ which is lower than our maximal computational limit of $2^{-12} \sim 2^{-14}$. Therefore, we switched to the split approach for finding the best 7-round related-key characteristic. We started with all possible 3.5-round characteristic (with the first 3.5 rounds and the last 3.5 rounds) with probability of at least 2^{-11} and tried to extend it to 7 rounds, thus we allowed a probability of 2^{-22}. The split approach found that the best related-key characteristic on 7 rounds has a probability of $2^{-20.38}$.

The results of the split search on 7 rounds can be used to find if 8-round characteristic with 2^{-22} exist, which in our case was negative. If we try to apply the related-key Matsui's approach for 8 rounds and allow $P_8^* = 2^{-22}$, then $P_8^*/\overline{P_5} = 2^{-22}/2^{-7.6} = 2^{-14.4}$, which is low. Hence, for 8 rounds we could not use neither Matsui's nor the split approach. However, we noted that the best characteristics of the first 7 rounds have a difference only in a few bits of the key. Hence, we ran a limited search for 8-round characteristic by allowing only a few bit difference in the key. The limited search gave us a characteristic with a probability $2^{-29.75}$ – better than the best single-key characteristic with $2^{30.8}$.

For higher rounds, the related-key Matsui's approach could not work because of the low probabilities $(P_n^*/\overline{P_{n-3}} < 2^{-12} \sim 2^{-14})$. However, if we assume that the 8-round characteristic found by the limited Matsui's approach is the best, then we can still run related-key Matsui's algorithm for the characteristic on 11 rounds. We found that if this holds, then the best related-key characteristics on 11 rounds is the best single-key characteristics.

For finding the best related-key characteristics on 9, 12, and 13 rounds we used our split approach. For 9 rounds, we allowed the 3-round characteristics to have at least $2^{-10.55}$ (because $(2^{-10.55})^3 = 2^{-31.65}$ and the best single-key on 9 rounds has $2^{-31.48}$). The search found that the best 9-round related-key characteristic is the best single-key characteristic. For 12 and 13 rounds, we allowed the starting 3-round characteristics with probability at least $2^{-11.85}$ (because $(2^{-11.85})^4 = 2^{-47.4}$ and the best single-key on 13 rounds has $2^{-47.22}$). Again, we obtained similar results – the best related-key characteristics on 12 and 13 rounds have no difference in the key, i.e. they are the single-key characteristics.

The result for the 13-round[4] related-key characteristic is especially interesting since Biham-Shamir analysis uses it for the attack on the whole DES. This means that *if the attacker uses related-key characteristics, he cannot improve the complexity of Biham-Shamir attack.*

The summary of our findings is presented in Tbl. 1. The related-key characteristics for 7 and 8 rounds are given in the Appendix (Fig. 1, 2).

5   The Case of DESL

DESL [6] uses a single S-box instead of eight different S-boxes as in DES. This S-box has a special design criteria to discard high probability (single-key) differential characteristics. Indeed, our initial analysis for single-key differential characteristics in DESL confirmed this result. Moreover, we could not find the best single-key differential characteristics (using the original Matsui's tool) for DESL for higher rounds (the absence of the probabilities for the best round-reduced single-key differential characteristics in the submission paper of DESL [6] seems

---

[4] We rerun the search for characteristics that cover rounds 1 to 12.

Table 1. Comparison of the probabilities of the best round-reduced differential single-key and related-key characteristics for DES

rounds	Single-key	Related-key	Method used
3	$2^{-4.0}$	2^0	RK Matsui's
4	$2^{-9.6}$	$2^{-4.61}$	RK Matsui's
5	$2^{-13.21}$	$2^{-7.83}$	RK Matsui's
6	$2^{-19.94}$	$2^{-12.92}$	RK Matsui's
7	$2^{-23.60}$	$2^{-20.38}$	Split
8	$2^{-30.48}$	$2^{-29.75} \leq \overline{P_8} < 2^{-22}$	Limited Matsui's
9	$2^{-31.48}$	$2^{-31.48}$	Split + Matsui's
10	$2^{-38.35}$	$\leq \overline{P_9}$	
11	$2^{-39.35}$	$2^{-39.35}$ if $\overline{P_8} = 2^{-29.75}$	RK Matsui's
12	$2^{-46.22}$	$2^{-46.22}$	Split + Matsui's
13	$2^{-47.22}$	$2^{-47.22}$	Split + Matsui's
14	$2^{-54.09}$	$\leq \overline{P_{13}}$	
15	$2^{-55.09}$	$2^{-55.09}$	RK Matsui's
16	$2^{-61.97}$	$\leq \overline{P_{15}}$	

to confirm our findings). Therefore, even the original Matsui's tool cannot be used (it is infeasible) for finding single-key characteristics, when they hold with low probabilities.

Our related-key Matsui's search algorithm, however, did find the best related-key characteristics for up to 7 rounds. Interestingly, the probabilities of these related-key characteristics are higher in DESL, than in DES (see Tbl. 2). For more rounds, we used the split approach as well. Nonetheless, for these characteristics, we were able to find only the upper bounds on their probabilities. For example, for 9-round related-key characteristic we used the split approach with 3-round probability of 2^{-10}. After running the search for the first, middle, and third three rounds, the algorithm did not return any characteristic. This means, there are no related-key characteristics on 9 rounds with probability at least 2^{-30}. Similarly, we used the split approach for finding the upper bound on the probability of the best characteristics for 12-rounds, and the related-key Matsui's approach for the bounds on 10,13, and 15 rounds. Our findings are presented in Tbl. 2.

The related-key characteristics that we have found can be used to launch boomerang attacks on the round-reduced cipher. For example, we can launch a related-key boomerang attack on 12 rounds (from round 4 to round 15), with two characteristics on 6 rounds – the first on rounds 4-9, the second on 10-15. The probability of the first characteristic is $2^{-14.68}$ (it is lower because we consider rounds 4-9), while the probability of the second is $2^{-12.09}$. Therefore, the probability of the whole boomerang is $2^{-2 \cdot 14.68 - 2 \cdot 12.09} = 2^{-53.54}$.

Table 2. Probabilities of the best round-reduced related-key differential characteristics for DESL

Round	Probability
3	2^0
4	$2^{-4.67}$
5	$2^{-7.24}$
6	$2^{-12.09}$
7	$2^{-19.95}$
8	$\leq \overline{P_7}$
9	$< 2^{-30}$
10	$< 2^{-31}$
11	$\leq \overline{P_{10}}$
12	$< 2^{-40}$
13	$< 2^{-41}$
14	$\leq \overline{P_{13}}$
15	$< 2^{-50}$
16	$< 2^{-51}$

6   The Case of s^2DES

Another variant of DES called s^2DES was proposed in [5]. The search for the best single-key differential characteristics in s^2DES was performed in [10]. For this purpose the authors used Matsui's tool. This analysis showed that the best round-reduced differential characteristics in s^2DES have higher probabilities than in DES.

We ran our search for related-key characteristics using only our related-key approached based on Matsui's algorithm. We noted that for each single-key characteristic on n-rounds, the value $\overline{P_n}/\overline{P_{n-3}}$ is at least $2^{-12.75}$ (for $n = 8$, see Tbl. 3), hence building all 3-round related-key characteristic might be feasible. However, the values $\overline{P_{n-3}}$ for different n could be updated, because they were the probabilities in the single-key scenario (the probability in the related-key scenario is not less than in the single-key). Indeed, the probabilities of the round-reduced related-key characteristics for the first 6 rounds, were higher than the probabilities of the single-key characteristics. This made $\overline{P_5}$ to be 2^{-8} instead of $2^{-9.22}$ as in the single-key case. Hence, for the related-key characteristic on 8 rounds, we had to allow $\overline{P_8}/\overline{P_5} = 2^{-22}/2^{-8} = 2^{-14}$ for the active S-boxes in the three rounds, instead of the previous $2^{-12.75}$. However, we were able to perform the search for this 7-round characteristic but with a significant computational cost – the search took around 3 weeks on 64 CPU cores.

After the sixth round, we found that all the best related-key characteristics have the same probability as the single-key (indeed they are single-key). The probabilities of the best single and related-key round-reduced characteristics are given in Tbl. 3.

Table 3. Comparison of the probabilities of the best round-reduce differential single-key and related-key characteristics for s^2DES

rounds	Single-key	Related-key
3	$2^{-4.39}$	2^0
4	$2^{-6.8}$	$2^{-5.19}$
5	$2^{-9.22}$	$2^{-8.0}$
6	$2^{-14.35}$	$2^{-12.61}$
7	$2^{-17.03}$	$2^{-17.03}$
8	$2^{-21.96}$	$2^{-21.96}$
9	$2^{-22.71}$	$2^{-22.71}$
10	$2^{-27.35}$	$2^{-27.35}$
11	$2^{-28.39}$	$2^{-28.39}$
12	$2^{-34.07}$	$2^{-34.07}$
13	$2^{-34.07}$	$2^{-34.07}$
14	$2^{-39.75}$	$2^{-39.75}$
15	$2^{-39.75}$	$2^{-39.75}$
16	$2^{-45.42}$	$2^{-45.42}$

7   Conclusions

We have presented the first algorithms for automatic search of the best round-reduced related-key differential characteristics in DES-like family of ciphers, DES, DESL, and s^2DES. We have shown that there is no significant difference between the probabilities of the best related-key and the best single-key characteristics on higher number of rounds of DES, and thus, the key schedule of DES has no notable weakness regarding differential attacks.

We believe our algorithms can be applied to similar 64-bit state and 64-bit key bit-oriented ciphers with linear key schedule. Moreover, our approaches can be used to search for high probability (up to 2^{-20}) related-key differential characteristics in any bit oriented ciphers with linear key schedule.

References

1. Aoki, K., Kobayashi, K., Moriai, S.: Best differential characteristic search of FEAL. In: Biham, E. (ed.) FSE 1997. LNCS, vol. 1267, pp. 41–53. Springer, Heidelberg (1997)
2. Biham, E., Shamir, A.: Differential cryptanalysis of DES-like cryptosystems. J. Cryptology 4(1), 3–72 (1991)
3. Biham, E., Shamir, A.: Differential cryptanalysis of the full 16-round DES. In: Brickell, E.F. (ed.) CRYPTO 1992. LNCS, vol. 740, pp. 487–496. Springer, Heidelberg (1993)
4. Biryukov, A., Nikolić, I.: Automatic Search for Related-Key Differential Characteristics in Byte-Oriented Block Ciphers: Application to AES, Camellia, Khazad and Others. In: Gilbert, H. (ed.) EUROCRYPT 2010. LNCS, vol. 6110, pp. 322–344. Springer, Heidelberg (2010)

5. Kim, K.: Construction of DES-like S-boxes based on boolean functions satisfyieng the SAC. In: Imai, H., Rivest, R.L., Matsumoto, T. (eds.) ASIACRYPT 1991. LNCS, vol. 739, pp. 59–72. Springer, Heidelberg (1991)

6. Leander, G., Paar, C., Poschmann, A., Schramm, K.: New lightweight DES variants. In: Biryukov, A. (ed.) FSE 2007. LNCS, vol. 4593, pp. 196–210. Springer, Heidelberg (2007)

7. Matsui, M.: On correlation between the order of S-boxes and the strength of DES. In: De Santis, A. (ed.) EUROCRYPT 1994. LNCS, vol. 950, pp. 366–375. Springer, Heidelberg (1995)

8. National Bureau of Standards. Data Encryption Standard. U.S. Department of Commerce, FIPS pub. 46 (January 1977)

9. Schneier, B., Kelsey, J., Whiting, D., Wagner, D., Hall, C., Ferguson, N.: The Twofish encryption algorithm: a 128-bit block cipher. John Wiley & Sons, Inc., New York (1999)

10. Tokita, T., Sorimachi, T., Matsui, M.: Linear cryptanalysis of LOKI and s^2DES. In: Pieprzyk, J., Safavi-Naini, R. (eds.) ASIACRYPT. LNCS, vol. 917, pp. 293–303. Springer, Heidelberg (1994)

A   Related-Key Characteristics for DES

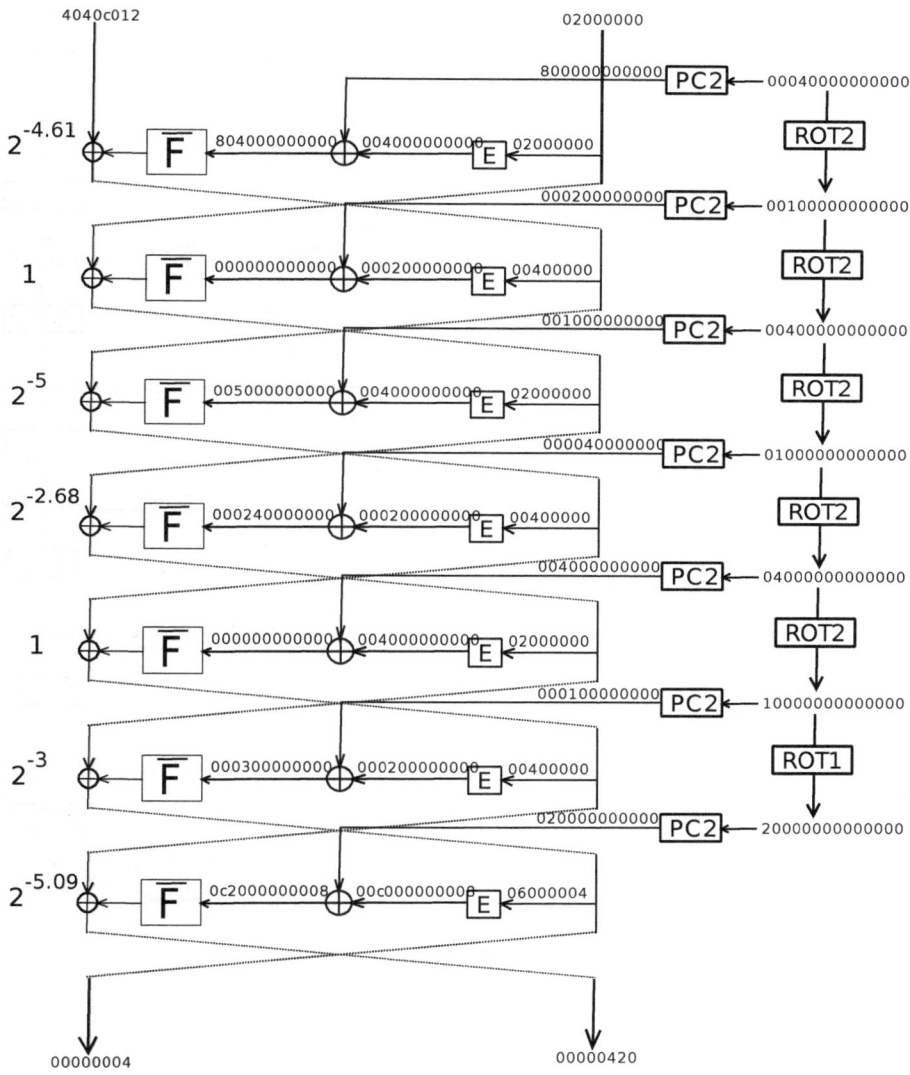

Fig. 1. The best related-key differential characteristic (with probability $2^{-20.38}$) on the last 7 rounds of DES

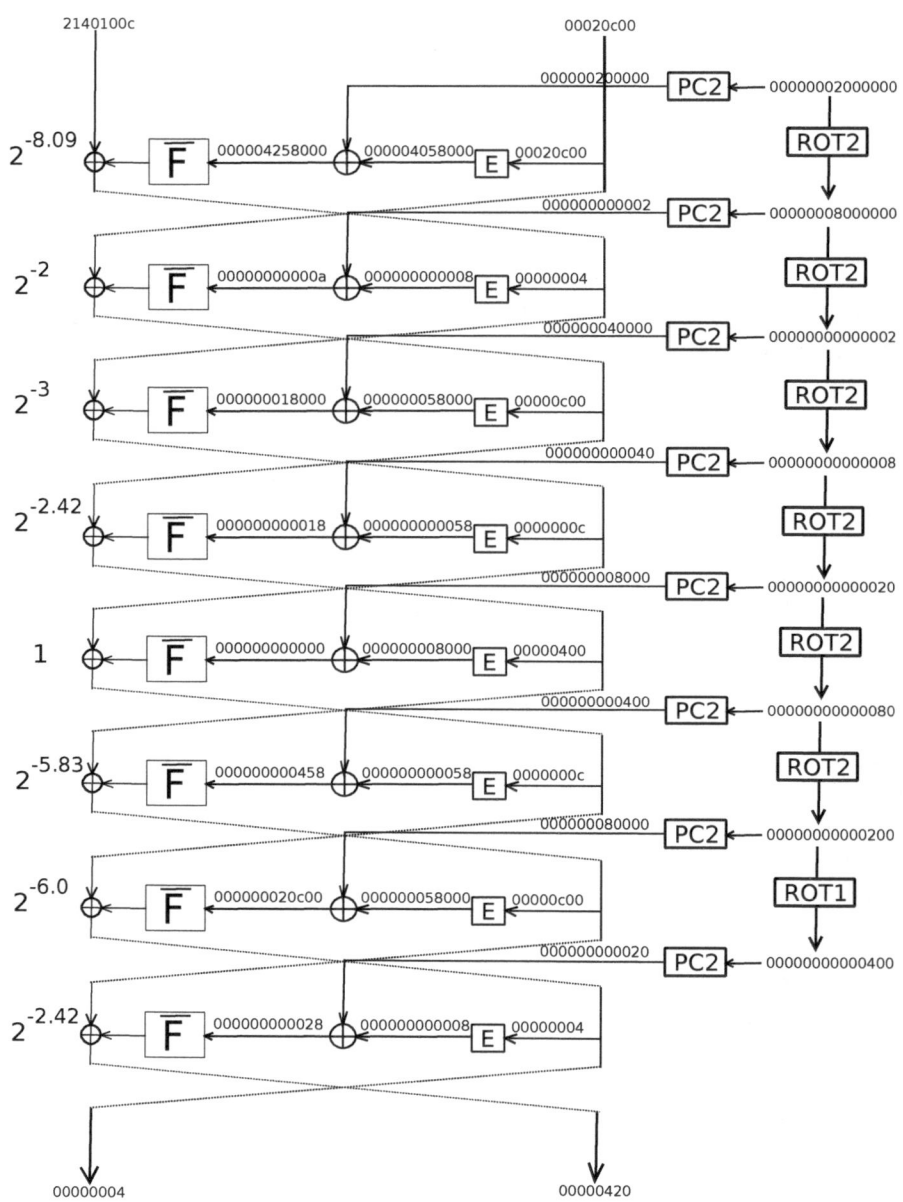

Fig. 2. Related-key differential characteristic (with probability $2^{-29.75}$) on the last 8 rounds of DES

Multiple Differential Cryptanalysis: Theory and Practice

Céline Blondeau and Benoît Gérard

SECRET Project-Team - INRIA Paris-Rocquencourt
Domaine de Voluceau - B.P. 105 - 78153 Le Chesnay Cedex - France
{celine.blondeau,benoit.gerard}@inria.fr

Abstract. Differential cryptanalysis is a well-known statistical attack
on block ciphers. We present here a generalisation of this attack called
multiple differential cryptanalysis. We study the data complexity, the
time complexity and the success probability of such an attack and we
experimentally validate our formulas on a reduced version of PRESENT.
Finally, we propose a multiple differential cryptanalysis on 18-round
PRESENT for both 80-bit and 128-bit master keys.

Keywords: iterative block cipher, multiple differential cryptanalysis,
PRESENT, data complexity, success probability, time complexity.

1 Introduction

Differential cryptanalysis has been introduced in 1990 by Biham and Shamir
[4,5] in order to break the *Data Encryption Standard* block cipher. This statistical cryptanalysis exploits the existence of a *differential*, *i.e.*, of a pair (α, β) of
differences such that for a given input difference α, the output difference after
encryption equals β with a high probability. This attack has been successfully
applied to many ciphers and has been extended to various different attacks, such
as truncated differential cryptanalysis, impossible differential cryptanalysis...

In the original version of differential cryptanalysis [4], a unique differential is
exploited. Then, Biham and Shamir have improved their attack by considering
together several differentials having the same output difference [5]. Truncated
differential cryptanalysis introduced by Knudsen [16] uses differentials with many
output differences that are structured as a linear space.

Here, we consider what we name *multiple differential cryptanalysis*. Similarly
to multiple linear cryptanalysis, multiple differential cryptanalysis is the general
case where the set of considered differentials has no particular structure, *i.e.*,
several input differences are considered together and the corresponding output
differences can be different from an input difference to another.

The problem of estimating the data complexity, time complexity and success
probability of a differential cryptanalysis is far from being simple. Since 1991, it
is widely accepted that the data complexity of a differential cryptanalysis is of
order p_*^{-1}, where p_* denotes the probability of the involved differential [5]. Theoretical studies based on hypothesis testing theory [2,3,7] confirm this statement

A. Joux (Ed.): FSE 2011, LNCS 6733, pp. 35–54, 2011.

and give more specific results. Concerning the success probability, a formula has been recently established by Selçuk in [23]. This formula, which is used in many recent papers on differential cryptanalysis, is derived from a Gaussian approximation of the binomial distribution. However, as already explained by Selçuk, the Gaussian approximation is not good in the setting of differential cryptanalysis. This was the motivation of the general framework presented in [9], that studies the complexity of any statistical cryptanalysis based on counters that follow a binomial distribution. But, this work does not apply to multiple differential cryptanalysis since the involved counters do not follow a binomial distribution in this case.

Our contribution. The main purpose of this paper is to provide a detailed analysis of the complexity of any multiple differential attack. It is worth noticing that it includes the variants of differential attacks such as classical differential cryptanalysis or truncated differential cryptanalysis. In Section 2, we introduce multiple differential cryptanalysis and study the complexity of this attack. We mainly provide formulas for the data complexity and the success probability of a multiple differential cryptanalysis. Then, in Section 3, we validate this theoretical framework by many experiments on a reduced version of the cipher PRESENT, namely SMALLPRESENT-[8]. Then, Section 4 focuses on the general problem of computing the involved probabilities. This problem arises in any statistical attack and is not directly related to the use of several differentials. Finally, to conclude this work, we propose a multiple differential cryptanalysis of 18-round PRESENT. This attack is not the best known attack on PRESENT since Cho has presented attacks up to 26 rounds [11]. Nevertheless, it improves the best previously known differential cryptanalysis on 16 rounds due to Wang [24].

2   Theoretical Framework

In this first section, we propose a framework for analysing multiple differential cryptanalyses. More precisely we provide estimates for the data complexity and the success probability of such differential attacks that use any number of differentials. The time and memory complexities of these attacks are also discussed.

2.1   Presentation and Notation

Let us start with some notation that will be used all along this paper. We consider an iterative block cipher parametrised by a key K.

$$E_K : \mathbb{F}_2^m \to \mathbb{F}_2^m$$
$$x \mapsto y = E_K(x),$$

where m is the block size. We denote by F the round function of this iterative cipher: $F_k(x)$ is the result of 1-round encryption of x using the subkey k. A multiple differential cryptanalysis aims at recovering the key K_* used to encipher the available samples. We consider here a last-round differential cryptanalysis on an iterative block cipher that recovers n_k bits of the last-round subkey that we will denote by k_* (this subkey is derived from the master key K_*). Such an attack belongs to the class of statistical cryptanalyses and thus follows the three following steps.

- *Distillation phase:* Extract the information on k_* obtained from the N available plaintext/ciphertext pairs.
- *Analysis phase:* From this information, compute the likelihoods of the candidates for the value of k_* and generate the list $\mathcal{L}$ of the best ℓ candidates.
- *Search phase:* Look down the list of candidates and test all the corresponding master keys until the good one is found.

Now, let us introduce the notation used for the differentials.

Definition 1. *[19] An r-round differential for a block cipher is a couple of differences $(\delta_0, \delta_r) \in \mathbb{F}_2^m \times \mathbb{F}_2^m$. The probability of the differential is defined by*

$$\Pr[\delta_0 \to \delta_r] \stackrel{\text{def}}{=} \Pr_{\mathbf{X},\mathbf{K}}[E_{\mathbf{K}}(\mathbf{X}) \oplus E_{\mathbf{K}}(\mathbf{X} \oplus \delta_0) = \delta_r].$$

In the setting of multiple differential cryptanalysis, the attacker exploits a collection Δ of differentials. The natural way of ordering these differentials is to gather the differentials with the same input difference. We denote by Δ_0 the set of all input differences involved in the set Δ

$$\Delta_0 \stackrel{\text{def}}{=} \{\delta_0, \exists \delta_r, (\delta_0, \delta_r) \in \Delta\}.$$

We number the input differences in Δ_0: $\Delta_0 = \{\delta_0^{(1)}, \ldots, \delta_0^{(|\Delta_0|)}\}$. Hence, for a fixed input difference $\delta_0^{(i)} \in \Delta_0$, we obtain a set $\Delta_r^{(i)}$ of the corresponding output differences:

$$\Delta_r^{(i)} \stackrel{\text{def}}{=} \{\delta_r \mid (\delta_0^{(i)}, \delta_r) \in \Delta\}.$$

Therefore, if we number thesse sets of output differences, the set of differentials Δ can be expressed as

$$\Delta = \left\{ \left(\delta_0^{(i)}, \delta_r^{(i,j)}\right) \; \middle| \; i = 1 \ldots |\Delta_0| \text{ and } j = 1 \ldots |\Delta_r^{(i)}| \right\}.$$

It is worth noticing that this definition is more general than truncated differential cryptanalysis since the set of output differences can be different from an input difference to another.

As in differential cryptanalysis, the algorithm used in multiple differential cryptanalysis consists in partially deciphering the N ciphertexts using all possible values for the last-round subkey and in counting the number of occurrences of the differentials in Δ^1. In other words, we count the number of plaintext pairs with a difference $\delta_0^{(i)}$ in Δ_0 that lead to an output difference in $\Delta_r^{(i)}$ after

---

[1] This way of combining differentials may not be optimal but it is the one used in all published attacks. Considering other techniques is out of the scope of this paper that aims at providing formulas for better estimating complexities of previous attacks.

r rounds. However, this attack (as it is) may not work because the cost of the partial decryption is prohibitive (there are too many pairs of ciphertexts and too many possible values for the subkey). In order to decrease this cost, a *sieving phase* is used[2] to discard some pairs for which we already know that the difference after r rounds cannot be in $\Delta_r^{(i)}$. This phase consists in precomputing the sets $\Delta_{r+1}^{(i)}$ of all δ_{r+1} in $\mathbb{F}_2^m$ such that there exists a j for which $\Pr\left[\delta_r^{(i,j)} \to \delta_{r+1}\right] \neq 0$ and in discarding every pair with an output difference not in $\Delta_{r+1}^{(i)}$. This set of differences is named a *sieve*. The multiple differential attack is summarised in Algorithm 1.

Algorithm 1. Multiple differential cryptanalysis

Input: N chosen plaintext/ciphertext pairs (x_i, y_i) with $y_i = E_{K_*}(x_i)$
Output: The key K_* used to encipher the samples

1 Initialise a table D of 2^{n_k} counters to 0.
2 **foreach** $\delta_0^{(i)} \in \Delta_0$ **do**
3   **foreach** *plaintext pair* (x_a, x_b) *such that* $x_b = x_a \oplus \delta_0^{(i)}$ **do**
4   **if** $y_a \oplus y_b \in \Delta_{r+1}^{(i)}$ **then**
5   **foreach** candidate k **do**
6   Compute $\delta = F_k^{-1}(y_a) \oplus F_k^{-1}(y_b)$;
7   **if** $\delta \in \Delta_r^{(i)}$ **then** $D[k] \leftarrow D[k] + 1$;
8 Generate a list $\mathcal{L}$ of the ℓ candidates with the highest values of $D[k]$;
9 **foreach** $k \in \mathcal{L}$ **do**
10   **foreach** possible master key K corresponding to k **do**
11   **if** $E_K(x) = y = E_{K_*}(x)$ **then return** K;

Such attacks are successful when the correct subkey is in the list $\mathcal{L}$ of candidates. Four important quantities have to be taken into consideration when quantifying the efficiency of a statistical cryptanalysis. The *success probability* P_S that is the probability of the correct subkey to be in the list of the best candidates,

$$P_S \overset{\text{def}}{=} \Pr\left[k_* \in \mathcal{L}\right],$$

the *data complexity* N that is the number of plaintext/ciphertext pairs used for the attack, the *time complexity* that heavily depends on the size ℓ of the list $\mathcal{L}$ and the *memory complexity*. The first three quantities are closely related since increasing N will increase P_S and increasing ℓ will also increase P_S together with the time complexity. We now study the time and memory complexities, while formulas for the data complexity and the success probability are provided in Section 2.4.

Remark. In a multiple differential attack, the *number of chosen plaintexts* N and the *number of samples* N_s are different quantities. The number of samples corresponds to the number of pairs with a difference in Δ_0 that we can form

---

[2] This is widely used in differential cryptanalysis.

with N plaintexts. In an attack with $|\Delta_0|$ input differences, we can choose the plaintexts such that the number of samples is $N_s = \frac{|\Delta_0| N}{2}$. This is done by choosing the plaintext set of the form $\bigcup_x \{x \oplus \delta, \delta \in Vect(\Delta_0)\}$ where $Vect(\Delta_0)$ is the linear space spanned by the elements of Δ_0. Such sets are classically named *structures*.

2.2   Time and Memory Complexities

In this section we discuss the details of Algorithm 1 in order to compute the time and the memory complexities of the multiple differential cryptanalysis defined in Algorithm 1.

In order to analyse the time complexity of this attack we introduce some notation. Let $S_r \overset{\text{def}}{=} \max_i\{|\Delta_r^{(i)}|\}$ and $S_{r+1} \overset{\text{def}}{=} \max_i\{|\Delta_{r+1}^{(i)}|\}$. We denote by p_{sieve} the maximum over all input differences in Δ_0 of the probability to pass the sieve *i.e.* $p_{sieve} = 2^{-m} S_{r+1}$.

When performing a multiple differential cryptanalysis, one needs to check many times if some difference belongs to a particular set A of differences. This step of the algorithm can be done with a time complexity logarithmic in $|A|$. On the other hand, this requires the use of $|A|$ memory blocks. Now let us consider each important step of the algorithm.

The total number of pairs to test is $N_s = |\Delta_0| N / 2$. For each pair we have to check if it passes the sieve. Thus the time complexity of this step is $N_s \log(S_{r+1})$. Nevertheless, one can decrease this complexity using the following simple trick. If there exists a set of positions in $\{1 \cdots m\}$ on which all elements in Δ_{r+1} vanish, then the plaintext/ciphertext pairs can be gathered depending on the values of the ciphertexts on these bits. Pairs formed by ciphertexts belonging to two different groups will not pass the sieve and thus only the pairs formed by ciphertexts in the same group must be considered. Using this trick together with plaintexts chosen to form structures, this step can take negligible time regarding the rest of the attack. Since the proposed cryptanalysis is a last-round attack, a partial inversion of the round function has to be performed for each pair that passes the sieve and for each last-round subkey. Therefore this step has a complexity of about $2^{n_k} N_s p_{sieve}$. Extracting the likeliest ℓ subkeys can be handled in linear time (regarding the number of candidates 2^{n_k}). The last part of the algorithm corresponds to an exhaustive search for the remaining bits of the master key. This step requires $\ell \cdot 2^{n_K - n_k}$ encryptions where n_K is the size of the master key.

Table 1 summarises the time complexities. The terms corresponding to steps with a small time complexity are neglected here, and it is assumed that the generation of the pairs has been done using the aforementioned trick.

The partial decryption cost can be seen as a the $1/(r+1)$-th of the cost of an encryption for an $(r+1)$-round cipher. The memory complexity of the attack is essentially due to the storage of the counters, of the plaintext/ciphertext pairs and of the sieves.

Table 1. Time complexity of a multiple differential cryptanalysis where S_r (resp. S_{r+1}) denote the maximal number of output differences for a given input difference in Δ_0 after r-rounds (resp. $(r + 1)$-rounds)

Encryptions	Partial decryptions	Comparisons
$O\left(\ell 2^{n_K - n_k}\right)$	$O\left(2^{n_k} N_s 2^{-m} S_{r+1}\right)$	$O\left(2^{n_k} N_s 2^{-m} S_{r+1} \log(1 + S_r)\right)$

2.3 Theoretical Framework

In this subsection we develop the theoretical framework used to analyse multiple differential cryptanalysis. In our context, the attacker obtains the ciphertexts corresponding to a set of N chosen plaintexts generated using structures.

The determination of the data complexity and the success probability of a multiple differential cryptanalysis requires the knowledge of the distribution of the counters used in Algorithm 1 and particularly the distribution of $D(k)$.

Definition 2. Let $D_x^{(i)}(k)$ be the basic counter corresponding to the set of differentials with $\delta_0^{(i)}$ as input difference and with output difference in $\Delta_r^{(i)}$. For a given plaintext and a given candidate k, $D_x^{(i)}(k)$ is defined as

$$D_x^{(i)}(k) \overset{\text{def}}{=} \begin{cases} 1 & \text{if } F_k^{-1}(E_{K_*}(x)) \oplus F_k^{-1}(E_{K_*}(x \oplus \delta_0^{(i)})) \in \Delta_r^{(i)}, \\ 0 & \text{otherwise.} \end{cases}$$

The counters $D_x^{(i)}(k)$ follow a Bernoulli distribution since, for a fixed input difference and a fixed plaintext, only one output difference can occur. For $k = k_*$, the value of $F_k^{-1}(x)$ corresponds to the value obtained after r rounds of the cipher and thus the distribution of $D_x^{(i)}(k_*)$ depends on the probability of the corresponding differential. On the other hand, for $k \neq k_*$, it is usually assumed that the value $F_k^{-1}(x)$ is uniformly distributed among all the possible values. This assumption is known as the *Wrong Key Randomisation Hypothesis* [15]. Most notably the distribution of the $D_x^{(i)}(k)$'s is the same for all wrong candidates k.

Hypothesis 1. (Wrong-Key Randomisation Hypothesis in the differential cryptanalysis setting).

$$\Pr\mathbf{x}\left[F_k^{-1}(E_{K_*}(X)) \oplus F_k^{-1}(E_{K_*}(X \oplus \delta_0^{(i)})) = \delta_r^{(i,j)}\right] = \begin{cases} p_*^{(i,j)} & \text{if } k = k_*, \\ p^{(i,j)} = \frac{1}{2^m - 1} & \text{for } k \neq k_*. \end{cases}$$

In the following of this paper we will take the value 2^{-m} instead of $\frac{1}{2^m - 1}$ for $p^{(i,j)}$. Then, using this hypothesis, we obtain that $D_x^{(i)}(k)$ follows a Bernoulli distribution with parameter $p_*^{(i)} \overset{\text{def}}{=} \sum_{j=1}^{|\Delta_r^{(i)}|} p_*^{(i,j)}$ if $k = k_*$ and $p^{(i)} \overset{\text{def}}{=} \sum_{j=1}^{|\Delta_r^{(i)}|} p^{(i,j)} \approx |\Delta_r^{(i)}| 2^{-m}$ otherwise. Then we define variables corresponding to sums of these basics counters.

Definition 3. *Let $D_x^{(i)}(k)$ be the basic counters defined in Definition 2. We define the sums of these counters over all input differences and the counter we are interested in that is the mark obtained by a subkey during the attack:*

$$D_x(k) \stackrel{\text{def}}{=} \sum_{i=1}^{|\Delta_0|} D_x^{(i)}(k) \quad \text{and} \quad D(k) \stackrel{\text{def}}{=} \frac{1}{2} \sum_x D_x(k).$$

The factor $1/2$ in the sum comes from the fact that for any i, x and any key k, the counters $D_x^{(i)}(k)$ and $D_{x \oplus \delta_0^{(i)}}^{(i)}(k)$ are equal. Hence, each statistical phenomenon is counted twice when summing over all possible values for x. Instead of putting such a factor $1/2$ it may be possible to sum over one half of the whole set of x in a way that each pair of plaintexts will be counted only once. For a fixed i, we consider only one input difference $\delta_0^{(i)}$ hence it is easy to split the set of plaintext in two. The problem is not so easy when we have to consider all input differences (*i.e.* for the sum $\sum_x D_x(k)$). Indeed, it may not be possible to find a set $\mathcal{X}$ containing $N/2$ plaintexts such that all pairs are counted once and only once in other words, a set such that $\sum_{x \in \mathcal{X}} D_x(k) = \frac{1}{2} \sum_x D_x(k)$. The existence of such a set $\mathcal{X}$ depends on the structure of the set of input differences Δ_0.

Definition 4. *The set of input differences Δ_0 is admissible if there exists a set $\mathcal{X}$ of $N/2$ plaintexts that fulfils the condition*

$$\forall \delta_0^{(i)} \in \Delta_0, \forall x \in \mathcal{X}, x \oplus \delta_0^{(i)} \notin \mathcal{X}. \tag{1}$$

An efficient way to test if a set Δ_0 is admissible is provided in Appendix A.1. From now, we consider that the set Δ_0 has been chosen to be admissible. Hence, each pair is only counted once, but some dependencies between counters still remain. Deriving a general formula for the distribution of a sum of dependent variables is not so easy. Moreover, the variables we consider have really small dependencies and hence, we will assume that they are independent.

Hypothesis 2. *For any subkey k (including k_*) and a set $\mathcal{X}$ that fulfils (1),*

- *For any x, the variables $(D_x^{(i)}(k))_{1 \leq i \leq |\Delta_0|}$ are independent.*
- *The variables $(D_x(k))_{x \in \mathcal{X}}$ are independent.*

This hypothesis is not so far to being true. The same kind of hypothesis is done in differential cryptanalysis. Indeed, in the differential setting, the random variables $D_x(k)$ follow a Bernoulli distribution of parameters p_* or p and the same kind of independence hypothesis is assumed when saying that the counters $D(k)$ follow a binomial distribution.

Assuming Hypothesis 2, the end of this section is now dedicated to the problem of finding good estimates for the distribution of the sum of M independent variables that follow Bernoulli distributions with different parameters. Actually, we aim at applying this estimate to the determination of the distributions of $D(k)$ and $D(k_*)$. In the following, we use $D(k)$ to instantiate some results but they obviously hold for $D(k_*)$, when p is replaced by p_*.

The first technique to find a good estimate of the distribution of $D(k)$ is to use the following theorem which states that the distribution of the counters $D_i(k)$ is close to a Poisson distribution.

Theorem 1. *[18] Let $D_x^{(i)}(k)$ be M independent Bernoulli random variables with parameters $p^{(i)}$. Let $D_x(k) \stackrel{\text{def}}{=} \sum_{i=1}^{M} D_x^{(i)}(k)$ and $\lambda = \sum_{i=1}^{M} p^{(i)}$. Then, for all $A \subset \{0, 1, \ldots, M\}$, we have*

$$\left| \Pr\left[D_x(k) \in A\right] - \sum_{a \in A} \frac{\lambda^a e^{-\lambda}}{a!} \right| < \sum_{i=1}^{M} \left(p^{(i)}\right)^2.$$

Hence, the distribution of $D_x(k)$ is close to a Poisson distribution of parameter $\sum_{i=0}^{|\Delta_0|} p^{(i)}$. Then, using the stability of the Poisson distribution under addition, we conclude that $\sum_{x \in \mathcal{X}} D_x(k)$ follows a Poisson distribution with parameter $\frac{N}{2} \cdot \sum_{i=0}^{|\Delta_0|} p^{(i)}$. We then introduce the following quantities that play a particular role in the analysis of the multiple differential cryptanalysis.

$$p_* \stackrel{\text{def}}{=} \frac{\sum_i p_*^{(i)}}{|\Delta_0|} = \frac{\sum_{i,j} p_*^{(i,j)}}{|\Delta_0|} \quad \text{and} \quad p \stackrel{\text{def}}{=} \frac{\sum_i p^{(i)}}{|\Delta_0|} = \frac{\sum_{i,j} p^{(i,j)}}{|\Delta_0|} \approx \frac{|\Delta| \cdot 2^{-m}}{|\Delta_0|}.$$

The bound on the error due to the use of the Poisson approximation is relatively small regarding probabilities of order 10^{-1} but it is not clear that this approximation is still accurate when considering tails of the distribution. Indeed, we have checked, using some experiments, that the cumulative function of the Poisson distribution is not a good estimate of the tails of the cumulative distribution function of the counters $D(k)$. For this reason, we have to use another result from large deviations theory to obtain a better estimate for the tails of the distribution of the $D(k)$'s.

Theorem 2. *[14, chapter 5.4] Let $D(k) = \sum_x D_x(k)$ be a sum of M discrete, independent and identically distributed random variables. Let $\mu(s)$ be the semi-invariant moment generating function of each of the $D_x(k)$. Then, for $s > 0$,*

$$\Pr\left[D(k) \geq M\mu'(s)\right] = e^{M[\mu(s) - s\mu'(s)]} \left[\frac{1}{|s|\sqrt{\pi 2 M \mu''(s)}} + o\left(\frac{1}{\sqrt{M}}\right)\right].$$

where μ' and μ'' denote the first and second-order derivatives of μ.

From this theorem, we can compute accurate formulas for the tails of the distribution of $D(k)$ by computing the semi-invariant moment generating function in the special case where all $D_x^{(i)}(k)$ follow a Bernoulli distribution. Details of the computation is done in [10, Appendix3] and leads us to Theorem 3. The result is expressed using the Kullback-Leibler divergence between two Brnoulli distributions of parameters x and y.

Definition 5. *Let $0 < x < 1$ and $0 < y < 1$ be two real numbers, the Kullback-Leibler divergence is defined by:*

$$D(x||y) \stackrel{\text{def}}{=} x \ln\left(\frac{x}{y}\right) + (1 - x) \ln\left(\frac{1 - x}{1 - y}\right).$$

Before giving the result obtained (Theorem 3), let us recall that $N_s = \dfrac{|\Delta_0|N}{2}$. This quantity appears naturally in the expression of the distribution tails.

Theorem 3. *Let $D(k)$ be a counter as defined in Definition 3 ($D(k)$ is a sum of $N/2$ independent and identically distributed variables and takes values in $\{0, 1, \ldots, N_s\}$). We define two functions of τ and q real numbers in $[0, 1]$ with $\tau \neq q$:*

$$G_-(\tau, q) \overset{\text{def}}{=} e^{-N_s D(\tau||q)} \cdot \left[\frac{q\sqrt{(1-\tau)}}{(q-\tau)\sqrt{2\pi\tau N_s}} + \frac{1}{\sqrt{8\pi\tau N_s}} \right], \tag{2}$$

$$G_+(\tau, q) \overset{\text{def}}{=} e^{-N_s D(\tau||q)} \cdot \left[\frac{(1-q)\sqrt{\tau}}{(\tau-q)\sqrt{2\pi N_s(1-\tau)}} + \frac{1}{\sqrt{8\pi\tau N_s}} \right]. \tag{3}$$

Then, the tails of the cumulative distribution function of $D(k)$ can be approximated by G_- and G_+, more precisely,

$$\Pr\left[D(k) \leq \tau N_s \right] = G_-(\tau, p) \left[1 + O\left(\frac{p-\tau}{p} \right) \right],$$

$$\Pr\left[D(k) \geq \tau N_s \right] = G_+(\tau, p) \left[1 + O\left(\frac{p-\tau}{p} \right) \right].$$

By combining the results of Theorem 1 and Theorem 3, we define the following estimate for the cumulative distribution function of the counters $D(k)$.

Proposition 1. *Let $G_{\mathcal{P}}(\tau, q)$ be the cumulative distribution function of the Poisson distribution with parameter qN_s. Let $G_-(\tau, q)$ and $G_+(\tau, q)$ as defined in Theorem 3. We define $G(\tau, q)$ as*

$$G(\tau, q) \overset{\text{def}}{=} \begin{cases} G_-(\tau, q) & \text{if } \tau < q - 3 \cdot \sqrt{q/N_s}, \\ 1 - G_+(\tau, q) & \text{if } \tau > q + 3 \cdot \sqrt{q/N_s}, \\ G_{\mathcal{P}}(\tau, q) & \text{otherwise.} \end{cases}$$

The cumulative distribution functions of the counters $D(k)$ and $D(k_*)$ can be approximated by G and G_*, where

$$G_*(\tau) \overset{\text{def}}{=} G(\tau, p_*) \quad \text{and} \quad G(\tau) \overset{\text{def}}{=} G(\tau, p),$$

with $p_* = \dfrac{\sum_{i,j} p_*^{(i,j)}}{|\Delta_0|}$ *and* $p = \dfrac{\sum_{i,j} p^{(i,j)}}{|\Delta_0|} \approx \dfrac{|\Delta|}{2^m |\Delta_0|}$ *from the wrong-key randomisation hypothesis.*

2.4   Data Complexity and Success Probability

For a set Δ_0 that is admissible and if Hypothesis 2 holds, the distributions of the counters are tightly estimated by Proposition 1 and are similar to the distributions involved in [9]. Therefore we can use the same framework to estimate the data complexity and the success probability of a multiple differential cryptanalysis. The results obtained are given in Corollary 1 and Corollary 2.

Corollary 1. *Using notation defined in Section 2.1, the data complexity of a multiple differential cryptanalysis with success probability close to 0.5 is*

$$N = -2 \cdot \frac{\ln(2\sqrt{\pi}\ell\, 2^{-n_k})}{|\Delta_0| D(p_*||p)},$$

where ℓ is the size of the list of the remaining candidates and n_k is the number of bits of the key we want to recover.

Proof. In [9], the authors approximate the tails of the binomial cumulative distribution function by $e^{-N_s \cdot D(\tau||p)} \frac{(1-p)\sqrt{\tau}}{(\tau-p)\sqrt{2\pi N(1-\tau)}}$ to obtain an estimate of the number of samples required to perform a statistical cryptanalysis. Here, the tails of the cumulative distribution function of the counters $D(k)$ are similar (see the definitions of G_- and G_+ given in Theorem 3). Therefore, we can use the same method to derive the required number of samples. We fix the relative threshold τ to p_* which corresponds to a success probability close to 0.5. Then, N_s is found by solving equation $1 - G(p_*) = \frac{\ell}{2^{n_k}}$ (recall that G depends on N_s). In differential cryptanalysis, p_* is quite larger than p hence, $G(p_*) = 1 - G_+(p_*, p)$. Therefore a good estimate of N_s can be found using a fixed point method for solving equation $G_+(p_*, p) = \frac{\ell}{2^{n_k}}$. As in [9], we here obtain that N_s is close to $-\frac{1}{D(p_*||p)}\left[\ln\left(\frac{\nu\ell 2^{-n_k}}{\sqrt{D(p_*||p)}}\right) + 0.5\ln(-\ln(\nu\ell 2^{-n_k}))\right]$ where $\nu \overset{\text{def}}{=} \frac{(p_*-p)\sqrt{8\pi(1-p_*)p_*}}{2p_*(1-p)+(p_*-p)\sqrt{1-p_*}}$. As proposed in [9], $\ln(2\sqrt{\pi D(p_*||p)})$ can be used as a good estimate of $\ln(\nu)$, implying that the number of samples N_s is close to $-\frac{\ln(2\sqrt{\pi}\ell\, 2^{-n_k})}{D(p_*||p)}$. The result finally follows from the fact that the number of plaintexts is $N = \frac{2N_s}{|\Delta_0|}$.   ◇

In [9] it is also conjectured that for a value N_s of the form $N_s = -c \cdot \frac{\ln(2\sqrt{\pi}\ell\, 2^{-n_k})}{|\Delta_0| D(p_*||p)}$, the success probability essentially depends on the value of the constant c.

   In Corollary 2, we provide an estimate for the success probability of a multiple differential cryptanalysis. This corollary can be proved using arguments similar to the one exposed in the proof of Theorem 3 in [9].

Corollary 2. *Let $G_*(x)$ (resp. $G(x)$) be the estimate of the cumulative distribution function of the counter $D(k_*)$ (resp. of $D(k)$) defined in Proposition 1. The success probability, P_S, of a multiple differential cryptanalysis is given by*

$$P_S \approx 1 - G_*\left[G^{-1}\left(1 - \frac{\ell - 1}{2^{n_k} - 2}\right) - 1\right] \tag{4}$$

where the pseudo-inverse of G is defined by $G^{-1}(y) = \min\{x | G(x) \geq y\}$.

2.5   Application to Known Differential Cryptanalyses

Intuitively speaking, exploiting more differentials should decrease the cost of the attack since we extract more information on the same key. Nevertheless, this intuition is not always true. Let n_k be the number of key-bits to recover and let us fix the size of the list to ℓ. Then, for a fixed c, taking N_s of the form $N_s = -c \cdot \frac{\ln(2\sqrt{\pi}\ell 2^{-n_k})}{|\Delta_0|D(p_*||p)}$, leads to the same success probability whatever is the set of differentials considered (and hence whatever are the values of $|\Delta_0|, p_*$ and p). That means that the greater the value $|\Delta_0|D(p_*||p)$ is, the more information we extract from the samples. This neither takes into account the time complexity for extracting information nor the time complexity for analysing it. More details on these complexities have been given in Section 2.2. We now focus on finding the set of differentials that provides the more information to the attacker.

A general statement on the best way to choose differentials is not so easy to make. Therefore, we will take a look at two particular cases.

Multiple inputs, single output. In [5], Biham and Shamir exploit several differentials to mount their attack on the DES. The differentials they use have all the same output difference but different input differences. In this case, we have several differentials $(\delta_0^{(i)}, \delta_r)$ with probabilities $p_*^{(i)}$ and corresponding random probabilities $p^{(i)} \approx 2^{-m}$ when a wrong candidate is used for deciphering. We also assume that differentials are sorted such that the $p_*^{(i)}$ are in decreasing order. The goal is to find a criterion to determine whether adding the best of the remaining differentials decreases the data complexity or not. For a fixed success probability and a fixed size of list, the data complexity decreases if and only if

$$|\Delta_0|D\left(\frac{\sum_{i=1}^{|\Delta_0|} p_*^{(i)}}{|\Delta_0|}\Big|\Big|2^{-m}\right) \leq (|\Delta_0|+1)D\left(\frac{\sum_{i=1}^{|\Delta_0|+1} p_*^{(i)}}{|\Delta_0|+1}\Big|\Big|2^{-m}\right). \quad (5)$$

This implies for instance that, if we have a set of differentials with several input differences and having the same probability, exploiting them will decrease the data complexity by a factor $|\Delta_0|$ compared to a simple differential attack that uses only one of them.

Single input, multiple outputs. Some truncated differential attacks [16] can be seen as multiple differential cryptanalyses with a single input and multiple outputs. Here we assume that we exploit several differentials $(\delta_0, \delta_r^{(j)})$ with probability $p_*^{(j)}$ for the correct subkey. We assume that the $p_*^{(j)}$ are sorted in decreasing order. Adding one more differential with the same input decreases the data complexity until

$$D\left(\sum_{j=1}^{|\Delta_r|} p_*^{(j)}\Big|\Big|2^{-m}|\Delta_r|\right) \leq D\left(\sum_{j=1}^{|\Delta_r|+1} p_*^{(j)}\Big|\Big|2^{-m}(|\Delta_r|+1)\right). \quad (6)$$

Moreover, by studying the derivative of the Kullback-Leibler divergence one can obtain that, if $a > b$ and $0 < \lambda \leq a^{-1}$, $D(\lambda a||\lambda b) > \lambda D(a||b)$. Therefore, if we have $|\Delta_r|$ differences with the same input difference and the same probabilities, taking this set of differentials decreases the data complexity by a factor greater than $|\Delta_r|$ compared to a simple differential.

Multiple inputs, multiple outputs. Both previous cases are particular cases of the general situation where the differentials are taken with several input differences and several output differences. Determining the optimal set of differentials that must be chosen to obtain the smallest data complexity is difficult. The reasons are that the differentials do not have the same probabilities and both previously defined criteria use the Kullback-Leibler divergence which is not so easy to study. For all attacks presented in the following sections, we have decided to first determine the optimal set of output differences for each input difference we consider. This has been done using the criterion defined in (6). Then, we have constructed the final set using (5) once the $p_*^{(i)}$'s have been obtained. We do not claim that the resulting set of differentials is optimal but it is an efficient way for choosing the differentials that provides good sets. Finding an algorithm to find the optimal set of differentials (in the sense that provides the more information to the attacker) is an interesting open problem.

3   Experimental Validation

In this section we experimentally validate the theoretical framework presented in Section 2. To confirm the tightness of the formulas for the data complexity and the success probability given by Corollary 1 and Corollary 2, we have mounted a multiple differential cryptanalysis on a reduced version of PRESENT namely SMALLPRESENT-[8].

3.1   Description of PRESENT and SMALLPRESENT-[s]

PRESENT is a 64-bit lightweight block cipher proposed at CHES 2007 [6]. It is a Substitution Permutation Network with 16 identical 4-bit S-boxes. PRESENT is composed of 31 rounds and is parametrised by a 80-bit or a 128-bit key. More details on the specification can be found in [6].

SMALLPRESENT-[s]. For relevant experiments, we need to be able to exhaustively compute the ciphertexts corresponding to all possible plaintexts for all possible keys. Therefore, we chose to work on a reduced version of PRESENT named SMALLPRESENT-[s] [17]. The family SMALLPRESENT-[s] has been designed to be used for such experiments. Parameter s corresponds to the number of S-boxes per round. The block size is then $4s$. Here, we present the results obtained on SMALLPRESENT-[8] *i.e.* on the version with 8 S-boxes and block size 32 bits. More details on the specification can be found in [17].

Adapting the key-schedule. In the reduced cipher presented in [17], the key-schedule is the same as for the full cipher PRESENT (*i.e.* with a 80-bit master key). But in the original PRESENT, most of the bits of a subkey are directly reused in the next-round subkey, while this is not the case anymore with SMALLPRESENT-[8] since the number of key bits is still 80 but each subkey only uses 32 bits. Then, we decided to modify the key-schedule for our experiments on SMALLPRESENT-[8]. This new key-schedule uses a 40-bit master key and is similar to the one of the full version.

The master key is represented as $K = k_{39}k_{38}\ldots k_0$. At round i, the 32-bit round subkey $K_i = k_{39}k_{38}\ldots k_8$ consists of the 32 leftmost bits of the current content of the register. After extracting the round key K_i, the key register is updated as follows: the key is rotated by 29 bit positions to the left, the leftmost four bits are passed through the PRESENT S-box, and the *roundcounter* value is XORed with bits $k_{11}k_{10}k_9k_8k_7$.

3.2 Experimental Validation of the Obtained Formulas

To validate the formulas for the data complexity and the success probability given in Corollary 1 and Corollary 2, we have mounted a toy attack on SMALLPRESENT-[8] using both the 40-bit and the 80-bit key-schedules. This attack uses differentials on 9 rounds and aims at recovering some bits of the last-two-round subkeys, *i.e.* it corresponds to an attack on 11 rounds of the cipher.

Design of the toy cryptanalysis. To empirically estimate the success probability of the attack, we have to experiment this multiple differential attack a large number of times. This implies that the number of key bits to recover has to be small enough (*i.e.* not more than 32). We took differentials with output differences of the form 0x????0000. This structure enable us to recover 16 bits of both last two subkeys. The set of input differences is $\Delta_0 = \{0x3, 0x5, 0x7, 0xB, 0xD, 0xF\}$. This set is admissible since we can split the set of plaintexts into two parts the even plaintexts and the odd plaintexts. This attack uses 55 differentials over 9 rounds of SMALLPRESENT-[8]. The probability of each differential for both key-schedule (40-bit and 80-bit) has been estimated by averaging over 200 keys. These 55 differentials are given in [10, Appendix A.4] with the estimation of their probabilities. The attack computes the list $\mathcal{L}$ of size $\ell = 2^{12}$ of the likeliest candidates for the last two round subkeys.

Validation of the formula given in Corollary 2. The theoretical success probability of the attack is $P_S = 1 - G_* \left[G^{-1} \left(1 - \frac{\ell-1}{2^{n_k}-2} \right) - 1 \right]$, where $G_*(x)$ and $G(x)$ are estimates of the cumulative distribution function of the counter $D(k_*)$ or of $D(k)$. In Fig. 1, we compare the experimental success probability with the theoretical success probabilities obtained using the Gaussian approximation [23], using a Poisson estimation of the distribution of the counters and using the hybrid cumulative function defined in Proposition 1. For both key-schedules, 250 cryptanalyses have been performed to obtain the empirical success rate. The curves obtained for 150, 200 and 250 experiments are quite similar thus we expect that using 250 experiments provides a good picture of the success probability behaviour. It is worth noticing that the theoretical results in both figures use empirical estimates for the probabilities of the differentials. It is clear from Fig. 1 that the Gaussian approximation used up to now for analysing

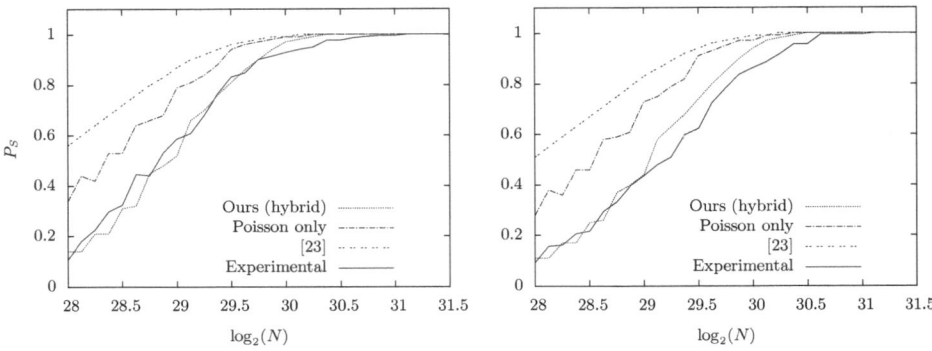

Fig. 1. Comparison of success probabilities for the 40-bit (left) and 80-bit (right) key-schedule

the complexity of differential cryptanalysis is not the most relevant, as already explained in [23]. Using the Poisson distribution (that provides good results in the case of simple differential cryptanalysis) is not here as good as using the hybrid cumulative function which results from large deviations theory to estimate the tails of the distributions. Since $\frac{\ell-1}{2^{n_k}-2}$ is small, the tightness of the estimate for $G^{-1}\left(1 - \frac{\ell-1}{2^{n_k}-2}\right)$ heavily depends on the accuracy of the tail estimate and thus the hybrid approach is the most relevant one. This result shows that the formula for the success probability given in Corollary 2 is a good approximation of the success probability of a multiple differential cryptanalysis.

Validation of the formula given in Corollary 1. Using the same experiments, we can also confirm the relevance of Corollary 1. It is conjectured in [9] that taking N of the form $N = -2 \cdot c \cdot \frac{\ln(2\sqrt{\pi}\ell\,2^{-n_k})}{|\Delta_0|D(p_*||p)}$ should lead to a success probability of about 50% for $c = 1$, 80% for $c = 1.5$ and 90% for $c = 2$. In Table 2 we give the empirical success rates corresponding to these three values of N for both attacks on the 40-bit and 80-bit versions of SMALLPRESENT-[8].

4   On the Estimations of the Probabilities p and p_*

We have shown that the formulas given by Corollary 1 and Corollary 2 are well-suited for multiple differential cryptanalysis. But all simulations have been performed on a toy example for which we were able to obtain good estimates of the probabilities of the differentials. However, one of the main difficulties in statistical attacks is the estimation of the underlying probabilities $p_*^{(i,j)}$.

Differential probabilities and trails probabilities. Computing the probability of a differential is, in general, intractable. Indeed, for an r-round differential (δ_0, δ_r), there exist many differential trails that have to be taken into account when computing the probability of this differential.

Table 2. Empirical success probabilities corresponding to values of N given by Corollary 1

Key-schedule	$c = 1.0$		$c = 1.5$		$c = 2.0$	
	40-bit	80-bit	40-bit	80-bit	40-bit	80-bit
N	$2^{28.92}$	$2^{29.06}$	$2^{29.50}$	$2^{29.65}$	$2^{29.92}$	$2^{30.06}$
P_S	0.55	0.47	0.83	0.75	0.92	0.88

Definition 6. *A differential trail β on r rounds of a cipher is a $(r+1)$-tuple $(\beta_0, \ldots, \beta_r)$ of elements of $\mathbb{F}_2^m$. Its probability is the probability that a plaintext pair with difference β_0 follows the differential path β when being encrypted: $p_\beta \stackrel{\text{def}}{=} \mathrm{Pr}_{\mathbf{X},\mathbf{K}} \left[\forall i, F_{\mathbf{K}}^i(\mathbf{X}) \oplus F_{\mathbf{K}}^i(\mathbf{X} \oplus \beta_0) = \beta_i \right].$*

The probability of a differential (δ_0, δ_r) can be computed by summing all the differential trails probabilities with input difference δ_0 and output difference δ_r. For recent ciphers, for a fixed differential, there is a lot of differential trails. This is the reason why, for most ciphers, it is impossible to estimate the exact probability of a differential. Using a branch & bound algorithm similar to the one used in linear cryptanalysis, it is possible to find all possible trails with given input and output differences up to a fixed probability. Summing the corresponding trail probabilities then provides a lower bound on the probability of the differential and thus on the efficiency of the attack.

Key dependence of the probabilities of the differentials. For Markov ciphers, introduced in [19], the classical way of estimating the probability of a differential trail is to use the following theorem.

Theorem 4. *[19] If an r-round iterated cipher is a Markov cipher and the r round keys are independent and uniformly random, then the probability of a differential trail $\beta = (\beta_0, \beta_1, \ldots, \beta_r)$ is*

$$p_\beta = \prod_{i=1}^r \mathrm{Pr}_{\mathbf{X},\mathbf{K}} \left[F_{\mathbf{K}}(\mathbf{X}) \oplus F_{\mathbf{K}}(\mathbf{X}') = \beta_i | \mathbf{X} \oplus \mathbf{X}' = \beta_{i-1} \right].$$

The point is that while many recent ciphers are Markov ciphers, their master key is not large enough to lead to independent and uniformly distributed round subkeys and thus, this theorem cannot be applied. Nevertheless, the independence of the round subkeys is generally assumed to obtain an estimate of a differential trail probability.

Hypothesis 3. (Round subkeys independence).
The round subkeys of the cipher E are independent and uniformly random.

Using Theorem 4, we define the theoretical probability of a differential trail $\beta = (\beta_0, \beta_1, \ldots, \beta_r)$ as $p_\beta^t \stackrel{\text{def}}{=} \prod_{i=1}^r \mathrm{Pr}_{\mathbf{X},\mathbf{K}} \left[F_{\mathbf{K}}(\mathbf{X}) \oplus F_{\mathbf{K}}(\mathbf{X}') = \beta_i | \mathbf{X} \oplus \mathbf{X}' = \beta_{i-1} \right]$. Hence, one may be able to estimate the probability $\mathrm{Pr}_{\mathbf{X},\mathbf{K}} \left[\delta_0 \to \delta_r \right]$ of a differential $\delta = (\delta_0, \delta_r)$ by summing the theoretical probabilities of the trails that compose it: $p_\delta^t \stackrel{\text{def}}{=} \sum_{\beta = (\delta_0, \beta_1 \ldots, \beta_{r-1}, \delta_r)} p_\beta^t$.

Now, another problem arises: the problem of fixed-key dependence. Theorem 4 can be used to estimate the probability of a differential $\delta = (\delta_0, \delta_r)$ but in an attack, the key is fixed and thus we are interested in the probabilities $p_\delta^K \stackrel{\text{def}}{=} \Pr_{\mathbf{X}} \left[E_K(\mathbf{X}) \oplus E_K(\mathbf{X} \oplus \delta_0) = \delta_r \right]$. Most of the analyses assume that this probability does not depend on the key $i.e.$, for two keys K and K', $p_\delta^K = p_\delta^{K'} = p_\delta^t$. This hypothesis is known as the *stochastic independence hypothesis*. It is actually far from being true since evidences show that the values of $2^{m-1}p_\delta^K$ are binomially distributed around $2^{m-1}p_\delta^t$ [13,8]. Nevertheless, in the setting of multiple differential cryptanalysis, this phenomenon seems to fade. The hypothesis we are using is then the following.

Hypothesis 4. (Stochastic equivalence in the multiple differential setting).
For any key K and for a set Δ of differences large enough, $\sum_{\delta \in \Delta} p_\delta^K = \sum_{\delta \in \Delta} p_\delta^t$.

Impact of the estimation of the probabilities of the differentials on the success probabilities. We have pointed out the problems related to the estimation of the probabilities of the differentials. They come from the large number of trails composing the differential and the fact that their probabilities depend on the key. In our attack on SMALLPRESENT-[8] with the 40-bit key-schedule, we have computed the success probability of the attack based on experimental values for the differential probabilities. We have also computed the theoretical values of the differential probabilities using trails up to probability 2^{-48}. The theoretical probabilities of the differentials are given in [10, Appendix A.4]. We observe that these values always underestimate the probability of the differentials. Using this estimation of the probability we have plot the success rate of the attack (Fig. 2) and we show how this underestimation of the probabilities of the differentials affects the estimation of the success probability of the attack

Estimation of p. In the analysis of the distribution of the counters, we have assumed that the $p^{(i,j)}$ were close to 2^{-m} (Hypothesis 1). The probability $p^{(i)}$ of a wrong-key counter to be incremented by a plaintext pair with difference $\delta_0^{(i)}$ has then been estimated by $|\Delta_r^{(i)}| 2^{-m}$. Thus, p that is the mean of the $p^{(i)}$'s has been estimated by $\frac{|\Delta|}{|\Delta_0|} 2^{-m}$. We use the results of the experiments on SMALLPRESENT-[8] (Section 3) to show that it is a good estimate for p. Let us recall that we took $|\Delta| = 55$ differentials with $|\Delta_0| = 7$ different input differences. Using the whole codebook we obtain $2^{31} \cdot |\Delta_0|$ samples and thus the expected value of the counters corresponding to wrong subkeys is $2^{31} \cdot |\Delta_0| \left(\frac{55}{|\Delta_0|} \right) 2^{-m} = 27.5$. The mean over the counters corresponding to wrong candidates has been computed for each performed attack and the results are in the range $[27.14; 28.15]$ (the mean value is 27.68). This confirms the relevance of the estimation $p \approx \frac{|\Delta|}{|\Delta_0|} 2^{-m}$.

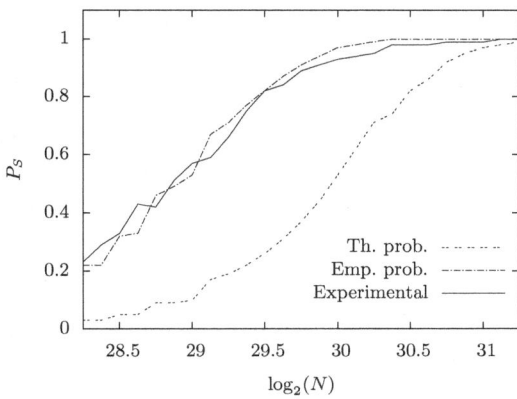

Fig. 2. Success probability of an attack on SMALLPRESENT-[8] with the 40-bit key-schedule

5   Application to PRESENT

There exists a lot of attacks on reduced versions of PRESENT. These attacks are summarised in Table 3. The best differential attack on PRESENT is due to Wang [24]. This attack, using 24 differentials on 14 rounds with same output difference, can break 16 rounds of PRESENT.

Table 3. Summary of the attacks on PRESENT

#rounds	version	type of attack	data	time	memory	reference
8	128	integral	$2^{24.3}$	$2^{100.1}$	$2^{77.0}$	[25]
16	80	differential	$2^{64.0}$	$2^{64.0}$	$2^{32.0}$	[24]
17	128	related keys	2^{63}	$2^{104.0}$	$2^{53.0}$	[22]
19	128	algebraic diff.	$2^{62.0}$	$2^{113.0}$	n/r	[1]
24	80	linear	$2^{63.5}$	$2^{40.0}$	$2^{40.0}$	[21]
24	80	statistical sat.	$2^{57.0}$	$2^{57.0}$	$2^{32.0}$	[12]
25	128	linear	$2^{64.0}$	$2^{96.7}$	$2^{40.0}$	[20]
26	80	multiple linear	$2^{64.0}$	$2^{72.0}$	$2^{32.0}$	[11]

We saw in Section 3 that experiments on SMALLPRESENT-[8] corroborate theoretical expectations. Assuming that this holds for the full cipher PRESENT too, we propose a multiple differential cryptanalysis for 18 rounds of PRESENT that improves the attack by Wang. This attack on 18 rounds uses 561 16-round differentials with 17 input differences forming the set $\Delta_0 = \{0x1001\} \cup \{0xY00Z, Y, Z \in \{2, 4, A, C\}\}$. This set Δ_0 is admissible (this can be checked using the method given in Appendix A.1). For each input difference the set of output differences is of size $|\Delta_r| = 33$ and each output differences is of the form 0x????????00000000. The sieves obtained after 18 rounds are similar for each input and of size $|\Delta_{r+2}^{(i)}| \approx 2^{32}$. The differential probabilities have been estimated by summing trails with probability up to 2^{-80} for each differential. The estimates obtained for the involved probabilities are $p_* = 2^{-58.50}$ and $p = 2^{-58.96}$. The number of active S-boxes is 8 for both final rounds, implying that the number of bits we recover is 64. In the case of the 80-bit key-schedule, there are 12 bits shared by both two-last-round subkeys and thus we actually recover $n_k = 52$ bits. Moreover, we can use the trick of decomposing the two rounds of the partial deciphering (see [24]). The sieves $\Delta_{r+1}^{(i)}$, that are the sets of possible differences after $r + 1$ rounds, are of size at most $2^{13.2}$ and thus, only $2^{32-16.8}$ last-round subkeys remain after deciphering one round. We give in Table 4 the complexities of the attack for different values of the data complexity, depending on the size of the list of remaining candidates.

Table 4. Different attacks on PRESENT with memory complexity 2^{32}

80-bit N	ℓ	P_S	time c.	128-bit N	ℓ	P_S	time c.
2^{60}	2^{51}	76%	$2^{79.00}$	2^{60}	2^{63}	76%	$2^{127.00}$
2^{62}	2^{47}	81%	$2^{75.04}$	2^{62}	2^{60}	88%	$2^{124.00}$
2^{64}	2^{36}	94%	$2^{71.72}$	2^{64}	2^{46}	90%	$2^{110.00}$

6   Conclusions

In this paper, we propose a general framework for analysing the complexity of multiple differential cryptanalysis. By studying the distributions of the counters involved in the attack, we obtain formulas for the data complexity, the time complexity and the success probability of such attacks. We have validated these theoretical results by mounting an attack on SMALLPRESENT-[8]. Using this framework we propose an attack on 18 rounds on PRESENT. This is not the best known attack on PRESENT since linear cryptanalysis seems to perform better on this cipher, but it improves the best previously known differential cryptanalysis of PRESENT [24].

References

1. Albrecht, M., Cid, C.: Algebraic techniques in differential cryptanalysis. In: Dunkelman, O. (ed.) FSE 2009. LNCS, vol. 5665, pp. 193–208. Springer, Heidelberg (2009)
2. Baignères, T., Junod, P., Vaudenay, S.: How far can we go beyond linear cryptanalysis? In: Lee, P.J. (ed.) ASIACRYPT 2004. LNCS, vol. 3329, pp. 432–450. Springer, Heidelberg (2004)
3. Baignères, T., Vaudenay, S.: The complexity of distinguishing distributions (Invited talk). In: Safavi-Naini, R. (ed.) ICITS 2008. LNCS, vol. 5155, pp. 210–222. Springer, Heidelberg (2008)
4. Biham, E., Shamir, A.: Differential cryptanalysis of DES-like cryptosystems. In: Menezes, A., Vanstone, S.A. (eds.) CRYPTO 1990. LNCS, vol. 537, pp. 2–21. Springer, Heidelberg (1991)

5. Biham, E., Shamir, A.: Differential cryptanalysis of DES-like cryptosystems. Journal of Cryptology 4, 3–72 (1991)
6. Bogdanov, A.A., Knudsen, L.R., Leander, G., Paar, C., Poschmann, A., Robshaw, M.J.B., Seurin, Y., Vikkelsoe, C.: PRESENT: An ultra-lightweight block cipher. In: Paillier, P., Verbauwhede, I. (eds.) CHES 2007. LNCS, vol. 4727, pp. 450–466. Springer, Heidelberg (2007)
7. Blondeau, C., Gérard, B.: On the data complexity of statistical attacks against block ciphers. In: Kholosha, A., Rosnes, E., Parker, M.G. (eds.) Workshop on Coding and Cryptography - WCC 2009, pp. 469–488 (2009)
8. Blondeau, C., Gérard, B.: Links between theoretical and effective differential probabilities: Experiments on present. In: TOOLS 2010 (2010),
   http://eprint.iacr.org/2010/261
9. Blondeau, C., Gérard, B., Tillich, J.-P.: Accurate estimates of the data complexity and success probability for various cryptanalyses. In: Charpin, P., Kholosha, S., Rosnes, E., Parker, M.G. (eds.) Designs, Codes and Cryptography, vol. 59(1-3). Springer, Heidelberg (2011)
10. Blondeau, C., Gérard, B.: Multiple Differential Cryptanalysis: Theory and Practice. Cryptology ePrint Archive, Report 2011/115 (2011),
   http://eprint.iacr.org/2011/115
11. Cho, J.Y.: Linear cryptanalysis of reduced-round PRESENT. In: Pieprzyk, J. (ed.) CT-RSA 2010. LNCS, vol. 5985, pp. 302–317. Springer, Heidelberg (2010)
12. Collard, B., Standaert, F.-X.: A statistical saturation attack against the block cipher PRESENT. In: Fischlin, M. (ed.) CT-RSA 2009. LNCS, vol. 5473, pp. 195–210. Springer, Heidelberg (2009)
13. Daemen, J., Rijmen, V.: Probability distributions of correlation and differentials in block ciphers. Journal of Mathematical Cryptology 1, 12–35 (2007)
14. Gallager, R.G.: Information Theory and Reliable Communication. John Wiley and Sons, Chichester (1968)
15. Harpes, C., Kramer, G.G., Massey, J.L.: A generalization of linear cryptanalysis and the applicability of matsui's piling-up lemma. In: Guillou, L.C., Quisquater, J.-J. (eds.) EUROCRYPT 1995. LNCS, vol. 921, pp. 24–38. Springer, Heidelberg (1995)
16. Knudsen, L.R.: Truncated and higher order differentials. In: Preneel, B. (ed.) FSE 1994. LNCS, vol. 1008, pp. 196–211. Springer, Heidelberg (1995)
17. Leander, G.: Small scale variants of the block cipher PRESENT. Cryptology ePrint Archive, Report 2010/143 (2010), http://eprint.iacr.org/2010/143
18. Cam, L.: An approximation theorem for the poisson binomial distribution. Pacific Journal of Mathematics 10, 1181–1197 (1960)
19. Lai, X., Massey, J.L.: Markov ciphers and differential cryptanalysis. In: Davies, D.W. (ed.) EUROCRYPT 1991. LNCS, vol. 547, pp. 17–38. Springer, Heidelberg (1991)
20. Nakahara Jr., J., Sepehrdad, P., Zhang, B., Wang, M.: Linear (Hull) and algebraic cryptanalysis of the block cipher PRESENT. In: Garay, J.A., Miyaji, A., Otsuka, A. (eds.) CANS 2009. LNCS, vol. 5888, pp. 58–75. Springer, Heidelberg (2009)
21. Ohkuma, K.: Weak keys of reduced-round PRESENT for linear cryptanalysis. In: Jacobson Jr., M.J., Rijmen, V., Safavi-Naini, R. (eds.) SAC 2009. LNCS, vol. 5867, pp. 249–265. Springer, Heidelberg (2009)
22. Özen, O., Varici, K., Tezcan, C., Kocair, Ç.: Lightweight block ciphers revisited: Cryptanalysis of reduced round PRESENT and HIGHT. In: Boyd, C., González Nieto, J. (eds.) ACISP 2009. LNCS, vol. 5594, pp. 90–107. Springer, Heidelberg (2009)

23. Selçuk, A.A.: On probability of success in linear and differential cryptanalysis. Journal of Cryptology 21, 131–147 (2008)
24. Wang, M.: Differential cryptanalysis of reduced-round PRESENT. In: Vaudenay, S. (ed.) AFRICACRYPT 2008. LNCS, vol. 5023, pp. 40–49. Springer, Heidelberg (2008)
25. Z'aba, M.R., Raddum, H., Henricksen, M., Dawson, E.: Bit-pattern based integral attack. In: Nyberg, K. (ed.) FSE 2008. LNCS, vol. 5086, pp. 363–381. Springer, Heidelberg (2008)

A   Appendix

A.1   Checking If a Set Δ_0 Is Admissible

For a set of input differences Δ_0 we want to determine whether this set is admissible, that mean we want to know if it is possible to obtain the value of the counter $D(k)$ by summing $N/2$ of the $D_x(k)$. This is possible if and only if there exists a set $\mathcal{X}$ containing $N/2$ plaintexts such that $\forall \delta_0^{(i)} \in \Delta_0, \forall x \in \mathcal{X}, x \oplus \delta_0^{(i)} \notin \mathcal{X}$. This is the case if $\mathcal{X}$ and its complement form the two parts of a bipartite graph where the edges correspond to the $\delta_0^{(i)}$. The existence of such a graph is equivalent to the non-existence of odd weight cycles (*i.e.* null sums of an odd number of $\delta_0^{(i)}$).

Testing this can be efficiently done if we now look at the problem in terms of coding theory. Let G be the matrix whose columns correspond to the binary decompositions of the differences in Δ_0. Then, saying that every odd combination of the columns is non-zero is equivalent to say that the dual of the code determined by G has only codewords with even Hamming weights. Also, this is equivalent to the fact that the dual of this dual code contains the all-one vector. Since the dual of the dual of a code is the original code, we deduce that the set Δ_0 is admissible if and only if the code determined by G contains the all-one vector. This can be tested in polynomial time using a Gaussian elimination. Indeed, putting the matrix G in the systematic form (i.e. $G' = (I||A)$ where I is the identity matrix), the following equivalence holds.

$$(1 \ldots 1) \cdot G' = (1 \ldots 1) \iff \Delta_0 \text{ is admissible.}$$

Fast Correlation Attacks: Methods and Countermeasures

Willi Meier

FHNW, Switzerland

Abstract. Fast correlation attacks have considerably evolved since their first appearance. They have lead to new design criteria of stream ciphers, and have found applications in other areas of communications and cryptography.

In this paper, a review of the development of fast correlation attacks and their implications on the design of stream ciphers over the past two decades is given.

Keywords: stream cipher, cryptanalysis, correlation attack.

1 Introduction

In recent years, much effort has been put into a better understanding of the design and security of stream ciphers. Stream ciphers have been designed to be efficient either in constrained hardware or to have high efficiency in software. A synchronous stream cipher generates a pseudorandom sequence, the keystream, by a finite state machine whose initial state is determined as a function of the secret key and a public variable, the initialization vector. In an additive stream cipher, the ciphertext is obtained by bitwise addition of the keystream to the plaintext.

We focus here on stream ciphers that are designed using simple devices like linear feedback shift registers (LFSRs). Such designs have been the main target of correlation attacks. LFSRs are easy to implement and run efficiently in hardware. However such devices produce predictable output, and cannot be used directly for cryptographic applications. A common method aiming at destroying the predictability of the output of such devices is to use their output as input of suitably designed non-linear functions that produce the keystream. As the attacks to be described later show, care has to be taken in the choice of these functions. Another well known method to destroy the linearity property of LFSRs is to use irregular clocking, where the output of an LFSR clocks one or more other LFSRs. All these are quite classical concepts. However they still form a valuable model for recent designs, as the hardware oriented finalists of the eSTREAM project illustrate, [41].

Several different cryptanalytic methods can be applied against stream ciphers. Amongst these methods, some only work for a specific cipher, whereas quite a number of other methods are more general, including correlation attacks, linear

A. Joux (Ed.): FSE 2011, LNCS 6733, pp. 55–67, 2011.

attacks, algebraic attacks, time/memory/data tradeoff attacks, and resynchronization attacks. We restrict here mainly to (fast) correlation attacks, and we comment on linear attacks. Beyond stream ciphers, methods similar to fast correlation attacks are of interest, e.g., in satellite communications, in the construction of a trapdoor stream cipher, [15], in digital watermarking [48], or for the learning parity with noise problem, [16], [28]. The appearance of correlation attacks has motivated various countermeasures in the form of criteria for Boolean functions that should be chosen in order to provide some correlation immunity.

This review is organized as follows. Section 2 describes the principles of correlation attacks. Section 3 forms the main part, and describes different types of fast correlation attacks. Sections 4 and 5 are aiming at countermeasures against these attacks: Section 4 discusses correlation immune functions and Bent functions, whereas Section 5 briefly deals with combiners with memory. In Section 6, linear attacks are discussed. They are viewed as a generalization of correlation attacks, and can be efficient in quite general stream cipher constructions. Finally, a few open problems are stated.

2   Correlation Attacks

The main targets of correlation attacks are filter generators and combiner generators. In a classical filter generator, the running device is a single binary LFSR. The keystream is generated as the output of a nonlinear Boolean function whose inputs are prespecified stages of a LFSR. The initial state of the LFSR is derived from the secret key and the initialization vector. In a nonlinear combiner generator, the keystream is generated as the output of a Boolean function whose inputs are the outputs of several LFSRs. In more detail, suppose the outputs $a_i^{(k)}$ of s LFSRs, $1 \leq k \leq s$, are used as input of a Boolean function f to produce keystream bits z_i for $i = 1, 2, \ldots,$

$$f(a_i^{(1)}, \ldots a_i^{(s)}) = z_i.$$

Then the keystream sequence may be correlated to the output sequence of one or more of the LFSRs.

Example 1. Let $s = 3$, and let f be the majority function,

$$y = f(x_1, x_2, x_3) = x_1 x_2 + x_1 x_3 + x_2 x_3.$$

Then $\text{Prob}(y = x_k) = 0.75$ for $k = 1, 2, 3$.

In general, if such correlations exist, decoding techniques may be used to determine the state of the LFSRs in a divide-and-conquer manner. This is the subject of correlation attacks.

The original correlation attack was proposed by Th. Siegenthaler in [45]. Hereby, it is assumed that some portion of the keystream is known. Suppose furthermore that the keystream sequence is correlated to the output of a LFSR, i.e., $P(a_i = z_i) \neq 0.5$, where a_i and z_i are the i-th output symbols of the LFSR and of the keystream generator, respectively. Besides the feedback connection of the LFSR, no further knowledge is required on the explicit structure of the generator.

Let the LFSR-length be n. For each of the 2^n possible initial states of the LFSR, the output sequence $\mathbf{a} = (a_1, a_2, .., a_L)$ for a suitable length $L > n$ is generated, and the value α, defined as $\alpha = L - d_H(\mathbf{a}, \mathbf{z})$ is computed. Here $d_H(\mathbf{a}, \mathbf{z})$ denotes the Hamming distance between $\mathbf{a}$ and $\mathbf{z}$, i.e., the number of positions in which $\mathbf{a}$ and $\mathbf{z}$ are different.

Then it is shown in [45], that α will take the largest value for the correct initial state with high probability, provided L in dependence of the correlation probability is sufficiently large.

This concept can be generalized to the situation where the keystream sequence is correlated to the outputs of a set of more than one LFSR: Assume that a keystream sequence is generated by a generator with several different LFSR's, and that a subset of LFSR-outputs are correlated to the keystream sequence. Then one can try to find the initial states of these LFSR's in a divide-and-conquer type of attack, and to guess the remaining LFSR-states in a separate phase.

Correlation attacks are often viewed as a decoding problem. For a LFSR of length n consider all possible output sequences of a fixed length $L > n$. This set of truncated output sequences can be viewed as a linear $[n, L]$ block code [29]. Thus the LFSR sequence $\mathbf{a} = (a_1, a_2, ..., a_L)$ is interpreted as a codeword in this code, and the keystream sequence $\mathbf{z} = (z_1, z_2, .., z_L)$ as the received channel output. The problem of the attacker can now be formulated as: Given a received word $\mathbf{z} = (z_1, z_2, ..., z_L)$, find the transmitted codeword. From coding arguments [44] it follows that the length L should be at least $L_0 = L/(1 - h(1 - p))$ for unique decoding, where $h(1 - p)$ is the binary entropy function, and $p = P(z_i = a_i)$ is the correlation probability.

3   Fast Correlation Attacks

A *fast correlation attack* is a correlation attack that is significantly faster than exhaustive search over the initial states of the target LFSR. In [32] two algorithms for fast correlation attacks are presented. Instead of exhaustive search as originally suggested in [45], the algorithms are based on using certain parity-check equations derived from the feedback polynomial of the LFSR. The algorithms have two different phases: in the first phase, a set of suitable parity-check equations is found. In the second phase, these equations are used in a fast decoding algorithm to recover the initial state of the LFSR. These algorithms have been demonstrated to be successful for quite long LFSR's ($n = 1000$ or longer), provided the number t of feedback taps is small ($t < 10$). However the algorithms fail if the LFSR has many taps. Due to these fast correlation attacks, one usually avoids using LFSR's with few feedback taps in stream cipher design. In [50], based on earlier work in [49], the linear syndrome method from coding theory is proposed for fast correlation attacks, with similar efficiency and limitations as the algorithms in [32].

The two algorithms in [32] are described here in order. In a preparation phase, parity check equations are determined by observing that for a given position j the digit a_j of the LFSR-sequence **a** satisfies a certain number m of linear relations involving a fixed number t of other digits of **a**. Here t denotes the number of taps of the LFSR. These linear relations are found by shifting and iterated squaring of the LFSR-relation.

Example 2. Consider the LFSR of length $n = 3$ with feedback relation

$$a_j = a_{j-1} + a_{j-3}, \ j \geq 3.$$

Then by squaring, the relation $a_j = a_{j-2} + a_{j-6}$ does hold as well. And by shifting, one gets three relations for the same digit a_j:

$$
\begin{aligned}
a_{j-3} + a_{j-1} + a_j &= 0 \\
a_{j-2} + a_j + a_{j+1} &= 0 \\
a_j + a_{j+2} + a_{j+3} &= 0
\end{aligned}
$$

The digits of the known output sequence **z** are substituted in the linear relations thus obtained. Some of the relations will still hold, some others will not. It has been observed that the more relations are satisfied for a digit z_j, the higher is the (conditional) probability that $z_j = a_j$. Denote by p^* the probability for $z_j = a_j$, conditioned on the number of relations satisfied.

Consider first a digit contained in one relation. Assume the digit $a^{(0)} = a_j$ at a given position j satisfies a linear relation involving t digits at some other positions of the LFSR-sequence **a**,

$$a^{(0)} + a^{(1)} + a^{(2)} + \cdots a^{(t)} = 0.$$

Denote by $z^{(0)}, z^{(1)}, \ldots z^{(t)}$ the digits in the same positions of the output sequence. Then

$$
\begin{aligned}
z^{(0)} &= a^{(0)} + b^{(0)} \\
z^{(1)} &= a^{(1)} + b^{(1)} \\
&\cdots\cdots\cdots\cdots \\
z^{(t)} &= a^{(t)} + b^{(t)},
\end{aligned}
$$

and for the perturbations, $\text{Prob}(b^{(0)} = 0) = \ldots = \text{Prob}(b^{(t)} = 0) = p$. Denote $s = \text{Prob}(b^{(1)} + \ldots + b^{(t)} = 0)$: $s = s(p, t)$. Then $s(p, t)$ can be computed recursively:

$$s(p, 1) = p, \ s(p, t) = ps \cdot (p, t - 1) + (1 - p)(1 - s(p, t - 1)) \text{ for } t > 1.$$

Next assume that a specified digit $a = a_j$ is contained in m relations, each involving t other digits. For a subset S of relations, denote by $E(S)$ the event that exactly the relations in S are satisfied. Then for $z = z_j$,

$$
\begin{aligned}
\text{Prob}((z = a), \text{and } E(S)) &= p \cdot s^h (1 - s)^{m-h}, \\
\text{Prob}((z \neq a) \text{ and } E(S)) &= (1 - p)s^{m-h}(1 - s)^h,
\end{aligned}
$$

where $h = |S|$ denotes the number of relations in S. Hence the conditional probability $p^* = \text{Prob}(z = a | E(S))$ is given by

$$p^* = \frac{p \cdot s^h (1-s)^{m-h}}{p \cdot s^h (1-s)^{m-h} + (1-p)s^{m-h}(1-s)^h}.$$

The probability distributions for the number h of satisfied relations are Binomial distributions. There are two cases. If the digit z is correct, i.e., if $z = a$,

$$p_1 = \binom{m}{h} s^h (1-s)^{m-h}.$$

Alternatively, if $z \neq a$,

$$p_0 = \binom{m}{h} s^{m-h}(1-s)^h.$$

It is intuitively clear that a digit a can be more reliably predicted the more the two distributions are separated. In [32] it is shown that the average number m of relations involving a that can be checked in the given output stream $\mathbf{z}$ is:

$$m = \log_2 \left(\frac{L}{2n} \right) (t + 1).$$

Example 3. Assume a correlation probability $p = 0.75$, a number $t = 2$ of taps, LFSR-length $n = 100$, and a length of $L = 5000$ known bits of $\mathbf{z}$. Then $m = 12$ relations are available (in average), and $s = 0.75^2 + 0.25^2 = 0.625$. The value of the probability p^* conditioned on the number h of relations satisfied is:

relations satisfied	probability
12	0.9993
11	0.9980
10	0.9944

Based on these considerations, two algorithms, Algorithms A and B for fast correlation attacks are described in [32].

Algorithm A essentially chooses a set I_0 of approximately n digits of the known output stream $\mathbf{z}$ that satisfy the most relations. The digits in I_0 are taken as a reference guess of $\mathbf{a}$ at the same positions. Thereafter, the initial state of the LFSR is found by solving a system of linear equations.

As the selected digits in I_0 are only correct with some probability, the correct guess of the initial state is found by testing modifications of I_0 of Hamming distance $1, 2, \ldots$ by correlation of the corresponding LFSR-sequence with the given sequence $\mathbf{z}$. Thus Algorithm A has exponential complexity, of order $O(2^{cn})$, $0 < c < 1$. The parameter c is a function of the correlation p, the number of taps t, and the ratio L/n.

Example 4. Let $p = 0.75$, $t = 2$, and $L/n = 100$. Then $c = 0.012$. The search complexity is of significantly reduced order $O(2^{0.012n})$ compared to $O(2^n)$ in case of exhaustive search.

Algorithm B is described step by step as follows:

Algorithm B

1. Assign the correlation probability p to every digit of **z**.
2. To every digit of **z** assign the new probability p^*. Iterate this step a number of times.
3. Complement those digits of **z** with $p^* < p_{thr}$ (for a suitable threshold p_{thr}).
4. Stop, if **z** satisfies the basic relation of the LFSR, else go to 1.

The number of iterations in 2. and the probability threshold in 3. have to be adequately chosen to obtain maximum correction effect. In 2. the formula for recomputing conditional probabilities has to be generalized to the case where assigned probabilities for each involved digit are different. After a few iterations, a strong separation effect can be observed between digits having probability p^* close to 0 or close to 1. Algorithms B is essentially linear in the LFSR-length n. The success of this algorithm has extensively been verified experimentally for various correlation probabilities, LFSR-lengths and numbers of taps $t <$ 10. Iterative methods similar to Algorithm B have been applied in decoding. In [17], R. G. Gallager has developed a decoding scheme, where the decoder computes all the parity checks and then changes any digit that is contained in more than some fixed number of unsatisfied parity-check equations. Using these new values, the parity-checks are recomputed, and the process is repeated. The method in [32] contrasts to this approach in that the process of assigning conditional probabilities to every digit is iterated rather than just changing digits according to the number of parity-check equations satisfied.

As these algorithms work only if the LFSR has few feedback taps, i.e., if the feedback polynomial is of low weight, the problem persisted, how to design algorithms that are efficient even if the number of taps is arbitrary.

A first approach is to look for polynomial multiples of the feedback polynomial: If the recursion is not of low weight, consider multiples of the feedback polynomial that are of low weight.

Example 5. ([30]) Consider the connection polynomial $g(x)$ over $GF(2)$ of degree 7 and of weight 5:

$$g(x) = x^7 + x^6 + x^4 + x + 1.$$

$g(x)$ has a polynomial multiple (a trinomial)

$$f(x) = g(x)m(x) = x^{21} + x^3 + 1$$

with a polynomial $m(x)$ of degree 14.

Fast correlation attacks can likewise be applied to the linear recursion of sparse polynomial multiples, [4]. There are quite different methods on how to find low weight polynomial multiples. These methods differ in the weight and degree of an attempted sparse multiple, and in the required memory and computing time, see, e.g., [18], [47]. In particular, a feedback polynomial of a LFSR of length n can be shown to have a polynomial multiple of weight 4 (i.e., with 3 taps) of expected length about $2^{n/3}$.

Low weight multiples of feedback polynomials are of more general interest, as they often allow for distinguishing attacks on LFSR-based stream ciphers, e.g., [23], [12], [13]. In these attacks, the primary aim is not to recover the key, but to distinguish the known keystream from random.

Apart from investigation of sparse multiples of the connection polynomial, there is vast literature dealing with improvements of the initial algorithms. A major improvement concerns fast correlation attacks on LFSR's with an arbitrary number of feedback taps. It appears that the algorithms as proposed in [24] and [35] are amongst the most efficient known thus far. Based on these methods, in [6], the algorithmic steps have been improved to accelerate the attacks in [24] or [35]. As to be expected, the complexity of these algorithms depends on the length n of the target LFSR as well as on the correlation probability p. A version of one of these algorithms is briefly sketched:

As opposed to other fast correlation attacks, the use of parity-checks is combined with a partial exhaustive search over a subset B of the initial state of the targeted LFSR. As predictions are true only with some probability, $D > n$ targeted bits of the LFSR-output are predicted by evaluating and counting a number of parity-check equations. As before, the parity-checks are found in a preprocessing phase. In [6], an elaborate match-and-sort algorithm is described how to generate many parity-checks. In an example case, the parity-checks involve a number of bits in the set B, the target bit a_i at position i of the LFSR-sequence $\mathbf{a}$ to be predicted, and two other bits at some positions j and m in the known output stream $\mathbf{z}$. The procedure is informally as follows:

- For each of the D target bits, evaluate a large number of parity-checks substituted into the output stream $\mathbf{z}$ and the guessed bits of B, and count the number of parity-checks satisfied, N_s, and the number of parity-checks N_u not satisfied.
- If the expression $|N_s - N_u|$ is larger than a threshold, predict $a_i = z_i$ if $N_s > N_u$, else $a_i = z_i + 1$.

Provided this majority poll is decisive for D target bits of the LFSR-sequence, the initial state can be easily recovered. Estimates of the complexity of this algorithm suggest that it is possible to attack LFSRs of length n about 100 in practice, provided p is not too close to 0.5. In [36] and [26], a large part of the state of the art in fast correlation attacks is found.

Fast correlation attacks have been applied successfully to concrete constructions: In [21], a fast correlation attack is applied to the summation generator. In [25], the stream cipher LILI-128 has been cryptanalysed by such methods.

More recently, in [2] the initial version of the eSTREAM finalist Grain with a key of 80 bits was broken. This motivated a careful tweak, Grain v1, which is an eSTREAM finalist, [41].

4   Towards Correlation Immunity

In many (fast) correlation attacks, the correlations are deduced as linear approximations of nonlinear output functions in stream ciphers. The existence of correlation attacks has thus led to new design criteria for Boolean functions used in stream ciphers, [46], [33]. In particular, combining (or filter) functions should have no statistical dependence between any small subset of inputs and the output.

More formally, let $X_1, X_2, \ldots X_n$ be independent binary variables which are balanced (i.e., each variable takes the values 0 and 1 with probability $\frac{1}{2}$). A Boolean function f of n variables is *m-th order correlation immune* if for each subset of m random variables $X_{i1}, X_{i2}, \ldots, X_{im}$ the random variable

$$Z = f(X_1, X_2, \ldots, X_n)$$

is statistically independent of the random vector $(X_{i1}, X_{i2}, \ldots, X_{im})$. There exists a tradeoff between the order of correlation immunity and the algebraic degree of Boolean functions, [46]. Low algebraic degree conflicts with security: Due to the Berlekamp-Massey algorithm, [31], and due to algebraic attacks, [8], the degree of output functions of combining or filter generators should not be low. Finally, to prevent good statistical approximations of the output function f by linear functions, f should have large distance to all affine functions. In this regard, early work by D. Chaum and J.-H. Evertse, [5], [14] on the cryptanalysis of the DES block cipher motivated a different trail concerning cryptographic properties of Boolean functions and S-boxes: In [33], a class of Boolean functions, coined perfect nonlinear functions, was studied, which turned out to coincide with the class of Bent functions, [42]. These functions have been used in the context of coding theory, [29]. Bent functions are not balanced, but otherwise they share a number of desirable properties: They have maximum nonlinearity, i.e., they have largest possible distance to affine functions, and they satisfy good correlation resistance. In addition, they have optimum differential properties. In a series of papers, K. Nyberg has studied Boolean functions and S-boxes related to Bent functions, starting with [39], [40]. A prominent example of such a vectorial Boolean function is the multiplicative inverse map in the finite field $GF(2^8)$ which is used in the S-box of the AES block cipher. The study of Boolean functions with good cryptographic properties has been an ongoing topic, see, e.g., the book [9].

5   Combiners with Memory

The tradeoff between correlation immunity and algebraic degree as noticed in [46] can be avoided if the combining function is allowed to have memory. Results on combiners with memory have first been published by R. Rueppel, [43].

A (k, m)-combiner with k inputs and m memory bits is a finite state machine which is defined by an output function

$$f : \{0,1\}^m \times \{0,1\}^k \to \{0,1\}$$

and a memory function

$$\varphi : \{0,1\}^m \times \{0,1\}^k \to \{0,1\}^m.$$

For a given stream $(X_1, X_2, \ldots)$ of inputs, $X_t \in \{0,1\}^k$, and an initial assignment $C_1 \in \{0,1\}^m$ of the memory, an output bitstream $(z_1, z_2, \ldots)$ is defined according to

$$z_t = f(C_t, X_t).$$

and

$$C_{t+1} = \varphi(C_t, X_t)$$

for all $t > 0$. For keystream generation, the stream of inputs $(X_1, X_2, \ldots)$ is produced by the output of k driving devices. The initial states are determined by the secret key. Often, the driving devices are LFSRs.

Example 6. The basic summation generator with $k = 2$ inputs is a combiner with $m = 1$ bit memory, which coincides with the usual carry of addition of integers: Write $X_t = (a_t, b_t)$. The functions f and φ are defined by

$$z_t = f(c_t, a_t, b_t) = a_t \oplus b_t \oplus c_t$$

and

$$c_{t+1} = \varphi(c_t, a_t, b_t) = a_t b_t \oplus a_t c_t \oplus b_t c_t.$$

The function f in this summation generator is 2nd-order correlation immune. Correlations in combiners with one bit memory have been studied in detail in [34].

Example 7. The stream cipher E_0 used in Bluetooth, [3], is a combiner with $k = 4$ inputs and $m = 4$ bit memory. The stream of inputs is produced by the outputs of 4 LFSRs of length 128 in total.

More recent (word-oriented) stream ciphers with memory are, e.g., SNOW, [11], the eSTREAM finalist SOSEMANUK, [41], or ZUC, [37]. A concept related to combiners with memory are feedback with carry shift registers (FCSRs) as introduced in [27]. A synthesis based on FCSRs enabled to cryptanalyze summation generators.

6   Linear Attacks

A correlation attack may be successful, if there are found linear relations that hold with nonnegligible probabilities, between single output bits and a subset of state bits of the LFSR's involved. A *linear attack* is more general, as it seeks for "good linear approximations" of the output stream, i.e., for correlations between linear functions of *several* output bits and linear functions of a subset of the LFSR-state bits involved. This type of attacks may be successful for both, key recovery as well as for distinguishing the output from random. Linear attacks have been developed by Golić, [19]. If there are strong enough correlations, a number of equations, each of which does hold with some probability, may be derived. There are fairly efficient methods (reminiscent to fast correlation attacks) to solve such systems of equations, provided the known output stream is long enough, i.e., provided there are many more equations than unknowns (see [22] for an attack of this type on the Bluetooth stream cipher algorithm). The distinction between correlations of a *single* output bit to linear functions of state bits of the LFSR's as opposed to correlations of linear functions of *several output bits* to linear functions of state bits of the LFSR's becomes relevant if the non-linear combining system contains m bit memory: Consider a block of M consecutive output bits, $Z_t^M = (z_t, z_{t-1}, ..., z_{t-M+1})$ as a function of the corresponding block of M consecutive inputs $X_t^M = (X_t, X_{t-1}, ..., X_{t-M+1})$ and the preceeding memory bits C_{t-M+1}. Here X_t denotes the bit vector at time t of the state bits of the LFSRs involved, and similarly, C_{t-M+1} denotes the bit vector of the m memory bits at time $t - M + 1$. Assume that X_t^M and C_{t-M+1} are balanced and mutually independent. Then, according to [20], if $M \geq m$, there *must* exist linear correlations between the output and input bits, but they may also exist if $M < m$. This shows that correlations cannot be easily defeated, even in presence of memory. Besides key recovery attacks, powerful distinguishing attacks using linear approximations of quite diverse stream cipher constructions have become known, e.g. a linear distinguisher on the initial version of the stream cipher SNOW, [7], or a distinguisher on the cipher Shannon, [1].

7   Open Problems

The topic of (fast) correlation attacks has considerably evolved over time. However, some open problems in this area are identified. A first question is how to devise fast correlation attacks in an algorithmically optimal way. Important steps in this direction have been taken in [6] and [10]. In a second direction, various word-oriented stream ciphers use LFSRs over a binary extension field of $GF(2)$ rather than over $GF(2)$. In this case, the established methods seem infeasible. It would be of interest to see fast correlation attacks for LFSRs, e.g, over $GF(2^{32})$. This question has been addressed initially in [26]. Finally, it was observed that correlations cannot be easily avoided in whatever construction is used in the design of a stream cipher. In a complexity-theoretic context, it has been shown that there exist pseudorandom generators with low computational requirements so that in a specified sense each linear distinguisher of the output stream has a bias that can provably be upper bounded, [38]. It would be interesting to come up with cryptographically secure constructions with similar properties.

Acknowledgments

This review owes much to useful discussions with María Naya-Plasencia and with attendees of FSE 2011. This work is supported by DTU Mathematics and by the Danish Center for Applied Mathematics and Mechanics as well as by the National Competence Center in Research on Mobile Information and Communication Systems (NCCR-MICS), a center of the Swiss National Science Foundation under grant number 5005-67322.

References

1. Ahmadian, Z., Mohajeri, J., Salmasizadeh, M., Hakala, R., Nyberg, K.: A practical distinguisher for the Shannon cipher. Journal of Systems and Software 83(4), 543–547 (2010)
2. Berbain, C., Gilbert, H., Maximov, A.: Cryptanalysis of grain. In: Robshaw, M.J.B. (ed.) FSE 2006. LNCS, vol. 4047, pp. 15–29. Springer, Heidelberg (2006)
3. Bluetooth SIG, Specification of the Bluetooth system, Version 1.1 (February 22, 2001), http://www.bluetooth.com/
4. Canteaut, A., Trabbia, M.: Improved Fast Correlation Attacks Using Parity-Check Equations of Weight 4 and 5. In: Preneel, B. (ed.) EUROCRYPT 2000. LNCS, vol. 1807, pp. 573–588. Springer, Heidelberg (2000)
5. Chaum, D., Evertse, J.-H.: Cryptanalysis of DES with a Reduced Number of Rounds, sequences of linear factors in block ciphers. In: Williams, H.C. (ed.) CRYPTO 1985. LNCS, vol. 218, pp. 192–211. Springer, Heidelberg (1986)
6. Chose, P., Joux, A., Mitton, M.: Fast Correlation Attacks: An Algorithmic Point of View. In: Knudsen, L.R. (ed.) EUROCRYPT 2002. LNCS, vol. 2332, pp. 209–221. Springer, Heidelberg (2002)
7. Coppersmith, D., Halevi, S., Jutla, C.S.: Cryptanalysis of stream ciphers with linear masking. In: Yung, M. (ed.) CRYPTO 2002. LNCS, vol. 2442, pp. 515–532. Springer, Heidelberg (2002), http://eprint.iacr.org/2002/020
8. Courtois, N., Meier, W.: Algebraic attacks on Stream Ciphers with Linear Feedback. In: Biham, E. (ed.) EUROCRYPT 2003. LNCS, vol. 2656, pp. 345–359. Springer, Heidelberg (2003)
9. Cusick, T., Stanica, P.: Cryptographic Boolean Functions and Applications. Academic Press, London (2009)
10. Edel, Y., Klein, A.: Computational aspects of fast correlation attacks (2010) (preprint)
11. Ekdahl, P., Johansson, T.: A New Version of the Stream Cipher SNOW. In: Nyberg, K., Heys, H.M. (eds.) SAC 2002. LNCS, vol. 2595, pp. 47–61. Springer, Heidelberg (2003)
12. Ekdahl, P., Meier, W., Johansson, T.: Predicting the Shrinking Generator with Fixed Connections. In: Biham, E. (ed.) EUROCRYPT 2003. LNCS, vol. 2656, pp. 330–344. Springer, Heidelberg (2003)
13. Englund, H., Johansson, T.: A New Simple Technique to Attack Filter Generators and Related Ciphers. In: Handschuh, H., Hasan, M.A. (eds.) SAC 2004. LNCS, vol. 3357, pp. 39–53. Springer, Heidelberg (2004)
14. Evertse, J.-H.: Linear structures in block ciphers. In: Price, W.L., Chaum, D. (eds.) EUROCRYPT 1987. LNCS, vol. 304, pp. 249–266. Springer, Heidelberg (1988)

15. Finiasz, M., Vaudenay, S.: When Stream Cipher Analysis Meets Public-Key Cryptography. In: Biham, E., Youssef, A.M. (eds.) SAC 2006. LNCS, vol. 4356, pp. 266–284. Springer, Heidelberg (2007)
16. Fossorier, M.P.C., Mihaljević, M.J., Imai, H., Cui, Y., Matsuura, K.: An Algorithm for Solving the LPN Problem and Its Application to Security Evaluation of the HB Protocols for RFID Authentication. In: Barua, R., Lange, T. (eds.) INDOCRYPT 2006. LNCS, vol. 4329, pp. 48–62. Springer, Heidelberg (2006)
17. Gallager, R.G.: Low-Density Parity-Check Codes. MIT Press, Cambridge (1963)
18. Golić, J.: Computation of low-weight parity-check polynomials. Electronic Letters 32(21), 1981–1982 (1996)
19. Golić, J.: Linear models for keystream generators. IEEE Trans. on Computers 45, 41–49 (1996)
20. Golić, J.: Correlation properties of a general binary combiner with memory. Journal of Cryptology 9, 111–126 (1996)
21. Golić, J., Salmasizadeh, M., Dawson, E.: Fast correlation attacks on the summation generator. Journal of Cryptology 13, 245–262 (2000)
22. Golić, J.D., Bagini, V., Morgari, G.: Linear Cryptanalysis of Bluetooth Stream Cipher. In: Knudsen, L.R. (ed.) EUROCRYPT 2002. LNCS, vol. 2332, pp. 238–255. Springer, Heidelberg (2002)
23. Hawkes, P., Rose, G.: Exploiting Multiples of the Connection Polynomial in Word-Oriented Stream Ciphers. In: Okamoto, T. (ed.) ASIACRYPT 2000. LNCS, vol. 1976, pp. 303–316. Springer, Heidelberg (2000)
24. Johansson, T., Jönsson, F.: Fast correlation attacks through reconstruction of linear polynomials. In: Bellare, M. (ed.) CRYPTO 2000. LNCS, vol. 1880, pp. 300–315. Springer, Heidelberg (2000)
25. Jönsson, F., Johansson, T.: A fast correlation attack on LILI-128. Inf. Process. Lett. 81(3), 127–132 (2002)
26. Jönsson, F.: Some results on fast correlation attacks, Thesis, Lund University, Sweden
27. Klapper, A., Goresky, M.: Feedback shift registers, 2-adic span, and combiners with memory. Journal of Cryptology 10, 111–147 (1997)
28. Levieil, É., Fouque, P.-A.: An Improved LPN Algorithm. In: De Prisco, R., Yung, M. (eds.) SCN 2006. LNCS, vol. 4116, pp. 348–359. Springer, Heidelberg (2006)
29. MacWilliams, F.J., Sloane, N.J.: The theory of error-correcting codes. North Holland, Amsterdam (1977)
30. Maitra, S., Gupta, K.C., Venkateswarlu, A.: Results on multiples of primitive polynomials and their products over GF(2). Theor. Comput. Sci. 341(1-3), 311–343 (2005)
31. Massey, J.L.: Shift-register synthesis and BCH decoding. IEEE Transactions on Information Theory 15, 122–127 (1969)
32. Meier, W., Staffelbach, O.: Fast correlation attacks on certain stream ciphers. Journal of Cryptology 1, 159–176 (1989)
33. Meier, W., Staffelbach, O.: Nonlinearity criteria for cryptographic functions. In: Quisquater, J.-J., Vandewalle, J. (eds.) EUROCRYPT 1989. LNCS, vol. 434, pp. 549–562. Springer, Heidelberg (1990)
34. Meier, W., Staffelbach, O.: Correlation properties of combiners with memory in stream ciphers. Journal of Cryptology 5, 67–86 (1992)
35. Mihaljević, M.J., Fossorier, M.P.C., Imai, H.: Fast correlation attack algorithm with list decoding and an application. In: Matsui, M. (ed.) FSE 2001. LNCS, vol. 2355, pp. 196–210. Springer, Heidelberg (2002)

36. Mihaljevic, M., Fossorier, M., Imai, H.: On decoding techniques for cryptanalysis of certain encryption algorithms. IEICE Trans. Fundamentals E84-A(4), 919–930 (2001)
37. Mobile Phone Security Algorithms - New Version, http://gsmworld.com/our-work/programmes-and-initiatives/
38. Mossel, E., Shpilka, A., Trevisan, L.: On ϵ-biased Generators in NC^0. Random Struct. Algorithms 29(1), 56–81 (2006)
39. Nyberg, K.: Constructions of Bent Functions and Difference Sets. In: Damgård, I.B. (ed.) EUROCRYPT 1990. LNCS, vol. 473, pp. 151–160. Springer, Heidelberg (1991)
40. Nyberg, K.: Perfect Nonlinear S-Boxes. In: Davies, D.W. (ed.) EUROCRYPT 1991. LNCS, vol. 547, pp. 378–386. Springer, Heidelberg (1991)
41. Robshaw, M., Billet, O.: New Stream Cipher Designs. LNCS, vol. 4986. Springer, Heidelberg (2008)
42. Rothaus, O.S.: On bent functions. Journal of Combinatorial Theory (A) 20, 300–305 (1976)
43. Rueppel, R.A.: Correlation immunity and the summation generator. In: Williams, H.C. (ed.) CRYPTO 1985. LNCS, vol. 218, pp. 260–272. Springer, Heidelberg (1986)
44. Shannon, C.E.: Communication theory of secrecy systems. Bell System Technical Journal 27, 656–715 (1949)
45. Siegenthaler, T.: Decrypting a class of stream ciphers using ciphertext only. IEEE Trans. on Computers C-34, 81–85 (1985)
46. Siegenthaler, T.: Correlation immunity of nonlinear combining functions for cryptographic applications. IEEE Trans. Inform. Theory 30, 776–780 (1984)
47. Wagner, D.: A generalized birthday problem. In: Yung, M. (ed.) CRYPTO 2002. LNCS, vol. 2442, pp. 288–303. Springer, Heidelberg (2002)
48. Wang, D., Lu, P.: Geometrically Invariant Watermark Using Fast Correlation Attacks. In: Proceedings of IIH-MSP 2006, pp. 465–468. IEEE Computer Society, Los Alamitos (2006)
49. Zeng, K., Huang, M.: On the linear syndrome method in cryptoanalysis. In: Goldwasser, S. (ed.) CRYPTO 1988. LNCS, vol. 403, pp. 469–478. Springer, Heidelberg (1990)
50. Zeng, K., Yang, C.H., Rao, T.R.N.: An improved linear syndrome algorithm in cryptanalysis with applications. In: Menezes, A., Vanstone, S.A. (eds.) CRYPTO 1990. LNCS, vol. 537, pp. 34–47. Springer, Heidelberg (1991)

Analysis of Reduced-SHAvite-3-256 v2*

Marine Minier[1], María Naya-Plasencia[2], and Thomas Peyrin[3]

[1] Université de Lyon, INRIA, CITI, F-69621, France
[2] FHNW, Windisch, Switzerland
[3] Nanyang Technological University, Singapore

Abstract. In this article, we provide the first independent analysis of the (2nd-round tweaked) 256-bit version of the `SHA-3` candidate `SHAvite-3`. By leveraging recently introduced cryptanalysis tools such as rebound attack or Super-Sbox cryptanalysis, we are able to derive chosen-related-salt distinguishing attacks on the compression function on up to 8 rounds (12 rounds in total) and free-start collisions on up to 7 rounds. In particular, our best results are obtained by carefully controlling the differences in the key schedule of the internal cipher. Most of our results have been implemented and verified experimentally.

Keywords: rebound attack, Super-Sbox, distinguisher, `SHAvite-3`, `SHA-3`.

1 Introduction

In cryptography hash functions are one of the most important and useful tools. An n-bit cryptographic hash function H is a function taking an arbitrarily long message as input and outputting a fixed-length hash value of size n bits. One wants such a primitive to be collision resistant and (second)-preimage resistant: it should be impossible for an attacker to obtain a collision (two different messages hashing to the same value) or a (second)-preimage (a message hashing to a given challenge) in less than $2^{n/2}$ and 2^n computations respectively. However, in many protocols hash functions are used to simulate the behavior of a random oracle [1] and the underlying security proof often requires the hash function H to be indistinguishable from a random oracle. This security notion naturally extends to a fixed-input length random oracle with a compression function.

In recent years, we saw the apparition of devastating attacks [23,22] that broke many standardized hash functions [20,14]. The National Institute of Standards and Technology (NIST) launched the `SHA-3` competition [16] in response

---

* This work was partially supported by the French National Agency of Research: ANR-06-SETI-013. The second author is supported by the National Competence Center in Research on Mobile Information and Communication Systems (NCCR-MICS), a center of the Swiss National Science Foundation under grant number 5005-67322. The third author is supported by the Singapore National Research Foundation under Research Grant NRF-CRP2-2007-03.

A. Joux (Ed.): FSE 2011, LNCS 6733, pp. 68–87, 2011.

to these attacks and in order to keep an appropriate security margin considering the increase of the computation power or potential further cryptanalysis improvements. The outcome of this competition will be a new hash function standard, to be selected in 2012. Among the 64 candidates originally submitted and the 14 selected for the 2nd round of the competition, one can observe that a non negligible proportion are AES-based proposals (reuse of some parts of the AES block cipher [15,5] or mimicry of its structure), as SHAvite-3 [3] for example.

This fact motivated the academic community to improve its knowledge concerning the security of the AES block cipher or AES-like permutations in the setting of hash functions [17,8,13,11,10,12,7]. For an attacker, one of the major distinction between cryptanalyzing a block cipher and a hash function is that in the latter he has full access to the internal computation and thus he can optimize its use of the freedom degrees. In particular, the recent SHA-1 attacks where made possible thanks to an improvement of the use of the freedom degrees. In the case of AES-based hash functions, the rebound attack [13,10], Start-from-the-middle [12] or Super-Sbox cryptanalysis [7] are very handy tools for a cryptanalyst.

During the first round of the SHA-3 competition, SHAvite-3 was first analyzed by Peyrin [18] who showed that an attacker could easily find chosen-counter chosen-salt collisions for the compression function. This weakness led to a tweaked version of the algorithm for the second round. Then, recently new cryptanalysis results [4,6] on the 512-bit version of SHAvite-3 were published, in particular a chosen-counter chosen-salt preimage attack on the full compression function of SHAvite-3-512.

Our contributions. In this paper, we give the first cryptanalysis results on the tweaked 256-bit version of SHAvite-3-256. By using the rebound attack or Super-Sbox cryptanalysis, we are able to derive distinguishers for a reduced number of rounds of the internal permutation of SHAvite-3-256. Those results can then be transformed into distinguishers or can be used to mount a free-start collision attack for reduced versions of the compression function. The number of rounds attacked can be further extended by authorizing the attacker to fully control the salt values. The results are summarized in Table 1. We emphasize that most of the attacks have been implemented and verified experimentally.

Table 1. Summary of results for the SHAvite-3-256 compression function

rounds	comput. complexity	memory complexity	type	section
6	2^{80}	2^{32}	free-start collision	sec. 3
7	2^{48}	2^{32}	distinguisher	sec. 3
7	2^{7}	2^{7}	chosen-related-salt distinguisher	sec. 4.1
7	2^{25}	2^{14}	chosen-related-salt free-start near-collision	sec. 4.2
7	2^{96}	2^{32}	chosen-related-salt semi-free-start collision	ext. vers.
8	2^{25}	2^{14}	chosen-related-salt distinguisher	sec. 4.2

2   The SHAvite-3-256 Hash Function

SHAvite-3-256 is the 256-bit version of SHAvite-3 [3], an iterated hash function based on the HAIFA framework [2]. We describe here the tweaked version of the algorithm. The message M to hash is first padded and then split into ℓ 512-bit message blocks $M_0\|M_1\|\ldots\|M_{\ell-1}$. Then, the 256-bit internal state (initialized with an initial value IV) is iteratively updated with each message block using the 256-bit compression function C_{256}. Finally, when all the padded message blocks have been processed, the output hash is obtained by truncating the internal state to the desired hash size n as follows:

$$h_0 = IV, \quad h_i = C_{256}(h_{i-1}, M_{i-1}, salt, cnt), \quad hash = trunc_n(h_i)$$

Internally, the 256-bit compression function C_{256} of SHAvite-3-256 consists of a 256-bit block cipher E^{256} used in classical Davies-Meyer mode. The input of the compression function C_{256} consists of a 256-bit chaining value h_{i-1}, a 512-bit message block M_{i-1}, a 256-bit salt (denoted $salt$) and a 64-bit counter (denoted cnt) that represents the number of message bits processed by the end of the iteration. The output of the compression function C_{256} is given by:

$$h_i = C_{256}(h_{i-1}, M_{i-1}, salt, cnt) = h_{i-1} \oplus E^{256}_{M_{i-1}\|salt\|cnt}(h_{i-1})$$

where $\oplus$ denotes the XOR function.

2.1   The Block Cipher E^{256}

The internal block cipher E^{256} of the SHAvite-3-256 compression function is composed of 12 rounds of a classical 2-branch Feistel structure. The chaining variable input h_{i-1} is first divided into two 128-bit chaining values (A_0, B_0). Then, for each round, the Feistel construction computes:

$$(A_{i+1}, B_{i+1}) = (B_i, A_i \oplus F_i(B_i)), \quad i = 0, \ldots, 11$$

where F_i is a non-linear function composed of three full AES rounds. More precisely, if one considers that $AESr$ denotes the unkeyed AES round (i.e. SubBytes SB, ShiftRows ShR and MixColumns MC functions in this order), then F_i is defined by (see Figure 1) :

$$F_i(x) = AESr(AESr(AESr(x \oplus k_i^0) \oplus k_i^1) \oplus k_i^2) \tag{1}$$

where k_i^0, k_i^1 and k_i^2 are 128-bit local keys generated by the message expansion of the compression function C^{256} (it can also be viewed as the key schedule of the internal block cipher E^{256}). We denote by $RK_i = (k_i^0, k_i^1, k_i^2)$ the group of local keys used during round i of the block cipher.

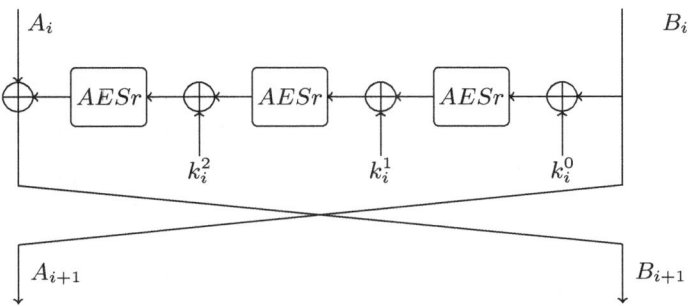

Fig. 1. Round i of the state update of SHAvite-3-256 compression function

2.2   The Message Expansion

The message expansion of C_{256} (the key schedule of E^{256}) takes a 512-bit message block M_i, the 256-bit salt ($salt$) and the 64-bit counter (cnt) as inputs. The 512-bit message block M_i is represented as an array of sixteen 32-bit words $(m_0, m_1, \ldots, m_{15})$, the 256-bit salt as an array of eight 32-bit words $(s_0, s_1, \ldots, s_7)$ and the counter as an array of two 32-bit words (cnt_0, cnt_1). 36 128-bit AES local subkeys k_i^j (with $0 \leq i \leq 11$ and $0 \leq j \leq 2$) are generated, seen as 144 words of 32 bits each (one word standing for one AES column), represented in an array $rk[0...143]$:

$$(k_i^0, k_i^1, k_i^2) = (rk[12 \cdot i], rk[12 \cdot i + 1], rk[12 \cdot i + 2], rk[12 \cdot i + 3]),$$
$$(rk[12 \cdot i + 4], rk[12 \cdot i + 5], rk[12 \cdot i + 6], rk[12 \cdot i + 7]),$$
$$(rk[12 \cdot i + 8], rk[12 \cdot i + 9], rk[12 \cdot i + 10], rk[12 \cdot i + 11])$$

The first 16 values of the array rk are initialized with the message block m_i, i.e. $rk[i] = m_i$ with $0 \leq i \leq 15$. Then, the rest of the array is filled by repeating four times the step described in Figure 2. During one step, sixteen 32-bit words are first generated using parallel AES rounds and a subsequent linear expansion step L_1 (the salt words are XORed to the internal state before applying the AES rounds). Note that during each step the two counter words cnt_0 and cnt_1 (or a complement version of them) are XORed with two particular 32-bits words at the output of the AES rounds. Then, sixteen more 32-bit words are computed using only another linear layer L_2. For more details on the message expansion, we refer to the submission document of the tweaked version of SHAvite-3 [3].

3   Rebound and Super-Sbox Analysis of SHAvite-3-256

Before describing our distinguishing and free-start collision attacks on reduced versions of the SHAvite-3-256 compression function, we first explain what are the main tools we are going to use.

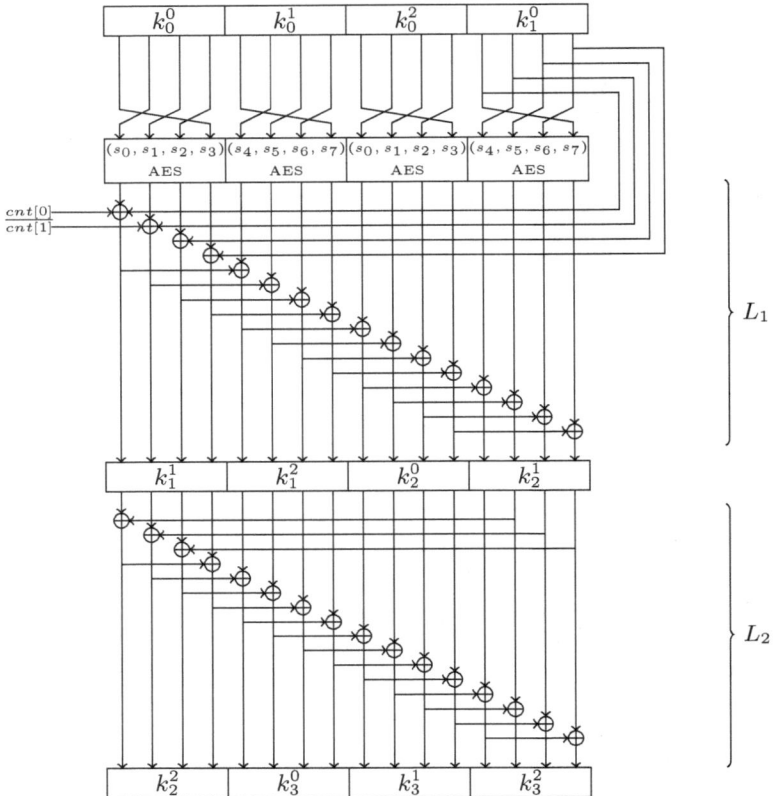

Fig. 2. The first step of the message expansion of the SHAvite-3-256 compression function. The salt words are XORed to the internal state before the parallel AES rounds application. The counters are XORed several times at different positions.

3.1   The Cryptanalyst Tool 1: The Truncated Differential Path

When cryptanalyzing AES-based hash functions (or more generally byte-oriented primitives), it has been shown [17] that it is very handy to look at truncated differences [9]: instead of looking at the actual difference value of a byte, one only checks if a byte contains a difference (active byte) or not (inactive byte). In addition to simplifying the analysis, the direct effect is that the differential behavior through the non-linear Sboxes becomes deterministic. On the other hand, the differential transitions through the linear MixColumns layer will be verified probabilistically.

More precisely, the matrix multiplication underlying the AES MixColumns transformation has the interesting property of being a Maximum Distance Separable (MDS) mapping: the number of active input and output bytes for one column is always greater or equal to 5 (unless there is no active input and output byte at all). When picking random input values, the probability of success for a differential transition that meets the MDS constraints through a MixColumns layer is determined by the number of active bytes in the output: if such a differential transition contains k active bytes in one column of the output, its probability of success will approximatively be equal to $2^{-8 \times (4-k)}$. For example, a $4 \mapsto 1$ transition for one column has success probability of approximatively 2^{-24}. Note that the same reasoning applies when dealing with the invert function as well.

In the following, we will use several different types of truncated differential masks for a given 128-bit AES state. We reuse the idea from [19] and we restrict ourselves to four types of byte-wise truncated differential words F, C, D and 1, respectively a fully active state, one fully active column only, one fully active diagonal only and one active byte only. Considering those 4 types of differential masks seems natural because of the symmetry and diffusion properties of an AES round.

We are especially interested in the truncated differential transitions through 3 rounds of the AES since it is the main basic primitive used in the round function of the SHAvite-3-256 compression function. We would like to know what is the probability to go from one truncated differential mask to another (both forward and backward) and the corresponding differential path. First, we can compute the approximate probability of success for a one-round transition between the four types of truncated differential states for both forward and backward directions. Those probabilities are simply obtained by studying the MixColumns transitions for one AES round. For example, one can easily check that when computing forward, going from D to F with the trail $D \mapsto 1 \mapsto C \mapsto F$ happens with probability 2^{-24} with randomly selected input values and active bytes difference values. The same probability holds for the inverse trail in the backward direction.

3.2 The Cryptanalyst Tool 2: The Freedom Degrees

The second very important tool for a hash function cryptanalyst are the freedom degrees. The rebound attack [13] uses a local meet-in-the-middle-like technique in which the freedom degrees are consumed in the middle part of the differential path, right where they can improve at best the overall complexity. More precisely, the rounds in the middle are controlled (the controlled rounds) and will be verified with only a few operations on average, while the rest of the path both in forward and backward direction is fulfilled probabilistically (the uncontrolled rounds). This method provides good results [11,10], but the controlled part is limited to two rounds only. In [12], this technique is generalized to start-from-the-middle attacks, allowing to control 3 rounds in the middle part, without increasing the complexity (i.e. only a few operations on average). However, this technique is more complex to handle and only works for differential paths for which the middle part does not contain too many active bytes. Finally, the Super-Sbox cryptanalysis (independently introduced in [10] and [7]) can also control 3 rounds in the middle of the differential trail with only a few operations on average and works for any differential path. The idea is that one can view two rounds of an AES-like permutation as the parallel application of a layer of big Sboxes, named Super-Sboxes, preceded and followed by simple affine transformations.

This technique can find several solutions for an average cost of 1, but there is a minimal cost to pay in any case: the complexity of the attack in the case of the AES permutation is $\max\{2^{32}, k\}$ computations and 2^{32} memory, where k is the number of solutions found verifying the controlled rounds.

3.3 Super-Sbox Attacks for Reduced SHAvite-3-256

Having introduced our cryptanalyst tools, we will derive distinguishing attacks for the 7-round reduced SHAvite-3-256 compression function, or even free-start collision attacks (collision for which the incoming chaining variable is fully controlled by the attacker) for the 6-round reduced version. We start with the 6-round truncated differential path built by removing the last round of the 7-round differential path depicted in Figure 3. First, one can check that this path is valid as it contains no impossible MixColumns transitions. Moreover, a simple analysis of the amount of freedom degrees (as it is done in [7]) shows that we have largely enough of them in order to obtain at least one valid solution for the whole differential path: we have a probability of about 2^{-48} that a valid pair for the differential path exists when the Δ and the subkeys values are fixed. Randomizing those values provides much more than the 2^{48} freedom degrees required.

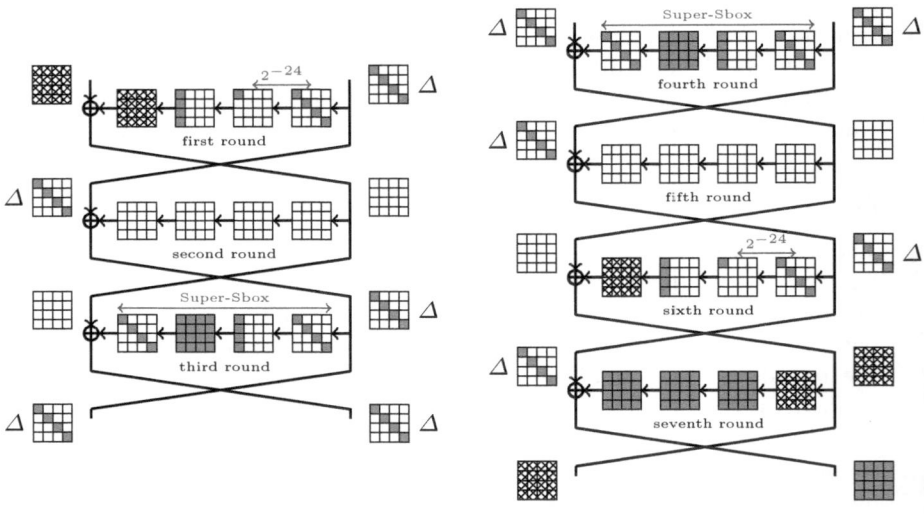

Fig. 3. The 7-round truncated differential path. The left part is the three first rounds and the right part the four last ones. Each gray cell stands for an active byte. A hatched state denotes a fully active state obtained by applying the MixColumns function on only one active byte per column. All D 128-bit words contain the difference Δ.

The most costly part is obviously located in the middle, during the third and fourth rounds where we have fully active AES states (F-type). Thus, we will use the available freedom degrees at those particular rounds precisely. Let Δ be one of the 2^{32} possible D-type difference values. With the Super-Sbox technique and by using the freedom degrees available on the message input, we will find a valid 128-bit pair for the fourth round, mapping the difference Δ to the very same difference through the 3 AES rounds. Then, we will do the same for the third round.

The subkeys used during the fourth round are k_3^0, k_3^1, k_3^2. We first choose a random value for k_3^2. Then, using the Super-Sbox technique, for a cost of $\max\{2^{32}, 2^{32}\} = 2^{32}$ computations and 2^{32} memory, one can generate 2^{32} pairs of 128-bit states verifying the truncated differential path for the 3 AES rounds in the right branch of the fourth round: D $\mapsto$ C $\mapsto$ F $\mapsto$ D. At the present time, we did not fix k_3^0 nor k_3^1, because we were only looking at truncated differences: for each 128-bit solution pair found, the 3 AES rounds truncated differential path will be verified whatever is the value of k_3^0 or k_3^1 (more precisely k_3^1 will only have an incidence on the exact D-type difference value on the input of the pairs, while k_3^0 will have no incidence on the difference at all). Note that the Super-Sbox technique allows us to directly force the exact difference value Δ at the output of the 3 AES rounds. Indeed, one can observe that the output of the 3 AES rounds is only a linear combination of the 4 Super-Sboxes outputs. However, the exact difference value on the four active bytes at the input of the 3 AES rounds is unknown because of the SubBytes layer of the first AES round. In order to get the desired difference value Δ on the input as well, for each solution pair we choose accordingly the value of k_3^1. More precisely, only the first column of k_3^1 must be accommodated (only the first column is active when incorporating k_3^1) and this can be done byte per byte independently. Exhausting all the AES Sbox differential transitions during a 2^{16} operations precomputation phase allows us to perform this step with only four table lookups. At this moment, for a cost of 2^{32} computations and memory, we found 2^{32} pairs of 128-bit states that map Δ to Δ through the 3 AES rounds of the fourth SHAvite-3-256 round. For each pair, the value of k_3^2 and the first column of k_3^1 are fixed, but the rest of the message is free to choose.

We perform exactly the same method in order to find 2^{32} pairs of 128-bit states that map Δ to Δ through the 3 AES rounds of the third SHAvite-3-256 round. The only difference is that we will have to fix k_2^2 and choose the first column of k_2^1. This can be done independently from the previously fixed values of k_3^2 and the first column of k_3^1 (by setting the second column of k_3^1, see Figure 2). We are left with two sets, each of 2^{32} pairs, verifying **independently** the third and fourth rounds. The subkey material that remains free is set to a random value and the second and fifth rounds of the differential path is verified with probability 1.

Then, the rest of the path (the uncontrolled rounds) is verified probabilistically: we have one D $\mapsto$ 1 $\mapsto$ C $\mapsto$ F transition in the first round and another one in the sixth round. As already demonstrated, this happens with probability $2^{-2*24} = 2^{-48}$. Overall, one can find a valid candidate for the whole 6-round truncated differential path with 2^{48} computations and 2^{32} memory.

If the very last branch switching of the Feistel structure is removed (at the end of the sixth round) and due to the Davies-Meyer construction, one can obtain a free-start collision if the active byte differences on the input of the first round are equal to the active byte differences on the output of the sixth round. This happens with probability 2^{-32} because we have the right 128-bit word of the internal state containing the difference Δ on both the input and output. The left one is fully active but its differences belong to a 4-byte subspace since only one MixColumns linear application away from a 4-byte difference (hatched states in Figure 3). Finally, by repeating the process, one can find a free-start collision for 6-round reduced SHAvite-3-256 with $2^{48+32} = 2^{80}$ computations and 2^{32} memory.

We can now move to the full 7-round path depicted in Figure 3. One solution for the entire path can still be obtained with 2^{48} computations and 2^{32} memory since the differential trail in the last round is verified with probability 1. A solution pair will have a difference Δ on its right word input and a difference maintained in a subspace of roughly 2^{32} elements on its left word input (as denoted with hatched cells in Figure 3). Concerning the output, the differences on the left 128-bit output word is kept in the same subspace of 2^{32} elements, while the right output word will have a random difference. Overall, after application of the feed-forward, we obtain a compression function output difference maintained in a subspace of at most 2^{160} elements, while the input difference is maintained in a subspace of 2^{32} elements (since the message/salt/counter inputs contain no difference and the value of Δ is fixed). This is equivalent to mapping a fixed 672-bit input difference (224 bits for the chaining variable, 256 bits for the non active incoming message chunk, 128 and 64 bits for the non active incoming salt and counter chunks respectively) to a fixed 96-bit output difference through a 704-bit to 256-bit compression function. According to the limited-birthday distinguishers [7], this should require 2^{64} computations in the ideal case.[1] Thus, one can distinguish a 7-round reduced version of the SHAvite-3-256 compression function with 2^{48} computations and 2^{32} memory.

4   Chosen-Related-Salt Distinguishers

While the previous section takes a full advantage of the Super-Sbox techniques in its analysis, this section presents an attack that uses rebound attack principle fully exploiting the message expansion. This leads to 7-round and 8-round chosen-related-salt distinguishers on the SHAvite-3-256 compression function with a complexity of 2^7 and 2^{25} operations respectively. Also, one can find chosen-related-salt semi-free-start collisions on 7 rounds of the SHAvite-3-256 compression function with a complexity of 2^{96} operations (see extended version of this article). In this section, we will insert differences in the message and in

---

[1] As shown in [7], if we denote by i (resp. j) the number of fixed-difference bits in the input (resp. in the output) and by t the total number of input bits of the compression function, the equivalent complexity to find such a structure for a random compression function is $2^{i+j-t} = 2^{672+96-704} = 2^{64}$ computations.

the salt input of the compression function. The main principle of those analyses relies on correcting the differences at the end of each round so that they are not spread afterwards. The 8-round differential path used is given in Figure 4, while the 7-round version is obtained by removing the seventh round. All the differences in the successive internal states will be canceled from state 1 to state 7. This could be done by considering differences in the first four 32-bit words of the *salt* denoted (s_0, s_1, s_2, s_3), in eight 32-bit message words (m_0, m_1, m_2, m_3) and $(m_8, m_9, m_{10}, m_{11})$ and differences in A_0, the left part of the initial chaining value, the other parameters being taken without any difference. The notations used are the ones of Section 2: A_i denotes the left part of the state at round i (where i goes from 0 to 8), and B_i the right part of the internal state at round i.

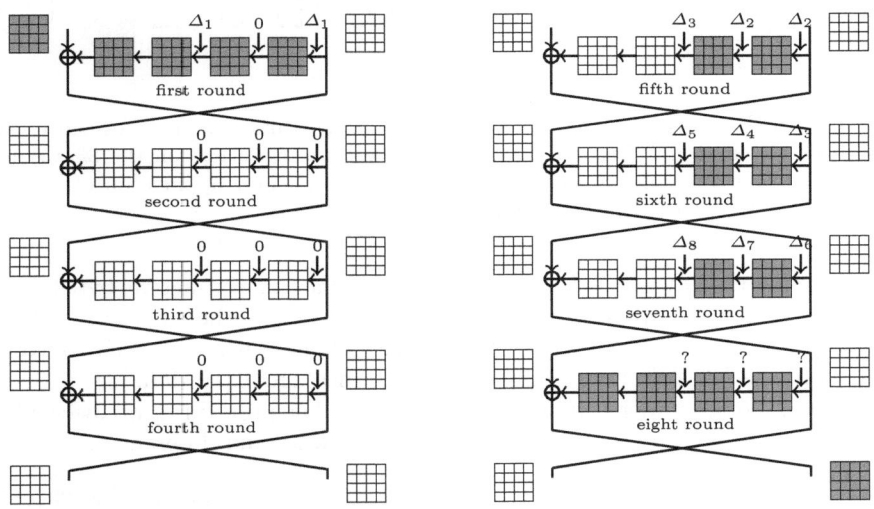

Fig. 4. The 8-round truncated differential path. The left part represents the four first rounds and the right part the four last ones. Each gray cell stands for an active byte. The Δ's in each round denote the differences incorporated by the subkeys. For the 7-round differential path, we remove the differences control in the seventh round.

The difference values in the salt words and in the message words are chosen to be identical (Δ_1), so that they cancel for the subkeys generated after the first round of the message expansion. Moreover, the subkeys involved in rounds 2, 3 and 4 will not contain any difference (as shown on Figure 5). We concentrated our analysis on the active rounds in the middle of the trail, i.e. rounds 5, 6 (and 7 in the case of the 8-round distinguisher). The probability of success for the rest of the differential path is one since in the first round the differences can spread freely.

We start by finding a valid pair that verifies the path for the rounds 5, 6 and 7. Let us remember that at the beginning, we have just established the truncated differential path, i.e. the actual difference values in the bytes are unknown. Thus,

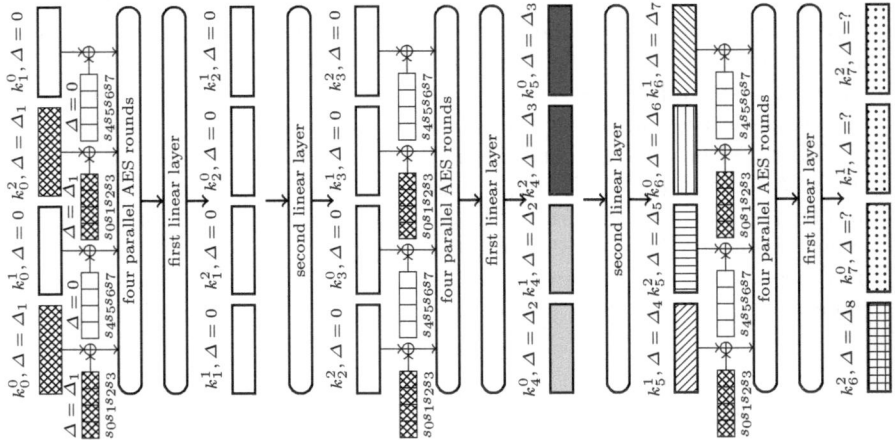

Fig. 5. Differential cancellation and differential equalities in the message expansion for the 7-round and 8-round chosen-related-salt distinguishers on the SHAvite-3-256 compression function. $\tilde{\Delta}$ represents the difference Δ after application of the column switching layer just before the salt incorporation.

when required, we will fix the difference values of the active bytes and also the values themselves. We will insert differences in the first four salt words (s_0, s_1, s_2, s_3), all their bytes being active. As before, when we refer to words, we denote the AES column 32-bit words in the message expansion.

For a better understanding of the distinguisher procedure, we discriminate 5 distinct kinds of ways in order to determine values and differences:

- Type a means that we directly choose the values and the differences. This can be done when there is no previous restriction.
- Type b are determined by the linear part of the message expansion. In this case, a linear relation of previously fixed differences or values completely determines the remaining ones.
- Type c are determined by the non-linear part of the message expansion. Here, a non linear relation of previously fixed differences and values determines the remaining ones.
- Type d are produced by some previous conditions on the Feistel path. That is, for example, if the value of $B_i \oplus k_{i+1}^0$ is fixed and then this subkey is determined, we automatically deduce the value of B_i from this equation. This will directly determine A_{i+1} since $A_{i+1} = B_i$.
- Type e are fixed by the AES rounds. This basically represents the conditions associated to the controlled rounds part.

From now on and for a better clarity, we indicate its type in brackets for each determination. When omitted, the default type is a. We will first describe the distinguisher on 7 rounds and then the one on 8 rounds.

4.1 7-Round Distinguisher with 2^7 Computations

As partially shown on Figure 4, the aim of the 7-round distinguisher is to find a pair of plaintext/ciphertext values for the internal block cipher of SHAvite-3-256 such that there is no difference on the right part of the plaintext and on the left part of the ciphertext. As shown in [7], the corresponding complexity to find such a structure for a random permutation is equal to 2^{64} computations. We show in this section how to find a pair of inputs that verifies the path with a time and memory complexity of 2^7 (by omitting the last branch swapping of the Feistel construction, the attack also applies to the compression function SHAvite-3-256 with the same practical and ideal complexities). In order to build this distinguisher, we would like to find values and differences of the subkeys such that the difference in the 3 AES rounds during SHAvite-3-256 rounds 5 and 6 are auto-erased. The differences generated by the subkeys in the seventh round are let completely free for the 7-round distinguisher and it will not be the case for the 8-round distinguisher. We first give in Table 2 how to fix the degrees of freedom in order to avoid any impossibility.

Table 2. Order and conditions for fixing values and differences. Δ_2 and Δ_3 are chosen such that Δ_2 and $\Delta_2 \oplus \Delta_3$ can be both generated from an AES layer with the same difference in the inputs (for randomly chosen Δ_2 and Δ_3, this is verified with very high probability). When going through all these steps followed in the given order, we are left with the degrees of freedom associated to the values of the words of the salt s_0, s_1, s_2, s_3. Thus, we can pick a random value for those remaining freedom degrees and finally compute the valid pair for the 7-round differential path. * represents the steps that are not executed for the distinguisher over 8 rounds.

instant	fixed (type a)	implies	type	cost
begin	$\Delta k_4^0 = \Delta_2$ $\Delta k_4^2 = \Delta_3$	$\Delta k_4^1 = \Delta_2$	b	1
		$\Delta k_4^0 = \Delta_3$	b	1
		$\Delta k_5^1 = \Delta_2 \oplus ((\Delta_3 \wedge (2^{96} - 1))\|((\Delta_3 >> 96) \oplus \Delta_2) \wedge (2^{32} - 1))$	b	1
		$\Delta k_5^2 = \Delta_2 \oplus ((\Delta k_5^1 \wedge (2^{96} - 1))\|((\Delta k_5^1 >> 96) \oplus \Delta_2) \wedge (2^{32} - 1))$	b	1
		$\Delta k_6^0 = \Delta_3 \oplus ((\Delta k_5^2 \wedge (2^{96} - 1))\|((\Delta k_5^2 >> 96) \oplus \Delta_3) \wedge (2^{32} - 1))$	b	1
		$\Delta k_6^1 = \Delta_3 \oplus ((\Delta k_6^0 \wedge (2^{96} - 1))\|((\Delta k_6^0 >> 96) \oplus \Delta_3) \wedge (2^{32} - 1))$	b	1
round 5	$B_4 \oplus k_4^0$	$k_4^1 : AES_r(AES_r(B_4 \oplus k_4^0) \oplus k_4^1) \oplus$ $AES_r(AES_r(B_4 \oplus k_4^0 \oplus \Delta_2) \oplus k_4^1 \oplus \Delta_2) = \Delta_3$	e	2^6
round 6	$B_5 \oplus k_5^0$	$k_5^1 : AES_r(AES_r(B_5 \oplus k_5^0) \oplus k_5^1) \oplus$ $AES_r(AES_r(B_5 \oplus k_5^0 \oplus \Delta_3) \oplus k_5^1 \oplus \Delta k_5^1) = \Delta k_5^2$	e	2^6
		$k_5^2 = k_4^1 \oplus ((k_5^1) \wedge (2^{96} - 1))\|((k_5^1 >> 96) \oplus k_4^1) \wedge (2^{32} - 1))$	b	1
		$k_4^0 \wedge (2^{32} - 1) = (k_5^1 \wedge (2^{32} - 1)) \oplus ((k_5^1) >> 96) \wedge (2^{32} - 1))$	b	1
		$B_4 \wedge (2^{32} - 1) = (k_4^0 \wedge (2^{32} - 1)) \oplus (k_4^0 \oplus B_4) \wedge (2^{32} - 1)$	b	1
end	$\delta s_0 \ldots \delta s_3$ $= \Delta_1$	$k_4^2 : AES_r^{-1}(k_4^2 \oplus k_4^1) \oplus$ $AES_r^{-1}(k_4^2 \oplus k_4^1 \oplus \Delta_2 \oplus \Delta_3) = \Delta_1$	c	1
		$k_6^0 = k_4^2 \oplus ((k_5^2 \wedge (2^{96} - 1))\|((k_5^2 >> 96) \oplus k_4^2) \wedge (2^{32} - 1))$	b	1
*	k_5^0	$(k_6^0 >> 32) \wedge (2^{96} - 1) = k_5^0 \wedge (2^{96} - 1)$	b	1
		$k_6^1 = k_5^0 \oplus (k_6^0 \wedge (2^{96} - 1))\|((k_6^0 >> 96) \oplus k_5^0) \wedge (2^{32} - 1))$	b	1
		$s_4 \ldots s_7 : AES_r^{-1}(k_4^0 \oplus cnt \oplus AES_r^{-1}(k_5^0 \oplus k_4^2) \oplus s_4\|s_5\|s_6\|s_7) \oplus$ $AES_r^{-1}(k_4^0 \oplus ct \oplus AES_r^{-1}(k_5^0 \oplus k_4^2) \oplus s_4\|s_5\|s_6\|s_7 \oplus \Delta_2) = \Delta_1$	c	1
		$B_4 = k_4^0 \oplus (k_4^0 \oplus B_4)$	d	1
		$B_5 = k_5^0 \oplus (B_5 \oplus k_5^0)$	d	1

Fifth round: As described in Figure 4, after the fourth round, the inputs of the fifth do not contain any active byte. The differences will be injected during the fifth round through the message expansion. The idea is to erase those differences directly during the fifth round. This will be achieved by carefully choosing the actual values of the bytes.

We choose (type a) the difference in k_4^0 to be Δ_2 and this sets (type b) the difference in k_4^1 to be Δ_2 as well. Then we pick a random difference value for k_4^2 that we denote Δ_3. Note that due to the message expansion, Δ_2 and $\Delta_3 \oplus \Delta_2$ must be 128-bit output differences that can be obtained from the same input difference after application of one AES round. Indeed, as shown in Figure 5, Δ_2 and $\Delta_3 \oplus \Delta_2$ are constructed from the Δ_1 difference inserted by the salt after application of one AES round. In fact, for randomly chosen Δ_2 and Δ_3 values, this is verified with very high probability. Those particular differences also fix (type b) the difference values of the 9 subkeys used in rounds 5, 6 and 7. However, note that k_6^2 is determined after one another AES non-linear round (as shown in Figure 5).

Once Δ_2 and Δ_3 differences fixed, we need to set the values themselves in this fifth round. To do so, we choose random value for $B_4 \oplus k_4^0$ (see Figure 6) and we can compute forward the differences just before the SubBytes layer of the second AES round. We can also propagate the differences backwards from the insertion of Δ_3 up to the output of this SubBytes layer. Due to the AES Sbox differential property, with a probability of 2^{-16}, the differences before and after the second SubBytes can be matched (2^{-1} per Sbox). Thus, we will have to try 2^{16} values of $B_4 \oplus k_4^0$ before finding a match. Note that this cost can be reduced to $4 \times 2^4 = 2^6$ by attacking all the columns independently. When such a solution is found, we directly pick (type e) a value of k_4^1 that makes this match happen.

Sixth round: We deal with the sixth round in a very same way. The differences in the subkeys k_5^0, k_5^1 and k_5^2 are already fixed by the message expansion. Thus, we would like to find the corresponding values that make the cancellation of differences possible in the sixth round computation. As before, we choose an appropriate value of $B_5 \oplus k_5^0$ such that we have a possible match between the input and output differences of the second SubBytes layer. When such a solution is found, we directly pick (type e) a value of k_5^1 that makes this match happen.

The final step: Now, if we randomly fix the values and differences $\delta s_0, \delta s_1, \delta s_2, \delta s_3$, and k_5^0, the values of $k_4^2, k_6^0, k_4^0, k_6^1, s_4, s_5, s_6, s_7, B_4$ and B_5 will also be determined (as shown in Table 2). At this point, we have obtained some coherent values that verify the differential path of rounds 5 and 6, and the only degrees of freedom left are the values of s_0, s_1, s_2 and s_3. We can just pick up one and compute backward and forward, and we obtain the whole path, where the inputs have just the left part of the state active, and the output, before the Davies-Meyer, the right one.

The total cost for the distinguisher is driven by the two first steps, that is $2 \times 2^6 = 2^7$ operations in order to find one valid candidate. This distinguisher has been implemented and verified experimentally. We provide in the extended version of this article an example of such a structured input/output pair (which should not be generated with less than 2^{64} operations in the ideal case).

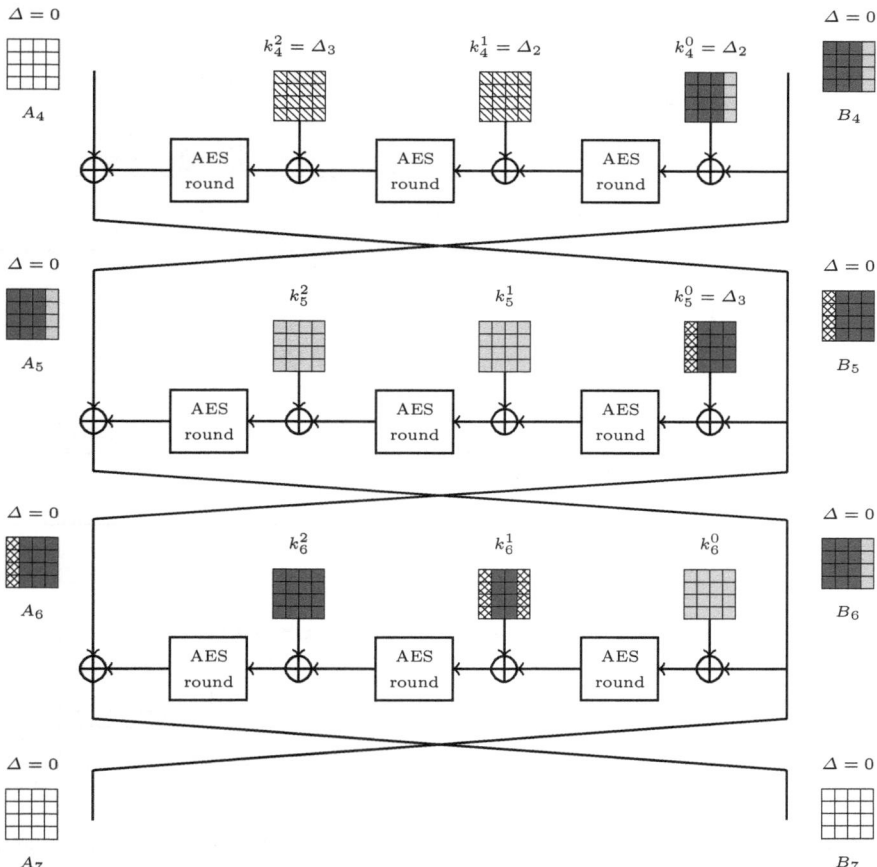

Fig. 6. Details of rounds 5, 6 and 7 of the 8-round chosen-related-salt distinguisher on the SHAvite-3-256 compression function. The bytes denoted with north-west lines are fixed during the fifth round. The light gray bytes are fixed during the sixth round. The dark gray bytes are fixed at the beginning of the seventh round whereas the bytes denoted with hatched cells are fixed at the end of the seventh round.

4.2 8-Round Distinguisher with 2^{25} Computations

In this section we describe the 8-round distinguisher. For rounds 5 and 6 the procedure is the same as the one for the 7-round distinguisher using the equations described in Table 2. However, now we would like also that the differences inserted during the seventh round cancel themselves, whereas the differences inserted during the eighth round can freely spread. In order to fulfill this requirement, after having handled the sixth round, instead of randomly choosing the value of k_5^0 one first chooses the values of s_0, s_1, s_2 and s_3, which will allow us to determine the difference in k_6^2.

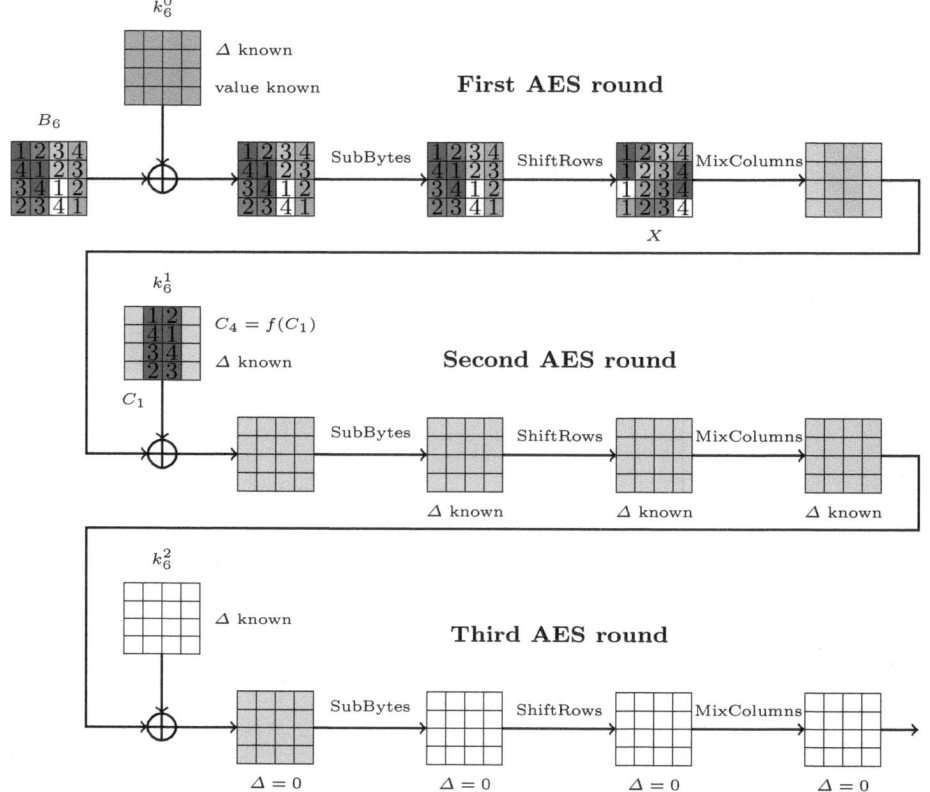

Fig. 7. The 3 AES rounds of the seventh round: the red bytes will determine the middle columns of k_6^1 (because of the message expansion and of previous conditions); the blue bytes will determine the differences in the 2 middle columns on the input of the second round; the orange bytes are fixed (due to round 6); the white and light gray bytes are free. The green bytes are the results of an XOR between blue bytes and orange bytes whereas pink bytes are the results of an XOR between blue bytes and red bytes.

Fixing the differences in the seventh round: We introduce the following notations: let S be a 128-bit AES state, we denote $(S)^i$ the i-th column of S and $(S*)^i$ the i-th column of $ShR(S)$ (i.e., the i-th diagonal of S), for $i \in [0..3]$. We give in Figure 7 a complete illustration of our attack.

Let us analyze the relations that link together the values already fixed and the values to be fixed. On the one hand, we have:

$$(B_6)^i \implies (A_5)^i = (B_4)^i \implies (k_4^0)^i \tag{2}$$

that one can read as "setting the value of $(B_6)^i$ will fix the value of $(A_5)^i$ (because the Feistel structure imposes $(A_5)^i = (B_4)^i$) which will in turn deduce the value of $(k_4^0)^i$". It is important to remark that the value of $(k_4^0)^3$ has already been fixed in round 6, as shown in Table 2. Therefore, because of Relation (2), $(B_6)^3$ will also be known. Then, from the message expansion, we can derive the following relations:

$$(k_4^0)^i \implies (k_5^0)^{i+1} \implies (k_6^1)^{i+1} \text{ for } i \in \{0,1\}, \tag{3}$$

$$(k_4^0)^2 \implies (k_5^0)^3 \implies (k_6^1)^3 = (k_5^0)^3 \oplus (k_6^1)^0. \tag{4}$$

One can check that the values of column 0 and column 3 of k_6^1 are associated by a linear relation. We recall that for the time being the difference of k_6^1 is already known, but not its value.

On the other hand, we have $(B_6*)^i \implies (X)^i$, where X is the state represented in Figure 7 just before the first MC. We can then write the following relation:

$$SB[(k_6^0*)^i \oplus (B_6*)^i] \oplus SB[(k_6^0*)^i \oplus \Delta(k_6^0*)^i \oplus (B_6*)^i] = \Delta(X)^i. \tag{5}$$

For the path to be verified, we need the difference $MC(\Delta(X)^i) \oplus \Delta(k_6^1)^i$ to be compatible with the difference fixed by k_6^2 after the second SB, and the differential transition must be possible by the values of k_6^1 (which are not fixed yet). From the previous equations (2), (3), (4) and (5) we obtain that for making the path to be verified over each column i, the following columns or diagonals intervene at the same time: for $i = 0$, $\{(B_6*)^0, (B_6)^3\}$; for $i = 1$, $\{(B_6*)^1, (B_6)^0\}$; for $i = 2$, $\{(B_6*)^2, (B_6)^1\}$; for $i = 3$, $\{(B_6*)^3, (B_6)^2\}$.

From the previous relations we can deduce that there are some bytes of B_6 that interact in more than one way with the relations for one column i, $((3,3)$ for $i = 0$, $(3,0)$ for $i = 1$, $(3,1)$ for $i = 2$, $(3,2)$ for $i = 3)$. Thus, since we have defined the main relations that must be verified, we can now describe how to find a valid pair. A conforming pair is a pair of values that verifies the differential path of the seventh round as well as the path of the previous rounds. This can be done with a complexity of 2^{25} in time and 2^{14} in memory with the following process:

• We consider 2^{24} values of the bytes of $(B_6*)^1$ (the ones at byte positions $(0,1); (1,2); (2,3); (3,0)$). As the byte $(2,3)$ has an already fixed value, because it belongs to $(B_6)^3$ which was determined in the previous steps, the 2^{24} values are all the possible ones. Thus, the second column of the state after the first MixColumns of the 7th round will be determined by the previous values (i.e. the bytes of $(B_6*)^1$). The values of $(B_6*)^1$ that give us the match we are looking for are such that the differences before and after the second SubBytes match for the second column. Those differences are influenced for the first one by the already fixed differences of k_6^1 and for the second one by the already fixed differences of k_6^2. In other words, at this step, the differences are mainly fixed but not the values. Due to the AES Sbox differential properties, the match between the two differences will happen with a probability equal to 2^{-4}.

We have then to consider that, when we have fixed the values for the bytes $((0,1); (1,2); (2,3); (3,0))$, the value of the byte $(3,1)$ of k_6^1 will also be determined due to equations (2) and (3).

So, when we find a match of differences, we also need that one of the two values for $(3,1)$ that makes the match possible, collides with the already fixed value for this byte. This will happen with a probability of 2^{-7}. We obtain: $2^{24} \cdot 2^{-4} \cdot 2^{-7} = 2^{13}$ values for $(B_6*)^1$ (bytes at positions $(0,1); (1,2); (2,3); (3,0)$) that make the differential path possible for the second column of the seventh round.

• We do the same thing with the bytes of $(B_6*)^2$ (bytes at positions $(0,2); (1,3); (2,0); (3,1)$), and we also obtain 2^{13} values that make the differential path possible for the third column of the seventh round.

• We now consider the interaction between the two previous steps in order to simultaneously find the solutions for the two middle columns. The byte at position $(0,1)$ of B_6 has already been fixed during the first step, and it determines the value of the byte $(0,2)$ of k_6^1, which belong to $(k_6^1)^2$ that affects the second step. Analogously the byte of the position $(2,0)$ from B_6 that was fixed at the second step determines the value of the byte $(2,1)$ of k_6^1, that belongs to $(k_6^1)^1$ and that affects the first step. Thus if we want to compute all the valid candidate values for the two columns in the middle before the second SB we will obtain: $2^{13} \cdot 2^{13} \cdot 2^{-7} \cdot 2^{-7} = 2^{12}$ possible values for fixing $(B_6*)^1$ and $(B_6*)^2$.

• We consider one of the previously determined values among the 2^{12} possibles. We consider now the yet unfixed bytes corresponding to positions $(1,0); (2,1)$ of B_6. Because of Relations (2), (3) and (4) they will determine the values of the associated bytes of k_6^1. Thus, since the differences are already fixed, each of these bytes has only 2 possible values that fulfill the difference match in the second SB for bytes $(1,1)$ and $(2,2)$. From $(B_6*)^3$ (i.e. the values from B_6 that will influence the fourth column in the second SB), we still have one byte, $(3,2)$, that has no influence on the already fixed parts. Therefore this byte can go through 2^8 distinct values. In total, for the fourth column after the first MC there are $2^{8+2} = 2^{10}$ possible values and differences that do not interfere with the differential path of the two middle columns in the second SB. We can compute how many of these 2^{12} values and differences for $(B_6*)^3$ could satisfy the path for the fourth column: $2^2 \cdot 2^8 \cdot 2^{-4} \cdot 2^4 = 2^{10}$ values for $(B_6*)^3$ make the fourth column on the second SubBytes also verify the path. The term 2^{-4} is present because one requires a possible match of differences and the term 2^4 comes from the fact that we can associate 2^4 values of the fourth column for one fixed difference before SubBytes.

• By applying the same method to the first column, we will obtain 2^{10} values for $(B_6*)^0$. We know from Relation (4) that $(k_6^1)^0$ and $(k_6^1)^3$ must satisfy a linear relation. This will occur for a fixed $(k_6^1)^0$ and a fixed $(k_6^1)^3$ with a probability of 2^{-32}. As we have 2^{10} possible values of $(k_6^1)^0$ from step 5 and 2^{10} possible values for $(k_6^1)^3$ from step 4, we obtain a valid couple $((k_6^1)^0, (k_6^1)^3)$ with probability 2^{-12}.

• We repeat step 5 about 2^{12} times with the different solutions of step 3, and we get a valid couple: we obtain all the valid values that verify the differential path, i.e. that have no differences after the 7th round.

Once those steps performed, all the desired values and differences that verify the path are determined except the ones in the first round. The values of s_4, s_5, s_6 and s_7 are not fixed yet and we can choose them as explained in Table 2.

Thus, Δ_1 will determine Δ_2 and we just compute backwards until we obtain the initial state that verifies the path (and consequently, also the first round). From this input, we get a corresponding output that contains no difference after seven rounds. If we apply the Davies-Meyer transformation, the right part of the state will not contain any difference. Thus, we have exhibited a free-start near-collision attack using chosen and related salts on a 7-round reduced version of the SHAvite-3-256 compression function used in the Davies-Meyer mode with 2^{25} computations and 2^{14} memory. The computation cost and memory requirements are quite far from the complexity corresponding to the ideal case (2^{64} operations).

Adding one more round at the end of the seventh round leads to a particular input/output structure (see Figure 4). More precisely, the subkeys differences of the eighth round are no more controllable, thus the right part A_8 of the state will contain differences but not the left side B_8 which is the seventh round's right part due to the Feistel structure. Therefore, after the Davies-Meyer transform, we will have the same difference in the left side of the input state and in the left side of the output state and we obtain a chosen-related-salt distinguisher on an 8-round reduced version of the SHAvite-3-256 compression function used in the Davies-Meyer mode with the same complexity as for the previous chosen-related-salt free-start near-collisions: 2^{25} computations and 2^{14} memory. Finding such a structure in the ideal case should require at least 2^{64} computations.

This 8-round distinguisher has been verified experimentally. We provide in the extended version an example of such a distinguisher.

5   Conclusion

In this paper, we have presented the first analysis of the (2nd-round tweaked) 256-bit version of the SHA-3 competition candidate SHAvite-3. As it is the case for many candidates based on the AES round function, we showed that the Super-Sbox cryptanalysis and the rebound attacks are very efficient when analyzing reduced-round versions of SHAvite-3-256. Namely, without using the salt or the counter inputs, one can attack up to seven rounds of the twelve rounds composing the SHAvite-3-256 compression function. We were even able to reach eight rounds when the attacker is assumed to be able to control the salt input. Despite the attacks being quite involved, all our practical complexity results were verified experimentally.

References

1. Bellare, M., Rogaway, P.: Random Oracles are Practical: A Paradigm for Designing Efficient Protocols. In: ACM Conference on Computer and Communications Security, pp. 62–73 (1993)
2. Biham, E., Dunkelman, O.: A Framework for Iterative Hash Functions - HAIFA. Cryptology ePrint Archive, Report 2007/278 (2007), http://eprint.iacr.org/2007/278 (accessed on Janaury 10, 2010)

3. Biham, E., Dunkelman, O.: The SHAvite-3 Hash Function. Submission to NIST (Round 2) (2009), http://www.cs.technion.ac.il/~orrd/SHAvite-3/Spec.15.09.09.pdf
4. Bouillaguet, C., Dunkelman, O., Leurent, G., Fouque, P.-A.: Attacks on Hash Functions based on Generalized Feistel - Application to Reduced-Round Lesamnta and SHAvite-3-512. Cryptology ePrint Archive, Report 2009/634 (2009), http://eprint.iacr.org/2009/634.pdf
5. Daemen, J., Rijmen, V.: The Design of Rijndael. In: Information Security and Cryptography. Springer, Heidelberg (2002) ISBN 3-540-42580-2
6. Gauravaram, P., Leurent, G., Mendel, F., Naya-Plasencia, M., Peyrin, T., Rechberger, C., Schläffer, M.: Cryptanalysis of the 10-round hash and full compression function of shavite-3-512. In: Bernstein, D.J., Lange, T. (eds.) AFRICACRYPT 2010. LNCS, vol. 6055, pp. 419–436. Springer, Heidelberg (2010)
7. Gilbert, H., Peyrin, T.: Super-Sbox Cryptanalysis: Improved Attacks for AES-like Permutations. In: Hong, S., Iwata, T. (eds.) FSE 2010. LNCS, vol. 6147, pp. 365–383. Springer, Heidelberg (2010), http://eprint.iacr.org/2009/531
8. Khovratovich, D.: Cryptanalysis of Hash Functions with Structures. In: Jacobson Jr., M.J., Rijmen, V., Safavi-Naini, R. (eds.) SAC 2009. LNCS, vol. 5867, pp. 108–125. Springer, Heidelberg (2009)
9. Knudsen, L.R.: Truncated and Higher Order Differentials. In: Preneel, B. (ed.) FSE 1994. LNCS, vol. 1008, pp. 196–211. Springer, Heidelberg (1995)
10. Lamberger, M., Mendel, F., Rechberger, C., Rijmen, V., Schläffer, M.: Rebound Distinguishers: Results on the Full Whirlpool Compression Function. In: Matsui, M. (ed.) ASIACRYPT 2009. LNCS, vol. 5912, pp. 126–143. Springer, Heidelberg (2009)
11. Matusiewicz, K., Naya-Plasencia, M., Nikolić, I., Sasaki, Y., Schläffer, M.: Rebound Attack on the Full LANE Compression Function. In: Matsui, M. (ed.) ASIACRYPT 2009. LNCS, vol. 5912, pp. 106–125. Springer, Heidelberg (2009)
12. Mendel, F., Peyrin, T., Rechberger, C., Schläffer, M.: Improved Cryptanalysis of the Reduced Grøstl Compression Function, ECHO Permutation and AES Block Cipher. In: Jacobson Jr., M.J., Rijmen, V., Safavi-Naini, R. (eds.) SAC 2009. LNCS, vol. 5867, pp. 16–35. Springer, Heidelberg (2009)
13. Mendel, F., Rechberger, C., Schläffer, M., Thomsen, S.S.: The Rebound Attack: Cryptanalysis of Reduced Whirlpool and Grøstl. In: Dunkelman, O. (ed.) FSE 2009. LNCS, vol. 5665, pp. 260–276. Springer, Heidelberg (2009)
14. National Institute of Standards and Technology. FIPS 180-1: Secure Hash Standard (April 1995), http://csrc.nist.gov
15. National Institute of Standards and Technology. FIPS PUB 197, Advanced Encryption Standard (AES). Federal Information Processing Standards Publication 197, U.S. Department of Commerce (November 2001)
16. National Institute of Standards and Technology. Announcing Request for Candidate Algorithm Nominations for a NewCryptographic Hash Algorithm (SHA-3) Family. Federal Register 27(212), 62212–62220 (2007), http://csrc.nist.gov/groups/ST/hash/documents/FR_Notice_Nov07.pdf (October 17, 2008)
17. Peyrin, T.: Cryptanalysis of GRINDAHL. In: Kurosawa, K. (ed.) ASIACRYPT 2007. LNCS, vol. 4833, pp. 551–567. Springer, Heidelberg (2007)
18. Peyrin, T.: Chosen-salt, chosen-counter, pseudo-collision on SHAvite-3 compression function (2009), http://ehash.iaik.tugraz.at/uploads/e/ea/Peyrin-SHAvite-3.txt

19. Peyrin, T.: Improved Differential Attacks for ECHO and Grøstl. In: Rabin, T. (ed.) CRYPTO 2010. LNCS, vol. 6223, pp. 370–392. Springer, Heidelberg (2010), http://eprint.iacr.org/2010/223.pdf
20. Rivest, R.L.: RFC 1321: The MD5 Message-Digest Algorithm (April 1992), http://www.ietf.org/rfc/rfc1321.txt
21. Wagner, D.: A Generalized Birthday Problem. In: Yung, M. (ed.) CRYPTO 2002. LNCS, vol. 2442, pp. 288–303. Springer, Heidelberg (2002)
22. Wang, X., Yin, Y.L., Yu, H.: Finding Collisions in the Full SHA-1. In: Shoup, V. (ed.) CRYPTO 2005. LNCS, vol. 3621, pp. 17–36. Springer, Heidelberg (2005)
23. Wang, X., Yu, H.: How to break md5 and other hash functions. In: Cramer, R. (ed.) EUROCRYPT 2005. LNCS, vol. 3494, pp. 19–35. Springer, Heidelberg (2005)

An Improved Algebraic Attack on Hamsi-256

Itai Dinur and Adi Shamir

Computer Science department
The Weizmann Institute
Rehovot 76100, Israel

Abstract. Hamsi is one of the 14 second-stage candidates in NIST's SHA-3 competition. The only previous attack on this hash function was a very marginal attack on its 256-bit version published by Thomas Fuhr at Asiacrypt 2010, which is better than generic attacks only for very short messages of fewer than 100 32-bit blocks, and is only 26 times faster than a straightforward exhaustive search attack. In this paper we describe a different algebraic attack which is less marginal: It is better than the best known generic attack for all practical message sizes (up to 4 gigabytes), and it outperforms exhaustive search by a factor of at least 512. The attack is based on the observation that in order to discard a possible second preimage, it suffices to show that one of its hashed output bits is wrong. Since the output bits of the compression function of Hamsi-256 can be described by low degree polynomials, it is actually faster to compute a small number of output bits by a fast polynomial evaluation technique rather than via the official algorithm.

Keywords: Algebraic attacks, second preimages, hash functions, Hamsi.

1 Introduction

The Hamsi family of hash functions [1] was designed by Özgül Küçük and submitted to the SHA-3 competition in 2008. In 2009 it was selected as one of the 14 second round candidates of the competition. Hamsi has two instances, Hamsi-256 and Hamsi-512, that support four output sizes 224, 256, 384 and 512.

Previous results on Hamsi include distinguishers [4] and [5], pseudo-preimage attacks [6] and near collision attacks [7]. However, these results do not break the core security properties of a hash function. More recently, Thomas Fuhr introduced the first real attack on Hamsi-256 [2]. The attack exploits linear relations between some input bits and output bits of the compression function in order to find pseudo preimages for the compression function of Hamsi-256 (a pseudo preimage of an arbitrary chaining value h_i^* under the compression function $\mathcal{F}$ is a message block $\bar{M}_i$ and a chaining value $\bar{h}_{i-1}$ such that $\mathcal{F}(\bar{M}_i, \bar{h}_{i-1}) = h_i^*$). The pseudo preimages can then be used in order to find a second preimage for a given message with complexity $2^{251.3}$, which is better than exhaustive search by a marginal factor of $2^{4.7} \approx 26$ (whose existence and exact size depends on how we measure the complexity of various operations). In addition, Fuhr's attack is better than a generic long message attack only for very short messages

A. Joux (Ed.): FSE 2011, LNCS 6733, pp. 88–106, 2011.

with up to 96 32-bit blocks[1](i.e. 384 bytes). Nevertheless, it is the first attack on Hamsi-256 that violates its core security claims.

In this paper, we develop new second preimage attacks on Hamsi-256 which are slightly less marginal. Our best attack on short messages of Hamsi-256 runs in time which is faster than exhaustive search by a factor of 512, which is about 20 times faster than Fuhr's attack. For longer messages, we develop another attack which is faster than the Kelsey and Schneier attack by a factor which is between 6 and 4 for all messages of practical size (i.e., up to 4 gigabytes). Our short message attack exploits some of the observations made in [2] regarding the Hamsi Sbox, but uses them in a completely different way to obtain better results: While Fuhr solved linear equations in order to speed up the search for pseudo preimages, our attacks use fast polynomial enumeration algorithms to quickly discard compression function inputs which cannot possibly yield the desired output.

Since the straightforward evaluation of the compression function of Hamsi-256, Fuhr's attack, and our attacks use different bitwise operations, comparing these attacks on Hamsi-256 cannot be done simply by counting the number of compression function evaluations. Instead, we compare the complexity of straightline implementations of the algorithms, counting the number of bit operations (such as AND, OR, XOR) on pairs of bits and ignoring bookkeeping operations such as moving a bit from one position to another (which only requires renaming of variables in straightline programs). In this model of computation, the best available implementation of one compression function evaluation of Hamsi-256 (given as part of the submission package in [1] and used as the reference complexity in this paper), requires about $10,500$ bit operations. Our best attack is about 512 times faster, and is thus equivalent to an algorithm than performs only 20 bit operations per message block.

Polynomial enumeration algorithms evaluate a polynomial function over all its possible inputs. Clearly, the complexity of such enumeration algorithms must be at least 2^n for n-bit functions and thus they may seem to provide little advantage over trivial exhaustive search. However, cryptographic primitives are usually heavy algorithms that require substantial computational effort per execution. Consequently, the complexity of exhaustive search (measured by the number of bit operations) can be much higher than 2^n. However, for low degree polynomials, the complexity of enumeration algorithms is higher than 2^n only by a small multiplicative factor. In order to attack Hamsi-256, we search for polynomials of low degree that relate some of the bits computed by the compression function: The variables of each polynomial are chosen from the inputs to Hamsi-256, and the output of each polynomial is either an output bit of Hamsi-256, a linear combination of output bits of Hamsi-256, or an intermediate state bit of Hamsi-256 from which an output bit (or output bits) can be easily computed.

---

[1] Since Hamsi-256 is built using the Merkle-Damgård construction and it has a 256-bit intermediate state, the best known generic second preimage attack on Hamsi-256 with long messages is the Kelsey and Schneier attack [3] that runs in time $k \cdot 2^{128} + 2^{256-k}$ for messages of length 2^k.

Our attack on short messages of Hamsi-256 is divided into two stages: In the first stage we find multiple pseudo preimages of a single target chaining value obtained by one of the invocations of the compression function during the computation of the hash of the given message. In the second stage we obtain a second preimage for the message by searching for a second preimage for one of the target pseudo preimages that are found in the first stage (this is done by traversing a tree-like structure of chaining values, as shown in figure 1). In both stages, we first efficiently enumerate a set of low degree polynomials for all the possible values of a carefully chosen set of variables which are input to Hamsi-256. We then run the compression function only for the inputs for which the polynomial evaluations match the values of the target (or targets). Since the compression function of Hamsi-256 mixes the chaining value less extensively than the message, in the first stage we find only pseudo preimages by selecting our set of input variables of the enumerated polynomials among the bits of the chaining value. In the second stage, we have to find second preimages and thus we have to select our set of input variables of the enumerated polynomials among the message bits. Therefore, the polynomials enumerated in the first stage have a lower degree than those enumerated in the second stage, implying that the first stage gives a better improvement factor than the second stage (compared to exhaustive search). We note that our two-stage process of finding a second preimage using an efficient pseudo preimage search algorithm is a variant of the well-known meet-in-the-middle algorithm, described in the appendix of the extended version of this paper [12]. The difference is that the second stage of meet-in-the-middle is performed using exhaustive search, whereas the second stage of our algorithm is optimized using efficient polynomial enumeration algorithms.

For longer messages, the generic attack of Kelsey and Schneier becomes increasingly better with the length, and quickly overperforms both Fuhr's attack and our enumeration-based attack. In this case, we develop another attack that directly plugs into and speeds up the algorithm of Kelsey and Schneier. The attack is based on the second stage of our short message attack, but uses different parameters since in this case we try to find a second preimage for a potentially huge number of targets.

The fact that our short message attack is faster than Fuhr's attack may seem surprising, as Fuhr's attack is based on very simple and efficient algorithms for solving linear equations, whereas our attack is based on exponential-time polynomial enumeration algorithms. However, linear equations are much more difficult to obtain than non-linear equations of relatively low degree. In particular, Fuhr can obtain useful linear equations in only 7 or 8 variables in the first stage. The complexity of interpolating and solving such a system is faster than exhaustive search (which requires 2^7 or 2^8 function evaluations) only by a small factor. In the second stage, [2] can not obtain any linear equations and proceeds by performing an exhaustive search, which makes the attack faster than Kelsey and Schneier's attack only for very short messages. Another reason why our attack is faster is that in the first stage we also exploit the weak diffusion of the input variables into some of the output bits. This allows our enumeration algorithms to evaluate some polynomials only over the possible values of small subsets of variables in order to obtain the values for the entire variable set.

2   Description of Hamsi-256

In this section we provide a brief description of the compression function of Hamsi-256. For more details, please refer to its specification [1].

The compression function of Hamsi-256 takes as an input a 32-bit message block M_i and a 256-bit chaining value h_i and outputs a new 256-bit chaining value h_{i+1}. The compression function first expands the 32-bit message to 8 blocks of 32 bits using a linear code over $GF(4)$: $E(M_i) = (m_0, m_1, ..., m_7)$. The expanded message is then concatenated to the chaining value to form a 512-bit state treated as a 4×4 matrix of 32-bit blocks as follows:

$$(m_0, m_1, ..., m_7, c_0, c_1, ..., c_7) \longrightarrow \begin{array}{|c|c|c|c|} \hline s_0 & s_1 & s_2 & s_3 \\ \hline s_4 & s_5 & s_6 & s_7 \\ \hline s_8 & s_9 & s_{10} & s_{11} \\ \hline s_{12} & s_{13} & s_{14} & s_{15} \\ \hline \end{array} = \begin{array}{|c|c|c|c|} \hline m_0 & m_1 & c_0 & c_1 \\ \hline c_2 & c_3 & m_2 & m_3 \\ \hline m_4 & m_5 & c_4 & c_5 \\ \hline c_6 & c_7 & m_6 & m_7 \\ \hline \end{array}$$

The concatenation is followed by a permutation defined by three rounds, where each round consists of three layers: In the first layer, the state bits are XORed with some constants. In the second layer, the 128 4-bit columns of the state undergo simultaneous applications of a 4×4 Sbox described in table 1.

Table 1. The Hamsi Sbox

x	0	1	2	3	4	5	6	7	8	9	A	B	C	D	E	F
$S[x]$	8	6	7	9	3	C	A	F	D	1	E	4	0	B	5	2

The third layer consists of several parallel applications of a linear transformation L on the state.

$$(s_0, s_5, s_{10}, s_{15}) := L(s_0, s_5, s_{10}, s_{15})$$
$$(s_1, s_6, s_{11}, s_{12}) := L(s_1, s_6, s_{11}, s_{12})$$
$$(s_2, s_7, s_8, s_{13}) := L(s_2, s_7, s_8, s_{13})$$
$$(s_3, s_4, s_9, s_{14}) := L(s_3, s_4, s_9, s_{14})$$

Finally, the second and fourth rows of the state are discarded and the initial chaining value h_i is XORed with the remaining state to form h_{i+1}. The last message block is processed differently, by applying 8 rounds of this permutation 8 (instead of the standard 3).

3 A Direct Attack on Hamsi-256

3.1 The Properties and Weaknesses of Hamsi-256 Which Are Exploited by the Attack

In the direct attack our goal is to consider the 32 message bits as variables and the 256-bit chaining value as a fixed input and analyze the degree of the state bits after each one of the three rounds of the Hamsi-256 compression function as polynomials in the message bits. Every Hamsi Sbox can be described as a polynomial of degree 3 in its 4 input variables. However, due to the way the expanded message bits and the chaining value bits are interleaved in Hamsi-256, after 1 round of the compression function, each state bit is a polynomial of reduced degree 2 in the message bits. This may seem insignificant, but after 2 rounds, the degree of each state bit as a polynomial in the message is at most 6 instead of the expected value of $3^2 = 9$. After the final compression function round, the degree is 18 instead of the expected 27. This low algebraic degree can be used to obtain distinguishers on Hamsi-256 (as already noticed in [4] and [5]), but even this reduced degree is too high for our algebraic attack. Instead, we exploit the low diffusion property of one round of Hamsi-256, namely, that several output bits of the compression function depend only on a small number of inputs from the second round.

3.2 Analysis of Polynomials of Degree 6 in 32 Variables

Since the attack on Hamsi-256 relies on a slightly improved version of exhaustive search, we have to use every possible saving and shortcut in the implementation of our algorithms, and can not ignore constants or low-order terms. In particular, we show how to efficiently interpolate and evaluate any polynomial of degree 6 in 32 variables for all the 2^{32} possible values of its inputs using fewer than $7 \cdot 2^{32}$ bit operations (instead of the 2^{64} complexity of the naive evaluation of the 2^{32} possible terms for each one of the 2^{32} possible inputs):

1. Given any black box implementation of the polynomial (e.g. in the form of the Hamsi-256 program), evaluate the output of the polynomial (which is a single bit value) only for input vectors of hamming weight ≤ 6 and store all the results.
2. Compute the coefficient of each term t_I of degree at most 6, where $I \subset \{0, 1, ...31\}, |I| \leq 6$ represents some subset of variables multiplied together. The coefficient of t_I is computed by summing all the outputs of the polynomial obtained from all inputs which assign 0 values to all variables that are not contained in I (where the variables that are contained in I are assigned all possible values).
3. Allocate an array of 2^{32} bits and copy all the coefficients of the polynomial into the array. The coefficient t_I is copied into the entry whose binary index is encoded by $b_0, b_1, ..., b_{31}$ where $b_i = 1$ if and only if $i \in I$. All the other entries of the array are set to 0.

4. Apply the Moebius transform [8] on the array and obtain an array which contains the evaluations of the polynomial for all 2^{32} possible input values.

Step 1 requires $\sum_{i=0}^{6} \binom{32}{i} \approx 2^{20}$ compression function evaluations. Step 2 requires $\sum_{i=0}^{6} 2^i \binom{32}{i} < 2^{26}$ bit operations. Step 3 can be combined with step 2 by writing the coefficients directly into the array and does not require additional work. A naive implementation of step 4 requires $32 \cdot 2^{31}$ bit operations, since the generic Moebius transform consists of 32 iterations, where in each iteration we add half of the array entries to the other half. However, in our case, the initial array can contain only about 2^{20} non zero values, whose locations are known and represent all the vectors with hamming weight of at most 6. In the first iteration of the Moebius transform, we split the array into 2 parts according to one variable and add only the entries whose index has a hamming weight of at most 6 in the remaining 31 variables (the others entries are left unchanged). The total complexity of the first iteration is thus $\sum_{i=0}^{6} \binom{31}{i}$ bit operations. In the second iteration, each half of the array is split into 2 parts according to another variable. Similarly, we add only the entries whose index has a hamming weight of at most 6 in the remaining 30 variables. The total complexity of the second iteration is thus $2 \sum_{i=0}^{6} \binom{30}{i}$. Generally, the complexity of the $j'th$ iteration where $0 \leq j \leq 25$ is $min(2^{31}, 2^j \sum_{i=0}^{6} \binom{31-j}{i})$. For $26 \leq j \leq 31$, the complexity is 2^{31}. Summing over all iterations, we get a total complexity of less than $7 \cdot 2^{32}$. We note that it is also possible to use the Gray-code based polynomial enumeration algorithm recently presented in [9] which is a bit more complicated than our Moebius-based transform, and has a similar time complexity.

Assuming that the straightforward evaluation of one Hamsi-256 compression function requires about $10,500$ bit operations, step 4 is the heaviest step and dominates the complexity of the algorithm. Note that when analyzing several polynomials which correspond to different output bits, step 1 needs to be performed only once since every compression function evaluation gives us the values of all the required polynomials. In addition, the bit locations XORed together during the Moebius transformation do not depend on the evaluated polynomial, and thus the evaluation of k unrelated polynomials can be achieved by XORing k bit words instead of single bits. This is particularly convenient when $k = 32$ or $k = 64$, which are standard word sizes in modern microprocessors.

3.3 Efficiently Eliminating Wrong Messages

Assume that we are given a target chaining value h_i^*, and a fixed chaining value h_{i-1}. We would like to efficiently find a single message block M_i such that

$\mathcal{F}(M_i, h_{i-1}) = h_i^*$, or decide that such a message does not exist for the given h_{i-1}, h_i^*. Due to the short message blocks of Hamsi-256, the probability to find a desired message for random chaining values h_{i-1} and h_i^* is about $2^{32-256} = 2^{-224}$. Hence, to succeed with high probability for a given h_i^*, we have to generate about 2^{224} random values for h_{i-1}. In order to obtain this number of chaining values, the target h_i^* must be located at least in block number 8 of the message (i.e. $i \geq 8$). This implies that we can apply our attack only when the given message contains at least 8 blocks.

The idea is to algebraically compute only a small set of output bits (indexed by N) for all the 2^{32} messages and compare their values to the values of those bits of the target chaining value. If the bits match, we run the compression function for the message to compute the whole 256-bit output, and compare it to the target h_i^*. Otherwise, we discard the message. Using the algorithm of section 3.2, we efficiently evaluate only the bits produced after 2 rounds of the compression function, which are required in order to determine the output bits specified by N. We then combine these values as specified by the last round, and obtain the values of the output bits indexed by N for all the possible 2^{32} messages.

In order to minimize the complexity of the attack, we need to select a set N that is big enough to eliminate a large number of messages with high probability. On the other hand, choosing N too big, will force us to evaluate many bits after two rounds, increasing the complexity of the attack. In fact we don't need to analyze all the second round bits that are required to compute the output bits N: We write the ANF form of the output bits N as a function of the second round bits and note that the sum of the second round variables which are not multiplied together (which we call the simple sum) is itself a polynomial of degree 6 in the message. Thus, the simple sum of each such output bit can be analyzed in the same way as the second round bits without computing separately each one of the summed bits. Note that in Step 1 of the analysis (interpolation), evaluating the polynomial means computing the sum of variables numerically from the output. The complexity of this computation is negligible compared to a compression function evaluation of Hamsi-256.

After choosing the set N of output bits and the set of second round bits of degree 6 $S(N)$ we have to evaluate, we do the following:

1. Given that $i \geq 8$: Choose an arbitrary message prefix of $i - 8$ blocks $M_1, M_2, ..., M_{i-8}$ and compute $h_{i-8} = \mathcal{F}(M_1, M_2, ..., M_{i-8}, IV)$ (which is fixed throughout the attack). Use DFS to traverse the tree of chaining values rooted at h_{i-8} by assigning values to the blocks $M_{i-7}, M_{i-6}, ..., M_{i-1}$ and computing the next chaining value h_{i-1} (as shown in figure 1). For each generated h_{i-1}:

2. Evaluate all the bits of $S(N)$ for all possible 32-bit message values using the algorithm of Section 3.2 and list them as $|S(N)|$ bit arrays of size 2^{32}.

3. Using the ANF form of the outputs bits of N as a function of the second round bits, calculate $|N|$ bit arrays of size 2^{32} representing the values of the $|N|$ output bits for all the messages.

4. Traverse the $|N|$ bit arrays and check whether the values of the output bits in N match the values of those bits in h_i^*, for each message M_i. Store all the messages for which there is a match.
5. For each message M_i stored in the previous step, evaluate the full Hamsi-256 compression function output by using its standard implementation and check whether $\mathcal{F}(M_i, h_{i-1}) = h_i^*$. If equality holds, output $M = M_1, M_2, ..., M_i$. Otherwise go to step 1.

The memory requirements of the algorithm can be reduced by calculating the bits of N iteratively and eliminating wrong messages according to the current calculated bit. We can then reuse some memory which was used for the calculation of the previous bits of $S(N)$ and which is not required anymore.

Given that $|N| = n_1$, $|S(N)| = n_2$, and the number of bit operations per message that is required to compute N from $S(N)$ is n_3 (calculated using the ANF form of the outputs bits), the complexity of the attack is about $2^{224}(n_3 2^{32} + 7n_2 2^{32} + 10500 \cdot 2^{32-n_1}) = 2^{256}(n_3 + 7n_2 + 10500 \cdot 2^{-n_1})$ bit operations. Compared to exhaustive search which requires about $10500 \cdot 2^{256}$ bit operations, this gives an improvement factor of about $(\frac{n_3 + 7n_2}{10500} + 2^{-n_1})^{-1}$.

We searched for sets of output bits N that optimize the complexity of the attack. The best set that we found is $N = \{5, 156, 184, 214, 221, 249\}$ whose 6 output bits depend just on 56 second round bits. We also have to add 6 bits for the simple sums of the second round variables, and the full list of $56 + 6 = 62$ bits is described in appendix A. For this parameter set we get that $n_2 = 62$, $n_1 = 6$ and the computation of all the evaluated output bits requires about $n_3 = 150$ bit operations per message. We calculated this number after a few optimizations which are based on finding some common parts in the ANF representation of the output bit polynomials (which can be calculated only once). The improvement factor (compared to exhaustive search) of this direct attack is therefore a very modest $(\frac{150 + 7 \cdot 62}{10500} + 2^{-6})^{-1} \approx 14$. The memory complexity is about $64 \cdot 2^{32} = 2^{38}$ bits and can be further reduced by iterative calculation of the output bits. We also found another interesting set of parameters: $N = \{5, 156, 214, 221, 249\}$ gives a slightly worse improvement factor of 13.

4   Improving the Direct Attack by Using Pseudo Preimages

While the direct attack seems to be worse than Fuhr's attack [2] , it can be the basis for substantial improvements. In this section we consider the generalized problem of finding a pseudo preimage, defined as a pair of a message block and chaining value $\bar{M}_i, \bar{h}_{i-1}$ such that $\mathcal{F}(\bar{M}_i, \bar{h}_{i-1}) = h_i^*$ for a given value h_i^*. Whereas in the direct attack described in the previous section we could only select our variables from the message, here we have the extra power of choosing our variables also from the chaining value bits which are mixed less extensively than

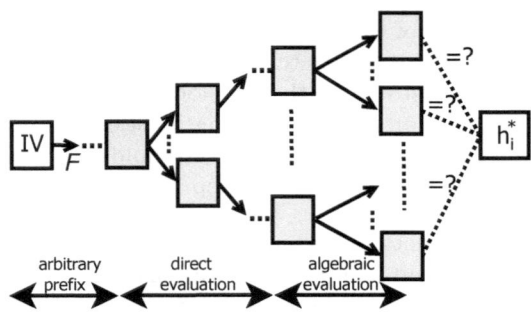

Fig. 1. A sketch of the second stage of the attack. After generating a prefix of chaining values using arbitrary message blocks, we start to traverse a tree-like structure of chaining values (shown as lightly filled boxes) using DFS: Each node is expanded into 2^{32} successor nodes by selecting the next value for the 32-bit message block in all possible ways .The tree has 8 levels so that the final level contains 2^{256} chaining values (which is roughly the number of chaining values that we need to generate in order to match the 256-bit target with high probability). The $7-th$ level nodes are not expanded by applying the Hamsi-256 compression function. Instead, we first efficiently evaluate only a small set of bits for all the 2^{32} possible message blocks. We then execute the compression function only for the messages for which the evaluation of those bits match the corresponding values of the target h_i^*. The attack succeeds once we find an $8-th$ level node that matches the target.

the message bits by the Hamsi-256 compression function. By carefully choosing these variables, we can lower the degree of the polynomials, allowing us to compute these outputs more efficiently compared to the direct attack.

Our improved attack exploits the very interesting observations made by Thomas Fuhr in section 3 of [2]. For a given message block M, we select our variables from the state that precedes the first Sbox layer as follows: Let $x^{(j)}$ denote the $j'th$ bit of the 32-bit word x. We define one variable bit $x^{(j)} \in \{0, 1\}$ for each j such that $s_{14}^{(j)} = 1$ and set $s_2^{(j)} = x^{(j)}$, $s_{10}^{(j)} = \overline{x^{(j)}}$. In addition, we define one variable bit $y^{(j)} \in \{0, 1\}$ for each j such that $s_1^{(j)} = 1$, $s_9^{(j)} = 0$ and set $s_5^{(j)} = y^{(j)}$, $s_{13}^{(j)} = \overline{y^{(j)}}$. According to [2], after 2 rounds of the compression function of Hamsi-256, the state bits depend linearly on our variable set: We chose our variable set such that after the first Sbox layer, only s_2, s_{13} depend linearly on our variable set. After the first round, only s_2, s_7, s_8, s_{13} depend linearly on our variable set. After the second Sbox layer, the dependency of the state on the variables remains linear and the second diffusion layer does not change this property. Note that for a random message, we expect $|V_1| = |\{x^{(j)}\}_{j \in j_x}| = 16$, $|V_2| = |\{y^{(j)}\}_{j \in j_y}| = 8$. We define $V = V_1 \cup V_2$ and for a random message we expect $|V| = 24$.

The observations of [2] allow us to select a relatively large set of variables such that the degree of all the output bits in those variables is 3. In addition, there are 28 specific output bits that depend only on 16 state bits after 2 rounds of the compression function. The indexes of these bits are $150 - 156, 182 - 188, 214 - 220, 246 - 252$. Moreover, each one of these output bits usually does not depend on all of our input variables. We calculate for each of these output bits the variables on which it actually depends, and efficiently enumerate (over all their possible values) only a certain subset of output polynomials (whose size we denote by α), called the *analyzed polynomials* or *analyzed bits*. We note that the dependencies of the 28 output bits on our variable set are influenced by the values of the message and the chaining value, but there are certain patterns that are common to most messages and chaining value pairs. For example, if our variable set contains 21 variables, there is usually an output bit which depends on only 12 of our variables, another 1 or 2 output bits that depend on 13 of our variables, another 2 or 3 output bits that depend on 14 of our variables and so forth.

4.1 The Polynomial Analysis Algorithm

All the 28 polynomials defined above have degree of at most 3 in the input variables. Given a message, a corresponding set of variables and a chaining value, we first interpolate the linear state bit polynomials of Hamsi-256 after 2 rounds. We can then use the optimized Moebius transform of section 3.2 (adapted to cubic polynomials) to efficiently evaluate any cubic polynomial, which corresponds to output bits $150 - 156, 182 - 188, 214 - 220, 246 - 252$ over all its inputs. These values are written into an array of size $2^{|S|}$, where $S \subseteq V$ is the set of input variables on which this polynomial depends. The enumeration algorithm starts with an initialization phase that interpolates the coefficients of the cubic polynomial by evaluating the function $\sum_{i=0}^{3} \binom{|S|}{i}$ times and performing $\sum_{i=0}^{3} 2^i \binom{|S|}{i}$ bit operations. The evaluation can be done by running the compression function, but we can use the second round polynomials in order to speed up this process: Given that we know the values of the 16 polynomials that are input to the third round, we can calculate the value of the output bit by evaluating 4 Sboxes and summing 4 of their outputs. An evaluation of one Hamsi Sbox output bit requires 8 bit operations (computing the 4 Sbox outputs requires 14 bit operations, but this number is reduced for individual bits), and the sum requires 3 more bit operations giving a total of 35 bit operations per evaluation. The 16 linear polynomials can be efficiently evaluated using a simple differential method which requires an average of 16 bit operations per evaluation. In total, one evaluation requires $16 + 35 = 51$ bit operations and the initialization step of the enumeration algorithm requires $\sum_{i=0}^{3} (51 + 2^i) \binom{|S|}{i}$ bit operations. After optimizations similar to the ones performed in section 3.2 (which exploit the sparseness of the coefficients in the array in most iterations of the algorithm), we get that the algorithm itself requires an additional number of $4 \cdot 2^{|S|}$ bit operations.

4.2   The Query Algorithm

Assume that we have already analyzed polynomials p_i for $1 \leq i \leq \alpha$ where p_i depends on a subset S_i of the variables. The output of the enumeration of each p_i is a table of size $2^{|S_i|}$. These α tables define a set of about $2^{|V|-\alpha}$ possible values for the variables such that when they are used as chaining value bits which are plugged into the compression function, the values of the α analyzed output bits match those of the target. Clearly, the remaining values of the variables that do not match the target can be safely discarded. However, these $2^{|V|-\alpha}$ solutions are only implicitly given by the tables and we have to efficiently obtain their explicit representation.

For example, assume that our set of variables is $V = \{v_1, v_2, v_3, v_4\}$, and we have analyzed polynomials p_1 that depends on $S_1 = \{v_1, v_2, v_4\}$, and p_2 that depends on $S_2 = \{v_2, v_3, v_4\}$. The table obtained after analyzing p_1 contains $2^3 = 8$ entries (an entry for each possible value of the variables of S_1). Out of these 8 entries, only entries $000, 010, 110, 111$ have a value that matches the value of the corresponding bit of the target. The other 4 entries have the complementary value (which does not match the value of the corresponding bit of the target). Note that each entry actually corresponds to two assignments of the 4 variables (For example, the point 001 corresponds to the assignments 0001 and 0011). Out of the 8 entries of the table obtained after analyzing p_2, the entries $000, 011, 110$ have a value that matches the value of the corresponding bit of the target. Our goal is to find the assignments to the 4 variables whose corresponding entries in both tables match the bits of the target. The explicit set of solutions in our example contains the points $0000, 0110, 1110$.

A naive approach in order to obtain an explicit representation of the solutions is to iterate the possible $2^{|V|}$ values for the variables, and check whether the value of the entry that corresponds to the value of the variables in each of one of the tables matches the value of the corresponding analyzed bit (note that we can discard a potential solution once the entry value in one of the tables does not match the value of the corresponding analyzed bit). This algorithm requires at least $2^{|V|}$ bit operations. We can easily save a factor of 2 by iterating only the values that match the target in one of the tables. However, we can do even better by considering the actual variable sets on which each analyzed output bit depends. The details and analysis of the improved query algorithm are specified in the appendix of the extended version of this paper [12]. Its expected complexity is $|S_1| + 2^{|S_1 \cup S_2|-1} + ... + 2^{|\bigcup_{i=1}^{\alpha} S_i|-\alpha+1}$ bit operations. Note that this complexity estimate is not symmetric with respect to the various S_i's, and thus different orders of analyzing the various tables will yield different complexities.

4.3   Post Filtering the Solutions

After the query algorithm, we are left with $2^{|V|-\alpha}$ solutions and we have to determine whether they indeed give a preimage which matches all the 256 bits of the given chaining value h_i^*. One option is to simply run the compression function and check whether the solutions match the target. However, it is more efficient to apply the following post filtering algorithm first.

- For each solution, evaluate the remaining $28 - \alpha$ output bits (that were not analyzed) one by one, and compare the output to the corresponding value of h_i^*. If the value of an output bit does not match the value of the corresponding target bit, drop the solution.

For each solution, we expect to evaluate 2 additional bits (we always evaluate one additional bit, a second bit is evaluated with probability 0.5, a third with probability 0.25, and so forth). Evaluating a bit requires evaluation of the 16 input linear polynomials up to round 2 plus 35 additional bit operations for the Sbox and XOR evaluations. A random linear polynomial in the $|V|$ input bits has about $\frac{|V|}{2}$ non zero coefficients, but this is not the case here. Our special choice of variables makes them diffuse slowly into the state of Hamsi-256, and as a result, our linear polynomials are very sparse and require about 3 bit operations per evaluation. The 2 evaluations thus require $2(35 + 3 \cdot 16) = 166$ bit operations. The post filtering requires in total about $166 \cdot 2^{|V|-\alpha}$ bit operations. The number of solutions that remain after the post filtering is about $2^{|V|-28}$ (i.e. we expect to have less than one solution per system on average if $|V| < 28$), and running the compression function after the post filtering requires negligible time.

4.4 Finding a Good Sequence of Analyzed Bits

In the previous sections we designed and calculated the complexities of the polynomial analysis algorithm, the query algorithm, and the post filtering algorithm. Given the sets $S_1, ..., S_{28}$ that correspond to the potential analyzed bits, we would like to find a good sequence of analyzed bits (of size α) which minimizes the complexity of the attack. Since there are many possible sequences, exhaustive search for the optimal sequence of analyzed bits is too expensive and thus we used a heuristic algorithm for this problem. A naive greedy algorithm which iteratively builds the sequence by selecting the next analyzed bit i that minimizes the added complexity $51|S|^3 + 3 \cdot 2^{|S|} + 2^{|\bigcup_{j=1}^{i} S_j|-i+1}$ seems to give reasonable results, but we got even better results by combining the greedy algorithm with exhaustive search over short sequences, as described next.

1. Given the dependencies of the 28 potential analyzed bits, exhaustively search for the optimal sequence of 3 analyzed bits that minimizes the sum of complexities of the query algorithm and their analysis.
2. Fill in the remaining $28 - 3 = 25$ bits of the sequence by iteratively searching for the next analyzed bit i that minimizes the added complexity $51|S|^3 + 3 \cdot 2^{|S|} + 2^{|\bigcup_{j=1}^{i} S_j|-i+1}$ (the post-filtering complexity is the same given the value of i).
3. Determine the length of the sequence α by calculating the total complexity of the attack for each possible value of $1 \leq \alpha \leq 28$ and truncate the sequence of 28 bits to size α.

The first step involves exhaustive search over $\frac{28!}{25!} < 2^{14.5}$ sequences, each requires a union of sets of at most $|V|$ variables represented as bit arrays, and an addition operation. The union requires $|V|$ bit operations and the addition a few more bit operations since the terms $51|S|^3 + 3 \cdot 2^{|S|}$ are computed only once and can be rounded in order to nullify the least significant bits. Assuming that $|V| < 2^5$, the complexity of the first step is about $2^{19.5}$ bit operations, which can be easily reduced to about $2^{18.5}$ by considering the sequences in a more clever way. The second and third steps take negligible time compared to the first step. Note that this algorithm is performed before analyzing the polynomials, although it is specified last.

4.5 Details of the Pseudo Preimage Attack on Hamsi-256

The details and analysis of the pseudo preimage attack on Hamsi-256 are specified below. The input of the algorithm is a chaining value h_i^*, and its output is a message block $\bar{M}_i$ and a chaining value $\bar{h}_{i-1}$ such that $\mathcal{F}(\bar{M}_i, \bar{h}_{i-1}) = h_i^*$.

1. Generate the next message block $\bar{M}_i$ (starting from the zero block, and incrementing its value each time this step is performed).
2. Compute the set of variables $V' = V_1 \cup V_2$ according to $\bar{M}_i$. If $|V'| < 21$ then discard the message and go to step 1. Otherwise, obtain the final set of 21 variables V for the current message block by dropping $|V'| - 21$ variables from V'. The variables that are dropped are arbitrarily chosen from the set V_1 (the variables are dropped from V_1 since the variables of V_2 tend to diffuse more slowly into the state of Hamsi-256, as noted in section 4 of [2]).
3. Generate the next partial chaining value $\bar{h}_{i-1}$, which does not assign values to the variables (starting from the zero partial chaining value each time step 1 is performed, and incrementing its value each time this step is performed). If no more partial chaining values exist, go to 1.
4. Given $\bar{M}_i$, V and $\bar{h}_{i-1}$, interpolate the linear state bit polynomials of Hamsi-256 after 2 rounds.
5. For each of the 28 output bits ($150 - 156, 182 - 188, 214 - 220, 246 - 252$), determine the variable subset on which it depends. This is done by retrieving the 16 linear second round state bits on which the output bit depends, and then performing a union over the variable subsets on which the 16 state bits depend.
6. Determine the heuristically best sequence of analyzed bits according to the algorithm in section 4.4.
7. Analyze the selected polynomials according to section 4.1.
8. Use the query algorithm of section 4.2 to determine the set of solutions.
9. Post filter the solutions according to section 4.3. If no solutions remain, go to step 3.
10. For each remaining solution, compute the compression function after assigning the value of the solution to the unspecified part of the partial chaining value, and check whether the output is equal to the target. If there is a solution for which equality holds, return the message and full chaining value. Otherwise, go to step 3.

In order to find at least one pseudo preimage with high probability, we must verify that we do not use too many degrees of freedom after throwing away messages and allocating the variables. We start with 32 degrees of freedom since the input to the Hamsi-256 compression function contains $256 + 32 = 288$ bits, (32 message bits and 256 chaining value bits) and the output of the compression function contains only 256 bits. We lose less than 0.5 degrees of freedom by throwing away messages for which the number of variables is too small. In addition, every variable sets one constraint on the input of the compression function and reduces the number of possible inputs to the compression function by a factor of 2. Thus, we lose a degree of freedom per allocated variable and less than 21.5 degrees of freedom overall. In total, we remain with a bit more than $32 - 21.5 = 10.5$ degrees of freedom which are expected to result in more than 2^{10} pseudo preimages for a random target.

We now estimate the complexity of the pseudo preimage attack: The complexity of some steps can be easily computed: For a given set of variables, step 4 of the algorithms requires 22 compression function evaluations of Hamsi-256 and $21 \cdot 512 < 2^{14}$ bit operations. Step 5 takes negligible time. Step 6 requires about $2^{18.5}$ bit operations. Step 10 requires $2^{|V|-28}$ compression function evaluations, which takes negligible time compared to the other steps of the attack. However, the complexity of the main steps of the attack $7 - 9$ cannot be easily computed since it depends on the message and the value of the chaining value used. Thus, we can only estimate the complexity of the attack by running simulations for randomly chosen messages and chaining values. In each simulation, we estimate the complexity of the attack by summing the complexities of the steps above with the complexity of steps $7 - 9$, as calculated in section 4.4. After thousands of simulations we found that for about 95% of messages and chaining values the attack is faster than exhaustive search by a factor which is at least 2^{13}. The average complexity of the attack is slightly better than $2^{256-13.5} = 2^{242.5}$ compression function evaluations.

Interestingly, the techniques of our pseudo preimage attack can also be used to speed up generic pseudo collision search algorithms on Hamsi-256 that are based on cycle detection algorithms (such as Floyd's algorithm [11]). The details of the pseudo collision attack are described in the appendix of the extended version of this paper [12].

5   Using Pseudo Preimages to Obtain Second Preimages for Hamsi-256

Given a message $M = M_1^*||M_2^*||...||M_\ell^*$ with $\ell \geq 9$, we can use the naive meet-in-the-middle algorithm (described in the appendix of the extended version of this paper [12]) in order to find an expected number of $2^{13.5/2} = 2^{6.75}$ pseudo preimages and use them as targets for the second preimage attack. This gives a total complexity of about $2^{256-5.75} = 2^{250.25}$ compression function evaluations. However, we can do better by using the result of section 3: Recall that our algorithm for finding pseudo preimages has more than 10 degrees of freedom left. We use 5 of the remaining degrees of freedom to set the input bits that correspond to the output bits of the set $N = \{5, 156, 214, 221, 249\}$ in all the pseudo preimages to some fixed value. As specified in section 3.3, the set N

represents the target bits for the direct second preimage attack on Hamsi-256 and this choice allows us to speed up the second phase by a factor of about $13 \approx 2^{3.7}$. In the first phase of the attack, the bits of N actually function as input bits to the pseudo preimage search algorithm. The details of the algorithm are specified below, where x is a numeric parameter:

1. Choose a target block with index of at least 9 (i.e. h_i^* with $i \geq 9$) and use the pseudo preimage search algorithm to find 2^x pseudo preimages in which the set of input bits $\{5, 156 + 128, 214 + 128, 221 + 128, 249 + 128\}$ is fixed to an arbitrary value. Note that the number 128 is added to some indexes of N due to the truncation of the output of the compression function.
2. Use the direct second preimage search algorithm to find a second preimage to one of the 2^x pseudo preimages found in the previous step.

We note that we still have $10 - 5 = 5$ degrees of freedom left, so we must choose $x \leq 5$ in the first step. The complexity of step 1 is about $2^{256-13.5+x} = 2^{242.5+x}$ compression function evaluations. The complexity of step 2 is about $2^{256-x-3.7}$ compression function evaluations. To optimize the attack, we choose $2^x = 30$, i.e $x \approx 4.9$ for which the total complexity of the attack is about $2^{248.4}$, which is about $2^{7.6} \approx 200$ times better than exhaustive search.

The algorithm presented above works for any message that contains at least 9 blocks. However, this restriction can be removed with little additional cost using an observation made by an anonymous referee of this paper: Since the 64-bit message length is encoded in the last two 32-bit blocks, we can find a pseudo preimage of the last intermediate chaining value with a non-zero message block. The 32 bits of the message block function as the most significant bits of the message length of our second preimage, which now contains enough blocks for the attack. The chaining value of the pseudo preimage gives us the target which we require for the algorithm above.

In addition, it is possible to improve the algorithm further by building a *layered hash tree*, similar to the one used in [10]. The optimized algorithm yields a less marginal improvement factor of $2^9 = 512$ over exhaustive search, which is about 20 times better than the attack published by Thomas Fuhr [2]. The details of this algorithm are specified in the appendix of the extended version of this paper [12].

6   Second Preimages for Longer Messages of Hamsi-256

The best known generic algorithm for finding second preimages for any Merkle-Damgård construction of hash functions is due to Kelsey and Schneier [3]. The algorithm needs to undergo a slight adaptation in order to be applied to the special structure of Hamsi-256 (see [2]). The complexity of the generic algorithm for Hamsi-256 is $k \cdot 2^{128} + 2^{256-k}$, where the message length satisfies $\ell \geq 4k + 2^k + 8$. Hence, the algorithm developed in the previous section is better than the generic algorithm only for $k \leq 9$, i.e. for messages that contain at most $4 \cdot 9 + 2^9 + 8 = 556$ blocks. For longer messages, we design a different algorithm that combines the techniques used in section 3 with a modified version of the Kelsey and Schneier algorithm. We elaborate only on the parts of the Kelsey and Schneier algorithm that are relevant to our modified attack.

Given an ℓ block message, in the the first phase of the Kelsey and Schneier algorithm, the attacker generates a (p, q) expandable message for $p = 4k$ and $q = 4k+2^k-1$ such that $q+8 \leq \ell-1$. This phase is left unchanged. We concentrate on the second phase of the Kelsey and Schneier algorithm, where we apply the compression function from a common chaining value and try to connect to one of the chaining values obtained by one of the invocations of the compression function during the computation of the hash of the given message. If the message is of size about 2^k, the complexity of this phase is 2^{256-k} compression function evaluations, which forms the bottleneck of the attack (assuming $k < 128$). Similarly to section 3, the idea is to speed up this phase simply by efficiently computing several bits of the output for all possible 2^{32} messages and filtering out messages which do not connect to any of the targets. Assuming that we efficiently compute the values of x output bits, then we still need to run the compression function a factor of 2^{-x+k} times for $x > k$ compared to the original algorithm.

Unlike section 3, a significant portion of the work here involves computing the output bits (almost) directly, and a smaller portion of the work involves analysis of the second round bits. The output bits are of degree 18 which is too high to be analyzed efficiently. However, we can exploit polynomials of a lower degree relatively easily. As in section 3.3, we use the ANF form of the output bits as a function of the second round bits. The symbolic representation is of degree 3 and we would like to get equations of degree 2. We remove all terms of degree lower than 3 in the ANF form. We then linearize the system of polynomials by assigning each distinct term of degree 3 a dedicated variable. We perform Gaussian Elimination on the linearized system and get a system in which about 120 rows contain only 1 variable and the rest of the rows contain 2 variables (each linearized variable represents 3 variables of round 2 multiplied together). This is of course not sufficient in order to reduce the degree. However, these linearized simple expressions (composed of 3 variables of degree 6 in the message bits) can be handled separately by the technique specified in section 3.2. We select a set of x linear combinations from the rows which contain only one linearized variables. The rest of the polynomial is of degree $2 \cdot 3 \cdot 2 = 12$ in the message bits, and is analyzed slightly differently. The analysis algorithm for such a linear combination is specified below. Its input is an arbitrary chaining value h and it outputs an array of size 2^{32} that contains the evaluations of the linear combination of the output bits for all possible 2^{32} message blocks.

1. Analyze the 3 second round variables that appear in the expression of the linearized variable of the linear combination, as specified in section 3.2.
2. Evaluate the remainder of the output bit combination on all input vectors of hamming weight ≤ 12 and store the results.
3. Interpolate the coefficients of the output bit combination: Place all its values in an array of size 2^{32}, where the values of entries of hamming weight ≥ 13

are set to zero. Then apply the Moebius transform [8] on the array and take only the coefficients of hamming weight ≤ 12 (the rest are known to be 0).

4. Apply the Moebius transform once more on the array and obtain the evaluations of the polynomial (not including the linearized variable) for all 2^{32} possible input values.

5. Add the values of the linearized variable to the array by computing it from the arrays produced in step 1.

Step 1 requires $3 \cdot 7 \cdot 2^{32} = 21 \cdot 2^{32}$ bit operations. Step 2 requires $\sum_{i=0}^{12} \binom{32}{i} \approx \frac{2^{32}}{10}$ compression function evaluations (which need to be performed once per chaining value). In addition, step 2 requires several bit operations to compute the value of the linear combination. Most of the linear combinations contain fewer than 40 additions and step 2 requires additional $40 \cdot 0.1 \cdot 2^{32} = 4 \cdot 2^{32}$ bit operations. Step 3 requires an application of the Moebius transform, which takes $16 \cdot 2^{32}$ bit operations. However, only about 0.1 of the entries of the array are relevant (the others are not accessed), hence the complexity is less than $2 \cdot 2^{32}$ bit operations. Step 4 requires the full $16 \cdot 2^{32}$ bit operations. Step 5 requires additional $3 \cdot 2^{32}$ bit operations. In total, the algorithm requires about $(21+4+2+16+3) \cdot 2^{32} = 46 \cdot 2^{32}$ bit operations in addition to the $\frac{2^{32}}{10}$ compression function evaluations that are performed globally.

The algorithm to find the second preimage is specified below, where x is a numeric parameter. It get as an input a message $M_1^* || M_2^* ||...|| M_\ell^*$ and outputs a message of the same length with the same Hamsi-256 hash value.

1. Generate a p, q expandable message for $p = 4k$ and $q = 4k + 2^k - 1$ such that $q + 8 \leq \ell - 1$.

2. Choose a set of x output bit combinations from the Gaussian elimination of the third round output bits in terms of the second round variables, such that each of these combinations contains a single expression of 3 second round bits multiplied together.

3. Compute and store all the values of the x output bit combinations of all the target chaining values h_i^* for $p + 8 \leq i \leq q + 8$.

4. Choose the common digest value of the expandable message h as a chaining variable and traverse the chaining value tree rooted at h using DFS by generating the next value for message blocks $M_1, M_2, ..., M_7$ (as shown in figure 1).

5. Compute the next chaining value $h_7 = \mathcal{F}(M_1, M_2, ..., M_7, h)$.

6. Analyze each one of the x output bit combinations as specified above for all possible 2^{32} values for the message block M_8, with the input chaining value h_7.

7. Traverse the x bit arrays and check whether the values of the output combinations match the values of the combinations of the target chaining values h_i^* for $p + 8 \leq i \leq q + 8$, for each possible value of the message block M_8. Store all the messages for which there is a match.

8. For each message block M_8 stored in the previous step, evaluate the full compression function and check whether $\mathcal{F}(M_8, h_7) = h_i^*$ for $p+8 \leq i \leq q+8$. If equality holds, output the message
$\mu_{i-8} || M_1 || M_2 || ... || M_8 || M_{i+1}^* || ... || M_\ell^*$, where μ_{i-8} is a message prefix of size $i - 8$ blocks (computed from the expandable message) such that $h = \mathcal{F}(\mu_{i-8}, IV)$. Otherwise, if there is no match, go to step 4.

We analyze the complexity of the algorithm per chaining value h_7 (i.e. steps $6 - 8$) in order to calculate the improvement factor of the attack over the generic algorithm. The Kelsey and Schneier algorithm requires 2^{32} compression function evaluations per chaining value, whereas we use only $\frac{2^{32}}{10}$ compression function evaluations. In addition, we require $46 \cdot x \cdot 2^{32}$ bit operations in step 6. However, we can optimize the complexity of this step for a group of combinations by taking combinations in which the linearized expressions share some common variables of the second round (which need to be analyzed only once). In particular, we can easily select a group of x combinations in which the x linearized expressions depend only on $2 \cdot x$ (instead of $3 \cdot x$) variables of the second round. This reduces the number of bit operations in step 4 to $39 \cdot x \cdot 2^{32}$. The improvement factor of the attack is thus $(\frac{1}{10} + \frac{39x}{10500} + 2^{-x+k})^{-1}$. By selecting an optimal value for x, we get a total improvement factor which is between 6 and 4 for all messages of practical length containing up to 2^{30} 32-bit blocks, whereas Fuhr's attacks [2] becomes worse than the generic attack for all messages which are longer than 96 blocks.

7   Conclusions

In this paper, we presented several second preimage attacks on Hamsi-256 that are based on polynomial enumeration algorithms. Our attacks are faster than Fuhr's attack for all message lengths, and unlike Fuhr's attack they are faster than the generic Kelsey and Schneier attack for all practical message sizes. Our new techniques can be applied in principle to other hash algorithms whose compression function can be described by a low degree multivariate polynomial, and demonstrate the potential vulnerability of such schemes to advanced algebraic attacks. In addition, our techniques can be used to speed up exhaustive search on secret key algorithms (such as block cipher, stream ciphers and MACs) that can be described by a low degree multivariate polynomial in the key bits. However, hash function designs with a stronger finalization function and an intermediate state that is bigger than the output (i.e "wide-pipe" designs), seem to better resist our attack.

Acknowledgements. The authors thank Orr Dunkelman and Nathan Keller for helpful discussions that led to this paper. The authors also thank the anonymous referees for their very helpful comments on this paper.

References

1. Küçük, Ö.: The hash function hamsi. Submission to NIST (updated) (2009)
2. Fuhr, T.: Finding Second Preimages of Short Messages for Hamsi-256. In: Abe, M. (ed.) ASIACRYPT 2010. LNCS, vol. 6477, pp. 20–37. Springer, Heidelberg (2010)

3. Kelsey, J., Schneier, B.: Second preimages on n-bit hash functions for much less than 2^n work. In: Cramer, R. (ed.) EUROCRYPT 2005. LNCS, vol. 3494, pp. 474–490. Springer, Heidelberg (2005)

4. Aumasson, J.-P., Käsper, E., Knudsen, L.R., Matusiewicz, K., Ødegård, R., Peyrin, T., Schläffer, M.: Distinguishers for the Compression Function and Output Transformation of Hamsi-256. In: Steinfeld, R., Hawkes, P. (eds.) ACISP 2010. LNCS, vol. 6168, pp. 87–103. Springer, Heidelberg (2010)

5. Boura, C., Canteaut, A.: Zero-sum Distinguishers for Iterated Permutations and Application to Keccak-f and Hamsi-256. In: Biryukov, A., Gong, G., Stinson, D.R. (eds.) SAC 2010. LNCS, vol. 6544, pp. 1–17. Springer, Heidelberg (2011)

6. Çalık, Ç., Turan, M.S.: Message Recovery and Pseudo-preimage Attacks on the Compression Function of Hamsi-256. In: Abdalla, M., Barreto, P.S.L.M. (eds.) LATINCRYPT 2010. LNCS, vol. 6212, pp. 205–221. Springer, Heidelberg (2010)

7. Wang, M., Wang, X., Jia, K., Wang, W.: New Pseudo-Near-Collision Attack on Reduced-Round of Hamsi-256. Cryptology ePrint Archive, Report 2009/484 (2009)

8. Joux, A.: Algorithmic Cryptanalysis, pp. 285–286. Chapman & Hall, Boca Raton

9. Bouillaguet, C., Chen, H.-C., Cheng, C.-M., Chou, T., Niederhagen, R., Shamir, A., Yang, B.-Y.: Fast Exhaustive Search for Polynomial Systems in F_2. In: Mangard, S., Standaert, F.-X. (eds.) CHES 2010. LNCS, vol. 6225, pp. 203–218. Springer, Heidelberg (2010)

10. Leurent, G.: MD4 is Not One-Way. In: Nyberg, K. (ed.) FSE 2008. LNCS, vol. 5086, pp. 412–428. Springer, Heidelberg (2008)

11. Joux, A.: Algorithmic Cryptanalysis, pp. 225–226. Chapman & Hall, Boca Raton

12. Dinur, I., Shamir, A.: An Improved Algebraic Attack on Hamsi-256. Cryptology ePrint Archive, Report 2010/602

A   Appendix: Parameters for the Direct Attack on Hamsi-256

The 56 second round bits on which the set $N = \{5, 156, 184, 214, 221, 249\}$ depends are listed below:

{3, 9, 18, 28, 44, 56, 63, 68, 79, 86, 91, 92, 93, 96, 107, 121, 131, 156, 184, 191, 196, 214, 219, 220, 221, 224, 249, 256, 259, 265, 274, 275, 284, 300, 312, 319, 324, 335, 342, 347, 348, 349, 352, 363, 377, 387, 412, 440, 447, 452, 470, 475, 476, 477, 480, 505}. The 6 simple sums for the sets $N = \{5, 156, 184, 214, 221, 249\}$ are given in the table below:

Table 2. The 6 simple sums of the second round variables denoted by x_i for $0 \leq i < 512$ for the output bits $N = \{5, 156, 184, 214, 221, 249\}$

Output Bit	Simple Sum
5	$x_9 + x_{18} + x_{19} + x_{63} + x_{86} + x_{92} + x_{121} + x_{137} + x_{146} + x_{147} + x_{214} + x_{249}$ $+ x_{265} + x_{274} + x_{275} + x_{319} + x_{342} + x_{393} + x_{402} + x_{403} + x_{470} + x_{476} + x_{505}$
156	$x_3 + x_{28} + x_{79} + x_{156} + x_{191} + x_{207} + x_{259} + x_{284} + x_{387} + x_{412} + x_{447} + x_{463}$
184	$1 + x_{56} + x_{63} + x_{107} + x_{235} + x_{319} + x_{347} + x_{447} + x_{475} + x_{491}$
214	$1 + x_9 + x_{86} + x_{121} + x_{137} + x_{v214} + x_{249} + x_{265} + x_{342} + x_{393} + x_{470} + x_{477} + x_{505}$
221	$1 + x_{18} + x_{63} + x_{68} + x_{92} + x_{96} + x_{146} + x_{224} + x_{274} + x_{319}$ $+ x_{324} + x_{352} + x_{402} + x_{452} + x_{476} + x_{477}$
249	$1 + x_{28} + x_{44} + x_{96} + x_{121} + x_{156} + x_{172} + x_{224} + x_{249} + x_{300}$ $+ x_{377} + x_{412} + x_{428} + x_{480} + x_{505}$

Practical Near-Collisions and Collisions on Round-Reduced ECHO-256 Compression Function

Jérémy Jean* and Pierre-Alain Fouque

Ecole Normale Supérieure
45 rue d'Ulm – 75230 Paris Cedex 05 – France
{Jeremy.Jean,Pierre-Alain.Fouque}@ens.fr

Abstract. In this paper, we present new results on the second-round SHA-3 candidate ECHO. We describe a method to construct a collision in the compression function of ECHO-256 reduced to four rounds in 2^{52} operations on AES-columns without significant memory requirements. Our attack uses the most recent analyses on ECHO, in particular the **Super-SBox** and **SuperMixColumns** layers to utilize efficiently the available freedom degrees. We also show why some of these results are flawed and we propose a solution to fix them. Our work improves the time and memory complexity of previous known techniques by using available freedom degrees more precisely. Finally, we validate our work by an implementation leading to near-collisions in 2^{36} operations for the 4-round compression function.

Keywords: Cryptanalysis, Hash Functions, SHA-3, ECHO-256, Collision attack.

1 Introduction

Recently, the National Institute of Standards and Technology (NIST) initiated an international public competition aiming at selecting a new hash function design [12]. Indeed, the current cryptanalysis of hash functions like SHA-1 and MD5 show serious weaknesses [18,19,20,21]. To study hash functions, one of the most powerful strategy is the differential cryptanalysis, which was introduced in [2] by Biham and Shamir to study the security of block ciphers. It consists in following the evolution of a message pair in the cipher by looking at the differences between the messages while they propagate through the encryption process. This type of analysis is particularly useful for studying hash functions where no secret-key is involved: in this known-key model [5], the attacker can thus follow the message pair at each step of the process. Knudsen generalized the idea in [4] with the concept of *truncated differentials*, aiming at following the *presence* of differences in a word, rather than their actual values. Initiated by the work of Peyrin on Grindhal [13], this kind of analysis leads to many other successful attacks against

* This author was partially supported by the French ANR project SAPHIR II and the French *Délégation Générale pour l'Armement* (DGA).

A. Joux (Ed.): FSE 2011, LNCS 6733, pp. 107–127, 2011.

block ciphers and hash functions, in particular those based on the AES [6,10] like ECHO. For the AES, since all differences are equivalent, only their presence matters.

Thanks to the SHA-3 contest, new kinds of attacks for AES-based permutations have been suggested in the past few years, in particular the rebound attack [10] and the start-from-the-middle attack [9]. In both cases, the novelty is to start searching for a message pair conforming to a given differential path in the middle of the trail. Doing so, we have the freedom of choosing values and differences where they can reduce greatly the overall cost of the trail.

The rebound technique uses these degrees of freedom to fulfill the most expensive part of the trail at very low average complexity whereas the remaining of the path is verified probabilistically. The number of controlled rounds in that case can not exceed two rounds. The start-from-the-middle technique improves the rebound attack in the sense that it uses the independence in the search process as much as possible. Consequently, it extends the number of controlled rounds to three, without any extra time.

In the present case of ECHO, Schläffer uses in [17] the idea of multiple inbound phases on two different parts of the whole path. Similar techniques have been introduced on Whirlpool [6] and on the SHA-3 proposal LANE [8]. In comparison to the rebound or the start-from-the-middle techniques, we are not limited to a controlled part located in the middle of the path. In the end, the partial message pairs are merged using remaining degrees of freedom. Schläffer's nice attacks permute some linear transformations of the ECHO round function to introduce the **SuperMixColumns** layer, which relies on a large matrix presenting non-optimal diffusion properties. It thus allows to build sparser truncated differential. In this paper, we show that the latest analyses of ECHO made by Schläffer fail with high probability at some point of the merging process: the attacks actually succeed with probability 2^{-128}. Nevertheless, we suggest an attack using degrees of freedom slightly differently to construct collisions and near-collisions in the compression function of ECHO-256 reduced to four rounds.

Our new techniques improve the rebound attack by using freedom degrees more precisely to get and solve systems of linear equations in order to reduce the overall time and memory complexity. We also describe a similar method as the one described by Sasaki et al. in [16] to efficiently find a message pair conforming to a truncated differential through the **SuperSBox** when not all input or output bytes are active. Both new techniques allow to repair some of the Schläffer's results to construct collisions in the compression function of ECHO-256. To check the validity of our work, we implement the attack to get a semi-free-start near-collisions in 2^{36} computations. That is, a chaining value h and a message pair (m, m') colliding on 384 bits out of 512 in the compression function f reduced to four rounds: $f(h, m) =_{384} f(h, m')$.

We summarize our results in Table 1.

The paper is organized as follows. In Section 2, we quickly recall the specifications of the ECHO hash function and the permutation used in the AES. In Section 3, we describe the differential path we use and present an overview of

Table 1. Summary of results detailed in this paper and previous analyses of ECHO-256 compression function. We measure the time complexity of our results in terms of operations on AES-columns. The notation $n/512$ describe the number n of bits on which the message pair collides in the near-collisions. Result from [17] have not been printed since flawed.

Rounds	Time	Memory	Type	Reference
3	2^{64}	2^{32}	free-start collision	[14]
3	2^{96}	2^{32}	semi-free-start collision *	[14]
4.5	2^{96}	2^{32}	distinguisher	[14]
4	2^{36}	2^{16}	semi-free-start near-collision 384/512	**This paper**
4	2^{44}	2^{16}	semi-free-start near-collision 480/512 †	**This paper**
4	2^{52}	2^{16}	semi-free-start collision	**This paper**

* With chosen salt
† This result is an example of other near-collisions that can be derived from the attack of this paper.

the differential attack to find a message pair conforming to this path. Then, in Section 4, we present the collision attack of ECHO-256 compression function reduced to four rounds. Finally, we conclude in Section 5. We validate our results by implementing the near-collision attack.

2   Description of ECHO

The hash function ECHO updates an internal state described by a 16×16 matrix of $GF\left(2^8\right)$ elements, which can also be viewed as a 4×4 matrix of 16 AES states. Transformations on this large 2048-bit state are very similar to the one of the AES, the main difference being the equivalent S-Box called **BigSubBytes**, which consists in two AES rounds. The diffusion of the AES states in ECHO is ensured by two *big* transformations: **BigShiftRows** and **BigMixColumns** (Figure 1).

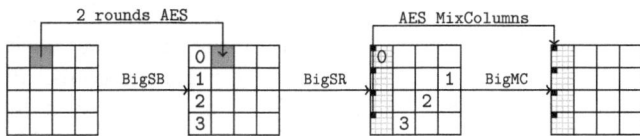

Fig. 1. One round of the ECHO permutation. Each of the 16 cells is an AES state (128 bits).

At the end of the 8 rounds of the permutation in the case of ECHO-256, the **BigFinal** operation adds the current state to the initial one (feed-forward) and adds its four columns together to produce the new chaining value. In this paper, we only focus on ECHO-256 and refer to the original publication [1] for more details on both ECHO-256 and ECHO-512 versions. Note that the keys used in the two AES rounds are an internal counter and the salt, respectively: they are mainly introduced to break the existing symmetries of the AES unkeyed permutation [7]. Since we are not using any property relying on symmetry and that adding constants does not change differences, we omit these steps.

Two versions of the hash function ECHO have been submitted to the SHA-3 contest: ECHO-256 and ECHO-512, which share the same state size, but inject messages of size 1536 or 1024 bits respectively in the compression function. Focusing on ECHO-256 and denoting f its compression function, H_i the i-th output chaining value, $M_i = M_i^0 \parallel M_i^1 \parallel M_i^2$ the i-th message block composed of three chunks of 512 bits each M_i^j and $S = [C_0 C_1 C_2 C_3]$ the four 512-bit ECHO-columns constituting state S, we have $(H_0 = IV)$:

$$C_0 \leftarrow H_{i-1} \qquad C_1 \leftarrow M_i^0 \qquad C_2 \leftarrow M_i^1 \qquad C_3 \leftarrow M_i^2$$

AES. We recall briefly one AES round on Figure 2 and refer as well to original publication [11] for further details. The **MixColumns** layer implements a Max-

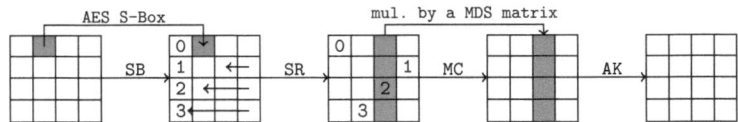

Fig. 2. One round of the AES permutation is the succession of four transformations: SubBytes (**SB**), ShiftRows (**SR**), MixColumns (**MC**) and AddKey (**AK**). Each of the 16 cells is an element of GF (2^8) (8 bits).

imum Distance Separable (MDS) code that ensures a complete diffusion after two rounds. It has good diffusion properties since its branch number, i.e. the sum of input and output active bytes, is always 0 or greater or equal than 5. As for the AES S-Box, it satisfies an interesting differential property: namely, a random differential transition exists with probability approximately 1/2. By enumerating each input/output difference pair, this result can be computed and stored in 2^{16} in the difference distribution table Δ. At the position (δ_i, δ_o), this table contains a boolean value whether the differential transition $\delta_i \rightarrow \delta_o$ exists. That is, if the equality $S(\lambda) + S(\lambda + \delta_i) = \delta_o$ holds for at least one element $\lambda \in$ GF (2^8), S being the AES S-Box. We note that this table can be slightly enlarged to 2^{19} to store one solution when possible.

Notations. Throughout this paper, we name each state of the ECHO permutation after each elementary transformation: starting from the first state S0, we end the first round after 8 transformations[1] in S8 and the four rounds in S32. Moreover, for a given ECHO-state Sn, we refer to the AES-state at row i and

---

[1] Transformations are: **SR** - **SB** - **MC** - **SB** - **SR** - **BSR** - **MC** - **BMC**.

column j by $Sn[i,j]$. Additionally, we introduce *column-slice* or *slice* to refer to a thin column of size 16×1 of the ECHO state. We use *ECHO-column* or simply *column* to designate a column of ECHO, that is a column of four AES states. Similarly, *ECHO-row* or *row* refer to a row of the ECHO state; that is, four AES states.

3   Differential Attack for Hash Functions

To mount a differential attack on a hash function, we proceed in two steps. First, we need to find a good differential path, in the sense that, being probabilistic, it should hold with a probability as high as possible. In the particular case of AES-based hash functions, this generally means a path with a minimum number of active S-Boxes. In comparison with the differential attacks where fixed differences chosen for their high probability go along with the differential path, for this particular design, all differences behave equivalently. Thus, the path is actually a truncated path, precising only whether a difference exists or not.

Second, we have to find a pair of messages following that differential path, which fixes values and differences. In the sequel, we present an equivalent description of the ECHO-permutation and then detail our choice of differential path, using the new round description. The part of the attack that finds a valid message pair for this path using the equivalent description is detailed in Section 3.3.

3.1   Reordering of Transformations in the ECHO Permutation

SuperSBox.   The concept of **SuperSBox** was independently introduced by Lamberger et al. in [6] and by Gilbert and Peyrin in [3] to study two AES rounds. By bringing the two non-linear layers together, this concept is useful to find a message pair conforming to a given differential path and leads to a new kind of cryptanalysis. The design of one AES round describes the sequence **SB-SR-MC** of transformations[2], but we can use the independence of bytes to reorder this sequence. Namely, dealing with the non-linear **BigSubBytes** layer of ECHO, we can permute the first **ShiftRows** with the first **SubBytes** without affecting the final result of the computation. We then glue the two non-linear layers into a unique **SB-MC-SB** non-linear transformation of the permutation. The so-called **SuperSBox** transformation is then viewed as a single non-linear layer operating in parallel on 32-bit AES-columns.

SuperMixColumns.   In a similar way, by permuting the **BigShiftRows** transformation with the parallel **MixColumns** transformations of the second AES round, a new *super* linear operation has been introduced by Schläffer in [17], which works on column-slices of size 16×1.

This super transformation called **SuperMixColumns** results of 16 parallel applications of **MixColumns** followed by the equivalent in ECHO, that is **Big-MixColumns**. This *super* transformation is useful for building particular sparse

---

[2] While omitting the key adding.

truncated differential. The matrix of the **SuperMixColumns** transformation is defined as the Kronecker product (or tensor product) of $\mathbf{M}$ with itself, $\mathbf{M}$ being the matrix of the **MixColumns** operation in the AES: $\mathbf{M}_{SMC} = \mathbf{M} \otimes \mathbf{M}$. Schläffer noted in [17] (in Section 3.3) that $\mathbf{M}_{SMC}$ is not a MDS matrix and its branch number is only 8, and not 17.

From this observation, it is possible to build sparse truncated differentials (Figure 3) where there are only 4 active bytes in both input and output slices of the transformation. The path $4 \rightarrow 16 \rightarrow 4$ holds with probability 2^{-24}, which reduces to 2^8 the number of valid differentials, among the 2^{32} existing ones. For a given position of output active bytes, valid differentials are actually in a subspace of dimension one. In particular, for slice s, $s \in \{0, 4, 8, 12\}$, to follow the truncated differential $4 \rightarrow 16 \rightarrow 4$ of Figure 3, we need to pick each slice of differences in the one-dimensional subspace generated by the vector v_s, where:

$$v_0 = [\text{E000 9000 D000 B000}]^{\mathrm{T}} \qquad v_4 = [\text{B000 E000 9000 D000}]^{\mathrm{T}}$$
$$v_8 = [\text{D000 B000 E000 9000}]^{\mathrm{T}} \qquad v_{12} = [\text{9000 D000 B000 E000}]^{\mathrm{T}}$$

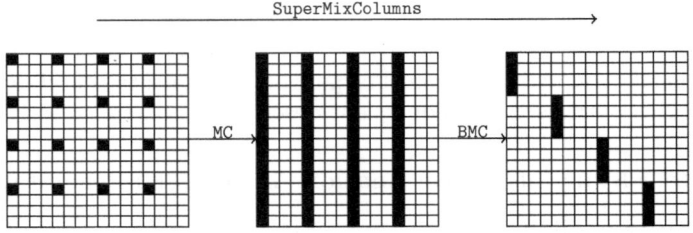

Fig. 3. The **SuperMixColumns** layer in the particular case of the truncated differential $4 \rightarrow 16 \rightarrow 4$

This new approach of the combined linear layers allows to build sparser truncated differentials but caused erroneous conclusions when it was used in [17] (in Section 4.1). Namely, at the end of the attack, where two partial solutions need to be merged to get a solution for the whole differential path, everything relies on this critical transformation: we need to solve 16 linear systems. We detail more precisely the problem in Section 3.4, where we study the merge process.

3.2 Truncated Differential Path

As in the more recent analyses of ECHO [15,17], we consider the path at the byte-level: this allows to build paths sparser than the ones we could obtain by considering only the AES-state level [1,3,9]. Our path is mostly borrowed from [17] and counts 418 active S-Boxes for the ECHO-permutation reduced to four rounds. In comparison to the path from [17], we increase significantly the number of active S-Boxes in the first round to decrease the time complexity of the attack. We note that the number of active S-Boxes is not directly correlated with the complexity of the attack. Moreover, in that case of an AES-based permutation, we can consider a truncated differential path because the actual differences are not really important since they are all equivalent: only their *presence* matters.

Figure 4 presents the truncated differential path used in this attack on the compression function reduced to four rounds. The attack being quite technical, colors have been used in order to improve the reader's understanding of the attack.

3.3 Finding a Message Pair Conforming to the Differential Path

Strategy. To find a message pair that follows the differential path of Figure 4, our attack splits the whole path into two distinct parts and merges them at the end. In the sequel, we refer to these two parts as *first subpart* and *second subpart*. The attack of Schläffer in [17] proceeds similarly but uses the rebound attack technique in the two subparts. We reuse this idea of finding message pairs conforming to partial truncated parts but most of our new techniques avoid the rebound attack on the **SuperSBox**. Both subparts are represented in the Figure 4: the first one starts in S7 and ends in S14 and fixes the red bytes of the two messages, whereas the second one starts at S16 until the end of the four rounds in S31 and fixes the yellow bytes. Additionally, the chaining value in the first round of the path are the blue bytes.

SuperSBox. In the differential path described on Figure 4, there are many differential transitions through the **SuperSBox** of the third round where input differences are reduced to *one* active byte. We are then interested in differential transitions such as the one described in Figure 5. For this kind of transition, the distribution difference table of the **SuperSBox** would work but requires 2^{64} to be computed and stored[3]. We show that we can compute a pair of columns satisfying this path in 2^{11} operations on one AES-column.

Let us consider the input difference to be $\Delta_i = [\delta_i, 0, 0, 0]^{\mathrm{T}}$ reduced to one active byte δ_i and the output difference $\Delta_o = [\delta_o^1, \delta_o^2, \delta_o^3, \delta_o^4]^{\mathrm{T}}$: we aim at finding a pair of AES-columns (c_1, c_2) conforming to those differences; that is: $c_1 + c_2 = \Delta_i$ and $\mathbf{SuperSBox}(c_1) + \mathbf{SuperSBox}(c_2) = \Delta_o$. In a precomputation phase of 2^{16}, we compute and store the differential distribution table of the AES S-Box.

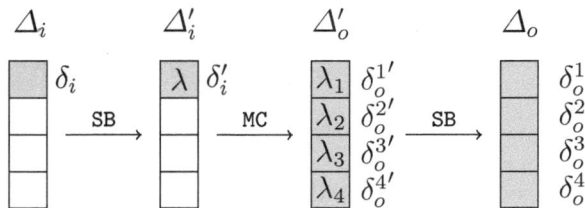

Fig. 5. A **SuperSBox** differential transition with only one active input byte

---

[3] In that case, we could compute and store smaller tables in 2^{40} for the four possible positions of active bytes.

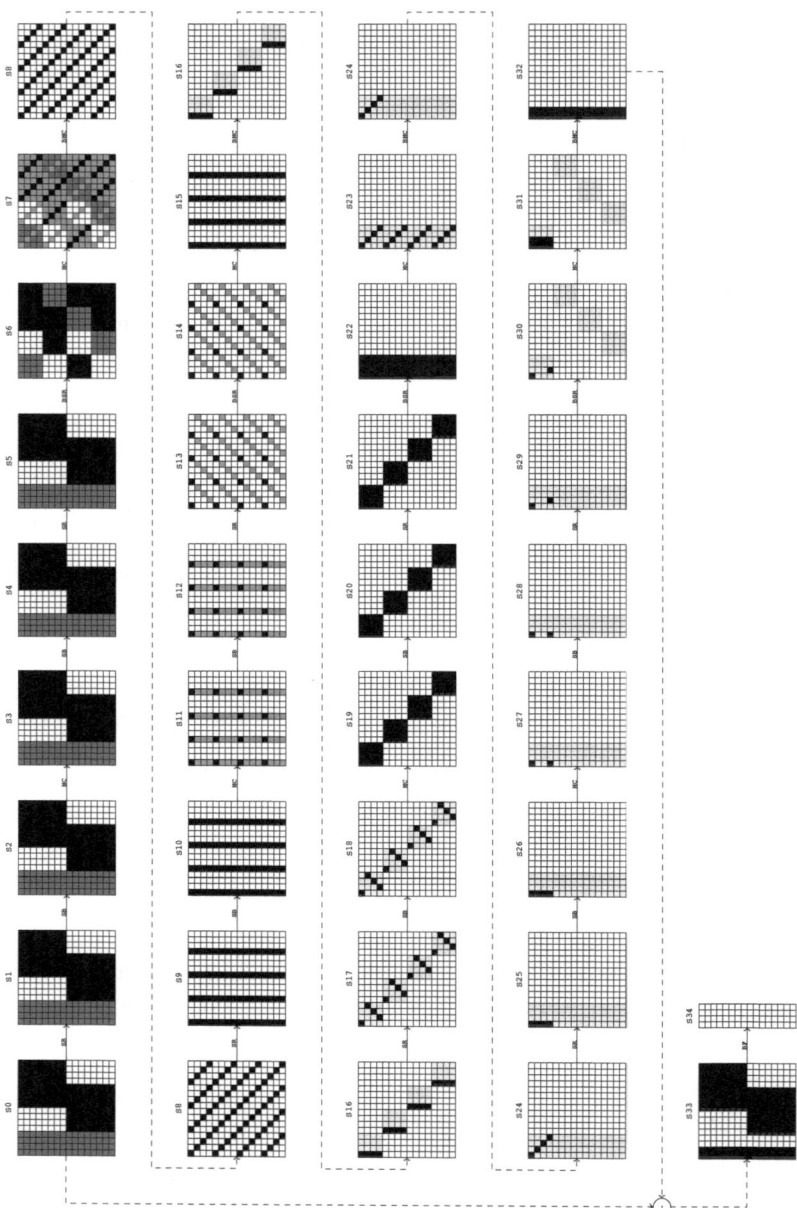

Fig. 4. The differential path used in this attack on the ECHO-256 compression function reduced to four rounds. To find a valid pair of messages, we split the path into two parts: the first subpart between S7 and S14 (red bytes) and the second subpart between S16 and S31 (yellow bytes). Black bytes are the only active bytes, blue bytes come from the chaining value and gray bytes in the first round are set to get a collision (or a near-collision) in the compression step.

The differential properties of the AES S-Box restrict the number of output differences of the first **SubBytes** layer to $2^7 - 1$ and for each one, the underlying values are set. Denoting δ_i' one of the output differences of this layer and λ the associated value such that $S^{-1}(\lambda) + S^{-1}(\lambda + \delta_i') = \delta_i$, we can propagate this difference $\Delta_i' = [\delta_i', 0, 0, 0]^T$ linearly to learn the four differences at the input of the second **SubBytes** layer. We note $\Delta_o' = \mathbf{MC}(\Delta_i') = [\delta_o^{1'}, \delta_o^{2'}, \delta_o^{3'}, \delta_o^{4'}]^T$ those differences. Here, both the input and the output differences are known and the four differential transitions $\delta_o^{i'} \rightarrow \delta_o^i$ exist with probability approximately 2^{-4}. Since we can restart with $2^7 - 1$ different δ_i', we get approximately 2^3 valid differential transitions. Each of these transitions fixes the underlying values, noted $\lambda_1, \lambda_2, \lambda_3, \lambda_4$. At this point, all intermediate differences conform to the path, but in terms of values, we need to ensure that λ is consistent with $\lambda_1, \lambda_2, \lambda_3, \lambda_4$. To check this, we exhaust the 2^4 valid vectors of values we can build by interchanging λ_i and $\lambda_i + \delta_o^{i'}$.

All in all, among the 2^{3+4} vectors of values we can build, only a fraction 2^{-8} will satisfy the 8-bit condition on λ. This means that the considered differential transition $\Delta_i \rightarrow \Delta_o$ through the **SuperSBox** occurs with probability 2^{-1} and if the transition exists, we can recover an actual AES-column pair in $2^7 \, 2^4 = 2^{11}$ operations.

3.4   Overview of the Attack

In this subsection, we describe the main steps used to find a message pair conforming to the differential path. We begin by the sensitive part of the attack, which caused erroneous statements in [17]: the merging phase of the two partial solutions.

Merging step.   Assume for a moment that we solved both subparts of the path, i.e. the red bytes between S7 and S14 are fixed as well as the yellow ones between S16 and S31: we have two partial solutions for the complete differential path. The truncated differential of Figure 4 is then partially verified but to merge the two parts, we need to set the white bytes so that the **SuperMixColumns** transition from S14 to S16 is possible.

Due to the particular construction of $\mathbf{M}_{SMC}$, some algebra considerations show that for the already-known values in S14 (red) and S16 (yellow), the **SuperMixColumns** transition will not be possible unless a 128-bit constraint is satisfied: the remaining degrees of freedom can not be used to satisfy this relation. Since all of the 16 column-slices of the considered matrices are independent, this leads to 16 constraints on 8 bits.

The flaw in [17] is to assume these relations are true, which holds only with probability 2^{-128}, whatever the value of unset bytes are. These equalities need to be true so that the 16 linear systems have solutions. The first system associated to the first slice is given by:

$$\mathbf{M_{SMC}} \begin{bmatrix} a_0 & x_0 & x_1 & x_2 & a_1 & x_3 & x_4 & x_5 & a_2 & x_6 & x_7 & x_8 & a_3 & x_9 & x_{10} & x_{11} \end{bmatrix}^T =$$
$$\begin{bmatrix} b_0 & b_1 & b_2 & b_3 & * & * & * & * & * & * & * & * & * & * & * & * \end{bmatrix}^T \tag{1}$$

where a_i and b_i are known values, x_i are unknowns and $*$ is any value in GF (2^8). This system has solutions if a particular linear combination of $[a_0, a_1, a_2, a_3]^T$ and $[b_0, b_1, b_2, b_3]^T$ lies in the image of some matrices: this constraints the already-known values to verify an 8-bit relation. The constraint comes from the fact that $\mathbf{M}_{SMC}$ is the Kronecker product $\mathbf{M} \otimes \mathbf{M}$. For example, in the following, we denote by a_i, $0 \leq i \leq 3$, the four known values of slice 0 of S14 coming from the first subpart (red) and b_i the known values for the same slice in S16, from the second subpart (yellow). With this notation, the first system will have solutions if and only if the following condition is satisfied:

$$2a_0 + 3a_1 + a_2 + a_3 = 14b_0 + 11b_1 + 13b_2 + 9b_3. \tag{2}$$

See Appendix A for the detailed proof. These constraints for each slice of the **SuperMixColumns** transition can also be viewed in a different way: consider all the b_i known for all slices, thus we can only pick 3 out of 4 a_i per slice in S14 and determine the last one deterministically. Alternatively, due to the **ShiftRows** and **BigShiftRows** transitions, we can independently determine slices 0, 4 and 8 in S12 so that slice 12 of S12 would be totally determined. This transfers the problem into the first subpart of the path.

Step 1. We begin by finding a pair of ECHO-columns satisfying the truncated path reduced to the first ECHO-column between S7 and S12. This is done with a randomized AES-state of the column, used to get and solve linear equations giving all differences between S7 and S9. Indeed, differences between S7 and S9 for the first column only depend on the four differences in $S7[2,0]^4$. Then, we search for valid differential transitions through the AES S-Box between S9 and S10 to finally deduce a good pair of ECHO-columns. This step can be done in 2^{12} operations on AES-columns (Section 4.1).

Step 2. Once we solved the first ECHO-column, we can deduce all differences between S12 and S16: indeed, the wanted **SuperMixColumns** transition imposes them as discussed in Section 3.1. This step is done in constant time (Section 4.2).

Step 3. Now that we have the differences in S16, we have a starting point to find a message pair for the second subpart of the whole truncated path: namely, states between S16 and S31 (yellow bytes). To do so, the idea is similar as in Step 1: since all differences between S20 and S24 only depend on the four differences of $S24^5$, we can use a randomized AES-column c in S18 to get four independent linear equations in S20 and thus deduce all differences between S20 and S24. Then, we search for input values for the 15 remaining **SuperSBoxes**, which have only one active byte at their input (Section 3.3). This succeeds with probability 2^{-15} so that we need to retry approximately 2^{15} new random c. The whole step can be done in 2^{26} operations on AES-columns (Section 4.2).

---

[4] Linear relations can be deduced by linearly propagating differences in $S7[2,0]$ forwards until S9.

[5] Linear relations can be deduced by linearly propagating the four differences of $S24[0,0]$ backwards until S20.

Being done, the truncated path is followed until the end of the four rounds in S32. Note that we can filter the **MixColumns** transition between S26 and S27 in a probabilistic way so that less slices would be active in S32.

Step 4. Getting back to the first subpart of the truncated path, we now find a valid pair of ECHO-columns satisfying the truncated path between S7 and S12 reduced to the second ECHO-column. This is basically the same idea as in Step 1. This can be done in 2^{12} operations on AES-columns as well (Section 4.3). Note that this step could be switched with Step 3.

Step 5. To construct a valid pair of ECHO-columns satisfying the truncated path between S7 and S12 reduced to the third ECHO-column, we proceed as before (steps 1 and 4), but we start by randomizing three AES states instead of one: indeed, differences between S7 and S9 at the input of the non-linear layer now depend on 12 differences, the ones in S7[0,2], S7[1,2] and S7[3,2]. Getting 12 linear systems then allow to learn those differences and we can finally search for four valid differential transitions through the AES S-Box in 2^4 operations on AES-columns (Section 4.3).

Step 6. The merging step in [17] fails with high probability, but we know how to get into the valid cases: since the three first ECHO-columns of the first subpart are now known, we can deduce the whole last ECHO-column allowing the 16 needed equations mentioned before. There is no freedom for that column, so we are left with a probabilistic behavior to check if it follows the column-reduced truncated differential. We then propagate the pair of deduced values backwards until S8 and check if the **invBigMixColumns** transition behave as wanted: namely, four simultaneous $4 \rightarrow 3$ active bytes, which occurs with probability $(2^{-8})^4$. Hence, we need to restart approximately 2^{32} times the previous Step 5 to find a valid pair of ECHO-columns satisfying both the path between S7 and 12 and the 128-bit condition imposed by the merging step. This step can be performed in 2^{36} operations on AES-columns (Section 4.3).

Step 7. To get a collision in the compression function, we then need to take care of the compression phase in the **BigFinal** operation: the feed-forward and the xor of the four ECHO-columns. The collision is reached when the sum of the two active AES-states per row in S0 equals the active one in S32. We have enough degrees of freedom to determine values in associated states of S7 (gray) to make this happens. Together with the probabilistic filter of Step 3, this step may impose the global time complexity of the attack; so, weakening the final objective (to get a near-collision, for instance) can make the whole attack practical (Section 4.4).

Step 8. The last step consists in filling all the remaining bytes by solving the 16 linear systems mentioned in Step 6, while taking care at the same time that the **invBigMixColumns** between S8 and S7 reaches the values determined by Step 7. Due to the particular structure of the solution sets, the systems can be solved in parallel in 2^{32} operations on AES-columns (Section 4.5).

4   Collision on the 4-Round Compression Function

4.1   Partial Message Pair for the First Subpart

This step aims at finding a pair of ECHO-columns satisfying the truncated differential of Figure 6. We consider the first column separately from the others in order to reach a situation where the merging process will be possible. Indeed, once we fix a slice, we can determine the differences at the beginning of the second subpart in S16.

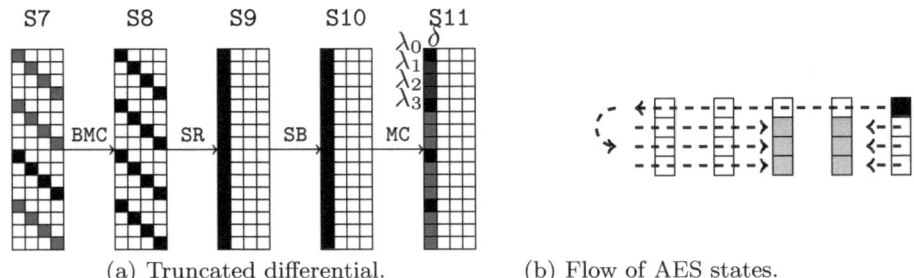

(a) Truncated differential.   (b) Flow of AES states.

Fig. 6. Truncated differential path (a) used for the first subpart of the attack for one ECHO-column. We represent on (b) the order in which AES states are randomized (black) or deduced by a small rebound attack (gray).

The previous method suggested in [17] (in Section 4.1) to find paired values following this truncated differential is a rebound attack working in time 2^{32} and using the differential distribution table of the **SuperSBox** of size 2^{64}. We show how we can find an ECHO-column pair conforming to this reduced path in 2^{12} operations on AES-columns without significant memory usage.

Rather than considering the whole column at once, we start by working on the top AES state in S11, that is S11[0,0]. We begin by choosing random values $(\lambda_0, \lambda_1, \lambda_2, \lambda_3)$ for the first AES-column of S11[0,0] (blue bytes), such that the active byte is set to difference δ, also chosen at random in $\mathrm{GF}\left(2^8\right) \setminus \{0\}$. Starting from S11[0,0] and going backwards, those values and differences are propagated deterministically until S8[0,0]. Since there is only one active byte per slice in the considered ECHO-column of S7, each of the associated four slices of S8 lies in a subspace of dimension one. Therefore, solving four simple linear systems leads to the determination of the 12 other differences of S8.

Therefore, in the active slice of S9 of Figure 6 at the input of the **SubBytes** layer, the four first paired bytes have values and differences known, whereas in the 12 other positions, only differences are set. Our goal now is to find good values for these byte pairs, which can be achieved by a small rebound attack on the AES S-Box where the output differences are propagated from S11 by choosing random differences. Thus, we iterate on the $(2^8)^3$ possible unset differences of S11 and propagate them linearly to S10. When both input and output differences of the 12 AES S-Boxes are known, we just need to ensure that these 12 differential transitions are possible. This is verified by the precomputed table[6] Δ. It ensures that the 12 transitions will occur simultaneously with probability 2^{-12}. Since we can try approximately $(2^8)^3$ output differences, we will have about $(2^8)^3 \; 2^{-12} \; (2^4)^3 \approx 2^{24}$ different vectors of values by trying them all. The factor $(2^4)^3$ comes from the possibility of interchanging the two solutions of the equations 12 times to get more vectors of values[7].

All in all, for any $\lambda_0, \lambda_1, \lambda_2, \lambda_3, \delta$ picked at random in $\mathrm{GF}\left(2^8\right)$ (with non-null difference), we can find 2^{24} ECHO-column pairs in S12 in 2^{12} such that the associated truncated differential from S12 to S7 is verified. We could thus build approximately $2^{24+8\times5} = 2^{64}$ pairs of ECHO-columns following the column-reduced truncated differential.

4.2   Finding a Message Pair for the Second Subpart

We now get a partial message pair conforming to the first subpart of the truncated path reduced to a single ECHO-column. Rather than completing this partial message pair for the three other active slices in S12, we now find a message pair conforming to the second subpart of the truncated path, located in the third round from S16 to S24 (yellow bytes).

Indeed, the mere knowledge of a single active slice pair of S12 in the first subpart is sufficient to get a starting point to find a message pair for the second subpart, i.e. yellow bytes. This is due to the desired transition through the **SuperMixColumns** transition: as explained in Section 3.1, differences in S14 lie in one-dimensional subspaces. Once a slice pair for the first slice of S12 is known and computed forwards to S14 (black and red bytes on Figure 7), there is no more choice for the other differences in S14. Finally, all differences between S12 and S17 have been determined by linearity of involved transformations.

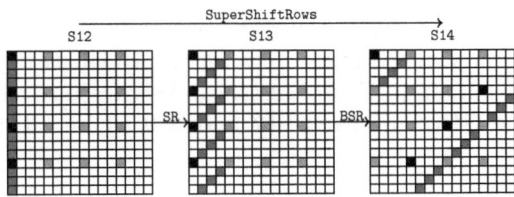

Fig. 7. The **SuperShiftRows** layer where only the values and differences of the first slice of S12 are known (black and red bytes)

At this point, only input differences of **SuperSBoxes** of the third round are known. We note that all operations between S20 and S24 are linear, so that all differences in those states only depend on the four differences of S24. We denote

---

[6] This is the differential distribution table of the AES S-Box, which required 2^{16} bits to be stored.

[7] There are cases where four solutions exist, but for simplicity, we consider only the more common two-solution case.

by k_i the non-null difference of column slice $i \in \{0, 1, 2, 3\}$ in state S24. By linearly propagating differences in S24 backwards to S20, we obtain constraints on the 64 output differences of the **SuperSBox** in S20. To find the actual differences, we need to find the four k_i and thus determine four independent linear equations. Considering arbitrarily the first AES-column of S20[0,0] (Figure 8), differences are: $[84k_0, 70k_3, 84k_2, 70k_1]^{\mathrm{T}}$ (black bytes).

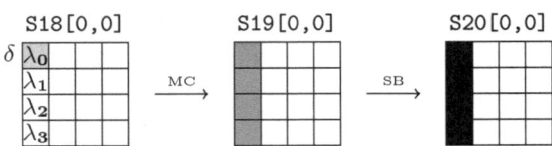

Fig. 8. The **MixColumns** and **SubBytes** transitions on the first AES-column between S18[0,0] and S20[0,0]

Starting from S18, let δ a random difference among the $2^7 - 1$ possible ones imposed by S17 for the considered columns (Figure 8). Any choice imposes the value associated to the differential transition as well: we denote it λ_0. At this step, we introduce more freedom by picking random values for the three remaining bytes of the column: $(\lambda_1, \lambda_2, \lambda_3)$. Note that we can choose $(2^8)^3 = 2^{24}$ of them and thus 2^{31} starting points in total. After this randomization the AES-column in S18, the same AES-columns in S19 and S20 are fully determined. We then need to link the four bytes with the differences provided by the right part of the path from S24 to S20: this is done by simple algebra by solving four linear equations in four variables, which are k_i, $0 \leq i \leq 3$.

After solving, we have the four differences k_i of state S24: we propagate them backwards from S24 to S20 and learn all the differences between S20 and S24. Only one pair of AES-columns out of the 16 was used in S18 to deduce differences k_i in S24, so we now try to find values for the 15 left (Figure 9).

Each of the remaining AES-columns, can be viewed as a differential transition through a **SuperSBox** between S17 and S20 where all differences have been

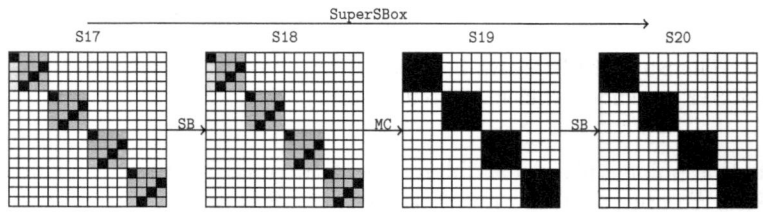

Fig. 9. Last step to get a message pair conforming to the second subpart of the path: finding the 15 remaining AES-columns using the **SuperSBox** properties. Black bytes are active and yellow bytes have already been defined in the previous step, as well as differences of the first AES-column of the first AES-state. Gray bytes are inactive and the target of this step.

previously set. As described in 3.3, we have 15 differential transitions through the **SuperSBox** with only one input active byte in each. The 15 transitions occur simultaneously with probability 2^{-15} and if so, we can recover the 15 AES-column pairs in parallel in 2^{11} using the technique previously described. Since there are 15 AES-columns to find in S17, we need to generate approximately 2^{15} new $(\delta, \lambda_0), \lambda_1, \lambda_2, \lambda_3$ and restart the randomization in S18[0,0].

Considering *one* message pair conforming to a single ECHO-column of the first subpart of the truncated path as starting point, the number of pairs we can build which follow the truncated path for this second subpart is: $2^7 \, 2^{8 \times 3} \, 2^{-15} \approx 2^{16}$. We note that we get one in 2^{26} operations in parallel.

In the collision attack on the compression function, we further extend this step by probabilistically filtering the active bytes in the **MixColumns** transition between S26 and S27. Among the 2^{16} message pairs we can build that follow the truncated path between S16 and S26, only one in average will verify the $4 \rightarrow 2$ transition through **MixColumns**. If such a pair is found then the pair conforms the truncated path until the end of the four rounds; otherwise, we need to find a new starting point, i.e. a new slice pair for slice 0 in S12. We reduce to two active bytes and not one or three because this is a best compromise we can make to lower the overall time complexity of the collision attack.

4.3 Completing the Partial Message Pair of the First Subpart

As discussed in Section 3.4, to solve the merging step, slice 12 of S12 is constrained by slices 0, 4 and 8 of S12. All values of slice pair 0 have been determined (Section 4.1) and used to fix yellow bytes and thus get a message pair conforming to the second subpart of the truncated path (Section 4.2).

Consequently, we only have freedom on the slice pairs 4 and 8 in S12. We determine values of slice pair 4 in the same way as slice 0 by considering the first subpart of the truncated path from S7 to S14 reduced to the second ECHO-column. There is a single active byte per slice in this ECHO-column of S7, so that we can build approximately 2^{60} valid columns[8] in that position in 2^{12} operations on AES-columns for a single one.

As soon as we have one, we use the remaining freedom of slice 8 to generate simultaneously slice pairs 8 and 12 of S12. We note that in the two last ECHO-columns of S7, there are three active bytes per slice (Figure 4). The method we suggest starts by finding a slice pair for slice 8 conforming to the truncated differential reduced to the third ECHO-column between S7 and S12. We proceed in the same way as we did for slices 0 and 4 and then, we deduce deterministically the slice pair 12 from the constraints imposed by the merge. Finally, we check whether that slice pair conforms the truncated differential reduced to the last ECHO-column until S7, namely the four simultaneous transitions $4 \rightarrow 3$ through **invMixColumns** between S8 and S7.

---

[8] Note that in Section 4.1, we could build 2^{64} of them because differences were chosen freely, whereas in the present case, differences are constrained by the AES S-Box differential properties to sets of size $2^7 - 1$. We thus loose 2^4 degrees of freedom.

The cost of 2^4 to construct a slice pair for slice 8 allows to repeat it 2^{32} times to pass the probability $\left(2^{-8}\right)^4$ of finding a valid slice pair for slice 12 conforming to both the linear constraints of the merge and the truncated differential through **invBigMixColumns**. Note that we have enough degrees of freedom to do so since we can find approximately $\left(2^7\right)^4 \left(2^8\right)^{3\times 3} = 2^{100}$ valid slice pairs for slice 8. However, only 2^{32} are needed, which completes this step in 2^{36} operations on AES-columns and fixes all the red bytes between S7 and S14.

4.4 Compression Phase in the Feed Forward

After four rounds, the round-reduced compression function applies the feed forward (S33 ← S0+S32) and XORs the four columns together (**BigFinal**). This operation allows to build the differential path such that differences would cancel out each other. As shown in the global path (Figure 4), states S0 and S32 XORed together lead to state S33 where there are three active AES-states in each row. In terms of differences, if each row sums up to zero, then we get a collision for the compression function in S34 after the **BigFinal**.

As we constructed the path until now, in both S0 and S32, we still have freedom on the values: only differences in S32 located in the two first slices are known from the message pair conforming to the second subpart of the truncated path. These differences thus impose constraints on the two other active pair states per row in S0. Namely, for each row r of S0 where active AES states are located in columns c_r and c'_r, we have $S0[r, c_r] + S0[r, c'_r] = S32[r, 0]$. Additionally, differences in S4 are known by linearly propagating the known differences from S7.

After the feed-forward, we cancel differences of each row independently: we describe the reasoning for an arbitrary row. We want to find paired values in the two active states of the considered row of S0, say (A, A') and (B, B'), such that they propagate with correct differences in S4, which are known, and with correct diagonal values (red bytes) in S7 after the **MixColumns**. In the sequel (Figure 10), we subscript the AES-state A by j to indicate that A_j is the AES-state A propagated until ECHO-state Sj with relevant transformations according to Figure 4.

The known differences of S4 actually sets the output differences of the **SuperSBox** layer: namely, $A_4 + A'_4 = \Delta_4$ and $B_4 + B'_4 = \Delta'_4$, where Δ_4 and Δ'_4 are the known differences in the considered row of S4. The constraint on the known diagonal values in A_7 and B_7 restricts the available freedom in the choice

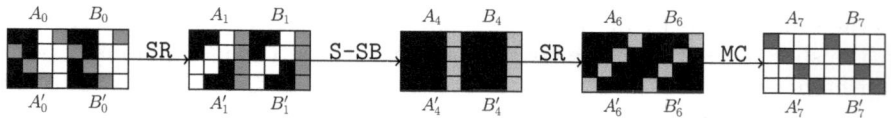

Fig. 10. Propagation of the pairs of AES-states (A_i, A'_i) and (B_i, B'_i) in a single ECHO-row in the first round. Non-white bytes represent active bytes; those in S7 (in red) are the known values and differences from the message pair conforming to the first subpart of the truncated path.

of the AES-columns of A_6 and B_6 (and linearly, to their equivalent A_6' and B_6' with diagonal values in A_7' and B_7') to reach the already-known diagonal values in S7 (red bytes). An alternative way of stating this is: we can construct freely the three first columns of (A_4, A_4') and (B_4, B_4') and deduce deterministically the fourth ones with the next **MixColumns** transition, since 4 out of 8 input or output bytes of **MixColumns** fix the 4 others. Furthermore, this means that if the three first columns of A_1, A_1', B_1 and B_1' are known, then we can learn the values of the remaining columns of S1 (bytes in gray).

We thus search valid input values for the three first **SuperSBoxes** of S1: to do so, we randomize the two differences per AES-column in this state and get valid paired values with probability 2^{-1} in 2^{18} computations with respect to output differences Δ_4 (Section 3.3). Consequently, we can deduce the differences of the same AES-columns in $B_1 + B_1'$ to get a zero sum with S32 after the **BigFinal**. This holds with the same 2^{-1} probability, with respect to Δ_4'. Once we have the three differential transitions for the three first AES-columns of both AES-states, all the corresponding values are then known and we propagate them in A_6, A_6', B_6 and B_6' (black bytes). Since in S7, diagonal values are known, we deduce the remaining byte of each column in A_6, A_6', B_6 and B_6' (gray) and propagate them backwards until S1.

The final step defines the nature of the attack: to get a collision, we check if those constrained values cancel out in the feed-forward, which holds with probability 2^{-32}. Restarting with new random values in S1 and in parallel on the four rows, we find a collision in $2^{18}\, 2^2\, 2^{32} = 2^{52}$ operations on AES-columns. Indeed, we need to repeat 2^{32} times the search of valid paired input values for the **SuperSBox**, which is done in time 2^{18} and succeeds with probability 2^{-2}.

4.5   Final Merging Phase

After we have found message pairs following both subparts of the truncated path so that the merge is possible, we need to finalize the attack by merging the two partial solutions.

In practice, this means finding values for each white bytes in the truncated path and in particular, at the second **SuperMixColumns** transition between S14 and S16. For each of the 16 slices, we get a system of linear equations like (1). In each solution set, each variable only depends on 3 others, and not on *all* the 11 others. This stems from the structured matrix $\mathbf{M}_{SMC}$. For example, in the first slice, we have:

$$L_0(x_0, x_3, x_6, x_9) = c_0 \tag{3}$$
$$L_1(x_1, x_4, x_7, x_{10}) = c_1 \tag{4}$$
$$L_2(x_2, x_5, x_8, x_{11}) = c_2 \tag{5}$$

where L_0, L_1, L_2 are linear functions and c_0, c_1, c_2 constants linearly deduced from the 8 known-values a_i and b_i, $0 \le i \le 3$, of the considered system.

In this phase of the merging process, we also need to set white bytes accordingly to the known values in S7 stemming from the feed-forward. We pick

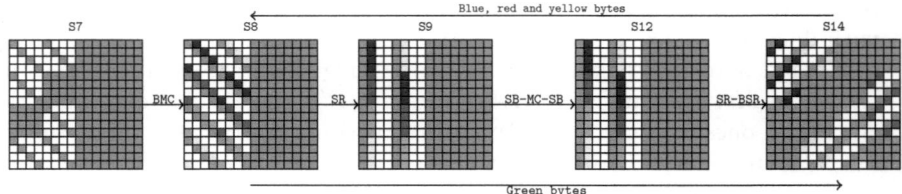

Fig. 11. After randomization of states S7[1,3] and S7[2,2], all values of gray bytes are known. Colors show the flow of values in one step of the merging process.

random values for unset bytes in S7[1,3] and S7[2,2] (Figure 11), such that all values in the two last ECHO-columns of S7 are set. Consequently, by indirectly choosing values for gray bytes in S14, we set the values of half of the unknowns per slice. For example, the system for the first slice becomes:

$$L'_0(x_0, x_3) = c'_0 \tag{6}$$
$$L'_1(x_1, x_4) = c'_1 \tag{7}$$
$$L'_2(x_2, x_5) = c'_2 \tag{8}$$

where L'_0, L'_1, L'_2 are linear functions and c'_0, c'_1, c'_2 some constants.

The three equations (6), (7), (8) are independent, which allows to do the merge in three steps: one on each pair of slices $(1,5)$, $(2,6)$ and $(3,7)$ of S12. Figure 11 represents in color only the first step, on the slice pair $(1,5)$ of S12. We show that each of the three steps can be done in 2^{32} computations and detail only the first step.

Because of the dependencies between bytes within a slice in S14, any choice of blue bytes in S12[0,0] determines blue bytes on S12[1,1] (and the same for yellow and red bytes, Figure 11). In total, we can choose $\left(2^{8 \times 4}\right)^3 = 2^{96}$ different values for the blue, yellow and red AES-columns of state S12. Since we are dealing with values, we propagate them backwards until S8. The **BigMixColumns** transition from S7 to S8 for these two slices imposes the 8 green values in S8[2,0] and S8[3,1]. Going forwards through the **SuperSBox**, we deduce green values in S14 and check whether the four pairs of green bytes satisfy the linear constraints in S14, which occur with probability $\left(2^{-8}\right)^4 = 2^{-32}$. We then have to restart with approximately 2^{32} new blue bytes and random yellow and red ones before satisfying the four constraints simultaneously.

After repeating this step for slices $(2,6)$ and $(3,7)$, we get a valid message pair that follows all the truncated path of Figure 4.

5   Conclusion

In this article, we introduce new results on ECHO-256 compression function reduced to four rounds by describing a collision attack. Our result is the first one which does not need to store the large difference distribution table of the

SuperSBox, which contributes in making the attack practical. We also prove that the latest results by Schläffer on ECHO are flawed and we suggest a way to correct it in some ways. We also improve the time and space complexity of the attack by taking into account more precisely the available degrees of freedom. We describe as well an efficient way to find paired input values conforming to particular truncated differentials through the **SuperSBox** where not all input bytes are active. Finally, we validate our claims by implementing a practical variant of the described attack. We believe this work can lead to new attacks: in particular, the collision attack by Schläffer on ECHO-256 might be corrected using our new techniques.

References

1. Benadjila, R., Billet, O., Gilbert, H., Macario-Rat, G., Peyrin, T., Robshaw, M., Seurin, Y.: SHA-3 proposal: ECHO. Submission to NIST (updated) (2009)
2. Biham, E., Shamir, A.: Differential Cryptanalysis of DES-like Cryptosystems. In: Feigenbaum, J. (ed.) CRYPTO 1991. LNCS, vol. 576. Springer, Heidelberg (1992)
3. Gilbert, H., Peyrin, T.: Super-Sbox Cryptanalysis: Improved Attacks for AES-Like Permutations. In: Hong, S., Iwata, T. (eds.) FSE 2010. LNCS, vol. 6147, pp. 365–383. Springer, Heidelberg (2010)
4. Knudsen, L.R.: Truncated and higher order differentials. In: Preneel, B. (ed.) FSE 1994. LNCS, vol. 1008, pp. 196–211. Springer, Heidelberg (1995)
5. Knudsen, L.R.. Rijmen, V.: Known-Key Distinguishers for Some Block Ciphers. In: Kurosawa, K. (ed.) ASIACRYPT 2007. LNCS, vol. 4833, pp. 315–324. Springer, Heidelberg (2007)
6. Lamberger, M., Mendel, F., Rechberger, C., Rijmen, V., Schläffer, M.: Rebound Distinguishers: Results on the Full Whirlpool Compression Function. In: Matsui, M. (ed.) ASIACRYPT 2009. LNCS, vol. 5912, pp. 126–143. Springer, Heidelberg (2009)
7. Van Le, T., Sparr, R., Wernsdorf, R., Desmedt, Y.G.: Complementation-Like and Cyclic Properties of AES Round Functions. In: Dobbertin, H., Rijmen, V., Sowa, A. (eds.) AES 2005. LNCS, vol. 3373, pp. 128–141. Springer, Heidelberg (2005)
8. Matusiewicz, K., Naya-Plasencia, M., Nikolić, I., Sasaki, Y., Schläffer, M.: Rebound attack on the full LANE compression function. In: Matsui, M. (ed.) ASIACRYPT 2009. LNCS, vol. 5912, pp. 106–125. Springer, Heidelberg (2009)
9. Mendel, F., Peyrin, T., Rechberger, C., Schläffer, M.: Improved Cryptanalysis of the Reduced Grøstl Compression Function, ECHO Permutation and AES Block Cipher. In: Jacobson Jr., M.J., Rijmen, V., Safavi-Naini, R. (eds.) SAC 2009. LNCS, vol. 5867, pp. 16–35. Springer, Heidelberg (2009)
10. Mendel, F., Rechberger, C., Schläffer, M., Thomsen, S.S.: The Rebound Attack: Cryptanalysis of Reduced Whirlpool and Grøstl. In: Dunkelman, O. (ed.) FSE 2009. LNCS, vol. 5665, pp. 260–276. Springer, Heidelberg (2009)
11. National Institute for Science, Technology (NIST):   Advanced E.D.F.-G.D.F.ncryption Standard (FIPS PUB 197) (November 2001)
    http://www.csrc.nist.gov/publications/fips/fips197/fips-197.pdf
12. National Institute of Standards, Technology: Announcing Request for Candidate Algorithm Nominations for a New Cryptographic Hash Algorithm (SHA-3) Family. Technical report, DEPARTMENT OF COMMERCE (November 2007)

126   J. Jean et al.

13. Peyrin, T.: Cryptanalysis of GRINDAHL. In: Kurosawa, K. (ed.) ASIACRYPT 2007. LNCS, vol. 4833, pp. 551–567. Springer, Heidelberg (2007)
14. Peyrin, T.: Improved Differential Attacks for ECHO and Grøstl. In: Rabin, T. (ed.) CRYPTO 2010. LNCS, vol. 6223, pp. 370–392. Springer, Heidelberg (2010)
15. Peyrin, T.: Improved differential attacks for ECHO and grøstl. In: Rabin, T. (ed.) CRYPTO 2010. LNCS, vol. 6223, pp. 370–392. Springer, Heidelberg (2010)
16. Sasaki, Y., Li, Y., Wang, L., Sakiyama, K., Ohta, K.: Non-full-active Super-sbox Analysis: Applications to ECHO and Grøstl. In: Abe, M. (ed.) ASIACRYPT 2010. LNCS, vol. 6477, pp. 38–55. Springer, Heidelberg (2010)
17. Schläffer, M.: Subspace Distinguisher for 5/8 Rounds of the ECHO-256 Hash Function (2010); To appear in Selected Areas in Cryptography 2010, Proceedings
18. Wang, X., Lai, X., Feng, D., Chen, H., Yu, X.: Cryptanalysis of the Hash Functions MD4 and RIPEMD. In: Cramer, R. (ed.) EUROCRYPT 2005. LNCS, vol. 3494, pp. 1–18. Springer, Heidelberg (2005)
19. Wang, X., Yin, Y.L., Yu, H.: Finding Collisions in the Full SHA-1. In: Shoup, V. (ed.) CRYPTO 2005. LNCS, vol. 3621, pp. 17–36. Springer, Heidelberg (2005)
20. Wang, X., Yu, H.: How to Break MD5 and Other Hash Functions. In: Cramer, R. (ed.) EUROCRYPT 2005. LNCS, vol. 3494, pp. 19–35. Springer, Heidelberg (2005)
21. Wang, X., Yu, H., Yin, Y.L.: Efficient Collision Search Attacks on SHA-0. In: Shoup, V. (ed.) CRYPTO 2005. LNCS, vol. 3621, pp. 1–16. Springer, Heidelberg (2005)

A   Merging Process in Detail

An instance of the problem to solve is the following: given a_0, a_1, a_2, a_3, b_0, b_1, b_2, $b_3 \in \mathrm{GF}\left(2^8\right)$, find x_0, x_1, x_2, x_3, x_4, x_5, x_6, x_7, x_8, x_9, x_{10}, $x_{11} \in \mathrm{GF}\left(2^8\right)$ such that:

$$\mathbf{M}_{\mathrm{SMC}} \begin{bmatrix} a_0 & x_0 & x_1 & x_2 & a_1 & x_3 & x_4 & x_5 & a_2 & x_6 & x_7 & x_8 & a_3 & x_9 & x_{10} & x_{11} \end{bmatrix}^{\mathrm{T}} = \tag{9}$$
$$\begin{bmatrix} b_0 & b_1 & b_2 & b_3 & * & * & * & * & * & * & * & * & * & * & * & * \end{bmatrix}^{\mathrm{T}}$$

where $*$ is any value in $\mathrm{GF}\left(2^8\right)$. Since we are only interested in the four first output values (the problem is similar for others slices), we do not take into consideration the lines other than the four first ones. Let $\mathbf{M}_{\mathrm{SMC}}|_{0,1,2,3}$ be that matrix. The system to be solved can be rewritten as ($\mathbf{M}_{\mathrm{SMC}}|_{0,1,2,3}^{j}$ is the matrix composed of rows $0,1,2,3$ and column j from $\mathbf{M}_{\mathrm{SMC}}$):

$$\mathbf{M}_{\mathrm{SMC}}|_{0,1,2,3}^{1,2,3,5,6,7,9,10,11,13,14,15} \begin{bmatrix} x_0 & x_1 & x_2 & x_3 & x_4 & x_5 & x_6 & x_7 & x_8 & x_9 & x_{10} & x_{11} \end{bmatrix}^{\mathrm{T}} =$$
$$\mathbf{M}_{\mathrm{SMC}}|_{0,1,2,3}^{0,4,8,12} \begin{bmatrix} a_0 \\ a_1 \\ a_2 \\ a_3 \end{bmatrix} + \begin{bmatrix} b_0 \\ b_1 \\ b_2 \\ b_3 \end{bmatrix}$$

$$\tag{10}$$

Now, we make the assumption that at least one solution to the problem exists. This means that the right-hand side of (10) lies in the image of the matrix $\mathbf{M}_{\mathrm{SMC}}|_{0,1,2,3}^{1,2,3,5,6,7,9,10,11,13,14,15}$ from the left-hand side. Because the matrix

$\mathbf{M_{SMC}}$ is a Kronecker product of $\mathbf{M}$ with itself, $\mathbf{M_{SMC}}|_{0,1,2,3}^{1,2,3,5,6,7,9,10,11,13,14,15}$,
$\mathbf{M_{SMC}}|_{0,1,2,3}^{9,10,11,13,14,15}$ and $\mathbf{M_{SMC}}|_{0,1,2,3}^{1,2,3,5,6,7}$ share the same image, described by:

$$S_0 = \{ \quad [t_0, t_1, t_2, L(t_0, t_1, t_2)], t_0, t_1, t_2 \in GF(2^8) \quad \} \tag{11}$$

where $L(t_0, t_1, t_2) = 247t_0 + 159t_1 + 38t_2$. Finally, if a solution exists, this means that:

$$\underbrace{\begin{bmatrix} 4 & 6 & 2 & 2 \\ 2 & 3 & 1 & 1 \\ 2 & 3 & 1 & 1 \\ 6 & 5 & 3 & 3 \end{bmatrix}}_{\mathbf{M_{SMC}}|_{0,1,2,3}^{0,4,8,12}} \begin{bmatrix} a_0 \\ a_1 \\ a_2 \\ a_3 \end{bmatrix} + \begin{bmatrix} b_0 \\ b_1 \\ b_2 \\ b_3 \end{bmatrix} \in S_0 \tag{12}$$

In other words, this means that the following equality is true:

$$14b_0 + 11b_1 + 13b_2 + 9b_3 = 2a_0 + 3a_1 + a_2 + a_3. \tag{13}$$

The given parameters a_0, a_1, a_2, a_3, b_0, b_1, b_2, b_3 are then constrained on an 8-bit condition. The converse is then: if this relation is not satisfied, then the problem has no solution.

We took the example of the very first slice, but the problem is similar for the 16 different slices in S14/S16. Namely, per slice, parameters need to satisfy the following equalities:

Slice	Condition
0	$14b_0 + 11b_1 + 13b_2 + 9b_3 = 2a_0 + 3a_1 + a_2 + a_3$
1	$11b_0 + 13b_1 + 9b_2 + 14b_3 = 2a_0 + 3a_1 + a_2 + a_3$
2	$13b_0 + 9b_1 + 14b_2 + 11b_3 = 2a_0 + 3a_1 + a_2 + a_3$
3	$9b_0 + 14b_1 + 11b_2 + 13b_3 = 2a_0 + 3a_1 + a_2 + a_3$
4	$14b_0 + 11b_1 + 13b_2 + 9b_3 = a_0 + 2a_1 + 3a_2 + a_3$
5	$11b_0 + 13b_1 + 9b_2 + 14b_3 = a_0 + 2a_1 + 3a_2 + a_3$
6	$13b_0 + 9b_1 + 14b_2 + 11b_3 = a_0 + 2a_1 + 3a_2 + a_3$
7	$9b_0 + 14b_1 + 11b_2 + 13b_3 = a_0 + 2a_1 + 3a_2 + a_3$
8	$14b_0 + 11b_1 + 13b_2 + 9b_3 = a_0 + a_1 + 2a_2 + 3a_3$
9	$11b_0 + 13b_1 + 9b_2 + 14b_3 = a_0 + a_1 + 2a_2 + 3a_3$
10	$13b_0 + 9b_1 + 14b_2 + 11b_3 = a_0 + a_1 + 2a_2 + 3a_3$
11	$9b_0 + 14b_1 + 11b_2 + 13b_3 = a_0 + a_1 + 2a_2 + 3a_3$
12	$14b_0 + 11b_1 + 13b_2 + 9b_3 = 3a_0 + a_1 + a_2 + 2a_3$
13	$11b_0 + 13b_1 + 9b_2 + 14b_3 = 3a_0 + a_1 + a_2 + 2a_3$
14	$13b_0 + 9b_1 + 14b_2 + 11b_3 = 3a_0 + a_1 + a_2 + 2a_3$
15	$9b_0 + 14b_1 + 11b_2 + 13b_3 = 3a_0 + a_1 + a_2 + 2a_3$

The main problem in the reasoning of [17] is to assume that a solution exists, while for some parameters, there is no solution.

In the end, if the condition is verified we can choose x_0, x_1, x_2, x_3, x_4, x_5, x_6, x_7, x_8 freely and determine x_9, x_{10}, x_{11} afterwards. If a solution exists, there are $(2^8)^9 = 2^{72}$ solutions to the problem. Taking any other slice leads to a very similar description of the set of solutions, with the same kind of dependencies between the variables.

On Cipher-Dependent Related-Key Attacks in the Ideal-Cipher Model

M.R. Albrecht[1], P. Farshim[2], K.G. Paterson[2], and G.J. Watson[3]

[1] SALSA Project -INRIA, UPMC, Univ. Paris 06
malb@lip6.fr
[2] Information Security Group, Royal Holloway, University of London
{pooya.farshim, kenny.paterson}@rhul.ac.uk
[3] Department of Computer Science, University of Calgary
gjwatson@ucalgary.ca

Abstract. Bellare and Kohno introduced a formal framework for the study of related-key attacks against blockciphers. They established sufficient conditions (output-unpredictability and collision-resistance) on the set of related-key-deriving (RKD) functions under which an ideal cipher is secure against related-key attacks, and suggested this could be used to derive security goals for real blockciphers. However, to do so requires the reinterpretation of results proven in the ideal-cipher model for the standard model (in which a blockcipher is modelled as, say, a pseudorandom permutation family). As we show here, this is a fraught activity. In particular, building on a recent idea of Bernstein, we first demonstrate a related-key attack that applies generically to a large class of blockciphers. The attack exploits the existence of a short description of the blockcipher, and so does not apply in the ideal-cipher model. However, the specific RKD functions used in the attack are provably output-unpredictable and collision-resistant. In this sense, the attack can be seen as a separation between the ideal-cipher model and the standard model. Second, we investigate how the related-key attack model of Bellare and Kohno can be extended to include sets of RKD functions that themselves access the ideal cipher. Precisely such related-key functions underlie the generic attack, so our extended modelling allows us to capture a larger universe of related-key attacks in the ideal-cipher model. We establish a new set of conditions on related-key functions that is sufficient to prove a theorem analogous to the main result of Bellare and Kohno, but for our extended model. We then exhibit non-trivial classes of practically relevant RKD functions meeting the new conditions. We go on to discuss standard model interpretations of this theorem, explaining why, although separations between the ideal-cipher model and the standard model still exist for this setting, they can be seen as being much less natural than our previous separation. In this manner, we argue that our extension of the Bellare–Kohno model represents a useful advance in the modelling of related-key attacks. In the full version of the paper, we also consider the topic of key-recovering related-key attacks and its relationship to the Bellare–Kohno formalism. In particular, we address the question of whether lowering the security goal by requiring the adversary to perform key-recovery excludes separations of the type exhibited by us in the Bellare–Kohno model.

Keywords: Related-key attack, Ideal-cipher model, Blockcipher.

A. Joux (Ed.): FSE 2011, LNCS 6733, pp. 128–145, 2011.
© International Association for Cryptologic Research 2011

1   Introduction

BACKGROUND. Related-key attacks were introduced by Biham and Knudsen [7,8,20], and have received considerable attention recently partly due to the discovery of various high-profile key-recovery attacks in this model ([9,10,14]). Some of these new attacks, in particular the family of attacks against AES, do not restrict key-derivation functions to either simple XORs or modular addition of constants. Instead non-linear key-derivation functions are used. This has sparked a debate as to whether these attacks should be considered valid, and in turn whether related-key attacks should be considered valid attacks on blockciphers in general. Part of the debate stems from the question of whether the job of preventing related-key attacks should fall to blockcipher designers or to designers of protocols making use of blockciphers. The latter group could put a stop to such attacks simply by avoiding the use of related keys within their protocols, and this in turn would remove any real incentive for cryptanalysts to consider ever more esoteric key relations. However, taking a pragmatic perspective, there are widely deployed real-world protocols which do make use of such related keys, so the study of related-key attacks holds relevance and interest both from cryptanalytic and theoretical perspectives. For example, key-derivation procedures leading to related-key scenarios seem to be widely used in the financial sector, with a public-domain example being the EMV specifications for card transactions [15, Appendix A1.3.1]. Other examples include the 3GPP confidentiality and integrity algorithms f8,f9 [19].

On the theoretical side, Bellare and Kohno [4] provided a thorough study of related-key attacks. Their main result established a general possibility result concerning security against related-key attacks, for certain classes of related-key-deriving (RKD) functions. Bellare and Kohno have as a thesis that the minimal requirement for a blockcipher security goal to be considered feasible is that it should be provably achievable for an ideal cipher. To this end, they showed that an ideal cipher is secure against related-key attacks involving any set of RKD functions that is both *collision-resistant* and *output-unpredictable*. However, to be usable in studying the security of real blockciphers, we need to be able to interpret such ideal-cipher-model results in the standard model, in which we might model a blockcipher as a pseudorandom permutation family. We note that [4] contains very little in the way of such interpretation.

However, the community's confidence in our ability to translate such results to the standard model has recently received a severe dent. In [18], Harris demonstrated that if the cipher itself is available for use during key derivation, then RKD functions can be constructed using which keys can be recovered for any cipher. Bernstein [6] presented a simple distinguishing attack on AES that also made use of the blockcipher itself in the RKD functions. Moreover, at least heuristically, the sets of RKD functions used in these attacks fulfil the conditions of collision-resistance and output-unpredictability needed to prove Bellare and Kohno's main result about security against related-key attacks. Researchers subsequently argued that, in view of these examples, the model for related-key attacks presented in [4] is broken, in the sense that, since any cipher can be broken in that model, then this model does not tell us anything about ciphers; rather it is simply too strong a model.

CONTRIBUTIONS. We begin by exploring the question of how to interpret the main result of Bellare and Kohno [4], restated here as Theorem 1, in the standard model. We provide two possible interpretations of this result, which vary only in the order in which they invoke certain quantifiers. We then formalise Bernstein's attack as a related-key attack that applies generically to a large class of blockciphers (those having equal-sized keys and messages). Moreover, we formally prove, under the standard assumption that the blockcipher is pseudorandom, that Bernstein's RKD functions meet the sufficient conditions of collision-resistance and output-unpredictability needed for the application of Theorem 1. We then explain how this attack can be seen as a separation between the ideal-cipher model and the standard model in the context of the second of our two interpretations of Theorem 1. We also justify why the first interpretation of Theorem 1 for the standard model is less interesting in practical contexts.

In an attempt to restore confidence in the Bellare–Kohno model, we extend the model to allow RKD functions which access the blockcipher itself. Since we are working with an ideal cipher, we model such access via oracle calls to the ideal cipher and its inverse. This allows us to do several things. Firstly, we can capture attacks like that due to Bernstein in our model (where it shows up as an attack against an ideal cipher for a particular set of RKD functions). Secondly, it allows us to prove the security of an ideal cipher for other sets of RKD functions which make use of the blockcipher during key derivation. Thirdly, it allows us to investigate analogues of Theorem 1 for the new setting. This leads to our main result, Theorem 4, in which we establish that an ideal cipher is secure against related-key attacks for sets of RKD functions that meet certain conditions. More precisely, we introduce oracle versions of collision-resistance and output-unpredictability, along with a new notion called *oracle-independence* of a set of RKD functions; we then show that these three conditions taken together are sufficient to establish the security of an ideal cipher against related-key attacks using that set of RKD functions. We go on to show that our main theorem is not vacuous by exhibiting non-trivial classes of practically relevant RKD functions meeting the new conditions. In particular, we show that RKD function sets like those used in the EMV standard meet the new conditions.

Given the problems we have identified with making standard model interpretations of Theorem 1, we then proceed to a careful discussion of how our main result, Theorem 4, can be translated into the standard model. When restricted to RKD sets which are independent of the blockcipher, our theorem becomes equivalent to that of Bellare and Kohno: its interpretation states that a reasonable blockcipher should resist related-key attacks when restricted to such an RKD set. On the other hand, for RKD sets which depend on the blockcipher, our theorem goes beyond that of Bellare and Kohno (which provides no guarantees) in the following way. Its interpretation asserts that if the dependency of the RKD functions is black box, and furthermore the set satisfies certain conditions, then a good blockcipher is expected to resist related-key attacks when restricted to such an RKD set. In particular, the RKD sets of Bernstein and Harris do not satisfy the required conditions. On the positive side, there exist cipher-dependent RKD sets which satisfy the required conditions.

Our final contribution is to ask whether the problems that arise in translating from the ideal-cipher model to the standard model in the context of related-key attacks can be avoided by lowering our sights. In particular, we consider the topic of related-key attacks that recover keys (rather than breaking the pseudorandomness of a cipher in the sense considered in [4]). This asks more of the adversary and therefore represents a weakening of the model. In turn, this opens up the possibility of excluding separation results like that we have shown. We can in fact show that the particular set of RKD functions used in Bernstein's attack *cannot* be used to mount a key-recovery attack. Unfortunately, we also have a negative result: using a modification of the attack of Harris, in the full version of the paper [1] we exhibit a specific set of RKD functions that does lead to a full key-recovery attack against real blockciphers, even though the functions satisfy the conditions for Theorem 1 to be applicable. Again, the RKD functions access the blockcipher itself, so the attack can be regarded as another separation between the ideal-cipher model and the standard model, but now for a weaker security notion than was originally considered in [4].

OTHER RELATED WORK. Bellare and Kohno also gave constructions of concrete blockciphers which are secure against adversaries which only partially transform the key. In subsequent work, Lucks [21] investigated RKA-secure blockciphers further, and gave improved security bounds for such partially key-transforming adversaries. In this work, the author also constructed a concrete blockcipher which is RKA-secure with respect to a rich set of related-key-deriving functions. Lucks's construction, however, was based on a non-standard, interactive number-theoretic assumption. The recent work of Goldenberg and Liskov [16] examines whether it is possible to build related-key-secure blockciphers from traditional cryptographic primitives. They show that while a related-key/secret pseudorandom bit is sufficient and necessary to build such a blockcipher, hard-core bits with typical security proofs are not related-secret secure. Very recently, Bellare and Cash [3] managed to construct PRFs and PRPs which are RKA secure with respect to key transformations which involve the action of a group element on the key. Their constructions are based on standard number-theoretic assumptions such as DDH. In yet another recent work, Applebaum, Harnik and Ishai [2] study related-key attacks for randomised symmetric encryption schemes. They also discuss the applications of such RKA-secure primitives to batch and adaptive oblivious transfer protocols.

KDM SECURITY. Key-dependent message (KDM) security [12,17] is a strong notion for primitives such as private/public-key encryption schemes and PRF/PRPs where one requires security in the presence of an adversary which can obtain the outputs of the encryption/function on points which depend, in known (or even chosen) ways, on secret keys. This setting is similar to related-key attacks in the sense that security games involve functions of an unknown key. However, while superficially similar in this sense, the RKA and KDM notions hand different capabilities to the adversary. A fuller discussion of the relations between these notions is beyond the scope of the present paper. In [17], the authors briefly define cipher-dependent KDM security, however their results are about relations which are *independent* of the cipher. We note that analogues of Bernstein's and Harris's attack in the context of KDM security were already noted in [17].

ORGANISATION. In the next section we settle notation and recall a number of definitions from [4]. Section 3 is concerned with the possible interpretations of the main result of [4] in the standard model. In Section 4 we extend the security model of Bellare and Kohno to include RKD sets that access the ideal cipher itself during key derivation. We also discuss some positive and negative results in this new model. We close by discussing the relevance of our results to practice.

2   Notation and Related-Key Attacks

NOTATION. We denote by $s \xleftarrow{\$} S$ the operation of sampling s uniformly at random from set S, and by $x \leftarrow y$ the assignment of value y to x. For a set S, $|S|$ denotes its size. We let $\mathrm{Perm}(\mathcal{D})$ denote the set of all permutations on $\mathcal{D}$. A blockcipher is a family of permutations $E : \mathcal{K} \times \mathcal{D} \to \mathcal{D}$, where $\mathcal{K}$ is the key space and $\mathcal{D}$ is the domain or message space.

We recall a number of definitions from [4].

Definition 1 (Pseudorandomness). *Let $E : \mathcal{K} \times \mathcal{D} \to \mathcal{D}$ be a family of functions. Let A be an adversary. Then*

$$\mathbf{Adv}_E^{\mathrm{prp}}(A) := \Pr\left[K \xleftarrow{\$} \mathcal{K} : A^{E(K,\cdot)} = 1 \right] - \Pr\left[G \xleftarrow{\$} \mathrm{Perm}(\mathcal{D}) : A^{G(\cdot)} = 1 \right]$$

is defined as the prp-advantage *of A against E.*

We let $\mathrm{Perm}(\mathcal{K}, \mathcal{D})$ denote the set of all blockciphers with domain $\mathcal{D}$ and key-space $\mathcal{K}$. Thus the notation $G \xleftarrow{\$} \mathrm{Perm}(\mathcal{K}, \mathcal{D})$ corresponds to selecting a random blockcipher. In more detail, it comes down to defining G via

$$\text{For each } K \in \mathcal{K} : G(K, \cdot) \xleftarrow{\$} \mathrm{Perm}(\mathcal{D}).$$

Given a family of functions $E : \mathcal{K} \times \mathcal{D} \to \mathcal{D}$ and a key $K \in \mathcal{K}$, we define the related-key oracle $E(\mathrm{RK}(\cdot, K), \cdot)$ as an oracle that takes two arguments, a function $\phi : \mathcal{K} \to \mathcal{K}$ and an element $x \in \mathcal{D}$, and returns $E(\phi(K), x)$. We shall refer to ϕ as a related-key-deriving (RKD) function. We let Φ be a set of functions mapping $\mathcal{K}$ to $\mathcal{K}$. We call Φ the set of allowed RKD functions and it will be a parameter of our definitions.

Definition 2 (Pseudorandomness with respect to related-key attacks). *Let $E : \mathcal{K} \times \mathcal{D} \to \mathcal{D}$ be a family of functions and let Φ be a set of RKD functions over $\mathcal{K}$. Let A be an adversary with access to a related-key oracle, and restricted to queries of the form (ϕ, x) in which $\phi \in \Phi$ and $x \in \mathcal{D}$. Then*

$$\mathbf{Adv}_{\Phi,E}^{\mathrm{prp\text{-}rka}}(A) := \Pr\left[K \xleftarrow{\$} \mathcal{K} : A^{E(\mathrm{RK}(\cdot, K), \cdot)} = 1 \right]$$
$$- \Pr\left[K \xleftarrow{\$} \mathcal{K}; G \xleftarrow{\$} \mathrm{Perm}(\mathcal{K}, \mathcal{D}) : A^{G(\mathrm{RK}(\cdot, K), \cdot)} = 1 \right]$$

is defined as the prp-rka-advantage *of A in a Φ-restricted related-key attack (RKA) on E.*

Therefore in a related-key attack an adversary's success rate is measured by its ability to distinguish values of the cipher on related-keys from those returned from a random blockcipher.

Definition 3 (RKA pseudorandomness in the ideal-cipher model). *Fix sets $\mathcal{K}$ and $\mathcal{D}$ and let Φ be a set of RKD functions over $\mathcal{K}$. Let A be an adversary with access to three oracles, and restricted to queries of the form (K', x) for the first two oracles and (ϕ, x) for the last, where $K' \in \mathcal{K}$, $\phi \in \Phi$ and $x \in \mathcal{D}$. Then*

$$\mathbf{Adv}_{\Phi,\mathcal{K},\mathcal{D}}^{\mathsf{prp\text{-}rka}}(A) := \Pr\left[K \xleftarrow{\$} \mathcal{K}; E \xleftarrow{\$} \mathrm{Perm}(\mathcal{K}, \mathcal{D}) : A^{E, E^{-1}, E(\mathsf{RK}(\cdot, K), \cdot)} = 1\right]$$

$$- \Pr\left[K \xleftarrow{\$} \mathcal{K}; E \xleftarrow{\$} \mathrm{Perm}(\mathcal{K}, \mathcal{D}); G \xleftarrow{\$} \mathrm{Perm}(\mathcal{K}, \mathcal{D}) : A^{E, E^{-1}, G(\mathsf{RK}(\cdot, K), \cdot)} = 1\right]$$

is defined as the prp-rka-*advantage of A in a Φ-restricted related-key attack on an ideal cipher with keys $\mathcal{K}$ and domain $\mathcal{D}$.*

This definition is simply an adaptation of Definition 2 to the ideal-cipher model by allowing oracle access to E and E^{-1}. To de-clutter the notation, we use E and E^{-1} as shorthand for $E(\cdot, \cdot)$ and $E^{-1}(\cdot, \cdot)$, respectively. An $E^{-1}(\mathsf{RK}(\cdot, K), \cdot)$ oracle can be added to the above definition to get *strong RKA pseudorandomness*. In this paper, however, we will work with the standard (i.e. non-strong) pseudorandomness. Our results can be extended to the strong setting.

Definition 4 (Output-unpredictability-2). *Let Φ be a set of RKD functions on the keyspace $\mathcal{K}$. Let $\mathcal{P}_K(\cdot)$ and $\mathcal{X}(\cdot)$ be a pair of oracles. The oracle $\mathcal{P}_K(\cdot)$ takes as input an element $\phi \in \Phi$ and the oracle $\mathcal{X}(\cdot)$ takes as input an element $K' \in \mathcal{K}$. Neither oracle returns a value. An adversary wins if it queries its $\mathcal{X}(\cdot)$ oracle with a key K' and if it queries its $\mathcal{P}_K(\cdot)$ oracle with a function ϕ such that $\phi(K) = K'$. We define the* up2-*advantage of an adversary A as*

$$\mathbf{Adv}_{\Phi}^{\mathsf{up2}}(A) := \Pr\left[K \xleftarrow{\$} \mathcal{K} : A^{\mathcal{P}_K(\cdot), \mathcal{X}(\cdot)} \text{ wins}\right].$$

The above definition captures the intuition that no adversary is able to predict the value $\phi(K)$, for a random K, with a high probability.

Definition 5 (Collision-resistance-2). *Let Φ be a set of functions on the key-space $\mathcal{K}$. Let $\mathcal{C}_K(\cdot)$ be an oracle that takes as input a function $\phi \in \Phi$ and that returns no value. An adversary wins if it queries its oracle with two distinct functions $\phi_1, \phi_2 \in \Phi$ such that $\phi_1(K) = \phi_2(K)$. We define the* cr2-*advantage of an adversary A as*

$$\mathbf{Adv}_{\Phi}^{\mathsf{cr2}}(A) := \Pr\left[K \xleftarrow{\$} \mathcal{K} : A^{\mathcal{C}_K(\cdot)} \text{ wins}\right].$$

The intuition here is that no adversary can trigger a collision between two different ϕ's with high probability. Note also that output-unpredictability is simply collision-resistance between a non-constant and a constant function. Throughout the paper we call an RKD set Φ output-unpredictable-2 or collision-resistant-2 if the corresponding advantage is "small" for efficient any adversary.

REMARK. Alternative and stronger notions of output-unpredictability and collision-resistance are also presented in [4]. However, the above definitions are enough for the main result there. We note that an attractive feature of the above definitions is their *non-interactiveness*. In fact, it is possible to simplify these definitions further by requiring an adversary which returns a single pair (K, ϕ) in the output-unpredictability-2 game, and two distinct RKD functions (ϕ_1, ϕ_2) in the collision-resistance-2 game. Using a standard reduction one can show that the simplified definitions are equivalent to the above definitions (respectively). In the first case a (multiplicative) security loss of $qq'/2$, where q and q' are, respectively, the number of queries to the $\mathcal{X}$ and $\mathcal{P}_K$ oracles, is introduced. In the second case, a loss of $q(q-1)/2$ is introduced, where q is the number of queries to the $\mathcal{C}_K$ oracle.

3   A Generic Cipher-Dependent Attack

Bellare and Kohno established the following theorem as their main result in [4].

Theorem 1 (Bellare and Kohno [4]). *Fix a key space $\mathcal{K}$ and domain $\mathcal{D}$. Let Φ be a set of RKD functions over $\mathcal{K}$. Let A be an ideal-cipher-model adversary that queries its first two oracles with a total of at most q' different keys and that queries its last oracle with a total of at most q different RKD functions from Φ. Then there are output-unpredictability-2 and collision-resistance-2 adversaries B and C such that*

$$\mathbf{Adv}^{\mathsf{prp\text{-}rka}}_{\Phi, \mathcal{K}, \mathcal{D}}(A) \leq \mathbf{Adv}^{\mathsf{cr2}}_{\Phi}(B) + \mathbf{Adv}^{\mathsf{up2}}_{\Phi}(C),$$

where B queries its $\mathcal{C}_K$ oracle q times, and C queries its $\mathcal{P}_K$ and $\mathcal{X}$ oracles q and q' times (respectively).

The above theorem states that for all Φ satisfying appropriate properties, an ideal cipher is secure against Φ-restricted related-key attacks. It is tempting to try to translate this ideal-cipher-model result to the standard model. Indeed, it is conceivable that a real blockcipher might also resist such Φ-restricted attacks under the same conditions. This statement can be interpreted in (at least) two ways.

1. For any Φ which is collision-resistant and output-unpredictable, there is a standard model blockcipher which resists Φ-restricted attacks; and
2. There is a standard model blockcipher E which resists all Φ-restricted attacks, as long as Φ is collision-resistant and output-unpredictable.

The essential difference between these two interpretations is their *order of quantifiers*. In the first interpretation, there may be no dependencies of Φ on E, whereas the blockcipher in the second interpretation should resist all Φ-restricted attacks, including those which depend on E. The theorem of Bellare and Kohno, on the other hand, does not allow the functions in Φ to depend on the ideal cipher itself, as the latter is chosen uniformly at random and independently from Φ. Therefore, the first interpretation is, in our opinion, a more accurate translation of the theorem to the standard model. In fact no natural counterexamples to this interpretation are yet known.[1] On the other hand, based

---

[1] Although artificial code-based separation results akin to that in [11] might be constructible.

on the aforementioned recent example of Bernstein, we show in the next theorem that the second interpretation is *invalid*. This result utilises RKD sets Δ_E, which depend on E, for each blockcipher E. The proof is given in the full version of the paper [1].

Theorem 2. *Let $E : \mathcal{K} \times \mathcal{D} \to \mathcal{D}$ be a family of functions with $\mathcal{K} = \mathcal{D}$. Let $0, 1 \in \mathcal{D}$ be any two distinct elements of $\mathcal{D}$ and consider the set of RKD functions*

$$\Delta_E := \{K \mapsto K, K \mapsto E(K, 0)\}.$$

Then there are a Δ_E-restricted related-key adversary A against E, and a prp adversary B against E such that

$$\mathbf{Adv}^{\mathsf{prp\text{-}rka}}_{\Delta_E, E}(A) \geq 1 - \mathbf{Adv}^{\mathsf{prp}}_E(B) - 2/|\mathcal{K}|.$$

Furthermore, for any collision-resistant-2 or output-unpredictable-2 adversary A, there is a prp adversary B such that

$$\mathbf{Adv}^{\mathsf{up2}}_{\Delta_E}(A) \leq \mathbf{Adv}^{\mathsf{prp}}_E(B) + 2q'/|\mathcal{K}| \text{ and } \mathbf{Adv}^{\mathsf{cr2}}_{\Delta_E}(A) \leq \mathbf{Adv}^{\mathsf{prp}}_E(B) + 1/|\mathcal{K}|,$$

where q' is the number of queries that an output-unpredictability-2 adversary A makes to its $\mathcal{X}$ oracle.

Hence if the blockcipher E is prp secure, then Δ_E is both collision-resistant-2 and output-unpredictable-2. This theorem therefore exhibits a class of ciphers E for which the second standard model interpretation of Theorem 1 does not hold. Note that we have in fact established a strong falsification of the second interpretation as it is enough to show that the inequality in the statement of Theorem 1 does not hold in the standard model. Note also that the inequalities in the above theorem can be somewhat simplified by observing that prp-rka security with respect to Δ_E-restricted adversaries implies prp security.

Note that in the Δ_E set, one can replace 0 with any $x \in \mathcal{D}$. Furthermore, there is no special role played by the identity function as a similar attack applies if the set was defined to be $\{K \mapsto E(K, 0), K \mapsto E(E(K, 0), 0)\}$. Note also that no efficiency requirements on RKD functions are made in Theorem 1, and the result holds even for ϕ that are infeasible to compute (in the ideal-cipher model). This allows us to define an RKD set containing a single function which allows an attacker to recover the key of a concrete cipher: $K \mapsto K'$ with K' such that $E(K', 0) = K$. We stress that this failure of the model, although technically allowed in the model of [4], is only of theoretical interest: the ability to compute this function would immediately break the prp-security of the cipher.

Harris [18] presents another cipher-dependent related-key attack which breaks every cipher in the standard model. In the full version of the paper [1] we formalise this attack and study its implications. In particular, the description of the RKD set is unclear in the original work, and depending on the interpretation, the set might or might not satisfy the collision-resistance-2 property. We clarify this issue by deriving accurate bounds for the advantage of a related-key adversary.

Theorem 2 can be seen as a *weak* separation between the standard model and the ideal-cipher model as it (only) rules out the second interpretation. It remains an open problem to prove or disprove the first interpretation. Disproving it would demonstrate a *strong* separation result as it also implies the weak separation.[2] We, however, do not consider this to be an important issue since, in a real-world attack, the attacker can choose its set of RKD functions after seeing the cipher, which relates more closely to the second interpretation. Put differently, at the core of the above attacks lies the dependence of the RKD set on E, which cannot be replicated in the Bellare–Kohno (BK) model.

Most concrete related-key attacks [9,10,14] lead to the recovery of a blockcipher's key. Hence one way to restore confidence in the BK model would be to raise the security bar for an attacker, and require it to recover keys. We formalise key-recovery in the full version of the paper [1] and show that this approach cannot succeed.

In the next section, we investigate how the model can be modified so as to capture such cipher-dependent related-key attacks.

4   RKD Functions with Access to E and E^{-1}

As discussed above, one weakness of the BK model lies in its inability to model related-key functions which depend on the blockcipher. In this section, we extend the BK model to address this issue and prove a result akin to Theorem 1 for this extended setting. In doing so, we treat the RKD functions as being oracle Turing machines and write each RKD function as $\phi^{\mathcal{O}_1,\mathcal{O}_2}$, where $\mathcal{O}_i$ are oracles. These oracles will be instantiated with a random blockcipher E and its inverse E^{-1} during security games. We denote a set of such oracle RKD functions by[3] $\Phi^{E,E^{-1}}$. We note that such oracle RKD functions are of interest in the ideal-cipher model only: a concrete blockcipher has a compact description and there is no need to grant access to the oracle.

We are now ready to define a refined notion of RKA pseudorandomness in the ideal-cipher model.

Definition 6 (Oracle RKA pseudorandomness in the ideal-cipher model). *Fix sets $\mathcal{K}$ and $\mathcal{D}$, and let $\Phi^{E,E^{-1}}$ be a set of oracle RKD functions over $\mathcal{K}$. Let A be an adversary with access to three oracles, and restricted to queries of the form (K', x) for the first two oracles and $(\phi^{E,E^{-1}}, x)$ for the last, where $K' \in \mathcal{K}$, $\phi^{E,E^{-1}} \in \Phi^{E,E^{-1}}$, and $x \in \mathcal{D}$. Then*

$$\mathbf{Adv}^{\mathrm{prp\text{-}orka}}_{\Phi^{E,E^{-1}},\mathcal{K},\mathcal{D}}(A) := \Pr\left[K \xleftarrow{\$} \mathcal{K}; E \xleftarrow{\$} \mathrm{Perm}(\mathcal{K},\mathcal{D}) : A^{E,E^{-1},E(\mathrm{RK}(\cdot,K),\cdot)} = 1\right]$$

$$- \Pr\left[K \xleftarrow{\$} \mathcal{K}; E \xleftarrow{\$} \mathrm{Perm}(\mathcal{K},\mathcal{D}); G \xleftarrow{\$} \mathrm{Perm}(\mathcal{K},\mathcal{D}) : A^{E,E^{-1},G(\mathrm{RK}(\cdot,K),\cdot)} = 1\right]$$

is defined as the prp-orka*-advantage of A in a $\Phi^{E,E^{-1}}$-restricted related-key attack on an ideal cipher with keys $\mathcal{K}$ and domain $\mathcal{D}$.*

---

[2] We remark that Proposition 9.1 of [4] demonstrates an intermediate result: it considers a restricted set of RKD functions which only alter the last few bits of the key, but may depend on the cipher.

[3] Although we use E and E^{-1} in the exponent, in the following security definitions they are chosen during each game.

Informally, we say that an ideal cipher is secure against $\Phi^{E,E^{-1}}$-restricted related-key attacks if the above advantage is "small" for any efficient A.

We now define a set of oracle RKD functions, which is the ideal-cipher model counterpart[4] of Δ_E defined in Theorem 2, as follows:

$$\Delta^E := \{K \mapsto K, K \mapsto E(K,0)\}.$$

The next theorem shows that this set can be used to break an ideal cipher in the sense of Definition 6.

Theorem 3. *Fix a set $\mathcal{K}$ and let $\mathcal{D} = \mathcal{K}$. Then there exists the ideal-cipher model Δ^E-restricted adversary A such that*

$$\mathbf{Adv}^{\mathsf{prp\text{-}orka}}_{\Delta^E, \mathcal{K}, \mathcal{D}}(A) \geq 1 - 2/|\mathcal{K}|.$$

Proof. The proof of this theorem is similar to that of Theorem 2. Adversary $A^{E,E^{-1},f(\mathsf{RK}(\cdot,K),\cdot)}$, whose goal is to decide whether $f = E$ or $f = G$, operates as shown in Figure 1.

Algorithm $A^{E,E^{-1},f}$:
Query RK oracle on $(K \mapsto K, 0)$ to get $x = f(K,0)$
Query RK oracle on $(K \mapsto E(K,0), 0)$ to get $y = f(E(K,0),0)$
Query oracle E on $(x,0)$ to get $z = E(x,0)$
Return $(z = y)$

Fig. 1. Δ^E-restricted adversary breaking an ideal cipher with $\mathcal{K} = \mathcal{D}$

When $f = E$, we have that $x = E(K,0)$, $y = E(E(K,0),0)$, and $z = E(E(K,0),0)$. Hence $z = y$ with probability 1. On the other hand, if $f = G$ then $x = G(K,0)$, $y = G(E(K,0),0)$, and $z = E(G(K,0),0)$. Similarly to the proof of Theorem 2 we have that $\Pr[E(G(K,0),0) = G(E(K,0),0)] \leq 2/|\mathcal{K}|$. The theorem follows. $\qquad\square$

As it can be seen, the proof is analogous to that of Theorem 2, reflecting the ability of our extended model to capture such cipher-dependent attacks. Another way to look at this result is to interpret it in terms of the event whose analysis underpins the proof of Theorem 1 in [4]. In that proof one needs to upper-bound the probability of:

Event D: A queries its related-key oracle with a function ϕ and queries its ideal cipher (in either the forward or backward directions) with a key K' such that $\phi(K) = K'$.

---

[4] Note that we use superscripts to denote an oracle access whereas subscripts are used to denote dependence. The former is of interest in the ideal-cipher model, and the latter in the standard model.

Looking at the code of A in Figure 1, it is easy to check that event D happens with probability 1: In A's attack $K' = E(K, 0)$ is a key queried to E and the RK oracle will have a key equal to K' when queried with $K \mapsto E(K, 0)$. This observation motivates the introduction of appropriately modified notions of output-predictability-2 and collision-resistance-2, as well as additional definitions which might enable a proof of oracle RKA pseudorandomness in the ideal-cipher model for $\Phi^{E,E^{-1}}$-restricted adversaries to be constructed.

Our first two conditions are modified versions of the collision-resistance-2 and output-unpredictability-2 notions of Bellare and Kohno as recalled in Section 2.

Definition 7 (Oracle-output-unpredictability-2). *Fix a key space $\mathcal{K}$ and domain $\mathcal{D}$ and let $\Phi^{E,E^{-1}}$ be a set of RKD functions on the key-space $\mathcal{K}$. Let $\mathcal{P}_K(\cdot)$ and $\mathcal{X}(\cdot)$ be a pair of oracles for each E. The oracle $\mathcal{P}_K(\cdot)$ takes as input an element $\phi^{E,E^{-1}} \in \Phi^{E,E^{-1}}$ and the oracle $\mathcal{X}(\cdot)$ takes as input an element $K' \in \mathcal{K}$. Neither oracle returns a value. An adversary wins if it queries its $\mathcal{X}(\cdot)$ oracle with a key K' and if it queries its $\mathcal{P}_K(\cdot)$ oracle with a function $\phi^{E,E^{-1}}$ such that $\phi^{E,E^{-1}}(K) = K'$. We define the oup2-advantage of an adversary A as*

$$\mathbf{Adv}^{\mathsf{oup2}}_{\Phi^{E,E^{-1}},\mathcal{K},\mathcal{D}}(A) := \Pr\left[K \xleftarrow{\$} \mathcal{K}; E \xleftarrow{\$} \mathrm{Perm}(\mathcal{K},\mathcal{D}) : A^{\mathcal{P}_K(\cdot),\mathcal{X}(\cdot)} \text{ wins}\right].$$

Definition 8 (Oracle-collision-resistance-2). *Fix a key space $\mathcal{K}$ and domain $\mathcal{D}$ and let $\Phi^{E,E^{-1}}$ be a set of functions on the key-space $\mathcal{K}$. Let $\mathcal{C}_K(\cdot)$ be an oracle for each E that takes as input a function $\phi^{E,E^{-1}} \in \Phi^{E,E^{-1}}$ and that returns no value. An adversary wins if it queries its oracle with two distinct functions $\phi_1^{E,E^{-1}}, \phi_2^{E,E^{-1}} \in \Phi^{E,E^{-1}}$ such that $\phi_1^{E,E^{-1}}(K) = \phi_2^{E,E^{-1}}(K)$. We define the ocr2-advantage of an adversary A as*

$$\mathbf{Adv}^{\mathsf{ocr2}}_{\Phi^{E,E^{-1}},\mathcal{K},\mathcal{D}}(A) := \Pr\left[K \xleftarrow{\$} \mathcal{K}; E \xleftarrow{\$} \mathrm{Perm}(\mathcal{K},\mathcal{D}) : A^{\mathcal{C}_K(\cdot)} \text{ wins}\right].$$

Once again we note that oracle-output-unpredictability can be seen as oracle-collision-resistance between a constant and a non-constant function.

Our third definition provides a sufficient condition on a set of oracle RKD functions $\Phi^{E,E^{-1}}$ to enable an analogue of Theorem 1 to be proved. Intuitively speaking, if an oracle RKD set is *oracle-independent*, then no collisions can take place between the explicit queries made by an adversary to one of its three oracles, and those made implicitly through the oracle RKD function, during its attack.

Definition 9 (Oracle-independence). *Fix a key space $\mathcal{K}$ and domain $\mathcal{D}$ and let $\Phi^{E,E^{-1}}$ be a set of functions on the key-space $\mathcal{K}$. Let $\mathcal{Q}_K(\cdot, \cdot)$ be an oracle for each E that takes as input a function $\phi^{E,E^{-1}} \in \Phi^{E,E^{-1}} \cup \mathrm{K}$, where K is the set of constant functions, and an $x \in \mathcal{D}$ and returns no value. An adversary wins if it queries its oracle with two (not necessarily distinct) oracle RKD functions $\phi_1^{E,E^{-1}}$ and $\phi_2^{E,E^{-1}}$, and a point $x_1 \in \mathcal{D}$ such that*

$$(\phi_1^{E,E^{-1}}(K), x_1) \in \{(K', x') : \phi_2^{E,E^{-1}}(K) \text{ queries } (K', x') \text{ to } E \text{ or } E^{-1}\}.$$

We define the oind*-advantage of an adversary A as*

$$\mathbf{Adv}^{\mathrm{oind}}_{\Phi^{E,E^{-1}},\mathcal{K},\mathcal{D}}(A) := \Pr\left[K \xleftarrow{\$} \mathcal{K}; E \xleftarrow{\$} \mathrm{Perm}(\mathcal{K},\mathcal{D}) : A^{\mathcal{Q}_K(\cdot,\cdot)} \text{ wins}\right].$$

Similarly to the remark at the end of Section 2, simpler alternatives of the above definitions can be formulated. For convenience, we call an oracle RKD set which satisfies the above three requirements *valid*.

Let us now state and prove the main result of this section.

Theorem 4. *Fix a key space $\mathcal{K}$ and a domain $\mathcal{D}$, and let $\Phi^{E,E^{-1}}$ be a set of RKD functions over $\mathcal{K}$. Let A be an ideal-cipher-model adversary that queries its first two oracles with a total of at most q' different keys and that queries its last oracle with a total of at most q different RKD functions from $\Phi^{E,E^{-1}}$. Then there exists an oracle-output-unpredictability-2 adversary B, an oracle-collision-resistance-2 adversary C, and an oracle-independence adversary D such that*

$$\mathbf{Adv}^{\mathrm{prp\text{-}orka}}_{\Phi^{E,E^{-1}},\mathcal{K},\mathcal{D}}(A) \leq \mathbf{Adv}^{\mathrm{ocr2}}_{\Phi^{E,E^{-1}},\mathcal{K},\mathcal{D}}(B) + \mathbf{Adv}^{\mathrm{oup2}}_{\Phi^{E,E^{-1}},\mathcal{K},\mathcal{D}}(C)$$
$$+ \mathbf{Adv}^{\mathrm{oind}}_{\Phi^{E,E^{-1}},\mathcal{K},\mathcal{D}}(D),$$

where B queries its $\mathcal{C}_K$ oracle q times, C queries its $\mathcal{P}_K$ and $\mathcal{X}$ oracles q and q' times (respectively), and D queries its $\mathcal{Q}_K$ oracle $q + q'$ times.

The intuition behind the proof of this theorem is similar to that for the proof of Theorem 1. The three conditions allow us to *separate* various oracle queries enabling us to simulate them by returning independently chosen random values. The output-unpredictability property is used to separate the ideal-cipher oracles from the related-key oracle. Collision-resistance is used to separate different ϕ's queried to the related-key oracle. The third condition is used in separating ϕ's oracles from those directly given to A; this was not necessary in the previous model when ϕ did not have access to any oracles. The proof of the above theorem can be found in the full version of the paper [1].

Note that if a set of RKD functions does not make any oracle calls to E or E^{-1} then the set automatically satisfies the oracle-independence criterion (with advantage 0). The oracle-collision-resistance-2 and oracle-output-unpredictability-2 conditions are identical to collision-resistance-2 and output-unpredictability-2 conditions of Bellare and Kohno respectively, and we also recover Theorem 1 in this case.

Let us now check why the attacks of Bernstein and Harris fail to satisfy the conditions required in Theorem 4 for an ideal cipher to be resistant to oracle related-key attacks. We have included a slightly modified and improved version of Harris's attack in the full version of the paper [1].

Theorem 5. *Let Δ^E and Ψ_i^E denote Bernstein's and Harris's set of oracle RKD functions, respectively. Then the oracle RKD sets Δ^E and Ψ_i^E do not satisfy the oracle-independence property.*

Proof. In Bernstein's attack, the function $K \mapsto E(K,0)$ queries E on K, as does the related-key oracle when queried on the identity function. For Harris's attack, note that

$H_{i,p}^E(K)$ queries E on $K \oplus p$ and $K \oplus p \oplus [1^k]_i$ and then, to compute the actual value of the related-key oracle when queried with this function, once again E is queried on one of these values.   □

So far we concentrated on ruling out attacks, and have not demonstrated how our choice of modelling can be used in a positive way. In other words, could it be the case that any non-trivial access to E or E^{-1} violates one of the three needed properties, rendering Theorem 4 meaningless. Fortunately, this is not the case. Our next two results demonstrate how one can model *new* cipher-dependent RKD functions which do not compromise security. The next theorem considers an RKD set from the EMV specification [15], and is proved in the full version of the paper [1].

Theorem 6. *Fix a key space $\mathcal{K}$ and let $\mathcal{D} = \mathcal{K}$. Define*

$$\Omega^E := \{K \mapsto E(K, x) : x \in \mathcal{D}\}.$$

Then for any adversary ocr2 *adversary A, any* oup2 *adversary B making at most q and q' queries to its $\mathcal{P}_K$ and $\mathcal{X}$ oracles (respectively), and any* oind *adversary C making at most q queries to its $\mathcal{Q}_K$ oracle, we have that*

$$\mathbf{Adv}_{\Omega^E,\mathcal{K},\mathcal{D}}^{\mathrm{ocr2}}(A) = 0, \mathbf{Adv}_{\Omega^E,\mathcal{K},\mathcal{D}}^{\mathrm{oup2}}(B) \le qq'/(2|\mathcal{K}|), \text{ and}$$

$$\mathbf{Adv}_{\Omega^E,\mathcal{K},\mathcal{D}}^{\mathrm{oind}}(C) \le q^2/(2|\mathcal{K}|).$$

The next theorem provides a possibility result in a scenario where the adversary has access to the identity function as well as other RKD functions.

Theorem 7. *Fix a key space $\mathcal{K}$ and let $\mathcal{D} = \mathcal{K}$. Define*

$$\Theta^E := \{K \mapsto K, K \mapsto E(K', K) : K' \in \mathcal{K}\}.$$

Then for any adversary ocr2 *adversary A making at most q queries to its $\mathcal{C}_K$ oracle, any* oup2 *adversary B making at most q and q' queries to its $\mathcal{P}_K$ and $\mathcal{X}$ oracles (respectively), and any* oind *adversary C making at most q queries to its $\mathcal{Q}_K$ oracle, we have that*

$$\mathbf{Adv}_{\Theta^E,\mathcal{K},\mathcal{D}}^{\mathrm{ocr2}}(A) \le q^2/(2|\mathcal{K}|), \mathbf{Adv}_{\Theta^E,\mathcal{K},\mathcal{D}}^{\mathrm{oup2}}(B) \le qq'/(2|\mathcal{K}|), \text{ and}$$

$$\mathbf{Adv}_{\Theta^E,\mathcal{K},\mathcal{D}}^{\mathrm{oind}}(C) \le q^2/(2|\mathcal{K}|).$$

The proof of this theorem is presented in the full version of the paper [1].

5   Interpretations in the Standard Model

Standard model interpretations of cryptographic results in an idealised model have always existed in the research community. The random oracle model and its real-world interpretations [13] provide a good example of the difficulties involved in attempting such translations. Another example is a result of Black [11], which gives a hash function construction provably secure in the ideal-cipher model, but insecure if the ideal cipher is instantiated with any concrete blockcipher. The result of [11] holds under related-key attacks as long as the RKD set under consideration contains the identity function (as in this case prp-rka security is at least as strong as the standard notion of prp security). This separation result, although theoretically valid, is *unnatural* as it is unlikely that a real-world hash function depends on the code of a blockcipher in the same artificial way as that used to derive the result in [11].

On the other hand, as shown in Theorem 2, in the related-key attack model of Bellare and Kohno, a *natural* (weak) separation result exists. This possibility seems to have been over-looked by the authors of [4], who did not discuss interpretations of their main result, Theorem 1, in the standard model. As pointed out in Section 3, this theorem can be interpreted in two different ways: the first interpretation, which was argued to be a more accurate translation, lacked a natural separation result, but was of smaller relevance to practice; the second interpretation, on the other hand, although relevant to practice, was shown to be invalid in Theorem 2.

Theorem 4 is an attempt to overcome the limitations of Theorem 1: oracle RKD functions enable modelling RKD functions which might depend on the cipher. For consistency and completeness, we should also investigate possible interpretations of this result in the standard model.

The first issue in interpreting this result arises when one attempts to relate an oracle RKD set $\Phi^{E,E^{-1}}$ to a concrete RKD set in the standard model. Our theorem concerns oracle RKD functions, that is RKD functions which use a blockcipher in a black-box way. Its relevance to the standard model is therefore restricted to RKD functions which use the cipher in a black-box (or a symbolic) way. Hence, given such an RKD set $\Phi_{E^\star}$ making subroutine calls to $E^\star$ and $E^{\star-1}$, one can rewrite it in the form of a natural oracle RKD set $\Phi^{E,E^{-1}}$, such that if it is instantiated at $E^\star$ (i.e. the oracle calls to E and E^{-1} are replaced with subroutine calls to $E^\star$ and $E^{\star-1}$), one recovers the original RKD set $\Phi_{E^\star}$.

Next, the validity of $\Phi_{E^\star}$ should be interpreted in terms of validity of $\Phi^{E,E^{-1}}$, as these sets are no longer the same (Note that this issue did not exist in interpreting Theorem 1 as the sets were identical). The minimum requirement on the set $\Phi_{E^\star}$ is that the associated oracle RKD set is valid, i.e. it satisfies the oracle-collision-resistant-2, oracle-output-unpredictable-2, and oracle-independent conditions. We additionally require that $\Phi^{E,E^{-1}}$ satisfies these conditions if E is no longer sampled uniformly at random in the games but is fixed to be $E^\star$. This latter condition is due to the fact that validity of the set for a random E is not enough to guarantee that at a specific $E^\star$ the set is also "reasonable".

We are now ready to present interpretations of our theorem that are analogous to those of Theorem 1. In the following, we let $\Phi^{E,E^{-1}}$ and $\Phi_{E^\star}$ be a pair of associated sets as discussed above. Our two interpretations, as before, concern the choice of order of quantifiers:

1. For all valid $\Phi^{E,E^{-1}}$, there exists a concrete blockcipher $E^\star$ which resists $\Phi_{E^\star}$-restricted attacks if $\Phi^{E,E^{-1}}$ is also valid at $E = E^\star$.
2. There is a concrete blockcipher $E^\star$, such that for all valid $\Phi^{E,E^{-1}}$, $E^\star$ resists $\Phi_{E^\star}$-restricted attacks if $\Phi^{E,E^{-1}}$ is also valid at $E = E^\star$.

In attempting to derive counterexamples to the above two interpretations a similar, but higher level, line of argument to that given for Theorem 1 applies. In Theorem 4, the strategy of dependence on E is fixed as E is chosen randomly and independently of $\Phi^{E,E^{-1}}$. In other words, for each E, each RKD function depends on E in the same way. This is exactly what is expressed by the first interpretation, and hence as in Theorem 1, we take this choice of order of quantifiers to be a more accurate interpretation. What is important here is that *unlike the first interpretation of the theorem of Bellare and Kohno, the RKD functions here may depend on E*, and hence this theorem has a greater relevance to practice than that provided by Theorem 1. As in Theorem 1, we do not expect there to be natural counterexamples to this interpretation.

Let us turn to the second interpretation. Due to the reversed order of quantifiers the strategy of dependence in a counterexample may itself depend on each cipher $E^\star$. In fact, Bernstein's attack still constitutes a counterexample to the second interpretation of Theorem 4, if one chooses the oracle RKD set to be identical to the concrete RKD set for each $E^\star$, i.e.

$$\Delta_{E^\star} := \{K \mapsto K, K \mapsto E^\star(K, 0)\} \text{ and } \Delta_{E^\star}^E := \{K \mapsto K, K \mapsto E^\star(K, 0)\}.$$

Note however that, as pointed out before, the dependency of $\Delta_{E^\star}^E$ on $E^\star$ is black box. This oracle RKD set may then be rewritten symbolically as

$$\Delta^E := \{K \mapsto K, K \mapsto E(K, 0)\},$$

and as we saw in Theorem 5, this set is not valid. A similar observation applies to Harris's RKD set. In general this dependency in a (natural) counterexample is likely to be black box, and the functions can be rewritten as an oracle RKD set with a *fixed* dependence strategy. This in turn would either constitute a counterexample to the first interpretation, which we have assumed to be unlikely, or the resulting new oracle RKD set will be invalid. On the other hand, a non-black-box dependency seems difficult to achieve. In conclusion, the second interpretation, for practical purposes, is the same as the first one.

Turning to positive results, Theorems 6 and 7 can be interpreted in the standard model in the following way. It is a "reasonable" goal to design a blockcipher which resist Ω_E- and Θ_E-restricted related-key attacks where

$$\Omega_E := \{K \mapsto E(K, x) : x \in \mathcal{D}\} \text{ and } \Theta_E := \{K \mapsto K, K \mapsto E(K', K) : K' \in \mathcal{K}\}$$

are respectively the RKD sets associated to Ω^E and Θ^E as defined in Theorems 6 and 7. These results may have applications in establishing the security of key hierarchies which use the cipher to derive new keys.

Let us look at the first standard model interpretations of Theorem 1 and Theorem 4 from a cryptanalytic perspective. Theorem 1 classifies a blockcipher E as broken if there exists a collision-resistance-2 and output-unpredictable-2 RKD set which can be used to break E in the related-key attack model *and furthermore* this set does not depend on E. This theorem provides no answers for RKD sets, such as Bernstein's set or that given in Theorem 7, which depend on E. Theorem 4, on the other hand, allowed dependency on E at the expense of an extra condition. This theorem classified a block-cipher E as broken in two cases: 1) the attack is independent of E and we are back at the conditions of Theorem 1; or 2) the attack is dependent on E in a black-box way, and the associated oracle RKD set is valid for a random E and also at $E^\star$.

According to the above cryptanalytic perspective, Bernstein's and Harris's attacks should not be seen as harmful. Attacks using RKD sets which involve a cipher's building blocks demonstrated by Biryukov et al. [9,10] on AES raise the following question: can Biryukov et al.'s set of RKD functions be simulated using calls to the full encryption and decryption routines of AES? If this is not the case, or if this is the case and the resulting oracle RKD set is valid, then the related-key attack against AES should be seen as interesting. Formalising such a natural dependency remains an open problem, and hence, in our opinion, Biryukov et al.'s attack should then be seen as a threat against AES in the related-key attack model (assuming the relevant RKD functions are available to the cryptanalyst). We note that our model might be further extended to consider RKD sets with oracle access to round functions so as to model Biryukov et al.'s results which exploit relations of this type.

REMARK. The requirements of Theorem 4 constitute a set of sufficient conditions for an ideal cipher to be secure in the prp-orka sense. These conditions, however, are quite strong and one might alternatively directly prove that an ideal cipher is prp-orka secure.[5] Validity at $E^\star$ now means resilience of an ideal cipher to related-key attacks when the oracle RKD set is instantiated with $E^\star$. Such proofs can then be used to conjecture the existence of a blockcipher resisting related-key attacks under the associated RKD set. From a cryptanalytic perspective, if these proofs exists, and the associated RKD set breaks a specific cipher, then this cipher should be seen as broken. Conversely, if an oracle RKD set can be used to break an ideal cipher, then the associated RKD set should not be seen as valid. This in turn means that one should neither expect there to be a blockcipher which resists such attacks, nor should such an attack be seen as harmful. These observations also apply to the two conditions used in Theorem 1.

IMPLICATIONS FOR PRACTICE. As well as considering standard-model interpretations of our main result, we also wish to reflect on what our results might mean for practice. Suppose we have an RKD set that is *invalid* for a blockcipher, so that the conditions of our main theorem are not met. Does this mean that there must be a related-key attack against the blockcipher? The answer is clearly no, since the possibility of a related-key attack depends on exactly how the blockcipher is used as a component in an overall system or protocol: if that environment does not make available to the attacker the relevant RK oracles, then the related-key attack will not be mountable in practice. A recent example of an interesting related-key attack which is not mountable in practice as is would be the attack of Dunkelman et al. [14] against KASUMI when used in the 3G network. On the other hand, even if we have an RKD set that is valid, then we can still not rule out related-key attacks altogether, because of the gap that exists between the ideal-cipher model and the standard model. Finally, we ask: what should a protocol designer do? The simple answer is to avoid the use of related keys in protocols

---

[5] This is the case for functions expressed in a contrived way such as $K \mapsto K \oplus 1 = K \oplus E(K, E^{-1}(K, 1))$ or $K \mapsto K \oplus E(0, 0)$.

altogether. If this is not possible, then our best advice is to only use related-keys in such a way that the relevant RKD set satisfies the conditions of Theorem 4. For example, the sets of functions exhibited in Theorems 6 and 7 would be suitable in this case.

Acknowledgements

The authors would like to thank Thomas Shrimpton and Martijn Stam for their detailed and insightful comments on earlier drafts of this paper. The work described in this paper has in part been supported by the Commission of the European Communities through the ICT program under contract ICT-2007-216676 (ECRYPT-II). M.R. Albrecht was supported by the Royal Holloway Valerie Myerscough Scholarship. The work of K.G. Paterson was supported by an EPSRC Leadership Fellowship, EP/H005455/1. The information in this document is provided as is, and no warranty is given or implied that the information is fit for any particular purpose. The user thereof uses the information at its sole risk and liability.

References

1. Albrecht, M.R., Farshim, P., Paterson, K.G., Watson, G.J.: On Cipher-Dependent Related-Key Attacks in the Ideal-Cipher Model. In: Cryptology ePrint Archive, Report 2011/??? (2011)
2. Applebaum, B., Harnik, D., Ishai, Y.: Semantic Security Under Related-Key Attacks and Applications. In: Cryptology ePrint Archive, Report 2010/544 (2010)
3. Bellare, M., Cash, D.: Pseudorandom Functions and Permutations Provably Secure Against Related-Key Attacks. In: Rabin, T. (ed.) CRYPTO 2010. LNCS, vol. 6223, pp. 666–684. Springer, Heidelberg (2010)
4. Bellare, M., Kohno, T.: A Theoretical Treatment of Related-Key Attacks: RKA-PRPs, RKA-PRFs, and Applications. In: Biham, E. (ed.) EUROCRYPT 2003. LNCS, vol. 2656, pp. 491–506. Springer, Heidelberg (2003)
5. Bellare, M., Rogaway, P.: The Security of Triple Encryption and a Framework for Code-Based Game-Playing Proofs. In: Vaudenay, S. (ed.) EUROCRYPT 2006. LNCS, vol. 4004, pp. 409–426. Springer, Heidelberg (2006)
6. Bernstein, D.J.: E-mail Discussion among the Participants of the Early Symmetric Crypto Seminar (2010)
7. Biham, E.: New Types of Cryptanalytic Attacks Using Related Keys (Extended Abstract). In: Helleseth, T. (ed.) EUROCRYPT 1993. LNCS, vol. 765, pp. 398–409. Springer, Heidelberg (1994)
8. Biham, E.: New Types of Cryptanalytic Attacks Using Related Keys. Journal of Cryptology 7(4), 229–246 (1994)
9. Biryukov, A., Khovratovich, D.: Related-Key Cryptanalysis of the Full AES-192 and AES-256. In: Matsui, M. (ed.) ASIACRYPT 2009. LNCS, vol. 5912, pp. 1–18. Springer, Heidelberg (2009)
10. Biryukov, A., Khovratovich, D., Nikolić, I.: Distinguisher and Related-Key Attack on the Full AES-256. In: Halevi, S. (ed.) CRYPTO 2009. LNCS, vol. 5677, pp. 231–249. Springer, Heidelberg (2009)
11. Black, J.: The Ideal-Cipher Model, Revisited: An Uninstantiable Blockcipher-Based Hash Function. In: Robshaw, M.J.B. (ed.) FSE 2006. LNCS, vol. 4047, pp. 328–340. Springer, Heidelberg (2006)

12. Black, J., Rogaway, P., Shrimpton, T.: Encryption-Scheme Security in the Presence of Key-Dependent Messages. In: Nyberg, K., Heys, H.M. (eds.) SAC 2002. LNCS, vol. 2595, pp. 62–75. Springer, Heidelberg (2003)
13. Canetti, R., Goldreich, O., Halevi, S.: The Random Oracle Methodology, Revisited. JACM 51(4), 557–594 (2004)
14. Dunkelman, O., Keller, N., Shamir, A.: A Practical-Time Attack on the A5/3 Cryptosystem Used in Third Generation GSM Telephony. In: Cryptology ePrint Archive, Report 2010/013 (2010)
15. EMV Integrated Circuit Card Specifications for Payment Systems, Book 2 Security and Key Management, Version 4.2 (June 2008)
16. Goldenberg, D., Liskov, M.: On Related-Secret Pseudorandomness. In: Micciancio, D. (ed.) TCC 2010. LNCS, vol. 5978, pp. 255–272. Springer, Heidelberg (2010)
17. Halevi, S., Krawczyk, H.: Security under Key-Dependent Inputs. In: ACM Conference on Computer and Communications Security, CCS 2007, pp. 466–475. ACM, New York (2007)
18. Harris, D.G.: Generic Ciphers are More Vulnerable to Related-Key Attacks than Previously Thought. In: WCC 2009 (2009)
19. Iwata, T., Kohno, T.: New Security Proofs for the 3GPP Confidentiality and Integrity Algorithms. In: Roy, B., Meier, W. (eds.) FSE 2004. LNCS, vol. 3017, pp. 427–445. Springer, Heidelberg (2004)
20. Knudsen, L.R.: Cryptanalysis of LOKI91. In: Zheng, Y., Seberry, J. (eds.) AUSCRYPT 1992. LNCS, vol. 718, pp. 196–208. Springer, Heidelberg (1993)
21. Lucks, S.: Ciphers Secure against Related-Key Attacks. In: Roy, B., Meier, W. (eds.) FSE 2004. LNCS, vol. 3017, pp. 359–370. Springer, Heidelberg (2004)
22. Razali, E., Phan, R.C.-W., Joye, M.: On the Notions of PRP-RKA, KR and KR-RKA for Block Ciphers. In: Susilo, W., Liu, J.K., Mu, Y. (eds.) ProvSec 2007. LNCS, vol. 4784, pp. 188–197. Springer, Heidelberg (2007)

On the Security of Hash Functions Employing Blockcipher Postprocessing

Donghoon Chang[1], Mridul Nandi[2], and Moti Yung[3]

[1] National Institute of Standards and Technology, USA
[2] C.R. Rao AIMSCS, Hyderabad, India
[3] Google Inc. and Department of Computer Science,
Columbia University, New York, USA
{pointchang,mridul.nandi}@gmail.com,
my123@columbia.edu

Abstract. Analyzing desired generic properties of hash functions is an important current area in cryptography. For example, in Eurocrypt 2009, Dodis, Ristenpart and Shrimpton [8] introduced the elegant notion of "Preimage Awareness" (PrA) of a hash function H^P, and they showed that a PrA hash function followed by an output transformation modeled to be a FIL (fixed input length) random oracle is PRO (pseudorandom oracle) i.e. indifferentiable from a VIL (variable input length) random oracle. We observe that for recent practices in designing hash function (e.g. SHA-3 candidates) most output transformations are based on permutation(s) or blockcipher(s), which are not PRO. Thus, a natural question is how the notion of PrA can be employed directly with these types of more prevalent output transformations? We consider the Davies-Meyer's type output transformation $OT(x) := E(x) \oplus x$ where E is an ideal permutation. We prove that $OT(H^P(\cdot))$ *is PRO if* H^P *is PrA, preimage resistant and computable message aware* (a related but not redundant notion, needed in the analysis that we introduce in the paper). The similar result is also obtained for 12 PGV output transformations. We also observe that some popular double block length output transformations can not be employed as output transformation.

Keywords: PrA, PRO, PRP, Computable Message Awareness.

1 Introduction

Understanding what construction strategy has a chance to be a good hash function is extremely challenging. Further, nowadays it is becoming more important due to the current SHA3 competition which is intended to make a new standard for hash functions. In TCC'04, Maurer *et al.* [17] introduced the notion of indifferentiability as a generalization of the concept of the indistinguishability of two systems [16]. Indifferentiable from a VIL (variable input length) random oracle (also known as PRO or pseudorandom oracle) is the appropriate notion of random oracle for a hash-design. Recently, Dodis, Ristenpart and Shrimpton [8] introduced a generic method to show indifferentiable or PRO security proof

A. Joux (Ed.): FSE 2011, LNCS 6733, pp. 146–166, 2011.

of a hash function, whose final output function $\mathcal{R}$ is a FIL (fixed input length) random oracle. More precisely, they defined a new security notion of hash function, called **preimage awareness** (PrA), and showed that $F(M) = \mathcal{R}(H^P(M))$ is PRO provided H^P is preimage aware (supposed to be a weaker assumption). The result is applied to prove the indifferentiable security of the Skein hash algorithm [2], a second round SHA3 candidate. Informally, a hash function H^P is called PrA if the following is true for any adversary A having access to P: For any y committed by A, if a preimage of y is not efficiently "computable" (by an algorithm called *extractor*) from the tuple of all query-responses of P (called *advise string*) then A should not be able to compute it even after making additional P queries. This new notion seems to be quite powerful whenever we have composition of a VIL hash function and a FIL output transformation.

Our Result. We start with a preliminary discussion about the different notions and interrelationship among them. We note that there are hash functions whose final output transformation *cannot be viewed as a random oracle* e.g. some SHA3 second round candidates. So one needs to extended results beyond that of Dodis *et al.* to cover the cases of hash functions with various output transformations which are in use and this becomes our major objective, since it is important to assure good behavior of these hash functions as well. As a good example of a prevalent transform for construction of hash functions, we choose Davies-Meyer [21] $OT(x) = E(x) \oplus x$ where E is a random permutation and study it in Section 3. We observe that the preimage awareness of H^P is not sufficient for the PRO security of F. In addition to PrA, if H^P is also preimage resistant (PI) and computable message aware (as we define in Section 3.1), then $F^{P,E}$ is PRO (proved in Theorem 1). Informally speaking, a hash function H^P is called **computable message aware** (or CMA) if there exists an efficient extractor (called computable message extractor) which can list the set of all computable messages whose H^P outputs are implied with high probability given the advise string of P. The main difference with PrA is that here, no adversary is involved and the extractor does not get any specific target (see Definition 2). We show that both preimage resistant and CMA are not implied by PrA and hence these properties can not be ignored. Our result can then be employed to prove that a close variant of Grøstl is PRO (see Section 4)[1]. We continue our research in finding other good output transformations. We found 12 out of 20 PGVs can be employed as output transformation OT and we require similar properties of H^P, i.e. PrA, PI and CMA, to have PRO property of $OT(H^P)$ (see Section 5). However these three properties are not sufficient for some DBL post processors. In section 6 we show PRO attacks when some popular double block length post processors are employed. It would be an interesting future research work to characterize the properties of the inner hash function H^P and the output transformation OT such that $OT(H^P)$ become PRO. In the appendix we review the results of [8,9].

---

[1] The indifferentiable security analysis of Grøstl has been studied in [1].

2   Preliminaries

A game is a tuple of probabilistic stateful oracles $G = (\mathcal{O}_1, \ldots, \mathcal{O}_r)$ where states can be shared by the oracles. The oracles can have access to primitives (e.g. random oracle) via black-box modes. It is reasonable to assume that all random sources of games come from the primitives. A probabilistic oracle algorithm A (e.g. an adversary) executes with an input x, its oracle queries being answered by the corresponding oracles of G. Finally it returns $y := A^G(x)$. An adversary A may be limited by different resources such as its runtime, number of queries to different oracles, size of its inputs or outputs, etc. If θ is a tuple of parameters describing the available resources of A then we say that A is a θ-*adversary*. In this paper H^P is an n-bit hash function defined over a message space $\mathcal{M}$ based on a primitive P which can be only accessed via a black-box.

Indifferentiability. The security notion of indifferentiability or PRO was introduced by Maurer *et al.* in TCC'04 [17]. In Crypto'05, Coron *et al.* adopted it as a security notion for hash functions [6]. Let F be a hash function based on ideal primitives $P = (P_1, ..., P_j)$ and $\mathcal{F}$ be a VIL random oracle, let $S^{\mathcal{F}} = (S_1^{\mathcal{F}}, ..., S_j^{\mathcal{F}})$ be a simulator (aimed to simulate $P = (P_1, \ldots, P_j)$) with access to $\mathcal{F}$, where S_i's can communicate to each other. Then, for any adversary A, the indifferentiability- or PRO-advantage of A is defined by $\mathbf{Adv}^{\text{pro}}_{F^P, S^{\mathcal{F}}}(A) = |\Pr[A^{F,P} = 1] - \Pr[A^{\mathcal{F},S} = 1]|$. When the value of the above advantage is negligible, we say that the hash function F is indifferentiable or PRO. Maurer *et al.* [17] also proved that if F is indifferentiable, then $\mathcal{F}$ (a VIL random oracle) used in any secure cryptosystem can be replaced by $F^{P_1, \ldots, P_j}$ with a negligible loss of security. In other words, F can be used as a VIL PRO.

Preimage-Awareness or PrA. Dodis, Ristenpart and Shrimpton defined a new security notion called *Preimage-Awareness* (or PrA) for hash functions [8,9] which plays an important role in analyzing indifferentiability of a hash function [2]. Given a game G^P (can be P itself), a tuple $\alpha = ((x_1, w_1), \ldots, (x_s, w_s))$ is called an *advise string* at some point of time in the execution of A^G, if w_i's are responses of all P-queries x_i's until that point of time. A PrA (q, e, t)-adversary A (making q queries and running in time t) commits $y_1, \ldots, y_e$ during the execution of A^P and finally returns M. We write $(y_1, \ldots, y_e) \leftarrow A^P_{\text{guess}}$, $M \leftarrow A^P$ and denote the advise string at the time A^P_{guess} commits y_i by α_i. The guesses and P-queries can be made in any order.

Definition 1. *The PrA-advantage of H^P with an extractor $\mathcal{E}$ is defined as* $\mathbf{Adv}^{\text{pra}}_{H,P,\mathcal{E}}(q, e, t) = \max_A \mathbf{Adv}^{\text{pra}}_{H,P,\mathcal{E}}(A)$ *where maximum is taken over all (q, e, t)-adversaries A and the PrA advantage of A, denoted $\mathbf{Adv}^{\text{pra}}_{H,P,\mathcal{E}}(A)$, is defined as*

$$\Pr[\exists i, \ H^P(M) = y_i, \ M \neq \mathcal{E}(y_i, \alpha_i) : M \leftarrow A^P; (y_1, \ldots, y_e) \leftarrow A^P_{\text{guess}}]. \quad (1)$$

H^P is called (q, e, t, t_e, ϵ)-PrA if $\mathbf{Adv}^{\text{pra}}_{H,P,\mathcal{E}}(q, e, t) \leq \epsilon$ for some extractor $\mathcal{E}$ with runtime t_e. In short, we say that a hash function is PrA or preimage-aware if there exists an "efficient" extractor such that for all "reasonable" adversaries A the PrA advantage is "small".

Implication among Collision Resistant, Preimage resistant (PI), PrA.
If a collision attacker B^P returns a collision pair (M, M'), then a PrA attacker
makes all necessary P queries to compute $H^P(M) = H^P(M') = y$ and finally
returns M if $\mathcal{E}(y, \alpha) = M'$, o.w. returns M'. So a PrA hash function must be
collision resistant. In [8] the authors consider a weaker version of PrA (called
weak-PrA) where an extractor can return a set of messages (possibly empty)
whose output is y. A PrA adversary wins this new weak game if it can find a
preimage of y different from those given by the extractor. They also have shown
that PrA is equivalent to collision resistant and weak-PrA. One can modify a
definition of a preimage-resistant hash function by introducing only one collision
pair. It still remains preimage resistant as the randomly chosen target in the
particular collision value has negligible probability. However, it is not preimage-
aware since a collision is known. On the other hand $H^P(x) = P^{-1}(x)$ or $H^P(x) = x$
are not preimage resistant but PrA.

3   Hash Function with Output Transformation $E(x) \oplus x$

In [8], hash functions have been analyzed for which the output transformation
can be modeled as a FIL random oracle. Generally, we can consider various kinds
of output transformations such as Davis-Meyer, PGV compression functions [20]
or some DBL (Double Block Length) compression functions [19,12,13,15] in
the ideal cipher model. Traditionally, the most popular known design of hash
function uses one of the above post-processors. It is well-known that all such
compression functions are not indifferentiably secure [6]. So, we need a separate analysis from [8]. In this section, we consider Davis-Meyer transformation
$OT(x) = E(x) \oplus x$, where E is a permutation modeled as a "random permutation." A simple example of H^P (e.g. the identity function) tells us that the
preimage awareness is not a sufficient condition to have PRO after employing Davis-Meyer post-processor. This suggests that we need something stronger
than PrA. We first observe that the preimage attack on identity function can
be exploited to have the PRO attack. So preimage resistant can be a necessary
condition. We define a variant of preimage resistant, called multipoint-preimage
(or mPI), which is actually equivalent to PI. The multipoint-preimage (or mPI)
advantage of a (q, t, s)-adversary A (i.e., adversary which makes q queries, runs
in time t and has s targets) for H^P is defined as

$$\mathbf{Adv}_{H^P}^{\mathrm{mPI}}(A) = \Pr_{h_1, \ldots, h_s \xleftarrow{\$} \{0,1\}^n} [\exists i, H^P(M) = h_i : M \leftarrow A^P(h_1, \ldots, h_s)] \quad (2)$$

When $s = 1$, it corresponds to the classical preimage advantage $\mathbf{Adv}_{H^P}^{\mathrm{PI}}(A)$.
Conversely, mPI advantage can be bounded above by preimage advantage as
described in the following. For any (q, t, s)-adversary A with multipoint-preimage
advantage ϵ against H^P, there is a $(q, t + O(s))$-adversary A' with preimage-
advantage ϵ/s. The adversary A' generates $(s-1)$ targets randomly and embeds
his target among these in a random position. So whenever an mPI adversary A
finds a multipoint preimage of these s targets, it is the preimage of A's target

with probability $1/s$ (since there is no way for A to know the position of the target for A'). W.l.o.g. one can assume that the targets are distinct and chosen at random. Otherwise we remove all repeated h_i's and replace them by some other random distinct targets. So we have the following result.

Lemma 1. *Let $h_1, \ldots h_s$ be distinct elements chosen at random (i.e. outputs of a random permutation for distinct inputs). Then, any (q,t)-adversary A^P can find one of the preimages of h_i's with probability at most $s \times \mathbf{Adv}_{H^P}^{\mathrm{PI}}(q,t)$.*

3.1   Computability

Next we show that preimage resistant and PrA are not sufficient to prove the PRO property. Consider the following example based on an n-bit one-way permutation f and random oracle P.

Example 1. $H^P(m) = P(f(m)) \oplus m$. Given $\alpha = (f(m), w)$ it is hard to find m and hence there is no efficient extractor to find the message m even though an adversary A knows m and its H^P-output. An adversary can compute $z = \mathcal{F}(m)$ and makes $E^{-1}(z \oplus w \oplus m)$ query. No feasible simulator can find message m from it with non-negligible probability and hence cannot return $w \oplus m$. However $w \oplus m$ is the response when A interacts with the real situation $(F^{P,E}, P, E, E^{-1})$. So A can have a PRO attack to F. It is easy to see that H^P is preimage resistant and PrA (given the advise string $\alpha = ((x_1, w_1), \ldots, (x_q, w_q))$ and the target x (that helps to find m back) the extractor finds i for which $f(w_i \oplus x) = x_i$ and then returns $w_i \oplus x$).

The above example motivates us to define a computable message given an advise string. A message M is called computable from α if there exists y such that $\Pr[H^P(M) = y | \alpha] = 1$. In other words, the computation of $H^P(M) = y$ can be made without making any further P-queries. We require the existence of an efficient extractor $\mathcal{E}_{\mathrm{comp}}$, called *computable message extractor*, which can list all computable messages given the advise string. We note that this is not same as weak-PrA as the extractor has to find all messages whose outputs can be computed to a value (unlike PrA, no such fixed target is given here). This notion does not involve any adversary.

Definition 2. *A pair $(H^P, \mathcal{E}_{\mathrm{comp}})$ is called (q, q_H, ϵ)-computable message aware or CMA if for any advise string α with q pairs, the number of computable messages is at most q_H and $\mathcal{E}_{\mathrm{comp}}(\alpha)$ outputs all these. Moreover, for any noncomputable messages M, $\Pr[H^P(M) = y | \alpha] \leq \epsilon$, $\forall y$.*

A hash function H^P is called (q, q_H, ϵ, t_c)-computable message aware or CMA if there is $\mathcal{E}_{\mathrm{comp}}$ with run time t_c such that $(H^P, \mathcal{E}_{\mathrm{comp}})$ is (q, q_H, ϵ)-CMA. In short we say that H^P is CMA if it is (q, q_H, ϵ, t_c)-computable message aware where for any feasible q, q_H and t_c are feasible and ϵ is negligible. We reconsider the above example $H^P(m) = P(f(m)) \oplus m$ for an one-way permutation f. We have seen that it is both PI and PrA. However, there is no efficient extractor that can not find all computable messages given the advise string say $(f(m), w)$. In fact,

m is computable but there is no way to know it by the extractor only from the advise string (extractor can know if the target $f(m) \oplus w = H^P(m)$ is given).

To be a computable message aware, *the list of computable message has to be small or feasible so that an efficient computable message extractor can exist*. For example, the identity function has a huge set of computable messages given any advise string which cannot be listed by any efficient algorithm even though we theoretically know all these messages.

3.2   PRO Analysis of a Hash Function with OT $E(x) \oplus x$

In this section we prove that $OT(H^P())$ is PRO whenever H^P is PrA, PI and CMA. We give an informal idea how the proof works. Note that for E-query, $E(x) \oplus x$ almost behaves like a random oracle and hence PrA property of H^P takes care the simulation. This would be similar to the random oracle case except that we have to deal with the fact there is no collision on E. The simulation of responses of P-queries will be same as P. The non-trivial part is to response E^{-1} query. If the E^{-1}-query y is actually obtained by $y = \mathcal{F}(M) \oplus H^P(M)$ then simulator has to find the M to give a correct response. Since simulator has no idea about the $\mathcal{F}(M)$ as he can not see $\mathcal{F}$-queries, the query y is completely random to him. However, he can list all computable messages and try to compute $H^P(M)$ and $\mathcal{F}(M)$. This is why we need CMA property. The simulator should be able to list all computable messages only from the P query-responses. If he finds no such messages then he can response randomly. The simulator would be safe as long as there is no preimage attack to the random output. Now we provide a more formal proof.

Let $F^{P,E}(M) = E(H^P(M)) \oplus H^P(M)$ and A be a PRO adversary making at most (q_0, q_1, q_2, q_3) queries to its four oracles with bit-size l_{max} for the longest $\mathcal{O}_0$-query. We assume that $H^P(\cdot)$ is preimage resistant and (q^*, q_H, ϵ)-computable message aware for an efficient computable message extractor $\mathcal{E}_{comp}$ where $q^* = q_1 + q_2\mathsf{NQ}[l_{max}]$. Let $q = q_H + q'$ and $q' = q_0 + q_1 + q_2 + q_3$. For any given PrA-extractor $\mathcal{E}$, we can construct a simulator $S^{\mathcal{F}} = (S_1, S_2, S_3)$ (defined in the oracles of C_A in Fig. 3) that runs in time $t^* = O(q^2 + q_3\mathsf{Time}(\mathcal{E}_{comp}))$. Given any indifferentiability adversary A making at most (q_0, q_1, q_2, q_3) queries to its four oracles with bit-size l_{max} for the longest $\mathcal{O}_0$-query, there exists a PrA $(q, q_2 + 1, t)$-adversary C_A with runtime $t = \mathsf{Time}(A) + O(q_2 \cdot \mathsf{Time}(\mathcal{E}) + q_0 + q_1 + (q_2 + q_0)\mathsf{NQ}[l_{max}])$. Now we state two lemmas which are useful to prove the main theorem of the section. Proof ideas of these lemmas are very similar to that of Lemma 8 and Lemma 10. The games $G4$ and $G5$ are defined in the Fig. 2. We use a simulation oracle simE which works given a runtime database E. A random element from the set $\{0,1\}^n \setminus Range(\mathsf{E})$ is returned for $\mathsf{simE}[1, y]$ whenever y is not in the domain of E. Similarly a random element from the set $\{0,1\}^n \setminus Domain(\mathsf{E})$ is returned in $\mathsf{simE}[-1, c]$ whenever c is not in the range of E. Whenever y or c are defined before the simulator oracle just returns the previous returned value. We use three such simulation oracles for E_0 (which keeps the input output behavior of E due to $\mathcal{O}_0$ queries only), E_1

Game	$G4$	and $G5$

Initialize : $\mathsf{E} = \mathsf{E}_0 = \mathsf{E}_1 = \phi$; $\mathsf{H} = \mathsf{H}' = \beta = \phi$, Bad $=$F;

300 On $\mathcal{O}_3$ - query c
301 $S = \{X_1, \ldots, X_r\} = \mathcal{E}_{\mathrm{comp}}(\beta)$;
302 For all $1 \leq i \leq r$ do
303 $\quad y_i = H^{\mathcal{O}_1}(X_i) := \mathsf{H}[X_i]$; $\mathcal{F}(X_i) = z_i$; $c_i = y_i \oplus z_i$;
304 If $\exists$ unique i s.t. $c_i = c$,
305 $\quad$ If $\mathsf{E}_1[y_i] = \perp$ then $y = y_i$;
306 $\quad$ Else if $\mathsf{E}_1[y_i] = c' \neq \perp$ then Bad $= T$; $y = \mathsf{simE}_1[-1, c]$;
307 Else if no i s.t. $c_i = c$, then $y = \mathsf{simE}_1[-1, c]$;
308 Else Bad $=$T; $y = \mathsf{simE}_1[-1, c]$;
309 If $\mathsf{E}_0^{-1}[c] \neq \perp$ and $\mathsf{E}_0^{-1}[c] \neq y_i$ then Bad $= T$; $\boxed{y = \mathsf{E}_0^{-1}[c];}$
310 If y is E_1-simulated and $\mathsf{E}_0[y] \neq \perp$ then Bad $= T$; $\boxed{y = \mathsf{sim\bar{E}}[-1, c];}$
311 $\mathsf{E}_1[y] := c$; $\mathsf{\bar{E}}[y] := c$; return y;

200 On $\mathcal{O}_2$ - query y
201 $X = \mathcal{E}(y, \beta)$; Ext $\overset{\cup}{\leftarrow} (y, X)$;
202 $y' = H^{\mathcal{O}_1}(X)$, $\mathsf{H} \overset{\cup}{\leftarrow} (X, y')$; $z = \mathcal{F}(X)$; $c = z \oplus y'$;
203 If $y' \neq y$ then
204 $\quad c' = \mathsf{simE}_1[1, y]$;
205 $\quad$ If $\mathsf{E}_0[y] \neq \perp$ then Bad $=$T; $\boxed{c' = \mathsf{E}_0[y];}$
206 $\quad$ Else if $c' \in \mathsf{Range}(\mathsf{E}_0)$ then Bad $=$T; $\boxed{c' = \mathsf{sim\bar{E}}[1, y];}$
207 $\quad \mathsf{E}_1[y] := c'$; $\mathsf{\bar{E}}[y] := c'$; return c';
208 If $y' = y$ and $c \in \mathsf{Range}(\mathsf{E}_1)$ then
209 $\quad c' = \mathsf{simE}_1[1, y]$;
210 $\quad$ If $\mathsf{E}_0[y] \neq \perp$, then Bad $=$T; $\boxed{c' = \mathsf{E}_0[y];}$
211 $\quad$ Else if $c' \in \mathsf{Range}(\mathsf{E}_0)$ then Bad $=$T; $\boxed{c' = \mathsf{sim\bar{E}}[1, y];}$
212 $\quad \mathsf{E}_1[y] := c'$; $\mathsf{\bar{E}}[y] := c'$; return c';
213 If $y' = y$ and $c \notin \mathsf{Range}(\mathsf{E}_1)$ then
214 $\quad$ If $\mathsf{E}_0[y] \neq \perp$, then Bad $=$T; $\boxed{c = \mathsf{E}_0[y];}$
215 $\quad$ Else if $c \in \mathsf{Range}(\mathsf{E}_0)$ then Bad $=$T; $\boxed{c = \mathsf{sim\bar{E}}[1, y];}$
216 $\quad \mathsf{E}_1[y] := c$; $\mathsf{\bar{E}}[y] := c$; return c;

100 On $\mathcal{O}_1$ - query u
101 $v = P(u)$; $\beta \overset{\parallel}{\leftarrow} (u, v)$;
102 return v;

000 On $\mathcal{O}_0$ - query M
001 $z = \mathcal{F}(M)$; $y = H^P(M)$; $c = z \oplus y$;
002 If $\exists M'$ s.t. $M \neq M'$ and $(M', y) \in \mathsf{H}'$ then
$\quad$ Bad $=$T; $\mathsf{H} \overset{\cup}{\leftarrow} (M, y)$; $\boxed{z = \mathcal{F}(M');}$ return z;
003 $\mathsf{H}' \overset{\cup}{\leftarrow} (M, y)$; $\mathsf{H} \overset{\cup}{\leftarrow} (M, y)$;
004 If $\mathsf{\bar{E}}[y] = \perp \wedge \mathsf{\bar{E}}^{-1}[c] = \perp$ then $\mathsf{\bar{E}}[y] := c$; $\mathsf{E}_0[y] := c$; return z;
005 If $\mathsf{\bar{E}}[y] = c$ then $\mathsf{E}_0[y] := c$; return z;
006 If $\mathsf{\bar{E}}[y] \neq \perp \wedge \mathsf{\bar{E}}[y] \neq c$ then Bad $=$T; $\boxed{z = \mathsf{\bar{E}}[y] \oplus y;}$ return z;
007 If $\mathsf{\bar{E}}[y] = \perp \wedge \mathsf{\bar{E}}^{-1}[c] \neq \perp$ then
$\quad$ Bad $=$T; $\mathsf{sim\bar{E}}[1, y] = c'$; $\mathsf{\bar{E}}[y] := c'$; $\mathsf{E}_0[y] := c'$; $\boxed{z = c' \oplus y;}$ return z;

Fig. 1. $G4$ executes with boxed statements whereas $G5$ executes without these. G_4 and G_5 perfectly simulate $(F^{P,E}, P, E, E^{-1})$ and $(\mathcal{F}, S_1, S_2, S_3)$, respectively. Clearly $G4$ and $G5$ are identical-until-Bad.

(which keeps the input output behavior of E due to $\mathcal{O}_2$ and $\mathcal{O}_3$ queries) and $\overline{E}$ (which keeps the input output behavior of E for all queries, i.e. it is the union of the previous two unions).

Lemma 2.
$G4 \equiv (F^{P,E}, P, E, E^{-1})$, $G5 \equiv (\mathcal{F}, S_1, S_2, S_3)$ and $G4, G5$ are identical-until-Bad.

Proof. It is easy to see from the pseudocode that $G4, G5$ are identical-until-Bad. The games $G5$ and the oracles simulated by C_A (same as $(\mathcal{F}, S_1, S_2, S_3)$) are actually identical. They have common random sources which are namely $\mathcal{F}$, P and the simulated oracle simE_0 (we can ignore the dead conditional statements which have boxed statements which are not actually executed in game $G5$). Now it remains to show that $G5$ is equivalent to a real game for a PRO attacker. Note that oracles $\mathcal{O}_2$ and $\mathcal{O}_3$ are statistically equivalent to a random permutation $\overline{E}$ and its inverse which are simulated runtime. Moreover $\mathcal{O}_0$ returns $\mathcal{F}(M)$ if $\overline{E}[y], \overline{E}^{-1}[c]$ are undefined or $\overline{E}[y] = c$ where $y = H^P(M)$ and $c = \mathcal{F}(M) \oplus y$. In all other cases $\mathcal{O}_C(M)$ either computes or simulates $c' = \overline{E}[y]$ and returns $c' \oplus y$. So $\mathcal{O}_0(M)$ is statistically equivalent to the oracle $\overline{E}(H^P()) \oplus H^P()$. Hence $G4$ is statistically equivalent to $(F^{P,E}, P, E, E^{-1})$.   ∎

The following result follows immediately from the fact that $\mathcal{F}$ and H^P are statistically independent and $\mathcal{F}$ is a random oracle.

Lemma 3. For any adversary $C^{P,\mathcal{F}}$ making q queries to the n-bit random oracle $\mathcal{F}$ we have $\Pr[\mathcal{F}(M) \oplus H^P(M) = \mathcal{F}(M') \oplus H^P(M'), M \neq M' : (M, M') \leftarrow C] \leq q(q-1)/2^{n+1}$.

Lemma 4. Whenever A^{G5} sets Bad true, C_A also sets one of the Bad events true. Moreover,

$$\Pr[C_A \text{ sets Bad true }] \leq \mathbf{Adv}^{\mathrm{pra}}_{H^P, P, \mathcal{E}}(C_A) + q_3 \times \mathbf{Adv}^{\mathrm{PI}}_{H^P}(q, t) +$$

$$q_0 q_3 \epsilon + \frac{2q_0 q_3 + q_2 q_0}{2^n - q_0 - q_2 - q_3} + \frac{(q_H + q_2 + q_0)^2}{2^{n+1}}.$$

Proof. The first part of the lemma is straightforward and needs to be verified case by case. We leave the details for readers to verify. It is easy to see that whenever Bad_{pra} sets true C is successful in a PrA attack. Now we estimate the probability of the other bad events, from which the theorem follows.

1. $\Pr[\mathsf{Bad}_{mPI} = T] \leq q_3 \times \mathbf{Adv}^{PI}_{H^P}(q^*, t^*)$. It is easy to see that whenever Bad_{mPI} sets true we have a preimage of some y_i which is generated from simE_1. Note that simE_1 responds exactly like a random permutation. So by lemma 1 we have the bound.

2. $\Pr[\mathsf{Bad}_{comp} = T] \leq q_0 q_3 \epsilon$. Whenever Bad_{comp} sets true we should have $H^P(M_i) = y_i$ where M_i is not computable (since it is not in the list given by $\mathcal{E}_{comp}$). So from computable message awareness definition we know that $\Pr[H^P(M_i) = y_i] \leq \epsilon$. The number of such M_i's and y_i's are at most $q_0 q_3$.

The Oracles $\mathcal{O}_2$ (or S_2) and $\mathcal{O}_3$ (or S_3)	The oracles $\mathcal{O}_0$ (or $\mathcal{F}$), $\mathcal{O}_1$ (or P) and Finalization
/The VIL random oracle $\mathcal{F}$ is simulated by C_A/	/The VIL random oracle $\mathcal{F}$ is simulated by C_A/
Initialize : $E_1 = \mathcal{L} = \mathcal{L}_1 = F' = H = \beta = \phi$;	100 On $\mathcal{O}_1$ - query u
Run A and response its oracles	101 $\quad v = P(u)$; $\beta \overset{\parallel}{\leftarrow} (u,v)$; return v;
300 On $\mathcal{O}_3$ - query c	
301 $\quad S = \{X_1, \ldots, X_r\} = \mathcal{E}_{\text{comp}}(\beta)$;	000 On $\mathcal{O}_0$ (or $\mathcal{F}$)- query M
302 $\quad$ For all $1 \le i \le r$ do	001 $\quad z = \mathcal{F}(M)$; $\mathcal{L} \overset{\cup}{\leftarrow} M$;
303 $\quad\quad y_i = H^{\mathcal{O}_1}(X_i) := H[X_i]$; $\mathcal{F}(X_i) = z_i$;	
304 $\quad\quad c_i = y_i \oplus z_i$; $F'[X_i] = c_i$; $\mathcal{L}_1 \overset{\cup}{\leftarrow} X$;	Finalization()
305 $\quad$ If $\exists$ unique i, $c_i = c$,	501 If collision in F' then Bad_{F1} =T;
306 $\quad\quad$ If $E_1[y_i] = \bot$ then $y = y_i$	502 If collision in H then Bad_{PrA} =T; Finish();
307 $\quad\quad$ Else if $\mathcal{O}_3(c') = y_i$ was queried and	503 For all $M \in \mathcal{L}$ do 504-518
308 $\quad\quad$ no i on that query then	504 $\quad z = \mathcal{F}(M)$, $H[M] = H^P(M) = y$;
309 $\quad\quad\quad \text{Bad}_{PI} = T$; $y = \text{simE}_1[-1, c]$;	505 $\quad F'[M] = c = y \oplus z$;
310 $\quad\quad$ Else if no i then $y = \text{simE}_1[-1, c]$;	506 If $F'[X] = c$, $X \ne M$ then Bad_{F1} =T;
311 $\quad\quad$ Else Bad_{F1} =T; $y = \text{simE}_1[-1, c]$;	507 If $H[X] = y$, $X \ne M$ then Bad_{PrA} =T; Finish();
312 $\quad E_1[y] = c$; return c;	508 If $\text{Ext}[y] \ne \bot$, M then Bad_{PrA} =T; Finish();
	509 If $\mathcal{O}_2(y) = c_i$, $y \ne y_i$
200 On $\mathcal{O}_2$ - query $y := y_i$, $i = i + 1$	510 $\quad$ then Bad_{E1} =T;
201 $\quad X = \mathcal{E}(y_i, \beta)$; $\text{Ext} \overset{\cup}{\leftarrow} (y, X)$;	511 Else if $\mathcal{O}_3(c_i) = y \ne y_i$ after M_i-query
202 $\quad y' = H^{\mathcal{O}_1}(X)$, $H \overset{\cup}{\leftarrow} (X, y')$; $z = \mathcal{F}(X)$;	512 $\quad$ then Bad_{comp} =T;
203 $\quad \mathcal{L}_1 \overset{\cup}{\leftarrow} X$; $c = z \oplus y$; $F'[X] = c$;	513 Else if $\mathcal{O}_3(c_i) = y \ne y_i$ before M_i-query
204 $\quad$ If $y' \ne y$	514 $\quad$ then Bad_{F2} =T;
205 $\quad\quad$ then $c = \text{simE}_1[1, y]$;	515 Else if $\mathcal{O}_3(c) = y_i$ after M_i-query, $c \ne c_i$
	516 $\quad$ then Bad_{E2} =T;
206 $\quad$ If $y' = y$, $\mathcal{O}_3(c)$ was queried	517 Else if $\mathcal{O}_3(c) = y_i$ before M_i-query, $c \ne c_i$
207 $\quad\quad$ then Bad_{F1} =T; $c = \text{simE}_1[1, y]$	518 $\quad$ then Bad_{PI} =T;
208 $\quad$ If $y' = y$, $\mathcal{O}_2(y'') = c$ was queried	519 return $\bot$;
209 $\quad\quad$ then $c = \text{simE}_1[1, y]$	
210 $\quad E_1[y] = c$; return c;	

Fig. 2. The oracles simulated by PrA adversary C_A to response an PRO adversary A. It has a finalization procedure which also sets some bad event true. Finish() which is defined similarly as in Fig. 4. It mainly completes the PrA attack. It is easy to see that whenever Finish() is being executed either we have a collision in H^P or there is some message M such that $H^P(M) = y, (y, M) \notin \text{Ext}$.

3. All other bad events hold due to either the special outputs of $\mathcal{F}(M)$ (when $\mathsf{Bad}_{F1} = T$ or $\mathsf{Bad}_{F2} = T$, we apply the lemma 3) or the special outputs of $\mathsf{simE}(c)$ (when $\mathsf{Bad}_{E1} = T$ and $\mathsf{Bad}_{E2} = T$). One can show the following:

$$\Pr[\mathsf{Bad}_{E1\vee E2\vee F1\vee F2} = T] \leq \frac{2q_0q_3 + q_2q_0}{2^n - q_0 - q_2 - q_3} + \frac{(q_H + q_2 + q_0)^2}{2^{n+1}}.$$

We have used Lemma 3 to bound the bad event Bad_{F1}. The other bad event probability calculations are straightforward. We leave details to readers. ∎

The main theorem of the section follows from the above lemmas.

Theorem 1. *For any indifferentiability adversary A making at most (q_0, q_1, q_2, q_3) queries to its four oracles with bit-size l_{max} for the longest $\mathcal{O}_0$-query, there exists a PrA $(q, q_2 + 1, t)$-adversary C_A with runtime $t = \mathsf{Time}(A) + O(q_2 \cdot \mathsf{Time}(\mathcal{E}) + q_0 + q_1 + (q_2 + q_0)\mathsf{NQ}[l_{max}])$ and*

$$\mathbf{Adv}_{F,S}^{\mathrm{pro}}(A) \leq \mathbf{Adv}_{H^P,P,\mathcal{E}}^{\mathrm{pra}}(C_A) + q_3 \times \mathbf{Adv}_{H^P}^{\mathrm{PI}}(q,t) + \delta,$$

where $\delta = q_0q_3\epsilon - \frac{2q_0q_3 + q_2q_0}{2^n - q_0 - q_2 - q_3} + \frac{(q_H + q_2 + q_0)^2}{2^{n+1}}$, $H^P(\cdot)$ is preimage resistant and (q^*, q_H, ϵ)-computable message aware for an efficient computable message extractor $\mathcal{E}_{\mathrm{comp}}$ where $q^* = q_1 + q_2\mathsf{NQ}[l_{max}]$.

4 Application of Theorem 1: PRO Analysis of a Variant of Grøstl

As an application of Theorem 1 we prove the PRO analysis of a variant of Grøstl hash function in which the output transformation is based on a permutation independent of the permutations used in iteration. The compression function $f^{P,Q}(z, m) = P(z \oplus m) \oplus Q(m) \oplus z$, where P and Q are invertible permutations in n-bit modeled to be independent random permutations (adversary can also have access to inverses). The hash function $H^{P,Q}$ of Grøstl without output transformation is Merkle-Damgård with strengthening (SMD) and the output transformation is $\mathrm{trunc}_s(P(x) \oplus x)$. In case of the variant of the hash function, the output transformation is same as the previous section, i.e. $OT(x) = E(x) \oplus x$ where E is a random permutation independent with P and Q. Since SMD preserves preimage awareness and preimage resistance of the underlying compression function, we focus on the proof of the compression function $f^{P,Q}$ to prove PrA and preimage resistance. The proof of the following lemmas are straightforward are given in the full version of the paper [5].

Lemma 5. *For any advise string α_P and α_Q of sizes $(q_1 + q_2)$ and $(q_3 + q_4)$ (for (P, P^{-1}) and (Q, Q^{-1}) respectively) the number of computable messages is at most $q_f \leq (q_1 + q_2)(q_3 + q_4)$ and for any non-computable message (z, m), $\Pr[f^{P,Q}(z, m) = c | \alpha_P, \alpha_Q] \leq \frac{1}{2^n - max(q_1 + q_2, q_3 + q_4)}$. Moreover there is an efficient computable message extractor $\mathcal{E}_{\mathrm{comp}}^f$ which can list all computable messages.*

Let $\mathbf{q} = (q_1, q_2, q_3, q_4)$. Now given the computable message extractor one can define a PrA extractor $\mathcal{E}^f$ as follows: $\mathcal{E}^f(y, \alpha_P, \alpha_Q) = (z, m)$ if there exists an unique computable message (from the list given by $\mathcal{E}^f_{comp}$) such that $f(z, m) = y$, otherwise it returns any arbitrary message.

Lemma 6. $\mathrm{Adv}^{\mathrm{PI}}_{H^{P,Q}}(\mathbf{q}, t) \leq \mathrm{Adv}^{\mathrm{PI}}_{f^{P,Q}}(\mathbf{q}, t) \leq \frac{(q_1+q_2)(q_3+q_4)}{2^n - max(q_1+q_2, q_3+q_4)}$ for any t.

Lemma 7. Let $\mathbf{q} = (q_1, q_2, q_3, q_4)$ and let $f^{P,Q} = P(h \oplus m) \oplus Q(m) \oplus h$, where P and Q are invertible ideal permutations. For any preimage awareness $(\mathbf{q}, e, t)$-adversary A making at most $\mathbf{q}$ queries to the oracles P, P^{-1}, Q, Q^{-1}, there exists an extractor $\mathcal{E}$ such that

$$\mathrm{Adv}^{\mathrm{pra}}_{f^{P,Q}, P, Q, \mathcal{E}}(A) \leq \frac{e(q_1+q_2)(q_3+q_4)}{2^n - max(q_1+q_2, q_3+q_4)} + \frac{(q_1+q_2)^2(q_3+q_4)^2}{2(2^n - max(q_1+q_2, q_3+q_4))},$$

Theorem 2. Let $Gr\o stl'(M) = P'(H^{P,Q}(M)) \oplus H^{P,Q}(M)$ where $H^{P,Q}$ is Grøstl without the output transformation and P, Q, P' are independent random permutations. Then for any adversary making at most q queries to all its oracles the PRO-advantage is bounded by $\ell_{max} q^2 q'^2 / 2^{n-2}$ if If $q' = q_1 + q_2 + q_3 + q_4 + l_{max} \leq 2^{n-1}$.

Proof. The result follows from Lemma 9, 10, 11 and 12, and Theorem 1. ∎

Remark 1. Our bound $\ell_{max} q^2 q'^2 / 2^{n-2}$ in the variant of Grøstl seems to be reasonable as we indeed have collision on the compression function in $2^{n/4}$ complexity i.e. the collision advantage is $q^4 / 2^n$. We believe the designers also noted that and this is why they consider at least double length hash function. We also strongly believe that the same bound can be achieved for original Grøstl. However to prove that we cannot apply Theorem 2 directly. This would be our one of the future research work.

5   PRO Analysis of Hash Functions with PGV Output Transformations

In the previous section, we considered the case that the finial output transformation is $OT(x) = E(x) \oplus x$. In this section, we consider 20 PGV compression functions shown in Table 1 as candidates of the final output transformation OT. Such 20 PGV hash functions based on them were proved to be collision resistant in the ideal cipher model [4]. More precisely, we will consider the case that $F^{P,E}(M) = OT(H^P(M_1), M_2)$, where E is an ideal cipher, $M = M_1 \| M_2$, $H^P(M_1)$ corresponds to h_{i-1} and M_2 corresponds to m_i in Table 1. Except for PGV 11, 13, 15-20 (See Example 2 and 3), Theorem 3 holds. The proof of the Theorem 3 is same as Davis-Meyer case. However we give the proof idea so that the reader could justify themselves.

Proof Idea for 1-10, 12 and 14: Like to the Davis-Meyer case we only need to worry about the E^{-1} query since we have chosen those PGV compression functions which behave like random oracle if adversary makes only E queries. Note that 5-10 and 12 and 14 PGV have w_i as keys. So given a $E_w^{-1}(y)$ query simulator can make the list of all h which can be computed, i.e. $H^P(M) = h$, and guess $m = h \oplus w$. Once simulator guesses m he can make $\mathcal{F}$ queries (M, m) and obtains responses z's. Now simulator can find a correct m if y is really obtained by some $\mathcal{F}(M, m)$. If there is no such m, simulator can response randomly and any bad behavior would be either bounded by the collision probability or by the preimage attack. The same argument works for PGV 1-4 since $H^P(M)$ is xor-ed with the the $E()$ output. So simulator can verify the correct h among all computable hash outputs.

Example 2. See PGV 15 in Table 1, which is $E_{m_i}(w_i) \oplus v$, where $w_i = h_{i-1} \oplus m_i$ and v is a constant. Now we want to give an indifferentiable attack on $F^{P,E}$ based on PGV 15, even though H^P is preimage aware, preimage resistant, and (q, q_H, ϵ)-computable message aware with feasible q and q_H and negligible ϵ. Let H^P be an Merkle-damgård construction with strengthening, where the underlying compression function is a preimage aware function based on the ideal primitive P. As shown in [8,9], SMD (Merkle-damgård construction with strengthening) preserves preimage awareness of the compression function. Also SMD preserves preimage resistance. So, H^P is also preimage aware and preimage resistant. We assume that H^P is (q, q_H, ϵ)-computable message aware with feasible q and q_H and negligible ϵ. Now we construct an indifferentiability adversary A for $F^{P,E}(M) = OT(H^P(M_1), M_2)$, where $OT(x, y) = E_y(x \oplus y) \oplus v$ is PGV 15, and v is a constant. First, A chooses a random query $M = M_1 \| M_2$ to $\mathcal{O}_1$, where $(\mathcal{O}_1, \mathcal{O}_2, \mathcal{O}_3, \mathcal{O}_4)$ is $(F^{P,E}, P, E, E^{-1})$ or $(\mathcal{F}, S_1^{\mathcal{F}}, S_2^{\mathcal{F}}, S_3^{\mathcal{F}})$ for any simulator $S^{\mathcal{F}} = (S_1^{\mathcal{F}}, S_2^{\mathcal{F}}, S_3^{\mathcal{F}})$. A gets its response z from $\mathcal{O}_1$. And A hands $(M_2, z \oplus v)$ over to $\mathcal{O}_4$. Then, A gets its response h from $\mathcal{O}_4$. A makes a new query $(M_2', h \oplus M_2 \oplus M_2')$ to $\mathcal{O}_3$. Then, A gets its response c from $\mathcal{O}_3$. Finally, A hands $(M_1 \| M_2')$ over to $\mathcal{O}_1$ and gets its response z'. If $(\mathcal{O}_1, \mathcal{O}_2, \mathcal{O}_3, \mathcal{O}_4)$ is $(F^{P,E}, P, E, E^{-1})$, $c \oplus v = z'$. On the other hand, since any simulator cannot know M_1, $c \oplus v \neq z'$ with high probability. Therefore, $F^{P,E}$ based on PGV 15 is not indifferentiable from a VIL random oracle $\mathcal{F}$. In the similar way, cases of PGV 11, 13, and 16 are not secure.

Example 3. See PGV 17 in Table 1, which is $E_{h_{i-1}}(m_i) \oplus m_i$. Firstly, we define a hash function $H^F(x) : \{0, 1\}^* \to \{0, 1\}^n$ as follows, where c is any n-bit constant and P is a VIL random oracle with n-bit output size.

$$H^P(x) = \begin{cases} c, & \text{if } x = c; \\ P(x), & \text{otherwise.} \end{cases}$$

In the similar way with the proofs of Section 6, we can prove that H^P is preimage aware, preimage resistant, $(q, q_H(= q + 1), 1/2^n)$-computable message aware, where q_H is the number of computable messages obtained from q input-output pairs of P. Now we want to give an indifferentiable attack on $F^{P,E}$ based on PGV 17. We construct an indifferentiability adversary A for $F^{P,E}(M) = OT(H^P(M_1), M_2)$, where $OT(x, y) = E_x(y) \oplus y$ is PGV 17. First, A chooses a query $M = c\|M_2$ to $\mathcal{O}_1$, where M_2 is a randomly chosen one-block message. A gets its response z from $\mathcal{O}_1$. And A hands $(c, z \oplus M_2)$ over to $\mathcal{O}_4$. Then, A gets its response m from $\mathcal{O}_4$. If $(\mathcal{O}_1, \mathcal{O}_2, \mathcal{O}_3, \mathcal{O}_4)$ is $(F^{P,E}, P, E, E^{-1})$, $m = M_2$. On the other hand, since any simulator cannot know M_2, $m \neq M_2$ with high probability. Therefore, $F^{P,E}$ based on PGV 17 is not indifferentiable from a VIL random oracle $\mathcal{F}$. In the similar way, cases of PGV 18-20 are not secure.

Theorem 3. *[PRO Construction via 12 PGVs]*
Let $F^{P,E}(M) = OT(H^P(M_1), M_2)$, where $M = M_1\|M_2$, OT is any PGV constructions except for PGV 11, 13, 15-20, E^{-1} is efficiently computable, and E is an ideal cipher. For any indifferentiability adversary A making at most (q_0, q_1, q_2, q_3) queries to its four oracles with bit-size l_{max} for the longest $\mathcal{O}_0$-query, there exists a PrA $(q, q_2 + 1, t)$-adversary C_A with runtime $t = \mathsf{Time}(A) + O(q_2 \cdot \mathsf{Time}(\mathcal{E}) + q_0 + q_1 + (q_2 + q_0)\mathsf{NQ}[l_{max}])$ and

$$\mathbf{Adv}^{\mathrm{pro}}_{F,S}(A) \leq \mathbf{Adv}^{\mathrm{pra}}_{H^P, P, \mathcal{E}}(C_A) + q_3 \times \mathbf{Adv}^{\mathrm{PI}}_{H^P}(q, t)$$

$$+ q_0 q_3 \epsilon + \frac{2q_0 q_3 + q_2 q_0}{2^n - q_0 - q_2 - q_3} + \frac{(q_H + q_2 + q_0)^2}{2^{n+1}},$$

where $H^P(\cdot)$ is preimage resistant and (q^*, q_H, ϵ)-computable message aware for an efficient computable message extractor $\mathcal{E}_{\mathrm{comp}}$ where $q^* = q_1 + q_2 \mathsf{NQ}[l_{max}]$.

Proof. The proof is very similar to that of Theorem 1. It will be referred to the full paper [5]. ∎

Table 1. 20 Collision Resistant PGV Hash Functions in the Ideal Cipher Model [4]. ($w_i = m_i \oplus h_{i-1}$)

Case	PGV	Case	PGV
1	$E_{m_i}(h_{i-1}) \oplus h_{i-1}$	11	$E_{m_i}(h_{i-1}) \oplus v$
2	$E_{m_i}(w_i) \oplus w_i$	12	$E_{w_i}(h_{i-1}) \oplus v$
3	$E_{m_i}(h_{i-1}) \oplus w_i$	13	$E_{m_i}(h_{i-1}) \oplus m_i$
4	$E_{m_i}(w_i) \oplus h_{i-1}$	14	$E_{w_i}(h_{i-1}) \oplus w_i$
5	$E_{w_i}(m_i) \oplus m_i$	15	$E_{m_i}(w_i) \oplus v$
6	$E_{w_i}(h_{i-1}) \oplus h_{i-1}$	16	$E_{m_i}(w_i) \oplus m_i$
7	$E_{w_i}(m_i) \oplus h_{i-1}$	17	$E_{h_{i-1}}(m_i) \oplus m_i$
8	$E_{w_i}(h_{i-1}) \oplus m_i$	18	$E_{h_{i-1}}(w_i) \oplus w_i$
9	$E_{w_i}(m_i) \oplus v$	19	$E_{h_{i-1}}(m_i) \oplus w_i$
10	$E_{w_i}(m_i) \oplus w_i$	20	$E_{h_{i-1}}(w_i) \oplus m_i$

6   PRO Attacks on Hash Functions with Some DBL Output Transformations

Here, we consider DBL (Double Block Length) output transformations. Unfortunately, many constructions with DBL output transformations are not indifferentiability secure, even though H^P satisfies all requirements as mentioned before.

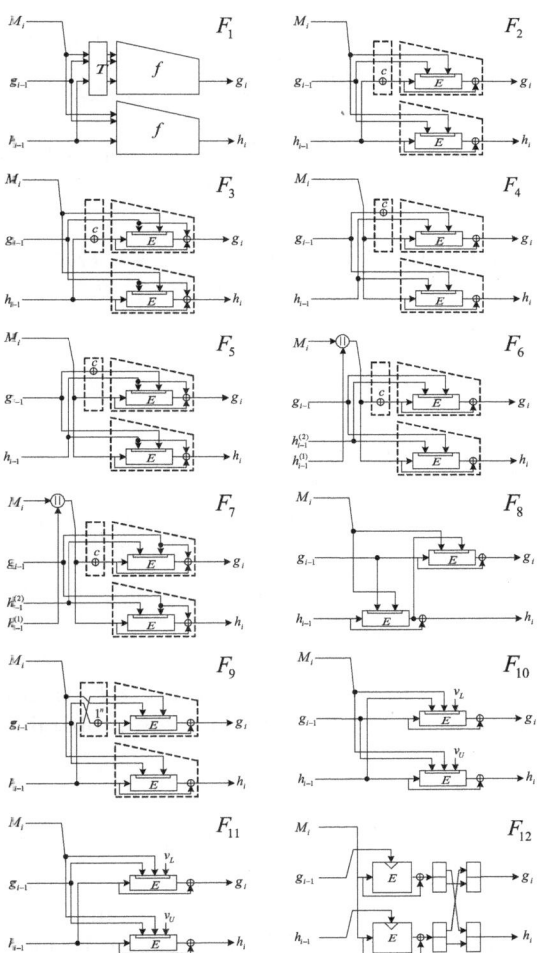

Fig. 3. Double Block Length constructions : F_i $i = 1$ to 12

6.1   The Case of $OT(x) = f(x)||f(x \oplus p)$

There are several DBL compression functions of the form $OT(x) = f(x)||f(x \oplus p)$ [19,13] where p is a non-zero constant and f is any function. See Fig. 3, where F_1 was proposed by Nandi in [19] and $F_2 \sim F_7$ were proposed by Hirose in [13]. In fact, the T in F_1 Fig. 3 is a permutation without any fixed point and $T^2 = id$. Here, we consider only $T(x) = x \oplus p$, where p is a non-zero constant. We define a hash function $H^P(x) : \{0,1\}^* \to \{0,1\}^n$ as follows, where c is any n-bit constant and P is a VIL random oracle with n-bit output size. 1^n and 0^n indicate the n-bit one and zero strings.

$$H^P(x) = \begin{cases} c \oplus p, & \text{if } x = 0^n; \\ c, & \text{if } x = 1^n; \\ P(x), & \text{otherwise.} \end{cases}$$

The following theorems show that H^P is preimage aware, q_H is small (more precisely computable-awareness) and preimage resistant. The proofs are given in the full version [5].

Theorem 4. *Let H^P be the above hash function. For any preimage awareness (q, e, t)-adversary A making at most q queries to the oracles P, there exists an extractor $\mathcal{E}$ such that*

$$\text{Adv}^{\text{pra}}_{H^P, P, \mathcal{E}}(A) \leq \frac{eq}{2^n} + \frac{(q+2)^2}{2^{n+1}}, \text{ and } q_H \leq q + 2.$$

Theorem 5. *Let H^P be the above hash function. Let q be the maximum number of queries to P. For any preimage-finding adversary A with q queries to P, $\text{Adv}^{\text{PI}}_{H^P}(A) \leq \frac{3+q}{2^n}$. For any n-bit y and any M not computable from any advise string α which consists of q query-response pairs of P, $\Pr[H^P(M) = y|\alpha] \leq 1/2^n$.*

Indifferentiability Attack on $F(M) = H^P(M)||(H^P(M) \oplus p)$.
Let $(\mathcal{O}_1, \mathcal{O}_2, \mathcal{O}_3)$ be $(F^{P,f}, P, f)$ or $(\mathcal{F}, S_1^{\mathcal{F}}, S_1^{\mathcal{F}})$ for any simulator S. Now we define an adversary A as follows. First, A makes query '0' and '1' to $\mathcal{O}_1$. Then, A obtains responses $(a_1||a_2)$ and $(b_1||b_2)$. If $\mathcal{O}_1 = F$, then $a_1 = b_2$ and $a_2 = b_1$. But, if $\mathcal{O}_1 = \mathcal{F}$, $a_1 = b_2$ and $a_2 = b_1$ with the probability $1/2^n$. So, F is not indifferentiably secure.

6.2   PRO Attack with $OT(x) = F_i(x)$ for $i = 8, 12$, (Fig. 3)

In the case of F_8 proposed by Lai and Massey in [15], which is called Tandem DM, there is the following structural weakness. If for any a $g_{i-1} = h_{i-1} = M_i = a$ in F_8 of Fig. 3, then $h_i \oplus g_i = a$. We can show an indifferentiability attack on $F(M) = F_9(H^P(M))$, where H^P is preimage aware and q_H is small. We define a hash function $H^P(x) : \{0,1\}^* \to \{0,1\}^n$ as follows, where c is any $n/2$-bit constant and P is a VIL random oracle with n-bit output size.

$$H^P(x) = \begin{cases} c||c, & \text{if } x = 0; \\ P(x), & \text{otherwise.} \end{cases}$$

We can easily show that H^P is preimage aware, preimage resistant and $(q_P, q_H(= q_P + 1), 1/2^n)$-computable message aware, where q_H is the number of computable messages obtained from q_P input-output pairs of P. and q_H is small.

Then, we show that $F(M_1||M_2) = F_8(H^Q(M_1), M_2)$ is not indifferentiable from a VIL random oracle $\mathcal{F}$ as follows. A makes a query '$(0||c)$' to $\mathcal{O}_1$ and get its response $z = (z_1||z_2)$. A checks if $z_1 \oplus z_2 = c$. If $\mathcal{O}_1$ is $\mathcal{F}$, $z_1 \oplus z_2 = c$ with the probability $1/2^{n/2}$. On the other hand, if $\mathcal{O}_1$ is F, $z_1 \oplus z_2 = c$ with probability 1. So F is not indifferentiability secure. In the case of F_{12}, which is called MDC-2, if the values of M_i and h_{i-1} are fixed, the half of bits of the output of F_{12} is also fixed regardless of what g_{i-1} is. Using this weakness, in a similar way as shown in above, we also can construct H^P such that $F(M_1||M_2) = F_{12}(H^P(M_1), M_2)$ is not indifferentiably secure.

7 Conclusion

In this paper we extend the applicability of preimage-awareness in those hash functions whose output transformation cannot be modeled as a random oracle. We choose Davis-Meyer as an output transformation based on a random permutation and show that the hash function is PRO if H^P is PrA, preimage resistant and computable message aware. The computable message awareness is a new notion introduced here similar to PrA. However this is not same as PrA as we can see the separation among these notions. As an application to our result we prove the PRO property of a variant of Grøstl hash function. We similarly prove that 12 PGV compression function out of 20 collision resistant PGV hash functions can be employed as output transformation with the similar assumption on H^P. However, some the popular double length hash function can not be used as we have shown PRO attacks. In summary, we study the choice of output transformation beyond the random oracle model and found both positive and negative results.

References

1. Andreeva, E., Mennink, B., Preneel, B.: On the Indifferentiability of the Grøstl Hash Function. In: Garay, J.A., De Prisco, R. (eds.) SCN 2010. LNCS, vol. 6280, pp. 88–105. Springer, Heidelberg (2010)
2. Bellare, M., Kohno, T., Lucks, S., Ferguson, N., Schneier, B., Whiting, D., Callas, J., Walker, J.: Provable Security Support for the Skein Hash Family, http://www.skein-hash.info/sites/default/files/skein-proofs.pdf
3. Bellare, M., Rogaway, P.: The Security of Triple Encryption and a Framework for Code-Based Game-Playing Proofs. In: Vaudenay, S. (ed.) EUROCRYPT 2006. LNCS, vol. 4004, pp. 409–426. Springer, Heidelberg (2006)
4. Black, J.A., Rogaway, P., Shrimpton, T.: Black-box analysis of the block-cipher-based hash-function constructions from PGV. In: Yung, M. (ed.) CRYPTO 2002. LNCS, vol. 2442, pp. 320–335. Springer, Heidelberg (2002)

5. Chang, D., Nandi, M., Yung, M.: On the Security of Hash Functions Employing Blockcipher Postprocessing, http://eprint.iacr.org/2010/629
6. Coron, J.S., Dodis, Y., Malinaud, C., Puniya, P.: Merkle-Damgard Revisited: How to Construct a Hash Function. In: Shoup, V. (ed.) CRYPTO 2005. LNCS, vol. 3621, pp. 430–448. Springer, Heidelberg (2005)
7. Damgård, I.B.: A design principle for hash functions. In: Brassard, G. (ed.) CRYPTO 1989. LNCS, vol. 435, pp. 416–427. Springer, Heidelberg (1990)
8. Dodis, Y., Ristenpart, T., Shrimpton, T.: Salvaging Merkle-Damgård for Practical Applications. In: Joux, A. (ed.) EUROCRYPT 2009. LNCS, vol. 5479, pp. 371–388. Springer, Heidelberg (2009)
9. Dodis, Y., Ristenpart, T., Shrimpton, T.: Salvaging Merkle-Damgård for Practical Applications. Full version of [6], Cryptology ePrint Archive: Report 2009/177
10. Ferguson, N., Lucks, S., Schneier, B., Whiting, D., Bellare, M., Kohno, T., Callas, J., Walker, J.: The Skein Hash Function Family. Submission to NIST (2008)
11. Gauravaram, P., Knudsen, L.R., Matusiewicz, K., Mendel, F., Rechberger, C., Schläffer, M., Thomsen, S.S.: Grøstl - a SHA-3 candidate. Submission to NIST (2008)
12. Hirose, S.: Secure Double-Block-Length Hash Functions in a Black-Box Model. In: Park, C.-s., Chee, S. (eds.) ICISC 2004. LNCS, vol. 3506, pp. 330–342. Springer, Heidelberg (2005)
13. Hirose, S.: How to Construct Double-Block-Length Hash Functions. In: Second Hash Workshop (2006)
14. Kelsey, J.: Some notes on Grøstl (2009), http://ehash.iaik.tugraz.at/uploads/d/d0/Grostl-comment-april28.pdf
15. Lai, X., Massey, J.L.: Hash Functions Based on Block Ciphers. In: Rueppel, R.A. (ed.) EUROCRYPT 1992. LNCS, vol. 658, pp. 55–70. Springer, Heidelberg (1993)
16. Maurer, U.: Indistinguishability of Random Systems. In: Knudsen, L.R. (ed.) EUROCRYPT 2002. LNCS, vol. 2332, pp. 110–132. Springer, Heidelberg (2002)
17. Maurer, U., Renner, R., Holenstein, C.: Indifferentiability, Impossibility Results on Reductions, and Applications to the Random Oracle Methodology. In: Naor, M. (ed.) TCC 2004. LNCS, vol. 2951, pp. 21–39. Springer, Heidelberg (2004)
18. Merkle, R.C.: One way hash functions and DES. In: Brassard, G. (ed.) CRYPTO 1989. LNCS, vol. 435, pp. 428–446. Springer, Heidelberg (1990)
19. Nandi, M.: Towards Optimal Double-Length Hash Functions. In: Maitra, S., Veni Madhavan, C.E., Venkatesan, R. (eds.) INDOCRYPT 2005. LNCS, vol. 3797, pp. 77–89. Springer, Heidelberg (2005)
20. Preneel, B., Govaerts, R., Vandewalle, J.: Hash Functions Based on Block Ciphers: A Synthetic Approach. In: Stinson, D.R. (ed.) CRYPTO 1993. LNCS, vol. 773, pp. 368–378. Springer, Heidelberg (1994)
21. Winternitz, R.: A Secure Hash Function built from DES. In: Proceedings of the IEEE Symp. on Information Security and Privacy, pp. 88–90. IEEE Press, Los Alamitos (1984)

Appendix A. Revisiting the Proof of "RO(PrA(·)) = PRO(·)"

In [8,9] it was proved that $F^{R,P}(M) = \mathcal{R}(H^P(M))$ is indifferentiable from a VIL random oracle $\mathcal{F}$, where $\mathcal{R} : \{0,1\}^m \rightarrow \{0,1\}^n$ is a FIL random oracle, P is an ideal primitive, and $H^P : \mathcal{M} \rightarrow \{0,1\}^m$ is preimage-aware. The result can be used to prove the indifferentiable security of any hash function which uses a post-processor defined independently from the underlying iteration function P. In the course of our studies we have found that the proof given in [8,9] is not completely correct (though the claims remain correct). We have reported this in a limited distribution abstracts on February and May 2010 (recently in October 2010, a correction on the e-print version has appeared by the original coauthors, further confirming our findings). Let us review the issues. There are two main flaws (to be described below) in the proof, and we need to provide alternative definitions of simulator and preimage-aware attacker to fix them. (We note that while somewhat technical, the revision is crucial). Let $NQ[l]$ be the number of P-queries required for the computation of $H^P(M)$ for $|M| = l$. We denote $\mathsf{Time}(\cdot)$ and $\mathsf{STime}(\cdot)$ to mean the run time and simulation run time of an algorithm. Now we restate the Theorem 4.1. in [8,9] (in terms of our PrA terminologies) and provide a sketch of the proof given in [8].

Theorem 4.1 of [8,9]
For any given efficient extractor $\mathcal{E}$, there exists a simulator $S = (S_1, S_2)$ with $\mathsf{Time}(S) = O(q_1 \cdot \mathsf{STime}(P) + q_2 \cdot \mathsf{Time}(\mathcal{E}))$. The simulator makes at most q_2 $\mathcal{F}$-queries. For any indifferentiability adversary $A^{\mathcal{O}_0, \mathcal{O}_1, \mathcal{O}_2}$ making at most (q_0, q_1, q_2) queries to its three oracles with bit-size l_{max} for the longest $\mathcal{O}_0$-query, there exists a $(q_1 + q_0 \cdot NQ[l_{max}], q_2 + 1, t)$-PrA adversary B_A^P with runtime $t = \mathsf{Time}(A) + O(q_0 \cdot NQ[l_{max}] + q_1 + q_2 \mathsf{Time}(\mathcal{E}))$ such that

$$\mathbf{Adv}_{F,S}^{\mathrm{pro}}(A) \le \mathbf{Adv}_{H^P, P, \mathcal{E}}^{\mathrm{pra}}(B_A).$$

Outline of Proof of Theorem 4.1. of [9]. Let $\mathcal{E}$ be an arbitrary extractor for H. Then $S = (S_1; S_2)$ works as follows. It maintains an internal advice string α (initially empty) that will consist of pairs $(u; v)$ corresponding to A's queries to P (via S_1). When A queries u to S_1, the simulator simulates $v \leftarrow P(u)$ appropriately, sets $\alpha \leftarrow \alpha \| (u; v)$, and returns v. For a query Y to S_2, the simulator computes $X \leftarrow \mathcal{E}(Y; \alpha)$. If $X = \perp$ then the simulator returns a random point. Otherwise it simulates $Z \leftarrow \mathcal{F}(X)$ and returns Z to the adversary. The games $R0, I1, G0. G1$ and B_A have been defined in [9] and the authors claimed the following:

(1) $G1 \equiv I1 \equiv (\mathcal{F}, S_1, S_2)$, (2) $G0 \equiv R0 \equiv (F^{P, \mathcal{R}}, P, \mathcal{R})$.

Due to the above claim the PRO-advantage of any adversary A is nothing but $|\Pr[A^{G0} = 1] - \Pr[A^{G1} = 1]|$. From the pseudocodes of games $G0$ and $G1$, it is easy to see that they are identical-until-Bad. Hence $\mathbf{Adv}_{F,S}^{\mathrm{pro}}(A) \le \Pr[A^{G1}$ sets Bad true]. The proof proceeds by defining a PRA-adversary B_A which makes preimage-aware attack successfully whenever B_A sets Bad true. Since B_A sets Bad true only if it finds a collision of H^P or finds a message M such that $\mathcal{E}(\alpha, Y) \ne M$ where $Y = H^P(M)$. So $\Pr[B_A$ sets Bad true] $\le \mathbf{Adv}_{H^P, P, \mathcal{E}}^{\mathrm{pra}}(B_A)$. The theorem follows immediately from the following claim:

(3) $\Pr[A^{G1}$ sets Bad true] $\le \Pr[B_A$ sets Bad true]. $\blacksquare$

7.1   Problems in Proof of Theorem 4.1. of [9]

In this section we explain the flaws we observe in the proof of Theorem 4.1. of [9]. To understand it one needs to go through the definitions of the games $G0$, $R0$, $G1$ and $G(B)$ (the tuple of three oracles simulated by B) described in [9].[2]

Flaw 1. $G0$ is not equivalent to $R0$.
If $\mathcal{O}_0$ in $G0$ has not been queried before (so the Bad event in $G0$ would not occur) then the output of $\mathcal{O}_2(Y)$ query is $\mathsf{F}(X)$ whenever $X = \mathcal{E}(Y, \alpha) \neq \bot$, otherwise it returns $\mathsf{R}(X)$ where F and R perfectly simulate two independent random oracles $\mathcal{F}$ and $\mathcal{R}$ respectively. We show that $\mathcal{O}_2$ cannot be equivalent to a random oracle. Suppose $\mathcal{E}$ is an extractor which returns a special message M^* whenever the advise string α is empty. If A makes first two successive distinct $\mathcal{O}_2$ queries $Y_{2,1}$ and $Y_{2,2}$ then $X_{2,1} = X_{2,2} = M^*$ and hence the outputs of $\mathcal{O}_2$ in game $G0$ are identical (same as $\mathsf{F}[M^*]$).

To get rid of the above problem, we can do the following steps in $\mathcal{O}_2$ (also in simulator S_2) immediately after it obtains $X = \mathcal{E}(Y, \alpha)$: Compute $H^P(X) = Y'$ and check whether it is the same as Y or not. If extractor returns a correct message, i.e. $Y = Y'$, then S_2 or $\mathcal{O}_2$ returns $\mathcal{F}(M)$. Otherwise, it returns randomly. To compute $H^P(X)$ one might need to simulate some P outputs (in case of $G0$) or make P-queries (in case of $\mathcal{O}_2$ of B).

Flaw 2. $G(B) \not\equiv G1$ and $G1, G(B)$ are not identical-until-Bad.
We first observe that the advise string α in $G1$ is not the same as that of B since the advice string α is updated whenever A has access to the oracle $\mathcal{O}_1$ in Game $G1$, but the advice string is updated whenever A has access to the oracle $\mathcal{O}_0$ and $\mathcal{O}_1$ of B in Fig 3 of [9]. For example, let $\mathcal{E}(Y, \alpha)$ return a message M whenever $H^P(M) = Y$ is "computable' from α otherwise return $\bot$. Any adversary which can guess $H^P(M)$ correctly and turn it to $\mathcal{O}_2$ query then $\mathcal{O}_2^B(Y|\tau)$ returns z. However, $\mathcal{O}_2^{G1}(Y|\tau)$ returns a random string $\mathsf{R}[Y]$ since α is the empty string in A^{G1}. So $G(B) \not\equiv G1$. One can similarly show that $G1, G(B)$ are not identical-until-Bad.

A possible attempt is to update the advise string for $\mathcal{O}_0$ queries in all games, in particular $G1$. However, if we do so then the simulator is not independent of $\mathcal{F}$-queries (since the advise string is updated whenever there is a $\mathcal{O}_0$-query and the advise string is used to define the response of S_2). On the other hand, we cannot ignore the $H^P(M)$ computation in B for $\mathcal{O}_0$ queries of A. This computation is essential to making PrA attack successfully. It seems impossible to handle the advise string so that it is updated in the same way for all games as well as $H^P(\cdot)$ computations are made for $\mathcal{O}_0$-queries. We can solve the problem if **we postpone the computation of H^P until all queries of A are made.** So we need a finalization procedure in B which essentially does all $H^P(M)$ computations of $\mathcal{O}_0(M)$-queries.

---

[2] We have defined the revised version of these games in the paper. We refer readers to [9] to see the original definitions to understand the flaws.

7.2   Revised Proof of Theorem 4.1 of [9]

We state the corrected version of theorem 4.1. below. The revised version of $B :=$ B_A, simulators and the games $G0$ and $G1$ are defined in Fig. 1. The adversary B_A has a subroutine called Finish() which is defined trivially. It mainly completes the PrA attack. It is easy to see that whenever Finish() is being executed either we have a collision in H^P or there is some message M such that $H^P(M) = y, (y, M) \notin$ Ext. For simplicity we ignore the details of the subroutine. Let $q = q_1 + (q_0 + q_2) \cdot$ NQ$[l_{max}]$.

Lemma 8. $G1 \equiv (\mathcal{O}_0^{B_A}, \mathcal{O}_1^{B_A}, \mathcal{O}_2^{B_A}) \equiv (\mathcal{F}, S_1, S_2)$. Games $G0$ and $G1$ are identical-until-Bad.

The Lemma is obvious from the games described in Fig. 1. We leave readers to verify. The following lemma essentially says that $G0$ is equivalent to $(\mathcal{R}(H^P), P, \mathcal{R})$. The proof of the lemma is easy to verify and we skip it for the full version.

Lemma 9. $G0 \equiv (\mathcal{R}(H^P), P, \mathcal{R})$, i.e. for any distinguisher A, the output distribution of A^{G0} and $A^{\mathcal{R}(H^P), P, \mathcal{R}}$ are identically distributed.

Lemma 10. Whenever A^{G1} sets Bad true, B_A sets Bad true and B_A makes PrA attack successful. So we have $\Pr[A^{G1}$ sets bad$] \leq \Pr[B_A$ sets bad$] \leq$ **Adv**$_{H^P, P, \mathcal{E}}^{pra}(B_A)$.

Proof. We already know from Lemma 8 that $G1$ is equivalent to the oracles simulated by B_A. However, the two games defined bad event in different manners. The game $G1$ sets bad during the computation of responses whereas the adversary B_A sets bad after all responses of the queries. A^{G1} sets bad true in line 209, 003 and in line 206, 005. We can see that if the conditional statements written in 209 and 003 in game $G1$ hold then we have a collision in H^P (there exist $M \neq X$ such that $H^P(M) = H^P(X)$). So we have PrA attack which is taken care of in 401 in the second step of B_A. For the lines 205 and 005 we have M such that $H^P(M) = y$ and Ext$[y] \neq M$, i which case PrA attack is possible due to incorrect guess of the extractor. This has been taken care of in 403. ∎

By using the above lemmas the theorem follows immediately

Theorem 6 (Ro domain extension via PrA). For any given extractor $\mathcal{E}$ we can construct a simulator $S = (S_1, S_2)$ with Time$(S) = O((q_1 + q_2 \cdot$ NQ$[l_{max}]) \cdot$ STime$(P) + q_2 \cdot$ Time$(\mathcal{E}))$. For any indifferentiability adversary $A^{\mathcal{O}_0, \mathcal{O}_1, \mathcal{O}_2}$ making at most (q_0, q_1, q_2) queries to its three oracles with bit-size l_{max} for the longest $\mathcal{O}_0$-query, there exists a $(q, q_2 + 1, t)$-adversary B with runtime $t = $ Time$(A) +$ $O(q_2 \cdot$ Time$(\mathcal{E}) + q_0 + q_1 + (q_2 + q_0)NQ[l_{max}])$ and

$$\mathbf{Adv}_{F,S}^{pro}(A) \leq \mathbf{Adv}_{H^P, P, \mathcal{E}}^{pra}(B),$$

Game $\boxed{G0}$ and $G1$	Adversary B_A^P and Simulator $S^{\mathcal{F}} = (S_1, S_2)$
Initialize : $\mathsf{H} = \mathsf{R}_2 = \mathsf{R}_0 = \phi$; $\mathcal{L} = \beta = \phi$, $i = 1$, Bad =F;	Initialize : $\mathsf{H} = \mathsf{R}_2 = \mathsf{R}_0 = \mathcal{L} = \beta = \phi$; $i = 1$, Bad =F; Run A and respond queries of A's as follows:
200 On $\mathcal{O}_2$ - query $y := y_i$, $i = i+1$ 201 $X = \mathcal{E}(y_i, \beta)$; Ext $\xleftarrow{\cup} (y, X)$; 202 $y' = H^P(X)$ and update β; 203 If $y' \neq y$ 204 then $z = \mathcal{R}(y)$; 205 If $y' \neq y \wedge (M, y) \in \mathsf{H}$ 206 then Bad =T; $\boxed{z = \mathsf{R}_0[y]};$ 207 If $y' = y$ 208 then $z = \mathcal{F}(X)$; 209 If $y' = y \wedge (M, y) \in \mathsf{H} \wedge M \neq X$ 210 then Bad =T; $\boxed{z = \mathsf{R}_0[y]};$ 211 $\mathsf{R}_2 \xleftarrow{\cup} (y, z)$; return z;	200 On $\mathcal{O}_2$ (or S_2)-query $y := y_i$, $i = i+1$ 201 $X = \mathcal{E}(y_i, \beta)$; Ext $\xleftarrow{\cup} (y, X)$; 202 $y' = H^P(X)$ and update β; 203 If $y' \neq y$ 204 then $z = \mathcal{R}(y)$; 207 If $y' = y$ 208 then $z = \mathcal{F}(X)$; 211 $\mathsf{R}_2 \xleftarrow{\cup} (y, z)$; return z;
100 On $\mathcal{O}_1$ - query u 101 $v = P(u)$; $\beta \xleftarrow{\|} (u, v)$; 102 return v;	100 On $\mathcal{O}_1$ (or S_1)-query u 101 $v = P(u)$; $\beta \xleftarrow{\|} (u, v)$; 102 return v;
000 On $\mathcal{O}_0$ - query M 001 $z = \mathcal{F}(M)$; $\mathcal{L} \xleftarrow{\cup} M$; 002 $y = H^P(M)$; $\mathsf{H} \xleftarrow{\cup} (M, y)$; 003 If $\mathsf{R}_0[y] \neq \perp$ 004 then Bad = T; $\boxed{z = \mathsf{R}_0[y]};$ 005 Else if $\mathsf{R}_2[y] \neq \perp \wedge (y, M) \notin \mathsf{E}$ 006 then Bad = T; $\boxed{z = \mathsf{R}_2[y]};$ 007 $\mathsf{R}_0 \xleftarrow{\cup} (y, z)$; return z;	000 On $\mathcal{O}_0$ (or $\mathcal{F}$)- query M 001 $z = \mathcal{F}(M)$; $\mathcal{L} \xleftarrow{\cup} M$; 002 $\mathsf{R}_0 \xleftarrow{\cup} (y, z)$; return z; 400 **Finalization:** (after A finishes queries.) 401 If $\exists M \neq M' \in \mathcal{L}$, $H^P(M) = H^P(M')$ 402 then bad =T, Finish(); 403 If $\exists M \in \mathcal{L}$, $(y, X) \in \mathsf{E}$, $X \neq M$, $H^P(M) = y$ 404 then bad =T, Finish(); 405 return $\perp$;

Fig. 4. $G0$ executes with boxed statements whereas $G1$ executes without these. Clearly $G0$ and $G1$ are identical-until-Bad and whenever $G1$ set bad true the adversary B_A^P set also bad true. In this case, Finish() subroutine executes which makes PrA successful. The tuple of simulated oracles of B_A is equivalent to $(\mathcal{F}, S_1, S_2)$.

Breaking Grain-128 with Dynamic Cube Attacks

Itai Dinur and Adi Shamir

Computer Science department
The Weizmann Institute
Rehovot 76100, Israel

Abstract. We present a new variant of cube attacks called a *dynamic cube attack*. Whereas standard cube attacks [4] find the key by solving a system of linear equations in the key bits, the new attack recovers the secret key by exploiting distinguishers obtained from cube testers. Dynamic cube attacks can create lower degree representations of the given cipher, which makes it possible to attack schemes that resist all previously known attacks. In this paper we concentrate on the well-known stream cipher Grain-128 [6], on which the best known key recovery attack [15] can recover only 2 key bits when the number of initialization rounds is decreased from 256 to 213. Our first attack runs in practical time complexity and recovers the full 128-bit key when the number of initialization rounds in Grain-128 is reduced to 207. Our second attack breaks a Grain-128 variant with 250 initialization rounds and is faster than exhaustive search by a factor of about 2^{28}. Finally, we present an attack on the full version of Grain-128 which can recover the full key but only when it belongs to a large subset of 2^{-10} of the possible keys. This attack is faster than exhaustive search over the 2^{118} possible keys by a factor of about 2^{15}. All of our key recovery attacks are the best known so far, and their correctness was experimentally verified rather than extrapolated from smaller variants of the cipher. This is the first time that a cube attack was shown to be effective against the full version of a well known cipher which resisted all previous attacks.

Keywords: Cryptanalysis, stream ciphers, Grain-128, cube attacks, cube testers, dynamic cube attacks.

1 Introduction

A well designed cipher is expected to resist all known cryptanalytic attacks, including distinguishing attacks and key recovery attacks. These two types of attacks are closely related since in many cases a distinguisher can be extended to a key recovery attack. Examples include many of the key-recovery attacks on iterated block ciphers such as differential cryptanalysis [1] and linear cryptanalysis [2]: First, the attacker constructs a distinguisher for a certain number of rounds of the iterated block cipher (usually one round less than the total number of rounds). Then, the attacker guesses part of the secret key and uses it to partially decrypt several ciphertexts in the final round. The distinguisher

A. Joux (Ed.): FSE 2011, LNCS 6733, pp. 167–187, 2011.

can be easily exploited to verify the guess: Under the correct guess, the partially decrypted ciphertexts are expected to exhibit the non-random property of the distinguisher. On the other hand, an incorrect guess is actually equivalent to adding another encryption round, and hence ciphertexts decrypted with an incorrect guess are expected to behave randomly. This is a very general technique for exploiting a distinguisher to recover the secret key of block ciphers, but it cannot be typically applied to stream ciphers, where partial decryption is not possible. Moreover, even when dealing with iterated block ciphers, more efficient key-recovery techniques often exist. In this paper we focus on the specific case of distinguishers obtained from cube testers (see [3]) and show how to use them in key recovery attacks.

Cube testers [3] are a family of generic distinguishers that can be applied to the black box representation of any cryptosystem. Cube attacks [4] are related to cube testers since both types of attacks sum the output of a cryptographic function over a subset of its input values. However, cube testers use the resultant sums to distinguish the cipher from a random function, whereas cube attacks use the sums to derive linear equations in the secret key bits. The success of cube testers and cube attacks on a given cryptosystem depends on subtle properties of the ANF (algebraic normal form) representation of the output function in the plaintext and key bits over GF(2). Although the explicit ANF representation is usually unknown to the attacker, cube testers and cube attacks can exploit a relatively low degree or sparse ANF representation in terms of some of its variables to distinguish the cipher from a random function and to recover the secret key.

Both cube attacks and cube testers are performed in two phases: The preprocessing phase which is not dependent on the key, and the online phase in which the key has a fixed unknown value. Whereas cube attacks are key recovery attacks and are thus stronger than cube testers, the preprocessing phase of cube attacks is generally more complex and has a lower chance of succeeding than the preprocessing phase of cube testers. The reason for this is that cube attacks require that the sum of the cipher's output function has a very specific property - it needs to be of low degree when represented as a polynomial in the key bits. Cube testers do not require such a specific property, but rather require that the value of the sum exhibits a property which is easily testable. An example of such a property is balance (i.e. whether the sum (modulo 2) is 0 and 1 with equal probabilities). Examples where cube testers succeed, while cube attacks seem to fail include scaled-down variants of the stream cipher Grain-128 (see [5]). Even in the case of scaled-down variants of the stream cipher Trivium, where cube attacks succeed ([4]), the preprocessing phase of cube attacks is much more time consuming than the one of cube testers. The challenge that we deal with in this paper is to extend cube testers to key recovery attacks in a new generic way. This combines the key recovery feature of cube attacks with the relatively low computational complexity of the preprocessing phase of cube testers.

We present a new attack called a dynamic cube attack that recovers the secret key of a cryptosystem by exploiting distinguishers given by cube testers.

The main observation that we use for the new attack is that when the inputs of the cryptosystem are not mixed thoroughly enough, the resistance of such a marginal cipher to cube testers usually depends on very few (or even one) non-linear operations that are performed at the latest stages of the encryption process. These few non-linear operations produce most of the relatively high degree terms in the ANF representation of the output function. If we manage to simplify the ANF representation of the intermediate encryption state bits that are involved in these non-linear operations (e.g. by forcing one of the two inputs of a multiplication operation to be zero), then the degree of the polynomial will be much lower, making it much more vulnerable to cube testers. In dynamic cube attacks, we analyze the cipher, find these crucial state bits and force them to be zero by using dedicated input bits called dynamic variables. Since the values of the state bits typically depend also on some key bits, we have to either guess them, or to assume that they have a particular value in order to apply the attack. For each guess, we use a different cube tester (that assigns the dynamic variables according to the guess) to distinguish the cipher from random. For the correct guess, the ANF representation of the crucial intermediate encryption state bits is simplified to zero, and the cube tester is likely to detect a strong non-random property in the output. On the other hand, for a large portion of wrong guesses, the cube tester is unlikely to detect this property. Thus, we can efficiently eliminate wrong guesses and thus recover parts of the secret key.

We applied the attack to two reduced variants of the stream cipher Grain-128 [6], and obtained the best known key recovery results for these variants. More significantly, we present an attack on the full version of Grain-128 which is faster than exhaustive search by a factor of 2^{15}, for a subset of 2^{-10} of all the possible keys. The attack can probably be optimized to break a larger set of weak keys of Grain-128, but even in its current form, it can break a practically significant fraction of almost one in a thousand keys. This is much better than other weak key attacks, which can typically break only a negligible fraction of keys.

The idea of assigning dynamic constraints (or conditions) to public variables and using them to recover key bits already appeared in previous work. In [15], the dynamic constraints were used to enhance differential and high order differential attacks by limiting the propagation of differences in the internal state of the cipher. This technique was applied to reduced variants of a few ciphers, and in particular was used to recover 2 key bits of Grain-128 when the number of initialization rounds is reduced from 256 to 213. In dynamic cube attacks, the dynamic constraints are used in a completely different way: The first and most crucial step of dynamic cube attacks is the careful analysis of the output function of the cipher. This analysis allows us to select constraints that weaken the resistance of the cipher to cube testers. Our carefully selected constrains, in addition to our novel algebraic key-recovery techniques, allow us to obtain much improved results on the cipher Grain-128: Our attack on a Grain-128 variant that uses 207 initialization rounds recovers the complete key (rather than a few key bits) with feasible complexity. In addition, we break a Grain-128 variant with 250 initialization rounds and a weak key set (containing a fraction of 2^{-10} of all the possible keys) of the full version of the cipher.

Next, we briefly describe the standard cube testers and cube attacks (for more details, refer to [3] and [4]). We then describe the new attack in detail and present our results on the cipher Grain-128 [6]. Finally, we conclude and list some open problems.

2   Cube Attacks and Cube Testers

2.1   Cube Attacks

In almost any cryptographic scheme, each output bit can be described by a multivariate master polynomial $p(x_1, .., x_n, v_1, .., v_m)$ over $GF(2)$ of secret variables x_i (key bits), and public variables v_j (plaintext bits in block ciphers and MACs, IV bits in stream ciphers). The cryptanalyst is allowed to tweak the master polynomial by assigning chosen values for the public variables, which result in derived polynomials, and his goal is to solve the resultant system of polynomial equations in terms of their common secret variables. The basic cube attack [4] is an algorithm for solving such polynomials, which is closely related to previously known attacks such as high order differential attacks [11] and AIDA [12].

To simplify our notation, we now ignore the distinction between public and private variables. Given a multivariate master polynomial with n variables $p(x_1, .., x_n)$ over $GF(2)$ in algebraic normal form (ANF), and a term t_I containing variables from an index subset I that are multiplied together, the polynomial can be written as the sum of terms which are supersets of I and terms that miss at least one variable from I:

$$p(x_1, .., x_n) \equiv t_I \cdot p_{S(I)} + q(x_1, .., x_n)$$

$p_{S(I)}$ is called the *superpoly* of I in p. Note that the superpoly of I in p is a polynomial that does not contain any common variable with t_I, and each term in $q(x_1, .., x_n)$ does not contain at least one variable from I. Moreover, compared to p, the algebraic degree of the superpoly is reduced by at least the number of variables in t_I.

The basic idea behind cube attacks is that the symbolic sum over $GF(2)$ of all the derived polynomials obtained from the master polynomial by assigning all the possible 0/1 values to the subset of variables in the term t_I is exactly $p_{S(I)}$ which is the superpoly of t_I in $p(x_1, .., x_n)$. A *maxterm* of p is a term t_I such that the superpoly of I in p is a linear polynomial which is not a constant.

The cube attack has two phases: the preprocessing phase, and the online phase. The preprocessing phase is not key-dependant and is performed once per cryptosystem. The main challenge of the attacker in the preprocessing phase is to find sufficiently many maxterms with linearly independent superpolys. Linear superpolys are not guaranteed to exist, and even when they exist, finding them can be a challenging preprocessing task. However, once sufficiently many linearly independent superpolys are found for a particular cryptosystem, we can repeatedly use them to easily find any secret key during the online phase.

2.2   Cube Testers

Similarly to cube attacks, cube testers [3] work by evaluating superpolys of terms of public variables. However, while cube attacks aim to recover the secret key, the goal of cube testers is to distinguish a cryptographic scheme from a random function, or to detect non-randomness by using algebraic property testing on the superpoly. One of the natural algebraic properties that can be tested is balance: A random function is expected to contain as many zeroes as ones in its truth table. A superpoly that has a strongly unbalanced truth table can thus be distinguished from a random polynomial by testing whether it evaluates as often to one as to zero. Other efficiently detectable properties include low degree, the presence of linear variables, and the presence of neutral variables.

In the preprocessing phase of cube testers, the attacker finds terms whose superpolys have some efficiently testable property.

3   A Simple Example of Dynamic Cube Attacks

Both standard (static) cube testers and dynamic cube attacks sum the output of the cipher over a given cube defined by a subset of public variables, which are called cube variables. In static cube testers, the values of all the public variables that are not summed over are fixed to a constant (usually zero), and thus they are called static variables. However, in dynamic cube attacks the values of some of the public variables that are not part of the cube are not fixed. Instead, each one of these variables (called dynamic variables) is assigned a function that depends on some of the cube public variables and some expressions of private variables. Each such function is carefully chosen, usually in order to zero some state bits in order to amplify the bias (or the non-randomness in general) of the cube tester. Dynamic cube attacks are clearly a generalization of standard cube testers, but also allow us to directly derive information on the secret key without solving any algebraic equations. Moreover, choosing the dynamic variables carefully may help to improve the time complexity of distinguishers obtained by using standard cube testers (we will need fewer cube variables to obtain a distinguisher). We note that the drawback of the new attack compared to basic cube attacks and cube testers, is that it requires a more complex analysis of the internal structure of the cipher.

To demonstrate the idea of the attack, we consider a polynomial P which is a function of the three polynomials P_1, P_2, and P_3:

$$P = P_1 P_2 + P_3$$

P_1, P_2, and P_3 are polynomials over five secret variables x_1, x_2, x_3, x_4, x_5 and five public variables v_1, v_2, v_3, v_4, v_5:

$P_1 = v_2 v_3 x_1 x_2 x_3 + v_3 v_4 x_1 x_3 + v_2 x_1 + v_5 x_1 + v_1 + v_2 + x_2 + x_3 + x_4 + x_5 + 1$
$P_2 = $ arbitrary dense polynomial in the 10 variables

$$P_3 = v_1v_4x_3x_4 + v_2x_2x_3 + v_3x_1x_4 + v_4x_2x_4 + v_5x_3x_5 + x_1x_2x_4 + v_1 + x_2 + x_4$$

Since P_2 is unrestricted, P is likely to behave randomly and it seems to be immune to cube testers (or to cube attacks). However, if we can set P_1 to zero, we get $P = P_3$. Since P_3 is a relatively simple function, it can be easily distinguished from random. We set $v_4 = 0$ and exploit the linearity of v_1 in P_1 to set $v_1 = v_2v_3x_1x_2x_3 + v_2x_1 + v_5x_1 + v_2 + x_2 + x_3 + x_4 + x_5 + 1$ which forces P_1 to zero. During the cube summation, the value of the dynamic variable v_1 will change according to its assigned function. This is in contrast to the static variable v_4, whose value will remain 0 during the cube summation. At this point, we assume that we know the values of all the secret expressions that are necessary to calculate the value of v_1: $x_1x_2x_3$, x_1, and $x_2 + x_3 + x_4 + x_5 + 1$. Plugging in the values for v_1 and v_4, we get:

$$P = v_2x_2x_3 + v_3x_1x_4 + v_5x_3x_5 + x_1x_2x_4 + x_2 + x_4 +$$
$$(v_2v_3x_1x_2x_3 + v_2x_1 + v_5x_1 + v_2 + x_2 + x_3 + x_4 + x_5 + 1) =$$
$$v_2v_3x_1x_2x_3 + v_2x_2x_3 + v_3x_1x_4 + v_5x_3x_5 + x_1x_2x_4 + v_2x_1 + v_5x_1 + v_2 + x_3 + x_5 + 1$$

After these substitutions, we can see that the simplified P is of degree 2 in the public variables, and there is only one term ($v_2v_3x_1x_2x_3$) of this degree. We have 3 free public variables (v_2, v_3, v_5) that are not assigned. We can now use them as cube variables: The sum over the big cube $v_2v_3v_5$ and two of its subcubes v_2v_5 and v_3v_5 is always zero. Moreover, the superpoly of v_2v_3 is $x_1x_2x_3$, which is zero for most keys. Thus, we can easily distinguish P from a random function using cube testers. However, the values of the expressions $x_1x_2x_3$, x_1, and $x_2 + x_3 + x_4 + x_5 + 1$ are unknown in advance, and it is not possible to calculate the dynamic values for v_1 without them. Thus, we guess the 3 values of the expressions (modulo 2). For each of the 8 possible guesses (there are actually 6 possible guesses since $x_1 = 0$ implies $x_1x_2x_3 = 0$, but this optimization is irrelevant at this point), we run the cube tester, and get 4 0/1 values - a value for each cube sum. The 7 wrong guesses will not zero P_1 throughout the cube summations. Hence the 4 cube sums for each wrong guess are likely to behave randomly, and it is unlikely that more than 1 wrong guess will give 4 zero cube sum values. On the other hand, the 4 cube sums for the correct guess will all equal to 0 with high probability. Hence, for most keys, we expect to remain with at most 2 possible guesses for the 3 expressions and we can recover the values for the expressions that are assigned a common value by these 2 guesses. This gives us a distinguisher for P and allows us to derive information regarding the secret key.

In the general decomposition of a polynomial P as $P = P_1P_2 + P_3$, we call P_1 (according to which we assign the dynamic variable) the *source polynomial*, P_2 the *target polynomial* and P_3 the *remainder polynomial*. There are many ways to express P is such a way, and the choice of source and target polynomials requires careful analysis of the given cipher.

4   Dynamic Cube Attacks on Grain-128

4.1   Description on Grain-128

We give a brief description of Grain-128, for more details one can refer to [6]. The state of Grain-128 consists of a 128-bit LFSR and a 128-bit NFSR. The feedback functions of the LFSR and NFSR are respectively defined to be

$s_{i+128} = s_i + s_{i+7} + s_{i+38} + s_{i+70} + s_{i+81} + s_{i+96}$

$b_{i+128} = s_i + b_i + b_{i+26} + b_{i+56} + b_{i+91} + b_{i+96} + b_{i+3}b_{i+67} + b_{i+11}b_{i+13} + b_{i+17}b_{i+18} + b_{i+27}b_{i+59} + b_{i+40}b_{i+48} + b_{i+61}b_{i+65} + b_{i+68}b_{i+84}$

The output function is defined as

$z_i = \sum_{j \in \mathcal{A}} b_{i+j} + h(x) + s_{i+93}$, where $\mathcal{A} = \{2, 15, 36, 45, 64, 73, 89\}$.

$h(x) = x_0 x_1 + x_2 x_3 + x_4 x_5 + x_6 x_7 + x_0 x_4 x_8$

where the variables x_0, x_1, x_2, x_3, x_4, x_5, x_6, x_7 and x_8 correspond to the tap positions b_{i+12}, s_{i+8}, s_{i+13}, s_{i+20}, b_{i+95}, s_{i+42}, s_{i+60}, s_{i+79} and s_{i+95} respectively.

Grain-128 is initialized with a 128-bit key that is loaded into the NFSR, and with a 96-bit IV that is loaded into the LFSR, while the remaining 32 LFSR bits are filled with the value of 1. The state is then clocked through 256 initialization rounds without producing an output, feeding the output back into the input of both registers.

4.2   Previous Attacks

Several attacks have been published on Grain-128 variants: [7] found a distinguisher when the number of initialization rounds is reduced from 256 to 192 rounds, [8] described shortcut key-recovery attacks on a variant with 180 initialization rounds, and [9] exploited a sliding property to speedup exhaustive search by a factor of two. Related-key attacks on the full cipher were presented in [10]. However, the relevance of related-key attacks is disputed and we concentrate on attacks in the single key model. Stankovski [16] presented a distinguishing attack on a variant that uses 246 initialization rounds, which works for less than half of the keys. The most powerful distinguishing attack on most keys of Grain-128 was given in [5], where cube testers were used in order to distinguish the cipher from random for up to 237 initialization rounds. Moreover, the authors claim that by extrapolating their experimentally verified results, one can argue that cube testers may be used in order to attack the full cipher. However, this conjecture has not been verified in practice due to the infeasibility of the attack. Note that [5] only gives a distinguisher, and leaves the problem of exploiting cube testers (or cube attacks) for key recovery open. More recently [15] used conditional differential cryptanalyses to recover 2 key bits of Grain-128 with 213 initialization rounds, which gives the best known key-recovery attack in the single key model up to this point.

4.3   Outline of the New Attacks on Grain-128

We present 3 attacks:

1. A feasible full key recovery attack on a Grain-128 variant that uses 207 initialization rounds, while utilizing output bits $208 - 218$.

2. An experimentally verified full key recovery attack on a Grain-128 variant with 250 initialization rounds.
3. An experimentally verified attack on the full Grain-128, which can recover a large subset of weak keys (containing 2^{-10} of all the possible keys).

We begin by describing the common steps shared by these three attacks. We then elaborate on each attack in more detail.

The preprocessing phase of the attacks consists of 2 initial steps:

Step 1. We first choose the state bits to nullify, and show how to nullify them by setting certain dynamic variables to appropriate values.

This is a complex process that cannot be fully automated and involves manual work to analyze the cipher: When applying the attack to Grain-128, we would like to decompose its output function into a source and target polynomials (representing intermediate state bits multiplied together), and a remainder polynomial which should be more vulnerable to cube testers than the original output. In our small example, this was easy since we could explicitly write down and analyze its ANF in terms of the public and private variables. However, the output function of Grain-128 is too complex to decompose and analyze in such a way. Our approach in this paper is to use the recursive description of the cipher's output function in order to find a good decomposition.

In the case of Grain-128, specific non-linear terms in the cipher's output stand out as being of higher degree than others and are good candidates to be nullified or simplified. The output function of Grain-128 is a multivariate polynomial of degree 3 in the state. The only term of degree 3 is $b_{i+12}b_{i+95}s_{i+95}$, and hence we focus on nullifying it. Since b_{i+12} is the state bit that is calculated at the earliest stage of the initialization steps (compared to b_{i+95} and s_{i+95}), it should be the least complicated to nullify. However, after many initialization steps, the ANF of b_{i+12} becomes very complicated and we were not able to nullify it when more than 230 initialization rounds are used (i.e. for $i > 230$). The compromise we make is to simplify (and not nullify) $b_{i+12}b_{i+95}s_{i+95}$: We write the most significant term of degree 3 that is used in the calculation of these state bits, which for b_{i+12} is $b_{i-128+12+12}b_{i-128+95+12}s_{i-128+95+12} = b_{i-104}b_{i-21}s_{i-21}$. The most significant term for both b_{i+95} and s_{i+95} is $b_{i-128+12+95}b_{i-128+95+95}s_{i-128+95+95} = b_{i-21}b_{i+62}s_{i+62}$. We can see that b_{i-21} participates in all terms, and thus nullifying it is likely to simplify the ANF of $b_{i+12}b_{i+95}s_{i+95}$ significantly.

The ANF of the earlier b_{i-21} is much easier to analyze compared to the one of b_{i+12}, but it is still very complex. Thus, we perform more iterations in which we simplify b_{i-21} further by using its recursive description to nullify previous state bits. When the ANF representation of b_{i-21} is simple enough, we select a linear public variable in its ANF and assign to it an expression which will make the whole expression identically zero. We elaborate on this multistage process for output bit 215 of Grain-128 (used in attack 1): We would like to zero $b_{215-21} = b_{194}$. However, we do not zero it directly. We first zero 4 other state bits in order to simplify its ANF representation. The details of how these bits were chosen are given in Appendix A.

Step 2. We choose a big cube and a set of subcubes to sum over during the online phase. We then determine the secret expressions that need to be guessed in order to calculate the values of the dynamic variables during the cube summations.

Some choices of the big cube give better results than other, and choosing a cube that gives good results is a crucial part of the preprocessing. One can use heuristics in order to find cubes that give better results (an example of a heuristic is given in [5]). However, it is difficult to predict in advance which cubes will give good results without actually executing the attack and calculating the results for many cubes.

The secret expressions that need to be guessed are calculated according to the symbolic expressions of the dynamic variables and the chosen big cube. This is a simple process that can be easily automated:

1. Given the symbolic form of a dynamic variable, look for all the terms which are combinations of variables from the big cube. In our simple example, the symbolic form of the single dynamic variable is $v_2 v_3 x_1 x_2 x_3 + v_2 x_1 + v_5 x_1 + v_2 + x_2 + x_3 + x_4 + x_5 + 1$. Our big cube is $v_2 v_3 v_5$. The terms which are combinations of variables from the big cube in the symbolic form are $v_2 v_3$, v_2, v_5 and the empty combination.

2. Rewrite the symbolic form as a sum of these terms, each one multiplied by an expression of secret variables. In our example, we write $v_2 v_3 x_1 x_2 x_3 + v_2 x_1 + v_5 x_1 + v_2 + x_2 + x_3 + x_4 + x_5 + 1 = v_2 v_3 (x_1 x_2 x_3) + v_2 (x_1 + 1) + v_5 (x_1) + (x_2 + x_3 + x_4 + x_5 + 1)$,

3. Add the expressions of secret variables to the set of expressions that need to be guessed. In the example, we add $x_1 x_2 x_3$, x_1 and $x_2 + x_3 + x_4 + 1$ (note that guessing the value of x_1 is the same as guessing the value of $x_1 + 1$, and we do not add it twice). In addition, we do not add expressions whose value can be deduced from the values of the expressions already in the set. For example, if x_1 and x_2 are is the set, we do not add $x_1 x_2$ or $x_1 + x_2$.

In steps $3 - 4$, the attacker uses the parameters obtained in the first two steps in order to derive information regarding the secret key. These steps constitute the online phase of the attack that is executed by the attacker after the secret key has been set. In addition, steps $3 - 4$ are simulated by the attacker in the preprocessing phase for several pseudo random keys, in order to verify his choices in steps $1 - 2$.

Step 3

1. For each possible value (guess) of the secret expressions, sum over the subcubes chosen in the previous step with the dynamic variables set accordingly, and obtain a list of sums (one sum per subcube).

2. Given the list of sums, calculate the guess score (which measures the non-randomness in the subcube summations). The output of this step is a sorted guess score list in which guesses are sorted from the lowest score to the highest.

Given that the dimension of our big cube is d, the complexity of summing over all its subcubes is bounded by $d2^d$ (this can be done using the Moebius transform [13]). Given that we need to guess the values of e expressions, the complexity of this step is bounded by $d2^{d+e}$. However, the data complexity of this step can by significantly lower than the time complexity: Assuming that we have only $y \leq e$ dynamic variables, the data complexity is bounded by 2^{d+y} (an output bit for every possible value of the cube and dynamic variables is sufficient).

After we obtain the summation values for each of the subcubes for a specific guess, we determine its score. The simple score function that we use in this paper measures the percentage of 1 values in the summations. The reason that we consider summation values which are biased towards 0 as non-random (but not summation values which are biased towards 1) is that the superpolys of the cubes in our attacks tend to be extremely sparse, and their ANF contains the constant 1 (or any other term) with very low probability. Such sparse polynomials evaluate to zero for almost all keys.

Step 4 Given the sorted guess score list, we determine the most likely values for the secret expressions, for a subset of the secret expressions, or for the entire key. The straightforward approach to calculate the values for the secret expressions is to simply take the values for the expressions from the guess that has the lowest score (or the least percentage of 1 values in its summations values), in the sorted guess list. However, this approach does not always work. Depending on the setting of the attack, there could be guesses that have a score that is at least as low as the correct guess score: In our small example, the correct guess score is expected to be 0, however there is a reasonable probability that there is another arbitrary guess with the score of 0. Therefore, the details of this step vary according to the attack and are specified separately for each attack.

4.4   Details of the First Attack

The first attack is a full key recovery attack on a Grain-128 variant that uses 207 initialization rounds, while utilizing output bits $208 - 218$. The key bits are recovered in small groups of size $1 - 3$, where each group is recovered using a different set of parameters that was obtained in the preprocessing phase.

One set of parameters for the attack is given in Table 1 in Appendix B. We now specify how the attack is carried out given the parameters of this table: First, we assign to each one of the dynamic variables in the table its symbolic value. Appendix B shows how to do this given the parameters of Table 1, and the assignment algorithm for the other tables in this paper is similar.

After all the dynamic variables are assigned, we determine the secret expressions that need to be guessed in order to fully calculate the values of the dynamic variables during the cube summations (step 2). Altogether, there are 7 expressions that need to be guessed, and since the big cube is of dimension 19, the total complexity of the step 3 of the attack with this specific set of parameters is about $19 \times 2^{19+7} < 2^{31}$. We will use only linear expressions for the full key recovery (step 4), hence we concentrate on retrieving their value from the sorted guess list. The two linear expressions actually contain only a single key bit and are listed in the "Expressions Retrieved" row. We sum on all subcubes of dimension at least $19 - 3 = 16$ of the big cube (of dimension 19), and the score for each guess is simply the fraction of 1 values among all the subcube sums. In step 4, we retrieve the value of the 2 expressions by taking the corresponding values from the best guess. We simulated the attack with the above parameters with hundreds of random keys. The attack failed to retrieve the correct values for the expressions $x_{127}, x_{122} + 1$ for about 10% of the keys. However for all the failed keys, the score of the best guess was at least 0.44 (i.e. the dynamic cube tester did not give a strong distinguisher), and thus we know when we fail by declaring the expressions as "undetermined" whenever we encounter a key for which the best guess score is at least 0.44 (this occurs for about 15% of the keys). This is important for the full key recovery attack that is described next.

We showed how to retrieve 2 key bits with high probability with one carefully chosen set of parameters. It is not difficult to find more sets of parameters that allow us to retrieve more key bits. Another example of such a set of parameters that uses the same output bit is given in table 2 in Appendix B. Note that we only changed the chosen big cube, which in turn changed the retrieved expressions. A different set of parameters that uses output bit 218 is given in table 3 in Appendix B. Altogether, we have 55 sets of parameters that ideally allow us to recover 86 of the 128 key bits. For each set of parameters, the score calculation method is identical to the one described above, i.e. we compute the percentage of 1 values in the cube sums for all cubes of dimension at least $d - 3$. The key recovery method is identical as well, i.e. we recover the values of the secret linear expressions from the guess with the best score, but only if its score is at least 0.44. We simulated the full attack on hundreds of random keys. On average, we could retrieve about 80 secret key bits per key. The remaining 48 key bits can be recovered with feasible complexity by exhaustive search.

We note that it is possible to retrieve more key bits in a similar way by using more output bits (e.g. output bits 219, 220, etc.), or using the same output bits with different sets of parameters. A more efficient key recovery method can try to determine values of non-linear secret expressions, some of which can be made linear by plugging in values for secret key bits which we already recovered. However, our main goal is to attack much stronger variants of Grain-128, as described next.

4.5   A Partial Simulation Phase

When attacking Grain-128, we perform the preprocessing steps $(1, 2)$ and then simulate the online steps of the attack $(3, 4)$ for several random keys. In this case, steps 3 and 4 are performed in order to estimate the success of the attack and are called the simulation phase. If we are not satisfied with the results, we can repeat steps 1 and 2 by choosing different parameters and performing another simulation phase. This process can be very expensive and its complexity is generally dominated by step 3. We can significantly reduce the complexity of the simulation phase by calculating the cube summations only for the correct guess and observing whether the correct guess exhibits a significant non-random property for most keys. This is unnecessary for the first attack in which we can run the full simulation phase and recover the secret key. However, in the second and third attacks, we try to attack variants of Grain-128 which are significantly stronger and the simulation phase becomes infeasible even for a single random key. In these cases, the observed non-randomness for the correct guess provides strong evidence that the stronger variants of Grain-128 can also be broken by the full key recovery version of the attack.

Given that we choose a big cube of size d and guess e expressions, the complexity of the cube summations when running the full simulation phase on one key is about $d2^{d+e}$ bit operations. However, the complexity of the simulation phase is actually dominated by the 2^{d+e} executions of the cipher: Assuming that each execution requires about b bit operations, the total complexity is about $b2^{d+e}$ (for Grain-128 $b > 2^{10} >> d$). Similarly, the partial simulation phase on one key requires $b2^d$ bit operations. Since the complexity does not depend on e, we can feasible verify the behavior of dynamic cube attacks even when their total complexity is infeasible when the dimension of the cube d is not too large. This ability to experimentally verify the performance of dynamic cube attacks is a major advantage over static cube attacks and cube testers.

4.6   A Generic Key Recovery Method

In the first attack, we run the full simulation phase and obtain the sorted guess list in step 3. Since we can do this many times and calculate the complexity of the attack, we tailored the key derivation algorithm used in step 4 such that it is very efficient for our chosen parameter sets. On the other hand, in the second and third attacks, we must perform the partial simulation phase as described above and we obtain only the score for the correct guess. Since we do not have the sorted guess list, we cannot calculate the exact complexity of the attack and we cannot customize the algorithm used in step 4 as in the first attack (for example, we cannot verify that the first guess in the sorted guess list assigns correct values for some expressions, as in the first attack). As a result, we use a key recovery method which is more generic in a sense that it is not tailored to a specific cipher, or to a specific set of parameters. The only property of the parameter sets for the attacks that it exploits, is that many guessed key expressions are linear. We now describe the details of this method as performed in real time (not in the simulation phase) and then estimate its complexity.

Assume that we have executed steps $1 - 3$ for Grain-128 with $n = 128$ secret key bits. Our big cube is of dimension d and we have e expressions to guess, out of which l are linear. Our sorted guess score list is of size 2^e and the correct guess is located at index g in the sorted list.

1. Consider the guesses from the lowest score to the highest: For each guess (that assigns values to all the expressions), perform Gaussian Elimination on the l linear expressions and express l variables as linear combinations of the other $n - l$ variables.

2. Exhaustively search the possible 2^{n-l} values for those $n - l$ variables: For each value, get the remaining part of the key from the linear expressions, execute the cipher with the key, and compare the result to the given data. If there is equality, return the full key.

Overall, we have 2^{n-l} iterations per guess. The evaluation of the linear expressions can be performed efficiently if we iterate over the 2^{n-l} values using Gray Codes. Hence, we assume that the evaluation of the linear expressions takes negligible time compared to the execution of the cipher. The total running time per guess is thus about 2^{n-l} cipher executions and the overall running time of step 4 is $g \times 2^{n-l}$. We can also to try improve the running time by using some of the $e - l$ non linear expressions which can be efficiently evaluated (compared to a single cipher execution): For each key we first check if the key satisfies these non linear equations before executing the cipher.

The complexity of the generic key recovery method is dependent on g which denotes the index of the correct guess in the sorted guess list. The expected value of g can be estimated for a random key by running several simulations of the attack on random keys. However, when the simulation phase is infeasible and we are forced to perform a partial simulation phase, we cannot estimate g this way since we do not have the guess list. A possible solution to this problem is to assume that all the incorrect guesses behave randomly (i.e. the subcube sums are independent uniformly distributed boolean random variables). Under this assumption, we run the partial simulation on an arbitrary key. If the cube sums for the correct guess detect a property that is satisfied by a random cipher with probability p, then we can estimate $g \approx max\{p \times 2^e, 1\}$.

The assumption that incorrect guesses behave randomly is clearly an oversimplification. In the first attack, we retrieve the value of a carefully chosen subset of the expressions by taking the corresponding values from the best guess. However for about half of the keys the best guess is not the correct guess, i.e. it does not assign the correct values for all the expressions, but rather to our chosen set of expressions. In other words, there are specific (non arbitrary) incorrect guesses that are likely to have a low score that can be at least as low as the score of the correct guess. These incorrect guesses usually assign a correct value to a fixed subset of the guessed expressions. In order to understand this, consider the following example: assume that $P = P_1 P_2 + P_3$, the source and target polynomials P_1 and P_2 are of degree 3, and the remainder polynomial is of degree 5 (all degrees are in terms of the public variables). We choose a dynamic variable to nullify P_1, and assume for the sake of simplicity that the degrees of P_2 and P_3 do not change after assigning this variable. We choose a cube of dimension 7, and sum on all its subcubes of dimension 6 and 7. Clearly, the correct guess will have a score of 0. However, any other which reduces the degree of P_1 to 1 or 0 will also have a score of 0.

To sum up, our estimation of g (and hence our estimation for the complexity of the attack) may not be completely accurate since incorrect guesses do not behave randomly. However, our simulations on Grain-128 variants on which the simulation phase is feasible, show that the effect of the incorrect guesses biased towards 0 is usually insignificant, and our estimation of g is reasonable. In addition, even incorrect non-uniform guesses are still likely to be highly correlated with the correct guess, and thus they can actually speed up the attack (this was experimentally verified in our first attack, which has a feasible complexity). Hence, our estimation of the complexity of step 4 of the attack is a reasonable upper bound.

4.7   Details of the Second Attack

In order to attack the almost full version of Grain-128 with 250 initialization rounds (out of 256), we nullify $b_{251-21} = b_{230}$. The parameters of the attack are specified in Table 4 in Appendix B. As in the first attack, most of the dynamic variables are used in order to simplify b_{230}. Note that we need many more dynamic variables compared to the previous attack. This is because it is much more difficult to nullify b_{230} than to nullify b_{194} or b_{197} (for example). In addition, we set v_{82} to the constant value of 1 so that v_{89} can function as a dynamic variable that nullifies b_{197}. Since the big cube is of dimension 37 and we have 24 dynamic variables, the data and memory complexity is $2^{37+24} = 2^{61}$. The number of expressions that need to be guessed seems to be 84. However, after removing many linearly dependent expressions, this number can be reduced to 59. Thus, the total complexity of the cube summations is about $37 \times 2^{37+59} < 2^{101}$, implying that we have to use the partial simulation phase. Out of the 59 expressions that need to be guessed, 29 contain only a single key bit on which we concentrate for generic key recovery.

During the partial simulation phase, we summed on all subcubes of dimension at least 35 of the big cube, calculating the percentage of 1 values separately for all the subcubes of each dimension (35, 36, or 37). We performed the partial simulation phase on dozens of random keys. For the sake of completeness, we also sampled a few random incorrect guesses for several keys and verified that they do not have a significant bias. For about 60% of the keys, the subcube sums for the correct guess contained only 0 values for the subcube of sizes 36 and 37, and less than 200 '1' values among the 666 subcubes of size 35. Assuming that incorrect guesses behave randomly, we expect the correct guess to be among the first guesses in the sorted guess list. The complexity of the unoptimized version of the attack (that ignores the non-linear expressions) is dominated by the exhaustive search for the remaining $128 - 29 = 99$ key bits per guess. Overall the complexity for about 60% of the keys is about 2^{100} cipher evaluations, and can almost surely be optimized further. For another 30% of the keys we tested, the non-randomness in the subcube sums was not as significant as in the first 60%, but still significant enough for the attack to be much faster than exhaustive search. For the remaining 10% of the keys, the non-randomness observed was not significant enough and the attack failed. However, we are certain that most of these problematic keys can still be broken by selecting different parameters for the attack.

4.8 Details of the Third Attack

In order to attack the full version of Grain-128 with 256 initialization rounds, we have to nullify $b_{257-21} = b_{236}$. However, the ANF of b_{236} is too complicated to zero using our techniques, and we had to make assumptions on 10 secret key bits in order nullify it. As a result, we could verify the correctness of our attack on the full version of Grain-128 only for a subset of about 2^{-10} of the possible keys in which 10 key bits are set to zero. Our current attack can thus be viewed as an attack on an unusually large subset of weak keys, but it is reasonable to assume that it can be extended to most keys with further improvements.

The parameters of the attack are specified in Table 5 in Appendix B. Since the big cube is of dimension 46 and we have 13 dynamic variables, the data and memory complexity is $2^{46+13} = 2^{59}$. After removing many linearly dependent expressions, the number of guessed expression is 61. Thus, the total complexity of the cube summations is about $46 \times 2^{46+61} < 2^{113}$ bit operations. Out of the 61 expressions that need to be guessed, 30 contain only a single key bit. Moreover, we can fix the values of 35 more variables such that 30 out of the remaining $61 - 30 = 31$ expression become linear. In order to recover the key efficiently, we use an extension of the generic key recovery method: Let the key be n, and denote the dimension of the big cube by d. Assume that given the values of c variables we can plug them into l (linear or non-linear) expressions such that they become linear, and perform Gaussian Elimination which makes it possible to express l variables as linear combinations of the remaining (unspecified) $n - l - c$ variables.

1. Consider the guesses from the lowest score to the highest: For each guess, iterate the $n-l$ variables using Gray Coding such that the c variables function as most significant bits (i.e their value changes every 2^{n-l-c} iterations of the remaining $n - l - c$ variables).
2. For each value of the c variables, perform Gaussian Elimination and express l variables as linear combinations of the remaining $n - l - c$ variables.
3. For each value of the remaining $n - l - c$ variables, compute the values of the l linear variables, execute the cipher with this derived key and compare the result to the given data. If there is equality, return the full key.

In our case, we have $n = 118$ (after fixing 10 key bits), $c = 35$, and $l = 60$. We call the second sub-step in which we perform Gaussian Elimination a big iteration and the third sub-step in which we do not change any value among the $c = 35$ variables, a small iteration. Note that big iterations are performed only every $2^{n-l-c} = 2^{23}$ small iterations. It is clear that computing the linear equations and performing Gaussian Elimination with a small number of variables in a big iteration takes negligible time compared to executing the cipher 2^{23} times in small iterations. Hence the complexity of the exhaustive search per guess is dominated by the small iterations and is about $2^{n-l} = 2^{58}$ cipher evaluations (as in the original generic key recovery method) . In order to complete the analysis of the attack, we need to describe the score calculation method and the estimate the index g of the correct guess.

During the partial simulation phase, we summed on all subcubes of dimension at least 44 of the big cube, calculating the percentage of 1 values separately for all the subcubes of each dimension. We performed simulations for 5 random keys (note that each simulation requires 2^{46} cipher executions, which stretched our computational resources): For 3 out of the 5 keys, we observed a significant bias towards 0 (which is expected to occur with probability less than 2^{-20} for a random cipher) in the subcubes of dimension 45 and 46. This implies that $g \approx 2^{61} \times 2^{-20} = 2^{41}$ and the total complexity of step 4 is about $2^{41} \times 2^{118-60} = 2^{99}$ cipher evaluations. Assuming that each cipher evaluation requires about 2^{10} bit operations, the total complexity of the attack remains dominated by step 3, and is about 2^{113} bit operations. This is better than exhaustive search by a factor of about 2^{15} even when we take into account the fact that our set of weak keys contains only 2^{-10} of the 2^{128} possible keys. For another key, the bias towards 0 in the subcubes of dimension 45 and 46 was not as strong and we also need to use the bias towards 0 in the subcubes of dimension 44. For this key, we were able to improve exhaustive search by a factor of about 2^{10}. For the fifth key, we also observed a bias towards 0, but it was not strong enough for a significant improvement compared to exhaustive search. As in the previous attack, we stress that it should be possible to choose parameters such that the attack will be significantly better than exhaustive search for almost all keys in the weak key set.

4.9   Discussion

Any attack which can break a fraction of 2^{-10} of the keys is sufficiently significant, but in addition we believe that our third attack can be improved to work on a larger set of weak keys of Grain-128. This can be done by making fewer assumptions on the key and optimizing the process of nullification of b_{236}. However, we do not believe that nullifying b_{236} will suffice to attack most keys of Grain-128. For such an attack, the most reasonable approach would be to choose a larger big cube to sum over, while nullifying fewer state bits at earlier stages of the cipher initialization process. The question whether a key recovery attack on most keys of Grain-128 can be feasibly simulated to yield an experimentally verified attack remains open.

5   Generalizing the Attack

In the previous section, we described in detail the dynamic cube attack on Grain-128. However, most of our techniques can naturally extend to other cryptosytems. In this section, we describe the attack in a more generic setting, emphasizing some important observations.

Step 1. As specified in the attack on Grain-128, choosing appropriate state bits to nullify and actually nullifying them is a complex process. In the case of Grain-128, specific non-linear terms in the cipher's output stand out as being of higher degree and enable us to decompose the output function to a source and target polynomials relatively easily. It is also possible to find good decompositions experimentally: We can tweak the cipher by removing terms in the output function. We then select various cubes and observe whether the tweaked cipher is more vulnerable to cube testers than the original cipher. If the tweaked cipher is indeed more vulnerable, then the removed terms are good candidates to nullify or simplify.

As in the case of Grain-128, there are several complications that may arise during the execution of this step and hence it needs to be executed carefully and repeatedly through a process of trial and error. One complication is that zeroing a certain group of state bits may be impossible due to their complex interdependencies. On the other hand, there may be several options to select dynamic variables and to zero a group of state bits. Some of these options may give better results than others. Another complication is that using numerous dynamic variables may overdeplete the public variables that we can use for the cube summations.

Step 2. The choice of a big cube, can have a major impact on the complexity of the attack. Unfortunately, as specified in the attack on Grain-128, in order to find a cube that gives good results we usually have to execute the attack and calculate the results for many cubes. After the big cube is chosen, the secret expressions that need to be guessed are calculated according to the simple generic process that is used for Grain-128.

Step 3. The only part of this step that is not automated is the score calculation technique for each guess from the subcube sums. We can use the simple method of assigning the guess its percentage of 1 values, or more complicated algorithms that give certain subcubes more weight in the score calculation (e.g. the sum of high dimensional subcubes can get more weight than the sum of lower dimensional ones, which tend to be less biased towards 0).

Step 4. Techniques for recovering information about the key differ according to the attack. It is always best to adapt the technique in order to optimize the attack as in the first attack on Grain-128. In this attack, we determined the values of some carefully chosen key expression from the guess with the best score. It is possible to generalize this technique by determining the value of a key expression (or several key expressions) according to a majority vote taken over several guesses with the highest score. We can also try to run the attack with different sets of parameters, but with some common guessed expressions. The values for those common guessed expressions can then be deduced using more data from several guess score lists.

When the simulation phase (steps 3 and 4) is not feasible we must use the partial simulation phase. The generic key recovery method and its extension in the third attack on Grain-128 can be used in case many of the guessed key expressions are linear, or can be made linear by fixing the values of some key bits.

6   Conclusions and Open Issues

Dynamic cube attacks provide new key recovery techniques that exploit in a novel way distinguishers obtained from cube testers. Our results on Grain-128 demonstrate that dynamic cube attacks can break schemes which seem to resist all the previously known attacks. Unlike cube attacks and cube testers, the success of dynamic cube attacks can be convincingly demonstrated beyond the feasible region by trying sufficiently many random values for the expressions we have to guess during the attack.

An important future work item that was discussed in section 4.9 is how to break most keys of Grain-128. In addition, the new techniques should be applied to other schemes. Preliminary analysis of the stream cipher Trivium [14] suggests that dynamic cube attacks can improve the best known attack on this cipher, but the improvement factor we got so far is not very significant.

References

1. Biham, E., Shamir, A.: Differential cryptanalysis of DES-like cryptosystems. In: Menezes, A., Vanstone, S.A. (eds.) CRYPTO 1990. LNCS, vol. 537, pp. 2–21. Springer, Heidelberg (1991)
2. Matsui, M.: Linear cryptanalysis method for DES cipher. In: Helleseth, T. (ed.) EUROCRYPT 1993. LNCS, vol. 765, pp. 386–397. Springer, Heidelberg (1994)
3. Aumasson, J.-P., Dinur, I., Meier, W., Shamir, A.: Cube Testers and Key Recovery Attacks on Reduced-Round MD6 and Trivium. In: Dunkelman, O. (ed.) FSE 2009. LNCS, vol. 5665, pp. 1–22. Springer, Heidelberg (2009)
4. Dinur, I., Shamir, A.: Cube attacks on tweakable black box polynomials. In: Joux, A. (ed.) EUROCRYPT 2009. LNCS, vol. 5479, pp. 278–299. Springer, Heidelberg (2009)
5. Aumasson, J.-P., Dinur, I., Henzen, L., Meier, W., Shamir, A.: Efficient FPGA Implementations of High-Dimensional Cube Testers on the Stream Cipher Grain-128. In: SHARCS - Special-purpose Hardware for Attacking Cryptographic Systems (2009)
6. Hell, M., Johansson, T., Maximov, A., Meier, W.: A stream cipher proposal: Grain-128. In: IEEE International Symposium on Information Theory, ISIT 2006 (2006)
7. Englund, H., Johansson, T., Sönmez Turan, M.: A framework for chosen IV statistical analysis of stream ciphers. In: Srinathan, K., Rangan, C.P., Yung, M. (eds.) INDOCRYPT 2007. LNCS, vol. 4859, pp. 268–281. Springer, Heidelberg (2007)
8. Fischer, S., Khazaei, S., Meier, W.: Chosen IV statistical analysis for key recovery attacks on stream ciphers. In: Vaudenay, S. (ed.) AFRICACRYPT 2008. LNCS, vol. 5023, pp. 236–245. Springer, Heidelberg (2008)
9. De Cannière, C., Kücük, Ö., Preneel, B.: Analysis of Grain's initialization algorithm. In: SASC 2008 (2008)
10. Lee, Y., Jeong, K., Sung, J., Hong, S.H.: Related-key chosen IV attacks on grain-v1 and grain-128. In: Mu, Y., Susilo, W., Seberry, J. (eds.) ACISP 2008. LNCS, vol. 5107, pp. 321–335. Springer, Heidelberg (2008)

11. Lai, X.: Higher order derivatives and differential cryptanalysis. In: Symposium on Communication, Coding and Cryptography, in Honor of James L. Massey on the Occasion of His 60'th Birthday, pp. 227–233 (1994)
12. Vielhaber, M.: Breaking ONE.FIVIUM by AIDA an algebraic IV differential attack. Cryptology ePrint Archive, Report 2007/413 (2007)
13. Joux, A.: Algorithmic Cryptanalysis, pp. 285–286. Chapman & Hall, Boca Raton
14. De Cannière, C., Preneel, B.: Trivium - a stream cipher construction inspired by block cipher design principles. estream, ecrypt stream cipher. Technical report of Lecture Notes in Computer Science
15. Knellwolf, S., Meier, W., Naya-Plasencia, M.: Conditional Differential Cryptanalysis of NLFSR-Based Cryptosystems. In: Abe, M. (ed.) ASIACRYPT 2010. LNCS, vol. 6477, pp. 130–145. Springer, Heidelberg (2010)
16. Stankovski, P.: Greedy Distinguishers and Nonrandomness Detectors. In: Gong, G., Gupta, K.C. (eds.) INDOCRYPT 2010. LNCS, vol. 6498, pp. 210–226. Springer, Heidelberg (2010)

A   Appendix: Zeroing State Bits of Grain-128

To demonstrate the process that we use to zero state bits of Grain-128, consider the problem of zeroing b_{194}. The ANF representation of b_{194} is a relatively small polynomial of degree 9 in the 128 secret variables and 96 public variables which contains 9813 terms. It is calculated by assigning all the key and IV bits a distinct symbolic variable, and calculating the symbolic value of the feedback to the NFSR after 67 rounds. It may be possible to choose a dynamic public variable and zero b_{194} directly. However, since the ANF representation of b_{194} is difficult to visualize, this has a few disadvantages: After we choose a cube to sum over, we need to guess all the secret expressions that are multiplied by terms of cube variables, and the complex ANF representation of b_{194} will force us to guess many expressions, which will unnecessarily increase the complexity of the attack. Moreover, since the ANF representation of b_{194} is of degree 9, many of the guessed expressions are expected to be non-linear, while ideally we would like to collect linear equations in order to be able to solve for the key bits efficiently. The process that we use to zero b_{194} is given below.

1. Use the description of Grain-128 to simplify the ANF representation of b_{194} by writing $b_{194} = b_{161}(b_{78}s_{161}) + P_{r1}$. In this form, b_{161} is the source polynomial, $b_{78}s_{161}$ is the target polynomial, and P_{r1} is some remainder polynomial with a simpler ANF representation compared to b_{194}.
2. The ANF representation of b_{161} is a simpler polynomial of degree 6 which contains 333 terms. Again, do not zero it directly, but write: $b_{161} = b_{128}(b_{45}s_{128}) + P_{r2}$, with b_{128} as the source polynomial with degree 3 and 26 terms. Choose v_0 as the dynamic variable and set it accordingly.
3. Now, the ANF representation of b_{161}, with v_0 set to its dynamic value is a polynomial of degree 2 which contains 47 terms. b_{161} can be zeroed by choosing v_{33} as a dynamic variable.

4. Recalculate the ANF of b_{194} with v_0 and v_{33} set to their dynamic values. It is now a polynomial of degree 5 which contains 1093 terms. Write $b_{194} = b_{134}b_{150} + P_{r3}$, and choose v_6 as the dynamic variable to zero b_{134}.

5. Write $b_{194} = b_{162} + P_{r4} = b_{129}(b_{46}s_{129}) + P_{r5}$ and choose v_1 as the dynamic variable to zero b_{129}.

6. Now, the symbolic form of b_{194} with v_0, v_{33}, v_6 and v_1 all set to their dynamic values, is a polynomial of degree 3 with 167 terms. Finally we choose v_{29} as the dynamic variable which can zero b_{194}.

B   Appendix: Parameters for Our Attacks on Grain-128

The parameter sets for the different attacks are given in the tables below. As an example, we demonstrate the process of assigning values to the dynamic variables in Table 1. The process for the other tables is similar.

The first index of the "Dynamic Variables" list in Table 1 is 0 (i.e v_0). It is used to nullify the first state bit in the "State Bits Nullified" list (b_{128}). The symbolic form of v_0 is calculated as follows:

1. Initialize the state of Grain-128 with all the key bits assigned a distinct symbolic variable and all the IV bits set to 0, except the IV bits in the "Cube Indexes" row and v_0 which are assigned a distinct symbolic variable.

2. Clock the state once and obtain the symbolic value of the bit fed back into the NFSR (note that v_0 is a linear variable of the polynomial).

3. Delete the term v_0 from the symbolic form of this polynomial and assign v_0 the symbolic sum of the remaining terms, i.e. set $v_0 = x_3x_{67} + x_{11}x_{13} + x_{17}x_{18} + x_{27}x_{59} + x_{40}x_{48} + x_{61}x_{65} + x_{68}x_{84} + x_0 + x_2 + x_{15} + x_{26} + x_{36} + x_{45} + x_{56} + x_{64} + x_{73} + x_{89} + x_{91} + x_{96}$.

Next, we determine the symbolic value of v_1 (second in the "Dynamic Variables" list), according to the second state bit in the "State Bits Nullified" list (b_{129}). It is calculated in a similar way to v_0, except that we set v_0 to the dynamic value calculated in the previous step and set v_1 to a distinct symbolic variable. Finally we assign v_1 the symbolic value that is fed back to the NFSR after 2 initialization rounds (again, removing the linear term of v_1 from the symbolic form). We iteratively continue assigning v_6, v_{33} and v_{29} according to the symbolic values fed back to the NFSR after $7, 34$ and 67 clocks respectively, each time setting the previously determined dynamic variables to their dynamic values.

Table 1. Parameter set No.1 for the attack on Grain-128, given output bit 215

Cube Indexes	$\{3,28,31,34,40,50,51,52,54,62,63,64,65,66,67,68,69,80,92\}$
Dynamic Variables	$\{0,1,6,33,29\}$
State Bits Nullified	$\{b_{128}, b_{129}, b_{134}, b_{161}, b_{194}\}$
Expressions Retrieved	$\{x_{127}, x_{122} + 1\}$

Table 2. Parameter set No.2 for the attack on Grain-128, given output bit 215

Cube Indexes	$\{5,19,28,31,34,40,50,51,52,54,62,63,64,65,66,67,68,80,92\}$
Dynamic Variables	$\{0,1,6,33,29\}$
State Bits Nullified	$\{b_{128}, b_{129}, b_{134}, b_{161}, b_{194}\}$
Expressions Retrieved	$\{x_{69}, x_{23}\}$

Table 3. A Parameter set for the attack on Grain-128, given output bit 218

Cube Indexes	$\{19,20,28,29,30,31,41,45,53,54,55,63,64,65,66,67,68,69,89,92\}$
Dynamic Variables	$\{2,3,4,9,1,36,7,32\}$
State Bits Nullified	$\{b_{130}, b_{131}, b_{132}, b_{137}, s_{129}, b_{164}, b_{170}, b_{197}\}$
Expressions Retrieved	$\{x_{98}, x_{49}\}$

Table 4. Parameter set for the attack on Grain-128, given output bit 251

Cube Indexes	$\{11,12,13,15,17,21,24,26,27,29,32,35,38,40,43,46,49,51,52,$ $53,55,57,58,63,64,65,66,74,75,77,78,79,81,84, 86,87,95\}$
Dynamic Variables	$\{8,9,10,14,0,1,39,2,72,3,4,5,80,25,90,92,41,7,36,37,88,23,89,54\}$
Public Variables Set to 1	$\{82\}$
State Bits Nullified	$\{b_{136}, b_{137}, b_{138}, b_{142}, b_{128}, b_{129}, s_{129}, b_{130}, s_{130}, b_{131}, b_{132}, b_{133},$ $b_{148}, b_{153}, b_{158}, b_{160}, s_{162}, b_{163}, b_{164}, b_{165}, b_{174}, b_{186}, b_{197}, b_{230}\}$

Table 5. Parameter set for the attack on a weak key subset of the full Grain-128, given output bit 257

Cube Indexes	$\{0,3,5,10,11,13,14,15,17,19,21,23,26,31,34,35,37,39,40,43,45,48,49,51,$ $53,54,55,56,57,59,63,65,66,67,68,71,77,78,79,81,85,91,92,93,94,95\}$
Dynamic Variables	$\{9,1,12,4,7,6,8,89,2,29,83,25,69\}$
State Bits Nullified	$\{b_{137}, b_{129}, s_{133}, b_{132}, b_{135}, b_{134}, b_{136}, s_{168}, b_{169}, s_{150}, b_{176}, b_{192}, b_{236}\}$
Key Bits Set to 0	$\{x_{48}, x_{55}, x_{60}, x_{76}, x_{81}, x_{83}, x_{88}, x_{111}, x_{112}, x_{122}\}$

Cryptanalysis of the Knapsack Generator

Simon Knellwolf and Willi Meier

FHNW, Switzerland

Abstract. The knapsack generator was introduced in 1985 by Rueppel and Massey as a novel LFSR-based stream cipher construction. Its output sequence attains close to maximum linear complexity and its relation to the knapsack problem suggests strong security. In this paper we analyze the security of practically relevant instances of this generator as they are recommended for the use in RFID systems, for example. We describe a surprisingly effective guess and determine strategy, which leads to practical attacks on small instances and shows that the security margin of larger instances is smaller than expected. We also briefly discuss a variant of the knapsack generator recently proposed by von zur Gathen and Shparlinski and show that this variant should not be used for cryptographic applications.

Keywords: knapsack, stream cipher, pseudorandom generator.

1 Introduction

Let $w_0, \ldots, w_{n-1}$ be n k-bit integers, and let $u_0, u_1, \ldots$ be a sequence of bits generated by a linear feedback shift register (LFSR) of order n over $\mathbb{F}_2$. At step i the *knapsack generator* computes

$$v_i = \sum_{j=0}^{n-1} u_{i+j} w_j \mod 2^k, \tag{1}$$

discards the ℓ least significant bits of v_i and outputs the remaining $k - \ell$ bits as part of the keystream. We call $u_0, u_1, \ldots$ the *control bits*, v_i the i-th *sum* and $w_0, \ldots, w_{n-1}$ the *weights* of the generator. The entire generator is defined by $n(2 + k)$ bits: n bits for the connection polynomial of the LFSR, n bits for the initial control bits (corresponding to the initial state of the LFSR) and kn bits for the weights. The connection polynomial should be primitive in order to achieve maximum period in the control sequence. Due to this special choice, it is natural to consider the connection polynomial as a public parameter. The remaining $n(1 + k)$ bits form the key of the generator. As a concrete example, a generator with $n = k = 64$ has a key length of 4160 bits.

Rueppel and Massey [16] introduced this generator in 1985. They addressed one of the main issues in the design of LFSR-based cryptosystems, which consists in breaking the linearity of the LFSR. The knapsack generator achieves this by the use of integer addition modulo 2^k which is a highly nonlinear operation when considered over $\mathbb{F}_2^k$, see [16,20] for a systematic analysis. Therewith it

A. Joux (Ed.): FSE 2011, LNCS 6733, pp. 188–198, 2011.

provides an interesting alternative to the use of nonlinear boolean filtering and combining functions. It avoids the tradeoff between high linear complexity and high correlation immunity which is inherent to nonlinear boolean functions, as it was shown in [19]. Besides the knapsack generator, Rueppel [14] also introduced the summation generator to avoid this tradeoff, but it turned out to be vulnerable to correlation attacks [5,11], algebraic attacks [10] and to attacks based on feedback with carry shift registers [9]. Compared to the summation generator, little cryptanalytic work has been published concerning the knapsack generator. To our knowledge it resisted the well known attack strategies for stream ciphers. Due to this absence of known security flaws and due to its ease of implementation, the authors of [2] recommend the knapsack generator for the use in RFID networks.

The name comes from the close relation to the knapsack problem (also known as the subset sum problem), which consists in finding a subset of given weights that add up to a given sum. The decisional version of this problem is known to be NP-complete and several attempts have been made to use it in cryptography. A prominent example is the Merkle-Hellman knapsack cryptosystem [13] which was broken by Shamir [17]. In a different direction, Impagliazzo and Naor [8] constructed provably secure pseudorandom generators and universal hash functions based on the knapsack problem.

Other than the above cryptosystems, the security of the knapsack generator is not directly related to the hardness of the knapsack problem. In the context of the knapsack problem the weights are known, whereas in the context of the knapsack generator they are not. For comparison, Howgrave-Graham and Joux [7] presented new generic algorithms for hard knapsack problems which allowed them to solve instances with $n = k = 96$ in practical time. These results have no implications on the security of the knapsack generator, and knapsack generators of much smaller size are not a priori insecure (even for $n = k = 32$, the key consists of 1056 bits).

Throughout the literature, for example in [2,12,15], it is recommended to choose $k = n$. We also focus on these cases and we do not always mention k explicitly in the following. The cases $n = 32, 64$ are of particular interest because they are favorable for software implementation. Besides n, the knapsack generator has an additional security parameter ℓ, which is the number of discarded bits per output. Intuitively, if ℓ is small, the output reveals more information about the control sequence, whereas if ℓ is large, the throughput of the generator gets low.

1.1   Previous Cryptanalytic Results

Rueppel [15] provided a first extensive analysis of the knapsack generator. He showed that the $\lceil \log n \rceil$ least significant bits of the sums do not achieve high linear complexity, and he provided some evidence that the other bits indeed do. This let him recommend to choose $\ell = \lceil \log n \rceil$. He further estimated the number of different boolean functions $\{0, 1\}^n \to \{0, 1\}$ mapping n control bits to the i-th bit of a sum as in (1), and he found that for $\lfloor \log n \rfloor \le i < n$ at least $2^{n(\lfloor \log n \rfloor - 1)}$

such functions can be specified by n weights. He stated this as a lower bound on the effective key length of the generator.

Von zur Gathen and Shparlinski [3] considered scenarios where either the control bits or the weights are known. In both cases they translate the task of finding the unknown parts of the key into a short vector problem which they solve by LLL lattice basis reduction algorithms. In the known control bit scenario, they can predict the generator if ℓ is not too small using about $n^2 - n$ outputs. It is difficult to estimate the practical time complexity of their strategy when extended to a guess and determine attack, and no empirical results are provided.

1.2 Contribution of This Paper

We describe a novel guess and determine attack which needs only a few more outputs than the number of weights n. Our analytical and empirical results show that the security level of the knapsack generator is not significantly higher than n bits. For a generator with $n = 32$ we implemented the full attack on a Desktop Computer.

Further, we analyze the faster variant of the knapsack generator recently proposed in [4], and show that it should not be used in cryptographic applications.

1.3 Road Map

In Section 2 we describe the knapsack generator as a system of modular equations and we introduce the notion of an approximation matrix, which is the basic concept of our analysis. In Section 3 we explain how to find good approximation matrices. In Section 4 we describe the full attack and illustrate its performance by empirical results, including a practical attack for $n = 32$. In Section 5 we briefly analyze the fast variant of the knapsack generator proposed in [4].

2 Problem Formalization

In this section we address the following problem: Given the control bits and s outputs of the knapsack generator, predict some bits of subsequent outputs with probability higher than $1/2$. Later, in Section 4, we extend this to a guess and determine attack when the control bits are not known.

We first formulate the knapsack generator as a system of modular equations and fix some notation.

2.1 A System of Modular Equations

In order to produce s outputs, the knapsack generator computes s sums according to (1). This can be written as

$$\mathbf{v} = U\mathbf{w} \mod 2^n, \tag{2}$$

where $\mathbf{v} = (v_0, \ldots, v_{s-1})$ are the sums, $\mathbf{w} = (w_0, \ldots, w_{n-1})$ are the weights and U is a $s \times n$ matrix whose coefficients are given by the control bits. We call U the *control matrix*. Its rows are the consecutive states of a binary LFSR, and the control matrix is entirely determined by one of its rows. We write $\mathbf{u}_i$ for the i-th row of U and, more generally, for the i-th state of the LFSR generating the control sequence. It is shown in [3] that n consecutive row vectors $\mathbf{u}_i$ are always linearly independent over the integers modulo 2^n. Hence, if U is known and $s \geq n$, the system described by (2) can be easily solved for $\mathbf{w}$. The challenge is to deal with the discarded bits. An attacker can only observe the $n - \ell$ most significant bits of each component of $\mathbf{v}$. Guessing the discarded bits is too expensive, since at least $n\ell$ bits would have to be guessed. The idea is to recover only the significant bits of each weight which might be sufficient to make a prediction.

2.2   Weight Approximation Matrices

We write the outputs as a vector $\mathbf{z} = (z_0, \ldots, z_{s-1})$ such that $z_i = v_i \gg \ell$ for $0 \leq i < s$. Here, $\gg$ denotes a right shift of n-bit integers, left shift is denoted by $\ll$, and when used for vectors, the shifting is applied componentwise. Since U has full rank modulo 2^n, there always exists a $n \times s$ matrix T with integer coefficients such that $TU = I_n \mod 2^n$, where I_n denotes the $n \times n$ identity matrix. We call such a T an *approximation matrix*. The name is motivated by the fact that

$$\mathbf{w} = T(\mathbf{z} \ll \ell) + T\mathbf{d} \mod 2^n$$

for some unknown vector $\mathbf{d} = (d_0, \ldots, d_{s-1})$ with $0 \leq d_i < 2^\ell$ for $0 \leq i < s$ (the d_i correspond to the discarded bits), which lets us hope to obtain *approximate weights* $\tilde{\mathbf{w}}$ by ignoring the discarded bits, that is, by computing

$$\tilde{\mathbf{w}} = T(\mathbf{z} \ll \ell) \mod 2^n. \tag{3}$$

The matrix T will be derived only from U (independently from $\mathbf{z}$). As soon as $s > n$, the choice of T is not unique. In the next paragraph we obtain a criterion for making a good choice.

2.3   Prediction with Approximate Weights

In order to predict $z_s = v_s \gg \ell$, we compute $\tilde{v}_s = \mathbf{u}_s \tilde{\mathbf{w}} \mod 2^n$, where $\mathbf{u}_s = (u_s, \ldots, u_{s+n-1})$ are the corresponding control bits. Substituting $\tilde{\mathbf{w}}$, we get $\tilde{v}_s = \mathbf{u}_s T(\mathbf{z} \ll \ell) \mod 2^n$. The generator actually computes

$$v_s = \mathbf{u}_s T(\mathbf{z} \ll \ell) + \mathbf{u}_s T\mathbf{d} \mod 2^n.$$

Intuitively, the significant bits of $\tilde{v}_s$ are likely to be correct if the integer summand $\mathbf{u}_s T\mathbf{d}$ is small in absolute value. We denote by p_λ the probability that at least λ significant bits of $\tilde{v}_s$ are correct,

$$p_\lambda = \Pr\left[(v_s \oplus \tilde{v}_s) \gg (n - \lambda) = 0\right].$$

The intuition is then formalized by the following lemma.

Lemma 1. *Let m be the smallest integer such that $|\mathbf{u}_s T\mathbf{d}| < 2^m$. Then, we have*

$$p_\lambda > 1 - \frac{1}{2^{\lambda-m}}$$

for all λ with $m \leq \lambda < n$.

Proof. For shorter notation we set $a = \mathbf{u}_s T(\mathbf{z} \ll \ell)$ and $b = \mathbf{u}_s T\mathbf{d}$. The difference $v_s \oplus \tilde{v}_s$ then writes as $(a + b) \oplus a$. Let's first assume that $b \geq 0$. Then, the sum $a+b$ can be recursively described by $(a+b)_j = a_j \oplus b_j \oplus c_{j-1}$, $c_j = a_j b_j \oplus a_j c_{j-1} \oplus b_j c_{j-1}$, where c_j denotes the carry bit, and $c_{-1} = 0$. For $j \geq m$ we have $b_j = 0$, and thus, $(a + b)_m \oplus a_m = c_{m-1}$ and for $j > m$, $(a + b)_j \oplus a_j = c_{m-1} \prod_{i=m}^{j-1} a_i$. The bound follows immediately under the assumption that the values of the bits a_i are independent and uniformly distributed for $m \leq i < n$. The case of $b < 0$ is very similar (using the recursive description of $a - b$ with a borrow bit instead of the carry bit). $\qquad\square$

Lemma 1 guarantees that we can correctly predict at least one bit per output with probability higher than $1/2$ if $m < n - 1$. Smaller m give more predictable bits. Hence, we are interested in an upper bound on m. Since the coefficients of $\mathbf{u}_s$ are restricted to be 0 or 1 and the coefficients of $\mathbf{d}$ are strictly smaller than 2^ℓ, we have $|\mathbf{u}_s T\mathbf{d}| < \|T\|2^\ell$, where we use $\|T\| = \sum_{i,j}|t_{ij}|$ as the *norm* of T. By the definition of m, this gives

$$m < \lceil \log\|T\| \rceil + \ell.$$

It follows that $\lceil \log\|T\| \rceil \leq n-\ell-1$ is a sufficient condition to predict at least one bit. In the next section we describe a method that finds approximation matrices with much lower norms than needed for typical values of ℓ.

3   Finding Good Approximation Matrices

The success of our attack essentially depends on the ability to find good approximation matrices, that is, matrices with small coefficients in absolute value. To compute such a matrix T we proceed row by row. We search for n row vectors $\mathbf{t}_i$ with small norm and such that $\mathbf{t}_i U = \mathbf{e}_i$, where $\mathbf{e}_i$ is the i-th unit vector of the standard basis of $\mathbb{F}_2^n$. This is a special case of the following problem:

Problem 1. Given an $s \times n$ integer matrix A and an integer column vector $\mathbf{b}$ of dimension n, find an integer row vector $\mathbf{x}$ such that $\mathbf{x}A = \mathbf{b}$ and such that the coefficients of $\mathbf{x}$ are small in absolute value.

Typically, in our scenario, there are many solutions to the equation $\mathbf{x}A = \mathbf{b}$ and it is not difficult to find one of them by linear algebra techniques. The difficult part is to find one with a small norm. We use an approach which is implemented in Victor Shoup's NTL [18]. The idea is the following: Given an arbitrary solution $\mathbf{x}'$, search for a vector $\mathbf{x}''$ in the kernel of A such that $\mathbf{x} = \mathbf{x}' - \mathbf{x}''$ has a small norm. This essentially corresponds to the approximate closest vector problem in the lattice spanned by the kernel vectors. Using Babai's algorithm [1] together with a LLL reduced kernel basis we can find very close vectors $\mathbf{x}''$ in practice. A nice introduction to Babai's algorithm and its use in combination with LLL can be found in [6].

In our specific application, the matrix A $(= U)$ has only 0 and 1 as coefficients and its rows are the successive states of an LFSR. It turns out that the small coefficients are favorable, that is, the average norm of the returned solution is smaller than for more general integer matrices. The particular structure of the control matrix (successive states of an LFSR) has no significant influence, the results are about the same as for random matrices with binary coefficients. In particular, the choice of the connection polynomial seems not to be important (as long as it is primitive).

Not surprisingly, the average norm of the returned solutions depends on s (it basically determines the kernel dimension of A). Figure 1 illustrates the performance of the method in function of s for $n = 64$. The graph indicates the average logarithmic norm as well as the lower and the upper quartile for samples of 100 approximation matrices obtained for $s = 68, 70, \ldots, 96$. Recall that s is the number of outputs used to compute the approximate weights.

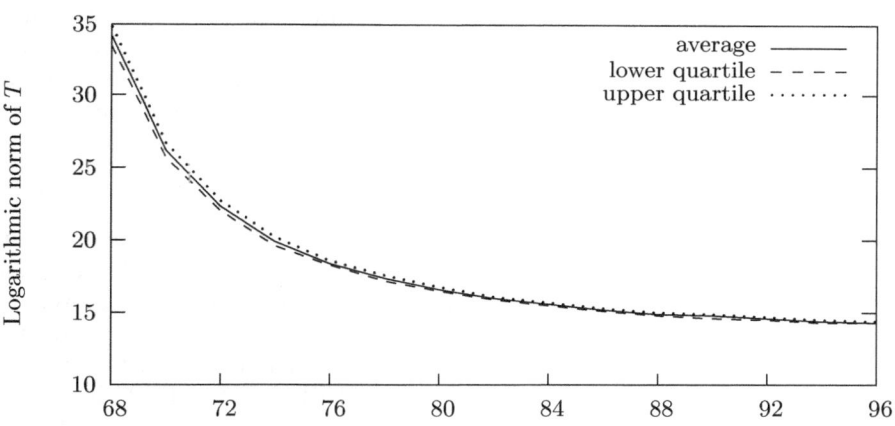

Fig. 1. Average logarithmic norm of T for $n = 64$ in function of s

4   Description of the Attack and Empirical Results

So far, we can predict the generator if the control bits are known. The empirical results in this section illustrate the effectiveness of the approach. But first, we describe the extension of our technique to an attack where the control bits are not known.

4.1 Description of the Attack

We assume a scenario where the attacker does not know the control bits nor the weights. Since the whole control sequence is determined by only n bits (typically, the initial state of the LFSR), the above approach naturally extends to a guess and determine attack:

1. Guess $u_0, \ldots, u_{n-1}$ and derive the $s \times n$ control matrix U.
2. Find an approximation matrix T based on U.
3. Use T and $z_0, \ldots, z_{s-1}$ to compute $\tilde{\mathbf{w}}$ as in (3).
4. Compute t predictions and check their λ most significant bits. If almost all of them are correct, the control bits have been guessed correctly. Otherwise, go back to step 1.

The parameters t and λ must be chosen such that the checking is reliable. At least, t should be chosen such that $n \geq t\lambda p_\lambda$. This is not a problem, because the norms of the approximation matrices are very low, and λ can be chosen such that p_λ is almost one. The attack then needs $s + t$ outputs: s outputs to approximate the weights and t outputs to check the predictions. The most expensive part of the attack is at step 2, where the approximation matrices are computed. Instead of computing 2^n such matrices, we can check several guesses by the same matrix T. Using $z_1, \ldots, z_s$ to compute $\tilde{\mathbf{w}}$ at step 3, we can check if $u_0, \ldots, u_{n-1}$ was the state of the LFSR after one clock and we can easily compute the initial state. In general, if $r \geq s + t$ outputs are available, only $2^n/(r - s)$ approximation matrices must be computed. Since this computation is independent of the observed outputs, it can even be done offline.

4.2 Practical Attack for $n = 32$

For $n = 32$ the above attack is practical on a desktop computer. In our experiments we assumed that 552 outputs could be observed. We used control matrices with $s = 40$ rows and the parameters for the checking part were $t = 20$ and $\lambda = 5$. Hence, only 2^{23} approximation matrices had to computed. A guess was accepted when less than 20 of the 100 predicted bits where wrong. On a Intel Core 2 Duo E8400 3.0 GHz Processor with 4 GB of RAM it took about three days to identify the correct initial control bits and about 870 bits of the weights. This allows an attacker to predict more than 22 bits per output (we used $\ell = 5$, hence an output has 27 bits).

4.3 Empirical Results for Larger n

For larger n the attack is not practical on a desktop computer, since we could not circumvent the guessing of the n bits. Hence, we assume in this paragraph that the control bits are known and that we have observed s outputs. This corresponds to a known control bits attack or to the checking part of our guess and determine attack. We are interested in the average number of significant bits per output that we can correctly predict. To be precise, let λ^* be the largest λ such that $(v_s \oplus \tilde{v}_s) \gg (n - \lambda) = 0$. We analyze the average size of λ^*. Note that if $\tilde{v}_s$ would be obtained by coin tossing, the expectation of λ^* would be $\sum_{i=1}^{n} i/2^{i+1} \approx 1$ bits per output. Table 1 contains the results for $n = 32, 64, 128$ with $\ell = \log n$ and different values of s. The samples were taken randomly from the key space of the generator (recall that the key consists of $n(1 + n)$ bits).

Table 1. Average number of correctly predicted bits per output (λ^*)

$s - n$	$n = 32$	$n = 64$	$n = 128$	$n = 256$
8	20.6	42.9	85.3	164.6
16	22.2	48.7	100.9	203.4
24	22.6	50.3	105.9	216.4
32	22.7	50.8	108.1	222.4

The results in Table 1 show that even for large n the security of the knapsack generator is not significantly higher than n bits. Computing an approximation matrix which allows to predict about 108 bits per output of a generator with $n = 128$ takes a few seconds and needs no more than 160 outputs.

5   Analysis of the Fast Knapsack Generator

In [4], von zur Gathen and Shparlinski describe a variant of the knapsack generator which achieves faster output generation. We call it the *fast knapsack generator*. They consider a slightly more general setting as we did in this paper by taking the weights in an arbitrary ring R (in this paper we just considered $R = \mathbb{Z}/m\mathbb{Z}$ with $m = 2^n$). The speedup is achieved by a special choice of the weights. Instead of randomly choosing each of the n weights, it is proposed to choose two elements $a, b \in R$ and to compute $w_j = ab^{n-j}$ for $0 \le j \le n - 1$. With these weights, the $(i + 1)$-th sum can be computed recursively from the i-th sum by

$$v_{i+1} = bv_i - ab^{n+1}u_i + abu_{i+n}, \text{ for } i \ge 0.$$

Hence, only one multiplication and two additions are needed for generating one output. If R is a prime field, the sequence (v_i) has provable properties concerning the uniformity of its distribution (see Theorem 3.5 in [4]). However, it was left open whether this specialization affects the cryptographic security of the generator. We show that the sequence does not provide cryptographic security if R is a prime field and we believe that the security is highly questionable if $R = \mathbb{Z}/m\mathbb{Z}$ for $m = 2^n$. Our attack is a guess and determine attack whose complexity essentially depends on the number of discarded bits (and not on n as in the case of the original generator). It specifically exploits the strong relation between the weights.

5.1   The Fast Generator over Prime Fields

Assume that R is a prime field, i.e. $R = \mathbb{F}_p$ for p prime. We think of the elements of $\mathbb{F}_p$ as $\lceil \log p \rceil$-bit integers and as above we denote by ℓ the number of discarded bits. Let's first suppose that the control bits are known. Then a and b can be determined as follows (here, all operations are modulo p):

1. Find i_0 such that $u_{i_0} = 0$ and $u_{i_0+n} = 0$
2. Guess the 2ℓ discarded bits of v_{i_0} and v_{i_0+1}
3. Compute $b = v_{i_0+1}/v_{i_0}$ and $a = v_{i_0}/\sum_{j=0}^{n-1} u_{i_0+j} b^{n-j}$
4. For some $i \neq i_0$ compute $\sum_{j=0}^{n-1} u_{i+j} a b^{n-j}$ and check if the significant bits agree with those of v_i.

The cost of this attack is about the cost of $2^{2\ell}$ times computing two modular inverses and two sums with n summands. If we drop the assumption that the control bits are known, we can not choose i_0 suitably in the first step. We have to guess it and hope that u_{i_0} and u_{i_0+n} are zero. Then, at the third step, we miss $n - 1$ control bits for computing a. Instead of guessing these control bits, we speculate on $u_{i_0+1} = 0$ and $u_{i_0+n+1} = 1$ such that $v_{i_0+2} = bv_{i_0+1} + ab$. So, we only have to guess the discarded bits of v_{i_0+2} for obtaining a. In order to check a and b we try to find $u_{i_0+2}, \ldots, u_{2n-1}$ such that the significant bits of $bv_i - ab^{n+1}u_i + abu_{i+n}$ agree with those of v_{i+1} for $i_0 + 2 \leq i < n$. If such control bits can be found, our guess is correct with high reliability. In average we need about 2^4 trials for finding a suitable i_0 (satisfying the conditions for the first and the third step) and for each trial we have to guess $2^{3\ell}$ discarded bits. Checking a guess costs at most $4n$ additions of three summands. The attack works with about $2^4 + n$ outputs and its total cost is about $2^{4+3\ell}$ times the cost of computing two modular inverses and at most $4n + 1$ additions.

5.2   The Fast Generator Modulo 2^n

The attack of the prime field case does not directly translate to the case $R = \mathbb{Z}/m\mathbb{Z}$ for $m = 2^n$ (or to other rings). The problem is that, in general, the elements of a ring do not have a unique inverse. Hence, the divisions at the third step are not well defined. Instead, we have to find a and b such that

$$bv_{i_0} = v_{i_0+1},$$
$$v_{i_0+2} = bv_{i_0+1} + ab.$$

For some guesses, no such a and b exist. These guesses can be easily ruled out. But for other guesses, many choices for a and b are possible and the checking of them will be more costly.

6   Discussion

It was already noticed by von zur Gathen and Shparlinski in [3] that the security of the knapsack generator is smaller than suggested by a naive estimate based on its key length. The results in this paper show that it is no more than n bits. Our approach applies to all relevant parameters ℓ, including the following:

n	32	64	128
ℓ	≤ 25	≤ 42	≤ 98

In particular, it applies to $\ell = \lceil \log n \rceil$. The full attack works with only a few more outputs than the number of weights and it is not difficult to translate our analysis to the cases where k and n are not equal.

However, we could not circumvent the guessing part of our attack, for example, using ideas from fast correlation attacks. Due to the high nonlinearity of integer addition with multiple inputs, it seems very unlikely to find approximate linear relations between outputs or correlations to the state of the LFSR.

So far, a knapsack generator of size n provides n-bit security. To guarantee this security level it needs a $n(n + 1)$ bit secret key, and the example of the fast knapsack generator shows that it is delicate to reduce the potential entropy of the weights. This has to be taken into account when evaluating the knapsack generator as an alternative to nonlinear boolean filtering and combining functions, or when using it in RFID applications.

Acknowledgements. We thank the anonymous reviewers for helpful comments. Especially, we thank the reviewer who pointed us to the use of LLL for computing the approximation matrices. This work was partially supported by European Commission through the ICT programme under contract ICT-2007-216676 ECRYPT II and by the Hasler Foundation www.haslerfoundation.ch under project number 08065.

References

1. Babai, L.: On Lovász' lattice reduction and the nearest lattice point problem. Combinatorica 6(1), 1–13 (1986)
2. Cole, P.H., Ranasinghe, D.C.: Networked RFID systems and lightweight cryptography: raising barriers to product counterfeiting. Springer, Heidelberg (2007)
3. von zur Gathen, J., Shparlinski, I.: Predicting Subset Sum Pseudorandom Generators. In: Handschuh, H., Hasan, M.A. (eds.) SAC 2004. LNCS, vol. 3357, pp. 241–251. Springer, Heidelberg (2004)
4. von zur Gathen, J., Shparlinski, I.: Subset sum pseudorandom numbers: fast generation and distribution. J. Math. Crypt. 3, 149–163 (2009)
5. Golic, J.D., Salmasizadeh, M., Dawson, E.: Fast Correlation Attacks on the Summation Generator. J. Cryptology 13(2), 245–262 (2000)
6. Hoffstein, J., Pipher, J., Silverman, J.H.: An introduction to mathematical cryptography. Springer, Heidelberg (2008)
7. Howgrave-Graham, N., Joux, A.: New Generic Algorithms for Hard Knapsacks. In: Gilbert, H. (ed.) EUROCRYPT 2010. LNCS, vol. 6110, pp. 235–256. Springer, Heidelberg (2010)
8. Impagliazzo, R., Naor, M.: Efficient Cryptographic Schemes Provably as Secure as Subset Sum. J. Cryptology 9(4), 199–216 (1996)

9. Klapper, A., Goresky, M.: Feedback Shift Registers, 2-Adic Span, and Combiners with Memory. J. Cryptology 10(2), 111–147 (1997)
10. Lee, D.H., Kim, J., Hong, J., Han, J.W., Moon, D.: Algebraic Attacks on Summation Generators. In: Roy, B., Meier, W. (eds.) FSE 2004. LNCS, vol. 3017, pp. 34–48. Springer, Heidelberg (2004)
11. Meier, W., Staffelbach, O.: Correlation Properties of Combiners with Memory in Stream Ciphers. In: Damgård, I.B. (ed.) EUROCRYPT 1990. LNCS, vol. 473, pp. 204–213. Springer, Heidelberg (1991)
12. Menezes, A.J., van Oorschot, P.C., Vanstone, S.A.: Handbook of Applied Cryptography. CRC Press, Boca Raton (2001)
13. Merkle, R., Hellman, M.: Hiding information and signatures in trapdoor knapsacks. IEEE Transactions Information Theory 24(5), 525–530 (1978)
14. Rueppel, R.A.: Correlation Immunity and the Summation Generator. In: Williams, H.C. (ed.) CRYPTO 1985. LNCS, vol. 218, pp. 260–272. Springer, Heidelberg (1986)
15. Rueppel, R.A.: Analysis and Design of Stream Ciphers. Springer, Heidelberg (1986)
16. Rueppel, R.A., Massey, J.L.: Knapsack as a nonlinear function. In: IEEE Intern. Symp. of Inform. Theory, vol. 46 (1985)
17. Shamir, A.: A Polynomial Time Algorithm for Breaking the Basic Merkle-Hellman Cryptosystem. In: CRYPTO, pp. 279–288 (1982)
18. Shoup, V.: NTL: A Library for doing Number Theory, www.shoup.net/ntl
19. Siegenthaler, T.: Correlation-immunity of nonlinear combining functions for cryptographic applications. IEEE Transactions on Information Theory 30(5), 776–780 (1984)
20. Staffelbach, O., Meier, W.: Cryptographic Significance of the Carry for Ciphers Based on Integer Addition. In: Menezes, A., Vanstone, S.A. (eds.) CRYPTO 1990. LNCS, vol. 537, pp. 601–613. Springer, Heidelberg (1991)

Attack on Broadcast RC4 Revisited

Subhamoy Maitra[1], Goutam Paul[2], and Sourav Sen Gupta[1]

[1] Applied Statistics Unit, Indian Statistical Institute,
Kolkata 700 108, India
{subho,souravsg_r}@isical.ac.in

[2] Department of Computer Science and Engineering,
Jadavpur University, Kolkata 700 032, India
goutam.paul@ieee.org

Abstract. In this paper, contrary to the claim of Mantin and Shamir (FSE 2001), we prove that there exist biases in the initial bytes (3 to 255) of the RC4 keystream towards zero. These biases immediately provide distinguishers for RC4. Additionally, the attack on broadcast RC4 to recover the second byte of the plaintext can be extended to recover the bytes 3 to 255 of the plaintext given $\Omega(N^3)$ many ciphertexts. Further, we also study the non-randomness of index j for the first two rounds of PRGA, and identify a strong bias of j_2 towards 4. This in turn provides us with certain state information from the second keystream byte.

Keywords: Bias, Broadcast RC4, Cryptanalysis, Distinguishing Attack, Keystream, RC4, Stream Cipher.

1   Introduction

RC4, designed by Ron Rivest for RSA Data Security in 1987, is the most popular commercial stream cipher algorithm. There are two components of the RC4 algorithm, namely, the Key Scheduling Algorithm (KSA) and the Pseudo-Random Generation Algorithm (PRGA), that are presented in Algorithm 1 and Algorithm 2 respectively. Given a secret key k of size l bytes (typically, $5 \leq l \leq 16$), an array K of size N bytes (typically, $N = 256$) is created to hold the key such that $K[y] = k[y \bmod l]$ for any $y \in [0, N-1]$. The KSA uses this secret key to scramble a permutation S of $\mathbb{Z}_N = \{0, 1, \ldots, N-1\}$, initialized as the identity permutation. After that, the PRGA generates keystream bytes to be bitwise XOR-ed with the plaintext. The indices i (deterministic) and j (pseudo-random) are used to point to the locations of S. All additions in the KSA and PRGA routines of RC4 algorithm are performed modulo N.

Since the advent of RC4 in 1987, it has faced rigorous analysis over the years due to its simple structure. Extensive research has been conducted to identify weaknesses of RC4 in terms of the KSA as well as the PRGA. There are several important results in cryptanalysis of RC4 where the initial bytes are not of concern. The most prominent recent works in this direction are the distinguisher proposed by Mantin [4] (based on the occurrence of strings of the pattern

A. Joux (Ed.): FSE 2011, LNCS 6733, pp. 199–217, 2011.

Input: Secret Key K. **Output**: S-Box S generated by K. **for** $i = 0, \ldots, N-1$ **do** $\quad \mid \quad S[i] = i;$ **end** Initialize counter: $j = 0;$ **for** $i = 0, \ldots, N-1$ **do** $\quad \mid \quad j = j + S[i] + K[i];$ $\quad \mid \quad$ Swap $S[i] \leftrightarrow S[j];$ **end**	**Input**: S-Box S, output of KSA. **Output**: Random stream Z $\qquad\qquad$ generated from S. Initialize the counters: $i = j = 0;$ **while** *TRUE* **do** $\quad \mid \quad i = i + 1;$ $\quad \mid \quad j = j + S[i];$ $\quad \mid \quad$ Swap $S[i] \leftrightarrow S[j];$ $\quad \mid \quad$ Output $Z = S[S[i] + S[j]];$ **end**

Algorithm 1. KSA $\qquad\qquad\qquad$ **Algorithm 2.** PRGA

$ABTAB$ with A, B bytes and T a string of bytes of small length), and the state recovery attack presented by Maximov and Khovratovich [6].

However, the major portion of the literature in RC4 cryptanalysis involves results related to initial keystream bytes of PRGA [2,5] (also see the references therein). To get rid of these problems, one may throw away some initial bytes of RC4 PRGA as suggested in [3,7]. But it may not be easy to modify the actual implementations immediately by throwing away some initial keystream bytes, since RC4 is already in use in many commercial applications. Thus the cryptanalytic results related to the initial bytes are still of importance. Moreover, these results are always of theoretical significance in terms of studying one of the most popular stream ciphers. The trend continues, including the most recent biases in this direction [8] that relates the initial keystream bytes, state variables and secret key of RC4. Recently, another paper [9] accepted at Eurocrypt 2011 exploited the known biases of RC4 (mostly involving the initial bytes) to provide distinguishers against WEP and WPA. Using related idea, this paper also proposes the best key recovery attack against WPA till date.

Notation. Let S_r, i_r, j_r, z_r denote the state, index i, index j, and the keystream byte respectively, after r (≥ 1) rounds of PRGA have been performed. Let S_0 denote the state just before the PRGA starts, i.e., right after the KSA ends. Further, let $p_{r,x}$ denote the probability $\Pr(S_r[x] = x)$, after r rounds of PRGA, where $r \geq 1$ and $0 \leq x \leq N - 1$.

Motivation and Contribution. In FSE 2001, Mantin and Shamir [5] published the best known distinguishing attack on RC4 based on the bias of the second byte towards zero. This result states that if the initial permutation is randomly chosen from the set of all ($N!$) permutations of $\mathbb{Z}_N$, then $\Pr(z_2 = 0) \approx \frac{2}{N}$ in RC4 keystream, whereas this should be $\frac{1}{N}$ in case of a random stream of bytes.

In [5, Section 3.2], after the description of the bias in the event ($z_2 = 0$), the following statement has been made:

"One could expect to see a similar (but weaker) bias towards 0 at all the other outputs z_t with $t = 0 \bmod n$, since in $1/N^2$ of these cases $S_t[2] = 0$ and $j = 0$, which would give rise to the same situation. However, extensive experiments have shown that this weaker bias at later rounds does not exist. By carefully analyzing this situation one can show that for any $j \neq 0$, the output is zero with a slight negative bias, and the total contribution of these negative biases exactly cancels the positive bias derived from $j = 0$. The only time we don't have this cancellation effect is at the beginning of the execution, when j always starts as 0 rather than as a uniformly distributed random value."

The main two claims implied by the above statement are as follows.

MS-Claim 1: $\Pr(z_r = 0) = \frac{1}{N}$ at PRGA rounds $3 \leq r \leq 255$.

MS-Claim 2: $\Pr(z_r = 0 \mid j_r = 0) > \frac{1}{N}$ and $\Pr(z_r = 0 \mid j_r \neq 0) < \frac{1}{N}$ for $3 \leq r \leq 255$. These two biases, when combined, cancel each other to produce no bias in the event $(z_r = 0)$ in rounds 3 to 255.

MS-Claim 2 was made to justify MS-Claim 1 in [5]. In the current work, contrary to MS-Claim 1, we show (Theorem 1) that $\Pr(z_r = 0) > \frac{1}{N}$ for all rounds r from 3 to 255. The immediate implications are that we find 253 new distinguishers of RC4, and that the validity of MS-Claim 2 is questionable. This motivates us to analyze the work of [5] to refute the aforementioned claims, and to study the (non)-randomness of j in PRGA. It is quite surprising that this issue has never been identified over the last decade.

The bias in the second byte was used in [5] to mount a distinguisher. We use our newly discovered biases to construct a class of 253 new distinguishers corresponding to the initial 253 keystream bytes z_r for $r \in \{3, 4, \ldots, 255\}$.

In addition, we study the non-randomness of index j rigorously to find a strong bias of j_2 towards 4. We can use this bias to guess the internal state variable $S_2[2]$ from the value of keystream byte z_2. Very recently, the results published in [8] claimed an exhaustive search for biases in all possible linear combinations of the state variables and the RC4 keystream bytes. However, our result concerning the bias of j_2 towards 4 is not covered in [8].

The literature of RC4 cryptanalysis, developed over more than two decades, is quite rich. In context of this paper, we have only referred to the publications which have direct relevance with our work. The reader may look into the references therein for a more detailed overview.

During the proposition and proof of our results in this paper, we shall require the following well known result in RC4 cryptanalysis from the existing literature. This appears in [3, Theorem 6.3.1], and we can restate the result as follows.

Proposition 1 ([3]). *At the end of KSA, for $0 \leq u \leq N-1$, $0 \leq v \leq N-1$,*

$$\Pr(S_0[u] = v) = \begin{cases} \frac{1}{N}\left[\left(\frac{N-1}{N}\right)^v + \left(1 - \left(\frac{N-1}{N}\right)^v\right)\left(\frac{N-1}{N}\right)^{N-u-1}\right] & \text{if } v \leq u; \\ \frac{1}{N}\left[\left(\frac{N-1}{N}\right)^{N-u-1} + \left(\frac{N-1}{N}\right)^v\right] & \text{if } v > u. \end{cases}$$

Remark 1. As Proposition 1 reveals, the underlying assumption of Mantin and Shamir [5] regarding the randomness of the initial permutation is violated in practice. This non-randomness in the permutation for the initial state of PRGA gives rise to the biases that we report in this paper.

2   Bytes 3 to 255 of PRGA are Biased to Zero

In this section we show that all the initial 253 bytes of RC4 keystream from round 3 to 255 are biased to zero. To prove the main theorem, we require the following technical result.

Lemma 1. *For $r \geq 3$, the probability that $S_{r-1}[r] = r$ is*

$$p_{r-1,r} \approx p_{0,r} \cdot \left[\left(\frac{N-1}{N} \right)^{r-1} - \frac{1}{N} \right] + \frac{1}{N}.$$

Proof. The event $S_{r-1}[r] = r$ may occur in the following two ways.

1. $S_0[r] = r$, and index r is not touched by any i or j during first $(r-1)$ PRGA rounds: The first event occurs with probability $p_{0,r}$. For the second one, note that index r is not touched by $i = 1, \ldots, r-1$ values, and the probability that none of j touches it either is approximately $(\frac{N-1}{N})^{r-1}$. Thus the contribution of this case is approximately $p_{0,r} \cdot (\frac{N-1}{N})^{r-1}$.

2. $S_0[r] \neq r$, and still $S_{r-1}[r]$ equals r by random association: The probability of the first event is $(1 - p_{0,r})$ and given this event, the second one is likely to occur only due to random association, thus with probability $\approx \frac{1}{N}$. Hence, the contribution of this case is approximately $(1 - p_{0,r}) \cdot \frac{1}{N}$.

Adding the two contributions calculated above, we get the result.   □

Remark 2. RC4 PRGA starts with $j_0 = 0$. For $r = 1$, we have $j_1 = j_0 + S_0[1] = S_0[1]$ which, due to Proposition 1, is not uniformly distributed. For $r = 2$, we have $j_2 = j_1 + S_1[2] = S_0[1] + S_1[2]$, whose probability distribution is more close to the uniform random distribution than that in case of j_1. In round 3, another pseudo-random byte $S_2[3]$ would be added to form j_3. From round 3 onwards, j can safely be assumed to be uniform over $\mathbb{Z}_N$. Experimental observations also confirm this. A detailed discussion on the randomness of j is presented in Section 4. In Item 1 of the proof of Lemma 1, the product

$$\Pr(j_1 \neq r) \cdot \Pr(j_2 \neq r) \cdots \Pr(j_{r-1} \neq r) = \Pr(j_1 \neq r) \cdot \Pr(j_2 \neq r) \cdot \left(\frac{N-1}{N} \right)^{r-3}$$

is approximated as $(\frac{N-1}{N})^{r-1}$, but one may always try the exact forms for the probabilities $\Pr(j_1 \neq r)$ and $\Pr(j_2 \neq r)$ to obtain further accuracy.

Now, we can state our main theorem on the bias of RC4 initial bytes.

Theorem 1. *For $3 \leq r \leq 255$, the probability that the r-th RC4 keystream byte is equal to 0 is*

$$\Pr(z_r = 0) \approx \frac{1}{N} + \frac{c_r}{N^2}.$$

where c_r is given by $\left[\left(\frac{N-1}{N} \right)^r + \left(\frac{N-1}{N} \right)^{N-r-1} - \left(\frac{N-1}{N} \right)^{N-1} \right] \cdot \left[\left(\frac{N-1}{N} \right)^{r-2} - \frac{1}{N-1} \right].$

Proof. We prove the result by decomposing the event $(z_r = 0)$ into two mutually exclusive and exhaustive cases[1], as follows.

$$\Pr(z_r = 0) = \Pr(z_r = 0 \ \& \ S_{r-1}[r] = r) + \Pr(z_r = 0 \ \& \ S_{r-1}[r] \neq r) \quad (1)$$

Now we consider the events $(z_r = 0 \ \& \ S_{r-1}[r] = r)$ and $(z_r = 0 \ \& \ S_{r-1}[r] \neq r)$ individually to calculate their probabilities. In this direction, note that

$$z_r = S_r[S_r[i_r] + S_r[j_r]] = S_r[S_r[r] + S_{r-1}[i_r]]$$
$$= S_r[S_r[r] + S_{r-1}[i_r]] = S_r[S_r[r] + S_{r-1}[r]].$$

This expression for z_r will be used in various effects throughout the paper.

Calculation of $\Pr(z_r = 0 \ \& \ S_{r-1}[r] = r)$: In this case $S_{r-1}[r] = r$, and thus we have the probability

$$\Pr(z_r = 0 \ \& \ S_{r-1}[r] = r)$$
$$= \Pr(S_r[S_r[r] + r] = 0 \ \& \ S_{r-1}[r] = r)$$
$$= \sum_{x=0}^{N-1} \Pr(S_r[x + r] = 0 \ \& \ S_r[r] = x \ \& \ S_{r-1}[r] = r)$$
$$= \sum_{x=0}^{N-1} \Pr(S_r[x + r] = 0 \ \& \ S_r[r] = x) \cdot \Pr(S_{r-1}[r] = r) \quad (2)$$

The last expression results from the assumption that the events $(S_r[x + r] = 0)$ and $(S_r[r] = x)$ are both independent from $(S_{r-1}[r] = r)$, as a state update has occurred in the process. Note that $S_{r-1}[r] = r$ is one of the values that gets swapped to produce the new state S_r (location $[r]$ denotes $[i_r]$ at this stage), and this is why we can claim the independence of $S_r[r]$ and $S_{r-1}[r]$. Otherwise, if a location $[s]$ is not same as $[i_r]$ or $[j_r]$, then $S_r[s]$ would be the same as $S_{r-1}[s]$, even after the state update.

Now, let us compute $\Pr(S_r[x + r] = 0 \ \& \ S_r[r] = x) = \Pr(S_r[x + r] = 0) \cdot \Pr(S_r[r] = x \mid S_r[x + r] = 0)$ independently. In this expression, if there exists any bias in the event $(S_r[x + r] = 0)$, then it must propagate from a similar bias in $(S_0[x + r] = 0)$, as was the case for $(S_{r-1}[r] = r)$ in Lemma 1. However,

---

[1] In the pre-proceedings version, we had considered the same cases, and had obtained the same expressions for $\Pr(z_r = 0)$ and c_r. However, the proof for Theorem 1 used Jenkin's bias [1] (Glimpse) in an intermediate step as a crude approximation. In this version, we present a rigorous analysis which does not require to use Jenkin's bias.

$\Pr(S_0[x+r] = 0) = \frac{1}{N}$ by Proposition 1, and thus we can safely assume $S_r[x+r]$ to be random as well. This provides us with $\Pr(S_r[x+r] = 0) = \frac{1}{N}$.

For $\Pr(S_r[r] = x \mid S_r[x+r] = 0)$, observe that when $x = 0$, the indices $[x+r]$ and $[r]$ in the state S_r point to the same location, and the events $(S_r[x+r] = S_r[r] = 0)$ and $(S_r[r] = x = 0)$ denote identical events. Thus in this case, $\Pr(S_r[r] = x \mid S_r[x+r] = 0) = 1$. In cases where $x \neq 0$, the indices $[x+r]$ and $[r]$ refer to two distinct locations in the permutation S_r, obviously containing different values. In this case,

$$\Pr(S_r[r] = x \mid S_r[x+r] = 0) = \Pr(S_r[r] = x \mid x \neq 0) = \frac{1}{N-1}.$$

For justifying the randomness of $S_r[r]$ for $x \neq 0$, one may simply observe that the location $[r] = [i_r]$ is the one that got swapped to generate state S_r from the previous state, and thus the randomness assumption of $S_r[r]$ is based on the randomness assumption of j_r, which is validated for $r \geq 3$ later in Section 4.

According to the discussion above, we obtain

$$\Pr\left(S_r[x+r] = 0 \ \& \ S_r[r] = x\right) = \begin{cases} \frac{1}{N} \cdot 1 = \frac{1}{N} & \text{if } x = 0, \\ \frac{1}{N} \cdot \frac{1}{N-1} = \frac{1}{N(N-1)} & \text{if } x \neq 0. \end{cases} \quad (3)$$

Substituting these probability values in Equation (2), we get

$$\Pr\left(z_r = 0 \ \& \ S_{r-1}[r] = r\right)$$
$$= \Pr\left(S_{r-1}[r] = r\right) \left[\sum_{x=0}^{N-1} \Pr\left(S_r[x+r] = 0 \ \& \ S_r[r] = x\right) \right]$$
$$= p_{r-1,r} \cdot \left[\frac{1}{N} + \sum_{x=1}^{N-1} \frac{1}{N(N-1)} \right]$$
$$= p_{r-1,r} \cdot \left[\frac{1}{N} + (N-1) \cdot \frac{1}{N(N-1)} \right] = p_{r-1,r} \cdot \frac{2}{N}. \quad (4)$$

Calculation of $\Pr\left(z_r = 0 \ \& \ S_{r-1}[r] \neq r\right)$: Similar to the previous case, we can derive the probability as follows:

$$\Pr\left(z_r = 0 \ \& \ S_{r-1}[r] \neq r\right)$$
$$= \sum_{y \neq r} \Pr(S_r[S_r[r] + y] = 0 \ \& \ S_{r-1}[r] = y)$$
$$= \sum_{y \neq r} \sum_{x=0}^{N-1} \Pr\left(S_r[x+y] = 0 \ \& \ S_r[r] = x \ \& \ S_{r-1}[r] = y\right)$$

An interesting situation occurs if $x = r - y$. In this case, on one hand, we obtain $S_r[x+y] = S_r[r] = 0$ for the first event, while on the other hand, we get

$S_r[r] = x = r - y \neq 0$ for the second event (note that $y \neq r$). This poses a contradiction (event with probability of occurrence 0), and hence we can write

$$\Pr(z_r = 0 \ \& \ S_{r-1}[r] \neq r)$$

$$= \sum_{y \neq r} \sum_{x \neq r-y} \Pr(S_r[x+y] = 0 \ \& \ S_r[r] = x \ \& \ S_{r-1}[r] = y)$$

$$= \sum_{y \neq r} \sum_{x \neq r-y} \Pr(S_r[x+y] = 0 \ \& \ S_r[r] = x) \cdot \Pr(S_{r-1}[r] = y), \qquad (5)$$

where the last expression results from the fact that the events $(S_r[x+y] = 0)$ and $(S_r[r] = x)$ are both independent from $(S_{r-1}[r] = y)$, as a state update has occurred in the process, and $S_{r-1}[r]$ got swapped during that update.

Similar to the derivation of Equation (3), we obtain

$$\Pr(S_r[x+y] = 0 \ \& \ S_r[r] = x) = \begin{cases} 0 & \text{if } x = 0, \\ \frac{1}{N(N-1)} & \text{if } x \neq 0. \end{cases} \qquad (6)$$

The only difference occurs in the case $x = 0$. In this situation, simultaneous occurrence of the events $(S_r[x+y] = S_r[y] = 0)$ and $(S_r[r] = x = 0)$ pose a contradiction as the two locations $[y]$ and $[r]$ of S_r are distinct (note that $y \neq r$), and they can not hold the same value 0 as the state S_r is a permutation. In all other cases $(x \neq 0)$, the argument is identical to that in the previous derivation.

Substituting the values above in Equation (5), we get

$$\Pr(z_r = 0 \ \& \ S_{r-1}[r] \neq r)$$

$$= \sum_{y \neq r} \Pr(S_{r-1}[r] = y) \left[\sum_{x \neq r-y} \Pr(S_r[x+y] = 0 \ \& \ S_r[r] = x) \right]$$

$$= \sum_{y \neq r} \Pr(S_{r-1}[r] = y) \left[0 + \sum_{\substack{x \neq r-y \\ x \neq 0}} \frac{1}{N(N-1)} \right]$$

$$= \sum_{y \neq r} \Pr(S_{r-1}[r] = y) \left[(N-2) \cdot \frac{1}{N(N-1)} \right]$$

$$= \frac{N-2}{N(N-1)} \sum_{y \neq r} \Pr(S_{r-1}[r] = y)$$

$$= \frac{N-2}{N(N-1)} \cdot (1 - \Pr(S_{r-1}[r] = r)) = \frac{N-2}{N(N-1)} \cdot (1 - p_{r-1,r}) \qquad (7)$$

Calculation for $\Pr(z_r = 0)$: Combining the probabilities from Equation (4) and Equation (7) in the final expression of Equation (1), we obtain the following.

$$\Pr(z_r = 0) = p_{r-1,r} \cdot \frac{2}{N} + \frac{N-2}{N(N-1)} \cdot (1 - p_{r-1,r})$$

$$= \frac{p_{r-1,r}}{N-1} + \frac{N-2}{N(N-1)} = \frac{1}{N} + \frac{1}{N-1} \cdot \left(p_{r-1,r} - \frac{1}{N} \right) \qquad (8)$$

Now, substituting the value of $p_{r-1,r}$ from Lemma 1 in Equation (8), we obtain

$$\Pr(z_r = 0) \approx \frac{1}{N} + \frac{1}{N-1} \cdot p_{0,r} \cdot \left[\left(\frac{N-1}{N} \right)^{r-1} - \frac{1}{N} \right]. \tag{9}$$

Further, we can use Proposition 1 to get the value of $p_{0,r}$ as

$$p_{0,r} = \Pr(S_0[r] = r) = \frac{1}{N} \left[\left(\frac{N-1}{N} \right)^r + \left(1 - \left(\frac{N-1}{N} \right)^r \right) \left(\frac{N-1}{N} \right)^{N-r-1} \right].$$

Substituting this expression for $p_{0,r}$ in Equation (9), we obtain the desired result $\Pr(z_r = 0) \approx \frac{1}{N} + \frac{c_r}{N^2}$ with the claimed value of c_r.   □

In Theorem 1, we have presented the bias in the probability $\Pr(z_r = 0)$ in terms of the parameter c_r, which in turn is a function of r. But we are more interested in observing the bias for specific rounds of RC4 PRGA, namely within the interval $3 \leq r \leq 255$. Thus, we are interested in obtaining numerical bounds on the bias for this specific interval. The next result is a corollary of Theorem 1 that provides exact numeric bounds on $\Pr(z_r = 0)$ within the interval $3 \leq r \leq 255$, depending on the corresponding bounds of c_r within the same interval.

Corollary 1. *For $3 \leq r \leq 255$, the probability that the r-th RC4 keystream byte is equal to 0 is bounded as follows*

$$\frac{1}{N} + \frac{0.98490994}{N^2} \geq \Pr(z_r = 0) \geq \frac{1}{N} + \frac{0.36757467}{N^2}.$$

Proof. We calculated all values of c_r (as in Theorem 1) for the range $3 \leq r \leq 255$, and checked that c_r is a decreasing function in r where $3 \leq r \leq 255$ (one may refer to the plot in Fig. 1 in this regard). Therefore we obtain

$$\max_{3 \leq r \leq 255} c_r = c_3 = 0.98490994 \quad \text{and} \quad \min_{3 \leq r \leq 255} c_r = c_{255} = 0.36757467.$$

Hence the result on the bounds of $\Pr(z_r = 0)$, depending on the bounds of c_r.   □

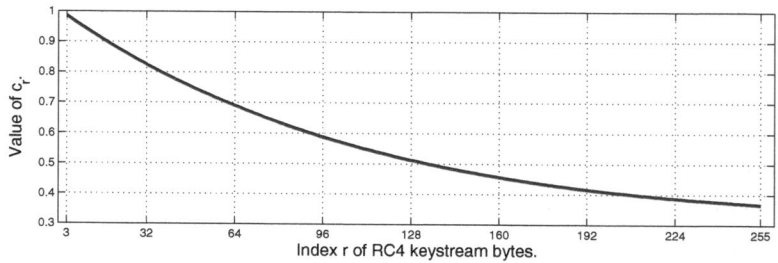

Fig. 1. Value of c_r versus r during RC4 PRGA ($3 \leq r \leq 255$)

Fig. 2 depicts a comparison between the theoretically derived vs. experimentally obtained values of $\Pr(z_r = 0)$ versus r, where $3 \leq r \leq 255$. The experimentation has been carried out with 1 billion trials, each trial with a randomly generated 16 byte key.

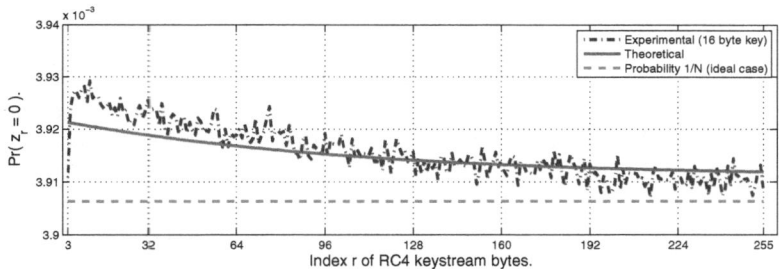

Fig. 2. $\Pr(z_r = 0)$ versus r during RC4 PRGA ($3 \leq r \leq 255$)

One may observe in Fig. 2 that the theoretical curve does not exactly coincide with the mean line of the experimental plot. This is algebraically expressed by the approximation in Theorem 1. The approximation arises due to the ideal randomness assumptions in the proof of Lemma 1, which do not hold in practice.

2.1 A Class of New Distinguishers

Theorem 1 immediately gives a class of distinguishers. In [5, Theorem 2], it is proved that if an event e happens with probabilities p and $p(1+\epsilon)$ in distributions X and Y respectively, then for p and ϵ with small magnitude, $O\left(p^{-1}\epsilon^{-2}\right)$ samples suffice to distinguish X from Y with a constant probability of success.

In our setting, let X and Y denote the distributions corresponding to *random stream* and *RC4 keystream* respectively, and e_r denote the event $(z_r = 0)$ for $r = 3$ to 255. From the formulation as in Equation (10), we can write $p = \frac{1}{N}$ and $\epsilon = \frac{c_r}{N}$. Thus, to distinguish RC4 keystream from random stream, based on the event $(z_r = 0)$, one would need number of samples of the order of

$$\left(\frac{1}{N}\right)^{-1}\left(\frac{c_r}{N}\right)^{-2} \sim O(N^3).$$

We can combine the effect of all these distinguishers by counting the number of zeros in the initial keystream of RC4, according to Theorem 2, as follows.

Theorem 2. *The expected number of 0's in RC4 keystream rounds 3 to 255 is approximately 0.9904610515.*

Proof. Let X_r be a random variable taking values $X_r = 1$ if $z_r = 0$, and $X_r = 0$ otherwise. Hence, the total number of 0's in rounds 3 to 255 is given by

$$C = \sum_{r=3}^{255} X_r.$$

We have $E(X_r) = \Pr(X_r = 1) = \Pr(z_r = 0)$ from Theorem 1. By linearity of expectation,

$$E(C) = \sum_{r=3}^{255} E(X_r) = \sum_{r=3}^{255} \Pr(z_r = 0).$$

Substituting the numeric values of the probabilities $\Pr(z_r = 0)$ from Theorem 1, we get $E(C) \approx 0.9904610515$. Hence the result.   □

For a random stream of bytes, this expectation is $E(C) = \frac{253}{256} = 0.98828125$. Thus, the expectation for RC4 is approximately 0.22% higher than that for the random case. The inequality of this expectation in RC4 keystream compared to that in a random stream of bytes may also be used to design a distinguisher.

2.2   A Critical Analysis of the Event $(z_r = 0)$ Given $j_r = $ or $\neq 0$

Recall the expression for $\Pr(z_r = 0)$ from Theorem 1:

$$\Pr(z_r = 0) = \frac{1}{N} + \frac{1}{N-1} \cdot \left(p_{r-1,r} - \frac{1}{N} \right) \approx \frac{1}{N} + \frac{c_r}{N^2}. \tag{10}$$

In the expression for $p_{r-1,r}$, as in Lemma 1, we see that $\left(\frac{N-1}{N} \right)^{r-1} > \frac{1}{N}$ for all $3 \leq r \leq 255$. Thus, there is always a *positive* bias in $p_{r-1,r}$, and in turn in $\Pr(z_r = 0)$. Further, for any $r \geq 1$, we can write

$$\Pr(z_r = 0) = \Pr(j_r = 0) \cdot \Pr(z_r = 0 \mid j_r = 0)$$
$$+ \Pr(j_r \neq 0) \cdot \Pr(z_r = 0 \mid j_r \neq 0). \tag{11}$$

One may note that MS-Claim 2 of Mantin and Shamir [5] essentially states that $\Pr(z_r = 0 \mid j_r = 0) = \frac{1}{N} + a_r$ and $\Pr(z_r = 0 \mid j_r \neq 0) = \frac{1}{N} - b_r$ for $3 \leq r \leq 255$, where both $a_r, b_r > 0$. Plugging these values in Equation (11), we have

$$\frac{1}{N} + \frac{c_r}{N^2} = \frac{1}{N} \left(\frac{1}{N} + a_r \right) + \left(1 - \frac{1}{N} \right) \left(\frac{1}{N} - b_r \right) \quad \text{for } 3 \leq r \leq 255.$$

Simplifying the above equation, we get $a_r = \frac{c_r}{N} + (N-1)b_r$. Thus, if MS-Claim 2 is correct, then we must have

$$\Pr(z_r = 0 \mid j_r = 0) = \frac{1}{N} + \frac{c_r}{N} + (N-1)b_r = \frac{1+c_r}{N} + (N-1)b_r,$$

where $0.98490994 \geq c_r \geq 0.36757467$ for $3 \leq r \leq 255$ (from Corollary 1). However, extensive experiments have confirmed that $\Pr(z_r = 0 \mid j_r = 0) \approx \frac{1}{N}$, thereby refuting MS-Claim 2 of Mantin and Shamir.

2.3   Guessing State Information Using the Bias in z_r

Mantin and Shamir [5] used the bias of the second byte of RC4 keystream to guess some information regarding $S_0[2]$, based on the following.

$$\Pr(S_0[2] = 0 \mid z_2 = 0) = \frac{\Pr(S_0[2] = 0)}{\Pr(z_2 = 0)} \cdot \Pr(z_2 = 0 \mid S_0[2] = 0) \approx \frac{1/N}{2/N} \cdot 1 = \frac{1}{2}.$$

Note that in the above expression, no randomness assumption is required to obtain $\Pr(S_0[2] = 0) = \frac{1}{N}$. This probability is exact and can be derived by substituting $u = 2, v = 0$ in Proposition 1. Hence, on every occasion we obtain $z_2 = 0$ in the keystream, we can guess $S_0[2]$ with probability $\frac{1}{2}$, and this is significantly more than a random guess with probability $\frac{1}{N}$.

In this section, we use the biases in bytes 3 to 255 (observed in Theorem 1) to extract similar information about the state array S_{r-1} using the RC4 keystream byte z_r. In particular, we try to explore the conditional probability $\Pr(S_{r-1}[r] = r \mid z_r = 0)$ for $3 \le r \le 255$, as follows.

$$\Pr(S_{r-1}[r] = r \mid z_r = 0) = \frac{\Pr(z_r = 0 \ \& \ S_{r-1}[r] = r)}{\Pr(z_r = 0)} \approx \frac{p_{r-1,r} \cdot \frac{2}{N}}{\frac{1}{N} + \frac{c_r}{N^2}}$$

In the above expression, c_r is as in Theorem 1. One may write

$$p_{r-1,r} = \frac{1}{N} + \frac{c_r}{N} - \frac{c_r}{N^2},$$

using Equation (8) from the proof of Theorem 1, and thereby obtain

$$\Pr(S_{r-1}[r] = r \mid z_r = 0) \approx \frac{\left(\frac{1}{N} + \frac{c_r}{N} - \frac{c_r}{N^2}\right) \cdot \frac{2}{N}}{\frac{1}{N} + \frac{c_r}{N^2}}$$

$$= 2 \cdot \left(\frac{1}{N} + \frac{c_r}{N} - \frac{c_r}{N^2}\right) \cdot \left(1 + \frac{c_r}{N}\right)^{-1} \approx \frac{2}{N} + \frac{2c_r}{N}.$$

From the expression for $\Pr(S_{r-1}[r] = r \mid z_r = 0)$ derived above, one can guess $S_{r-1}[r]$ with probability more than twice of the probability of a random guess, every time we obtain $z_r = 0$ in the RC4 keystream. In Fig. 3, we plot the theoretical probabilities

$$\Pr(S_{r-1}[r] = r \mid z_r = 0) = 2 \cdot \left(\frac{1}{N} + \frac{c_r}{N} - \frac{c_r}{N^2}\right) \cdot \left(1 + \frac{c_r}{N}\right)^{-1}$$

against r for $3 \le r \le 255$, and the corresponding experimental values observed by running the RC4 algorithm 1 billion times with randomly selected 16 byte keys. It clearly shows that all the experimental values are also greater than $\frac{2}{N}$, as desired. The crisscross nature of the curves in Fig. 3 originates from a similar behavior observed in the curves of Fig. 2.

3   Attacking the RC4 Broadcast Scheme

Let us now revisit the famous attack of Mantin and Shamir [5] on broadcast RC4. As mentioned in their paper,

> "A classical problem in distributed computing is to allow N Byzantine generals to coordinate their actions when up to one third of them can be traitors. The problem is solved by a multi-round protocol in which each general broadcasts the same plaintext (which initially consists of either "Attack" or "Retreat") to all the other generals, where each copy is encrypted under a different key agreed in advance between any two generals."

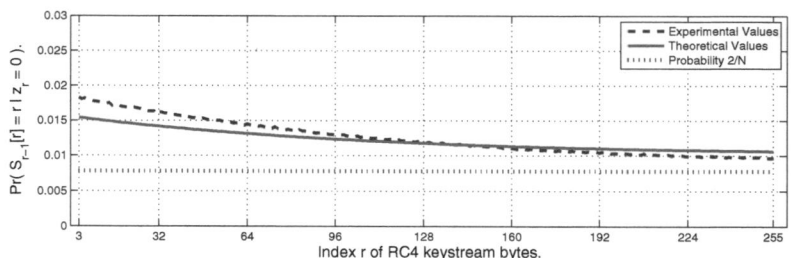

Fig. 3. $\Pr(S_{r-1}[r] = r \mid z_r = 0)$ versus r during RC4 PRGA ($3 \leq r \leq 255$)

In [5], the authors propose a practical attack against an RC4 implementation of the broadcast scheme, based on the bias observed in the second keystream byte. They prove that an enemy that collects $k = \Omega(N)$ number of ciphertexts corresponding to the same plaintext M, can easily deduce the second byte of M, by exploiting the bias in z_2.

In a similar line of action, we may exploit the bias observed in bytes 3 to 255 of the RC4 keystream to mount a similar attack on RC4 broadcast scheme. Notice that we obtain a bias of the order of $\frac{1}{N^2}$ in each of the bytes z_r where $3 \leq r \leq 255$. Thus, roughly speaking, if the attacker obtains about N^3 ciphertexts corresponding to the same plaintext M (from the broadcast scheme), then he can check the frequency of occurrence of bytes to deduce the r-th ($3 \leq r \leq 255$) byte of M.

The most important point to note is that this technique will work for each r where $3 \leq r \leq 255$, and hence will reveal *all the 253 initial bytes* (number 3 to 255 to be specific) of the plaintext M. We can formally state our result (analogous to [5, Theorem 3]) as follows.

Theorem 3. *Let M be a plaintext, and let $C_1, C_2, \ldots, C_k$ be the RC4 encryptions of M under k uniformly distributed keys. Then if $k = \Omega(N^3)$, the bytes 3 to 255 of M can be reliably extracted from $C_1, C_2, \ldots, C_k$.*

Proof. Recall from Theorem 1 that $\Pr(z_r = 0) \approx \frac{1}{N} + \frac{c_r}{N^2}$ for all $3 \leq r \leq 255$ in the RC4 keystream. Thus, for each encryption key chosen during broadcast, the r-th plaintext byte $M[r]$ has probability $\frac{1}{N} + \frac{c_r}{N^2}$ to be XOR-ed with 0.

Due to the bias of z_r towards zero, $\frac{1}{N} + \frac{c_r}{N^2}$ fraction of the r-th ciphertext bytes will have the same value as the r-th plaintext byte, with a higher probability. When $k = \Omega(N^3)$, the attacker can identify the most frequent character in $C_1[r], C_2[r], \ldots, C_k[r]$ as $M[r]$ with constant probability of success. □

The attack on broadcast RC4 is applicable to many modern Internet protocols (such as group emails encrypted under different keys, group-ware multi-user synchronization etc.). Note that Mantin and Shamir's attack [5] works at the byte level. It can recover only the second byte of the plaintext under some assumptions. On the other hand, our attack can recover additional 253 bytes (namely, bytes 3 to 255) of the plaintext.

4   Non-randomness of j in PRGA

During the PRGA round of RC4 algorithm, two indices are used; the first is i (deterministic) and the second is j (pseudo-random). Index i starts from 0 and increments by 1 (modulo N) at the beginning of each iteration, whereas j depends on the values of i and $S[i]$ simultaneously. The pseudo-randomness of the internal state S triggers the pseudo-randomness in j. In this section, we attempt to understand the pseudo-random behavior of j more clearly.

In RC4 PRGA, we know that for $r \geq 1$, $i_r = r \mod N$ and $j_r = j_{r-1}+S_{r-1}[i_r]$, starting with $j_0 = 0$. Thus, we can write the values assumed by j at different rounds of PRGA as follows.

$$j_1 = j_0 + S_0[i_1] = 0 + S_0[1] = S_0[1],$$
$$j_2 = j_1 + S_1[i_2] = S_0[1] + S_1[2],$$
$$j_3 = j_2 + S_2[i_3] = S_0[1] + S_1[2] + S_2[3],$$
$$\vdots \quad \vdots \quad \vdots$$
$$j_r = j_{r-1} + S_{r-1}[i_r] = S_0[1] + S_1[2] + \cdots + S_{r-1}[r] = \sum_{x=1}^{r} S_{x-1}[x],$$

where $1 \leq r \leq N - 1$, and all the additions are performed modulo N, as usual.

4.1   Non-randomness of j_1

In the first round of PRGA, $j_1 = S_0[1]$ follows a probability distribution which is determined by S_0, the internal state array after the completion of KSA. According to Proposition 1, we have

$$\Pr(j_1 = v) = \Pr(S_0[1] = v) = \begin{cases} \frac{1}{N} & \text{if } v = 0; \\[2mm] \frac{1}{N}\left(\frac{N-1}{N} + \frac{1}{N}\left(\frac{N-1}{N}\right)^{N-2}\right) & \text{if } v = 1; \\[2mm] \frac{1}{N}\left(\left(\frac{N-1}{N}\right)^{N-2} + \left(\frac{N-1}{N}\right)^{v}\right) & \text{if } v > 1. \end{cases}$$

This clearly tells us that j_1 is *not* random. This is also portrayed in Fig. 4.

4.2   Non-randomness of j_2

In the second round of PRGA however, we have $j_2 = S_0[1]+S_1[2]$, which demonstrates better randomness, as discussed next. Note that we have the following in terms of probability for j_2.

$$\Pr(j_2 = v) = \Pr(S_0[1] + S_1[2] = v)$$
$$= \sum_{w=0}^{N-1} \Pr(S_0[1] = w) \cdot \Pr((S_1[2] = v - w) \mid (S_0[1] = w)) \quad (12)$$

In the above expression, $(v - w)$ is performed modulo N, like all arithmetic operations in RC4. The following cases may arise with respect to Equation (12).

Case I. Suppose that $j_1 = S_0[1] = w = 2$. Then, we will have $S_1[i_2] = S_1[2] = S_1[j_1] = S_0[i_1] = S_0[1] = 2$. In this case,

$$\Pr((S_1[2] = v - 2) \mid (S_0[1] = 2)) = \begin{cases} 1 & \text{if } v = 4, \\ 0 & \text{otherwise.} \end{cases}$$

Case II. Suppose that $j_1 = S_0[1] = w \neq 2$. Then $S_0[2]$ will not get swapped in the first round, and hence we will have $S_1[2] = S_0[2]$. In this case,

$$\Pr((S_1[2] = v - w) \mid (S_0[1] = w \neq 2)) = \Pr(S_0[2] = v - w).$$

Let us substitute the results obtained from these cases to Equation (12) to obtain

$$\Pr(j_2 = v) = \begin{cases} \Pr(S_0[1] = 2) + \displaystyle\sum_{\substack{w=0 \\ w \neq 2}}^{N-1} \Pr(S_0[1] = w) \Pr(S_0[2] = v - w), & \text{if } v = 4; \\[2em] \displaystyle\sum_{\substack{w=0 \\ w \neq 2}}^{N-1} \Pr(S_0[1] = w) \Pr(S_0[2] = v - w), & \text{if } v \neq 4. \end{cases}$$

$$(13)$$

Equation (13) completely specifies the exact probability distribution of j_2, where each of the probabilities $\Pr(S_0[x] = y)$ can be substituted by their exact values from Proposition 1. However, the expression suffices to exhibit the non-randomness of j_2 in the RC4 PRGA, having a large bias for $v = 4$. We found that the theoretical values corresponding to the probability distribution of j_2 (as in Equation (13)) match almost exactly with the experimental data plotted in Fig. 4. For the sake of clarity, we do not show the theoretical curve in Fig. 4.

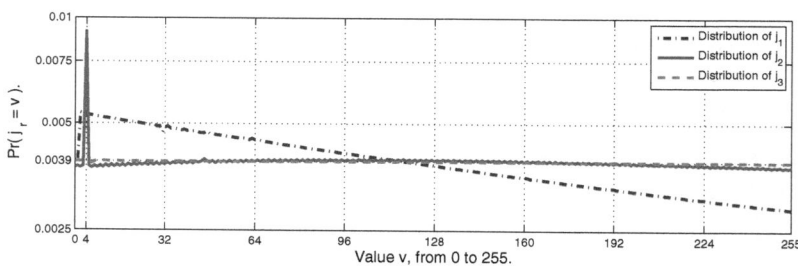

Fig. 4. Probability distribution of j_r for $1 \leq r \leq 3$

Calculation of $\Pr(j_2 = 4)$. Let us now evaluate $\Pr(j_2 = 4)$ independently:

$$\Pr(j_2 = 4)$$

$$= \Pr(S_0[1] = 2) + \sum_{\substack{w=0 \\ w \neq 2}}^{N-1} \Pr(S_0[1] = w) \cdot \Pr(S_0[2] = 4 - w)$$

$$= \frac{1}{N}\left[\left(\frac{N-1}{N}\right)^{N-2} + \left(\frac{N-1}{N}\right)^2\right] + \sum_{\substack{w=0 \\ w \neq 2}}^{N-1} \Pr(S_0[1] = w) \cdot \Pr(S_0[2] = 4 - w)$$

Following Proposition 1, the summation term in the above expression evaluates approximately to $\frac{0.965268}{N}$ for $N = 256$. Thus, we get

$$\Pr(j_2 = 4) \approx \frac{1}{N}\left[\left(\frac{N-1}{N}\right)^{N-2} + \left(\frac{N-1}{N}\right)^2\right] + \frac{0.965268}{N} \approx \frac{7/3}{N}.$$

This verifies our experimental observation, as depicted in Fig. 4.

Guessing State Information Using the Bias in j_2. It is also feasible to use this bias of j_2 to guess certain information about the RC4 state S_2. In particular, we shall focus on the event $(S_2[i_2] = 4 - z_2)$ or $(S_2[2] = 4 - z_2)$, and prove a bias in the probability of occurrence of this event, as follows.

Proposition 2. *After completion of the second round of RC4 PRGA, the state variable $S_2[2]$ equals the value $4 - z_2$ with probability*

$$\Pr(S_2[2] = 4 - z_2) \approx \frac{1}{N} + \frac{4/3}{N^2}.$$

Proof. First, note that we can write z_2 in terms of the state variables as follows

$$z_2 = S_2[S_2[i_2] + S_2[j_2]] = S_2[S_1[j_2] + S_1[i_2]] = S_2[S_1[j_2] + S_1[2]].$$

Thus, we can write the probability of the target event $(S_2[2] = 4 - z_2)$ as follows

$$\Pr(S_2[2] = 4 - z_2) = \Pr(S_2[i_2] = 4 - S_2[S_1[j_2] + S_1[2]])$$
$$= \Pr(S_1[j_2] = 4 - S_2[S_1[j_2] + S_1[2]])$$
$$= \Pr(S_1[j_2] + S_2[S_1[j_2] + S_1[2]] = 4)$$

Now, the idea is to exploit the bias in the event $(j_2 = 4)$ to obtain the bias in the probability mentioned above. Thus, we decompose the target event into two mutually exclusive and exhaustive cases[2], as follows.

$$(S_1[j_2] + S_2[S_1[j_2] + S_1[2]] = 4) = (S_1[j_2] + S_2[S_1[j_2] + S_1[2]] = 4 \,\&\, j_2 = 4)$$
$$\cup\, (S_1[j_2] + S_2[S_1[j_2] + S_1[2]] = 4 \,\&\, j_2 \neq 4)$$

---

[2] In the pre-proceedings version, we had considered the same cases, and had obtained the same expression for $\Pr(S_2[2] = 4 - z_2)$. However, the proof used Jenkin's bias [1] (Glimpse) in an intermediate step as a crude approximation. In this version, we present a rigorous analysis which does not require to use Jenkin's bias.

First event $(S_1[j_2] + S_2[S_1[j_2] + S_1[2]] = 4$ & $j_2 = 4)$: The probability for the first event can be calculated as follows.

$$\Pr(S_1[j_2] + S_2[S_1[j_2] + S_1[2]] = 4 \text{ \& } j_2 = 4)$$
$$= \Pr(S_1[4] + S_2[S_1[4] + S_1[2]] = 4 \text{ \& } j_2 = 4)$$
$$= \sum_{y=0}^{N-1} \Pr(S_1[4] + S_2[y] = 4 \text{ \& } S_1[4] + S_1[2] = y \text{ \& } j_2 = 4)$$
$$= \sum_{y=0}^{N-1} \Pr(S_1[4] + S_2[y] = 4 \text{ \& } S_1[4] + S_1[2] = y) \cdot \Pr(j_2 = 4)$$
$$= \Pr(j_2 = 4) \sum_{y=0}^{N-1} \Pr(S_1[4] + S_2[y] = 4 \text{ \& } S_1[4] + S_1[2] = y)$$

In the last expression, the values taken from S_1 are independent of the value of j_2, and thus the events $(S_1[4] + S_2[y] = 4)$ and $(S_1[4] + S_1[2] = y)$ are both independent of the event $(j_2 = 4)$. Also note that if $y = 4$, we obtain

$$S_1[4] + S_2[y] = S_1[4] + S_2[4] = S_1[4] + S_2[j_2] = S_1[4] + S_1[i_2] = S_1[4] + S_1[2],$$

which results in the events $(S_1[4] + S_2[y] = 4)$ and $(S_1[4] + S_1[2] = y)$ being identical. In all other cases, we have $S_1[4] + S_2[y] \neq S_1[4] + S_1[2]$ and thus the values are chosen distinctly independent at random. Hence, we obtain

$$\Pr(S_1[4] + S_2[y] = 4 \text{ \& } S_1[4] + S_1[2] = y) = \begin{cases} \frac{1}{N} & \text{if } y = 4; \\ \frac{1}{N(N-1)} & \text{if } y \neq 4. \end{cases}$$

The probabilities in the above expression are verified through experimentation by running the RC4 algorithm 1 billion times, choosing a 16 byte key uniformly at random in each run. The probability for the first event turns out to be

$$\Pr(S_1[j_2] + S_2[S_1[j_2] + S_1[2]] = 4 \text{ \& } j_2 = 4)$$
$$= \Pr(j_2 = 4) \cdot \left[\frac{1}{N} + \sum_{y \neq 4} \frac{1}{N(N-1)} \right]$$
$$= \frac{7/3}{N} \cdot \left[\frac{1}{N} + (N-1) \cdot \frac{1}{N(N-1)} \right] = \frac{7/3}{N} \cdot \frac{2}{N}.$$

Second event $(S_1[j_2] + S_2[S_1[j_2] + S_1[2]] = 4$ & $j_2 \neq 4)$: For the second event, the probability calculation can be performed in a similar fashion, as follows.

$$\Pr(S_1[j_2] + S_2[S_1[j_2] + S_1[2]] = 4 \text{ \& } j_2 \neq 4)$$
$$= \sum_{x \neq 4} \Pr(S_1[x] + S_2[S_1[x] + S_1[2]] = 4 \text{ \& } j_2 = x)$$
$$= \sum_{x \neq 4} \sum_{y=0}^{N-1} \Pr(S_1[x] + S_2[y] = 4 \text{ \& } S_1[x] + S_1[2] = y \text{ \& } j_2 = x)$$

Note that the case $y = x$ poses an interesting situation. On one hand, we obtain $S_1[x] + S_2[y] = S_1[x] + S_2[x] = S_1[x] + S_2[j_2] = S_1[x] + S_1[i_2] = S_1[x] + S_1[2] = 4$, while on the other hand, we get $S_1[x] + S_1[2] = x \neq 4$. We rule out the case $y = x$ from the probability calculation due to this contradiction, and get

$$\Pr\left(S_1[j_2] + S_2[S_1[j_2] + S_1[2]] = 4 \ \& \ j_2 \neq 4\right)$$
$$= \sum_{x \neq 4} \sum_{y \neq x} \Pr(S_1[x] + S_2[y] = 4 \ \& \ S_1[x] + S_1[2] = y \ \& \ j_2 = x)$$
$$= \sum_{x \neq 4} \sum_{y \neq x} \Pr(S_1[x] + S_2[y] = 4 \ \& \ S_1[x] + S_1[2] = y) \cdot \Pr(j_2 = x).$$

As before, in the last expression, the values taken from S_1 are independent of the value of j_2, and thus the events $(S_1[x] + S_2[y] = 4)$ and $(S_1[x] + S_1[2] = y)$ are both independent of the event $(j_2 = x)$.

Another interesting case occurs if $y = 4$ in the above calculation. In this case, one one hand, we have $S_1[x] + S_2[4] = 4$, while one the other hand we get $S_1[x] + S_1[2] = 4$. One may notice that $S_1[4]$ is a value that does not get swapped to obtain the state S_2. This is because the only two values to get swapped at this stage are from the locations $[i_2] = [2]$ and $[j_2] = [x] \neq [4]$. Thus, $S_2[4] = S_1[4]$ and we get $S_1[x] + S_1[4] = 4$ and $S_1[x] + S_1[2] = 4$, indicating $S_1[4] = S_1[2]$. As S_1 is a permutation, this situation is not possible, and all other cases deal with two distinct locations of the permutation S_1. Therefore, we obtain

$$\Pr(S_1[x] + S_2[y] = 4 \ \& \ S_1[x] + S_1[2] = y) = \begin{cases} 0 & \text{if } y = 4; \\ \frac{1}{N(N-1)} & \text{otherwise.} \end{cases}$$

In turn, we obtain the probability of the second event as follows.

$$\Pr\left(S_1[j_2] + S_2[S_1[j_2] + S_1[2]] = 4 \ \& \ j_2 \neq 4\right)$$
$$= \sum_{x \neq 4} \Pr(j_2 = x) \sum_{y \neq x} \Pr(S_1[x] + S_2[y] = 4 \ \& \ S_1[x] + S_1[2] = y)$$
$$= \sum_{x \neq 4} \Pr(j_2 = x) \left[0 + \sum_{\substack{y \neq x \\ y \neq 4}} \frac{1}{N(N-1)} \right]$$
$$= \sum_{x \neq 4} \Pr(j_2 = x) \left[(N-2) \cdot \frac{1}{N(N-1)} \right]$$
$$= \frac{N-2}{N(N-1)} \sum_{x \neq 4} \Pr(j_2 = x)$$
$$= \frac{N-2}{N(N-1)} \cdot (1 - \Pr(j_2 = 4)) = \frac{N-2}{N(N-1)} \cdot \left(1 - \frac{7/3}{N^2} \right).$$

Calculation for $\Pr(S_2[2] = 4 - z_2)$**:** Combining the probabilities for the first and second events, we obtain the final probability as

$$\Pr(S_2[2] = 4 - z_2) = \frac{7/3}{N^2} \cdot \frac{2}{N} + \frac{N-2}{N(N-1)} \cdot \left(1 - \frac{7/3}{N^2}\right) \approx \frac{1}{N} + \frac{4/3}{N^2}.$$

Hence the desired probability for the event $(S_2[2] = 4 - z_2)$.   □

Thus, one can guess the value of $S_2[i_2] = S_2[2]$ with probability greater than that of a random guess (probability $\frac{1}{N}$). For $N = 256$, the result matches with our experimental data generated from 1 billion runs of RC4 with randomly selected 16 byte keys.

4.3   Randomness of j_r for $r \geq 3$

Along the same line of analysis as in the case of j_2, it is possible to compute the explicit probability distributions of $j_r = \sum_{x=1}^{r} S_{x-1}[x]$ for $3 \leq r \leq 255$ as well. We do not present the expressions $\Pr(j_r = v)$ for $r \geq 3$ to avoid complication. However, it turns out that $j_r = \sum_{x=1}^{r} S_{x-1}[x]$ becomes closer to be random as r increase. The probability distributions of j_1, j_2 and j_3 are shown in Fig. 4, where the experiments have been run over 1 billion trials of RC4 PRGA, with randomly generated keys of size 16 bytes.

One may note that the randomness in j_2 is more than that of j_1 (apart from the case $v = 4$), and j_3 is almost uniformly random. This trend continues for the later rounds of PRGA as well. However, we do not plot the graphs for the probability distributions of j_r with $r \geq 4$, as these distributions are almost identical to that of j_3, i.e., almost uniformly random in behavior.

5   Conclusion

In this paper, we revisit the attack on broadcast RC4 introduced in FSE 2001 by Mantin and Shamir [5], and refute some claims made in that paper. Mantin and Shamir claimed that amongst the initial bytes of RC4 keystream, only the second one shows a bias to zero, and none of the other initial bytes has any bias (even weaker). Contrary to this claim, we prove that all the other initial keystream bytes (3 to 255 to be specific) *also* exhibit a bias to zero. It comes as a surprise to us that this observation has escaped the scrutiny of the RC4 research community for a long time.

The above biases can distinguish RC4 keystream reliably from a random stream of bytes. Further, these biases can also be exploited to mount an attack against broadcast RC4. In addition to the second plaintext byte recovery as in [5], our technique can retrieve the bytes 3 to 255 of the plaintext. The bias shown by these initial bytes also allow us to guess some state information from the RC4 keystream ($S_{r-1}[r]$ given $z_r = 0$ for $3 \leq r \leq 255$).

Further, we study the non-randomness of index j in RC4 PRGA that reveals a strong bias of j_2 towards 4. This bias in turn helps in guessing the state value $S_2[2]$ from the second keystream byte.

We would like to make a small note on a related observation. The probability calculation for event $(z_r = 0)$ in this paper was triggered by the observation that the event $(S_{r-1}[r] = r)$ is biased in the first place. There exist similar biases (though in a much weaker magnitude) in the event $(S_r[u] = v)$ for other values of u, v as well. These biases may in turn lead to corresponding biases in events $(z_r = k)$ for $k \neq 0$, but we do not study these in the scope of this paper.

Another observation that caught our attention during this work was the noticeable negative bias in $\Pr(z_1 = 0)$. Similar issues of non-random behavior in the first keystream byte z_1 has been reported earlier in [7, Section 6]. But neither [7] nor we could provide a satisfactory proof of this bias. We would like to pose this as an open problem to conclude our paper:

Open problem: Compute $\Pr(z_1 = 0)$ *explicitly to support the observations made in [7] and the negative bias observed in the line of our work.*

Acknowledgment. The authors are thankful to the anonymous reviewers for their comments and suggestions that helped in improving technical and editorial details of the paper. The authors would also like to express their gratitude towards Dr. Mridul Nandi and Mr. Santanu Sarkar, who have helped improve the technical content of the paper through discussions regarding some of the probability computations.

References

1. Jenkins, R.J.: ISAAC and RC4 (1996), `http://burtleburtle.net/bob/rand/isaac.html`
2. Maitra, S., Paul, G.: New Form of Permutation Bias and Secret Key Leakage in Keystream Bytes of RC4. In: Nyberg, K. (ed.) FSE 2008. LNCS, vol. 5086, pp. 253–269. Springer, Heidelberg (2008)
3. Mantin, I.: Analysis of the stream cipher RC4. Master's Thesis, The Weizmann Institute of Science, Israel (2001), `http://www.wisdom.weizmann.ac.il/~itsik/RC4/Papers/Mantin1.zip`
4. Mantin, I.: Predicting and Distinguishing Attacks on RC4 Keystream Generator. In: Cramer, R. (ed.) EUROCRYPT 2005. LNCS, vol. 3494, pp. 491–506. Springer, Heidelberg (2005)
5. Mantin, I., Shamir, A.: A Practical Attack on Broadcast RC4. In: Matsui, M. (ed.) FSE 2001. LNCS, vol. 2355, pp. 152–164. Springer, Heidelberg (2002)
6. Maximov, A., Khovratovich, D.: New State Recovery Attack on RC4. In: Wagner, D. (ed.) CRYPTO 2008. LNCS, vol. 5157, pp. 297–316. Springer, Heidelberg (2008)
7. Mironov, I.: (Not So) Random Shuffles of RC4. In: Yung, M. (ed.) CRYPTO 2002. LNCS, vol. 2442, pp. 304–319. Springer, Heidelberg (2002)
8. Sepehrdad, P., Vaudenay, S., Vuagnoux, M.: Discovery and Exploitation of New Biases in RC4. In: Biryukov, A., Gong, G., Stinson, D.R. (eds.) SAC 2010. LNCS, vol. 6544, pp. 74–91. Springer, Heidelberg (2011)
9. Sepehrdad, P., Vaudenay, S., Vuagnoux, M.: Statistical Attack on RC4 Distinguishing WPA. Accepted at EUROCRYPT 2011 (2011)

Boomerang Attacks on BLAKE-32

Alex Biryukov, Ivica Nikolić*, and Arnab Roy

University of Luxembourg
{alex.biryukov,ivica.nikolic,arnab.roy}@uni.lu

Abstract. We present high probability differential trails on 2 and 3 rounds of BLAKE-32. Using the trails we are able to launch boomerang attacks on up to 8 round-reduced keyed permutation of BLAKE-32. Also, we show that boomerangs can be used as distinguishers for hash/ compression functions and present such distinguishers for the compression function of BLAKE-32 reduced to 7 rounds. Since our distinguishers on up to 6 round-reduced keyed permutation of BLAKE-32 are practical (complexity of only 2^{12} encryptions), we are able to find boomerang quartets on a PC.

Keywords: SHA-3 competition, hash function, BLAKE, boomerang attack, cryptanalysis.

1 Introduction

The SHA-3 competition [6] will soon enter the third and final phase, by selecting 5 out of 14 second round candidates. The hash function BLAKE [2] is among these 14 candidates, and it is one of the few functions that has not been tweaked from the initial submission in 2008. Being an addition-rotation-xor (ARX) design, BLAKE is one of the fastest functions on various platforms in software. Indeed, among the fastest candidates, BLAKE has the highest published security level, i.e. the best published attacks work only on a small fraction of the total number of rounds. Few attacks, however, were published on the round-reduced compression function and keyed permutation of BLAKE-32 (which has 10 rounds). In [3] Ji and Liangyu present collision and preimage attacks on 2.5 rounds of the compression function of BLAKE-32. Su et al. [7] give near collisions on 4 rounds with a complexity of 2^{21} compression function calls. However, one can argue that the message modification they use, requires an additional effort of 2^{64} (see Sec. 5). Aumasson et al. in [1], among other, present near collisions on 4 rounds of the compression function with 2^{56} complexity, and impossible differentials on 5 rounds of the keyed permutation.

Our Contribution. We show various boomerang distinguishers on round-reduced BLAKE-32. Our analysis is based on the fact that BLAKE-32, being a keyed permutation, has some high probability differential trails on two

---

* This author is supported by the Fonds National de la Recherche Luxembourg grant TR-PHD-BFR07-031.

A. Joux (Ed.): FSE 2011, LNCS 6733, pp. 218–237, 2011.

and three rounds (2^{-1} on two and 2^{-7} on three rounds). Moreover, we can extend the three round trail to four rounds. First, we use these trails to build boomerang distinguishers for the round-reduced keyed permutation of BLAKE-32 on up to 8 rounds. Then we extend the concept of boomerang distinguishers to hash functions. As far as we know, this is the first application of the standard boomerangs to hash function. An amplified boomerang attack applied to hash functions was presented in [4], however it was used in addition to a collision attack. Our boomerang attacks, on the other hand, are standalone distinguishers, and work in the same way as for block ciphers – by producing the quartet of plaintexts and ciphertexts (input chaining values and output chaining values). We also show how to obtain simpler zero-sum distinguisher from the boomerang and present such distinguishers for 4, 5, 6 rounds of BLAKE-32. Our final result is a boomerang distinguisher for 7 rounds of the compression function of BLAKE-32. The summary of our results is given in Table 1.

Although in this paper we focus on BLAKE-32, our attacks can be easily extended to the other versions of BLAKE (with similar complexities and number of attacked rounds). The attacks do not contradict any security claims of BLAKE.

Table 1. Summary of the attacks on the compression function (CF) and the keyed permutation (KP) of BLAKE-32

Attack	CF/KP	Rounds	CF/KP calls	Reference
Free-start collisions	CF	2.5	2^{112}	[3]
Near collisions[a]	CF	4	2^{21}	[7]
Near collisions	CF	4	2^{56}	[1]
Impossible diffs.	KP	5	-	[1]
Boomerang dist.	CF	4	2^{67}	Sec. 5
Boomerang dist.	CF	5	$2^{71.2}$	Sec. 5
Boomerang dist.	CF	6	2^{102}	Sec. 5
Boomerang dist.	CF	6.5	2^{184}	Sec. 5
Boomerang dist.	CF	7	2^{232}	Sec. 5
Boomerang dist.	KP	4	2^{3}	Sec. 6
Boomerang dist.	KP	5	$2^{7.2}$	Sec. 6
Boomerang dist.	KP	6	$2^{11.75}$	Sec. 6
Boomerang dist.	KP	7	2^{122}	Sec. 6
Boomerang dist.	KP	8	2^{242}	Sec. 6

[a] The attack assumes that message modification can be used anywhere in the trail.

2   Description of BLAKE32

The compression function of BLAKE-32 processes a state of 16 32-bit words represented as 4×4 matrix. Each word in BLAKE-32 has 32 bits. In the *Initialization* procedure, the state is loaded with a chaining value $h_0, \ldots, h_7$, a salt $s_0, \ldots, s_3$, constants $c_0, \ldots, c_7$, a counter t_0, t_1 as follows:

$$\begin{pmatrix} v_0 & v_1 & v_2 & v_3 \\ v_4 & v_5 & v_6 & v_7 \\ v_8 & v_9 & v_{10} & v_{11} \\ v_{12} & v_{13} & v_{14} & v_{15} \end{pmatrix} \longleftarrow \begin{pmatrix} h_0 & h_1 & h_2 & h_3 \\ h_4 & h_5 & h_6 & h_7 \\ s_0 \oplus c_0 & s_1 \oplus c_1 & s_2 \oplus c_2 & s_3 \oplus c_3 \\ t_0 \oplus c_4 & t_0 \oplus c_5 & t_1 \oplus c_6 & t_1 \oplus c_7 \end{pmatrix}$$

After the *Initialization*, the compression function takes 16 message words $m_0, \ldots, m_{15}$ as inputs and iterates 10 rounds. Each round is composed of eight applications of G function. A column step:

$$\mathsf{G}_0(v_0, v_4, v_8, v_{12}), \mathsf{G}_1(v_1, v_5, v_9, v_{13}), \mathsf{G}_2(v_2, v_6, v_{10}, v_{14}), \mathsf{G}_3(v_3, v_7, v_{11}, v_{15})$$

followed by the diagonal step:

$$\mathsf{G}_4(v_0, v_5, v_{10}, v_{15}), \mathsf{G}_5(v_1, v_6, v_{11}, v_{12}), \mathsf{G}_6(v_2, v_7, v_8, v_{13}), \mathsf{G}_7(v_3, v_4, v_9, v_{14})$$

where $\mathsf{G}_i (i \in \{0, \ldots, 7\})$ depend on their indices, message words $m_0, \ldots, m_{15}$, constants $c_0, \ldots, c_{15}$ and round index r. At round r, $\mathsf{G}_i(a, b, c, d)$ is described with following steps:

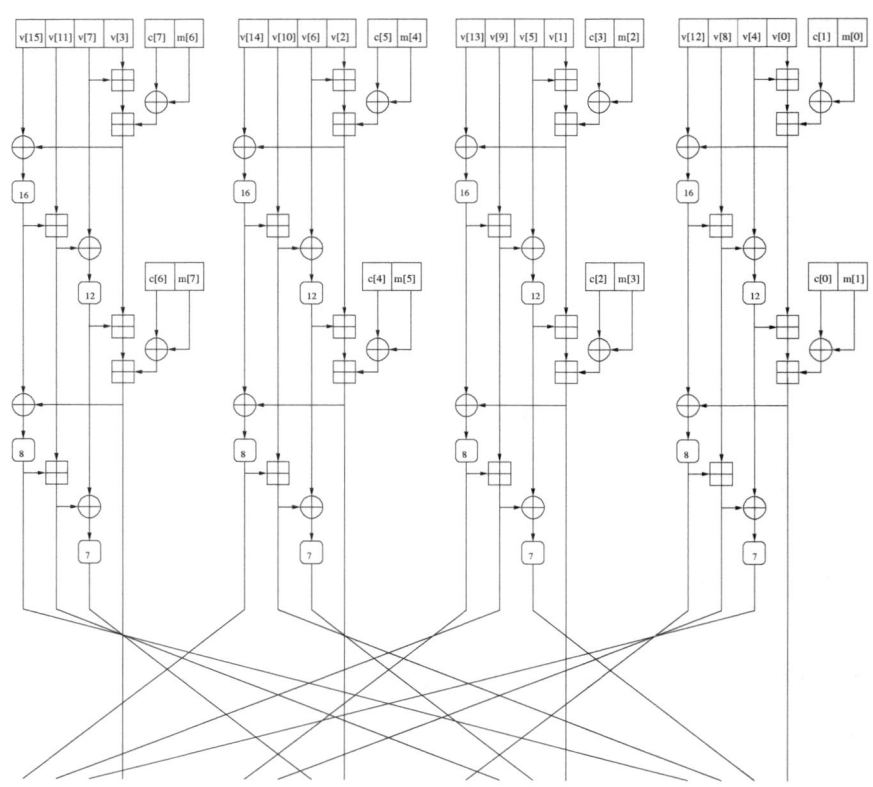

Fig. 1. Column step of round-0

$$1 : a \leftarrow a + b + (m_{\sigma_r(2i)} \oplus c_{\sigma_r(2i+1)})$$
$$2 : d \leftarrow (d \oplus a) \ggg 16$$
$$3 : c \leftarrow c + d$$
$$4 : b \leftarrow (b \oplus c) \ggg 12$$
$$5 : a \leftarrow a + b + (m_{\sigma_r(2i+1)} \oplus c_{\sigma_r(2i)})$$
$$6 : d \leftarrow (d \oplus a) \ggg 8$$
$$7 : c \leftarrow c + d$$
$$8 : b \leftarrow (b \oplus c) \ggg 7$$

where σ_r belongs to the set of permutations as specified in [2]. The *Finalization* procedure in BLAKE-32 is depicted as:

$$h_0' \leftarrow h_0 \oplus s_0 \oplus v_0 \oplus v_8$$
$$h_1' \leftarrow h_1 \oplus s_1 \oplus v_1 \oplus v_9$$
$$h_2' \leftarrow h_2 \oplus s_2 \oplus v_2 \oplus v_{10}$$
$$h_3' \leftarrow h_3 \oplus s_3 \oplus v_3 \oplus v_{11}$$
$$h_4' \leftarrow h_4 \oplus s_0 \oplus v_4 \oplus v_{12}$$
$$h_5' \leftarrow h_5 \oplus s_1 \oplus v_5 \oplus v_{13}$$
$$h_6' \leftarrow h_6 \oplus s_2 \oplus v_6 \oplus v_{14}$$
$$h_7' \leftarrow h_7 \oplus s_3 \oplus v_7 \oplus v_{15}$$

where $h_0, \ldots, h_7$ is the initial chaining value and $v_0, \ldots, v_{15}$ is the state value after the ten rounds, and $h_0', \ldots, h_7'$ are the words of the new chaining value.

3 Boomerang Attacks on Block Ciphers and Compression Functions

The boomerang attack [8] is a differential-type attack that exploits high probability differential trails in each half of a cipher E. When successful, it outputs a quartet of plaintexts and corresponding ciphertexts with some fixed particular differences between some of the pairs. This property can be used to distinguish the cipher from a random permutation, and in some cases, to recover the key.

Let us decompose the initial cipher E into two ciphers E_0, E_1, i.e. $E = E_1 \circ E_0$. Let $\Delta \rightarrow \Delta^*$ be some differential trail for E_0 that holds with probability p and $\nabla \rightarrow \nabla^*$ be a trail for E_1 with probability q. We start with a pair of plaintexts $(P_1, P_2) = (P_1, P_1 \oplus \Delta)$ and produce a pair of corresponding ciphertexts $(C_1, C_2) = (E(P_1), E(P_2))$. Then we produce a new pair of ciphertext $(C_3, C_4) = (C_1 \oplus \nabla^*, C_2 \oplus \nabla^*)$, decrypt this pair, and get the corresponding pair of plaintexts $(P_3, P_4) = (E^{-1}(C_3), E^{-1}(C_4))$. The difference $P_3 \oplus P_4$ is Δ with probability at least $p^2 q^2$: 1)the difference $E_0(P_1) \oplus E_0(P_2)$ is Δ^* with probability p; 2) the differences $E_1^{-1}(C_1) \oplus E_1^{-1}(C_3)$, $E_1^{-1}(C_2) \oplus E_1^{-1}(C_4)$ are both ∇ with probability q^2; 3)when 1), 2) hold, then the difference $E_1^{-1}(C_3) \oplus E_1^{-1}(C_4)$ is Δ^* (with probability pq^2) and $E^{-1}(C_3) \oplus E^{-1}(C_4)$ is Δ with probability $p^2 q^2$.

We would like to address a couple of issues. First, the boomerang distinguisher can be used even in the case when it returns a pair (P_3, P_4) with a difference

$P_3 \oplus P_4$ specified only in certain bits (instead of the full plaintext). When the difference is specified in t bits ($t < n$), then the probability of the boomerang (in order to be used as a distinguisher) should be higher than 2^{-t}, i.e. $p^2 q^2 > 2^{-t}$. Second, the real probability of the boomerang is $\hat{p}^2 \hat{q}^2$, where $\hat{p}, \hat{q}$ are so-called amplified probabilities, defined as:

$$\hat{p} = \sqrt{\sum_{\Delta^*} P[\Delta \to \Delta^*]^2}, \hat{q} = \sqrt{\sum_{\nabla} P[\nabla \to \nabla^*]^2} \tag{1}$$

Since finding these values is hard, in some cases, we try to get experimental results for the probability of the boomerang. We run a computer simulation, start the boomerang with a number of pairs with some prefixed difference Δ, and count the number of returned pairs that have the same difference Δ. Obviously the ratio of the returned pairs to the launched pairs is the probability of the boomerang.

The main obstacle for applying the boomerang attack to compression functions, is that in general, the compression functions are non-invertible. Hence, after obtaining the pairs (C_3, C_4) from (C_1, C_2), one cannot go backwards and obtain the pair (P_3, P_4). One way to deal with this is to switch to amplified boomerang attacks [5]. However, this type of boomerangs usually has lower probability, and more importantly, since it requires internal collisions, in the case when the underlying compression functions are double pipes, the attack complexity becomes higher than in a trivial attack.

Indeed, the standard boomerang attack can be used as a differential distinguisher for a compression function F. The idea is to start the attack in the middle of F and then go forward and backwards to obtain the quartets, thus escaping the feedforward. Let $F(H)$ be obtained from some invertible function $f(H)$ with a feedforward, for example Davies-Meyer mode $F(H) = f(H) \oplus H$. As in the attack on block ciphers, first step is to decompose f into two functions f_0, f_1 and to find two differential trails for f_0 and f_1 (further we use the same notation as in the attacks on block ciphers). We start with four states S_1, S_2, S_3, S_4 at the end of the function f_0 (beginning of f_1) such that $S_1 \oplus S_2 = S_3 \oplus S_4 = \Delta^*$ and $S_1 \oplus S_3 = S_2 \oplus S_4 = \nabla$. From these states we obtain the initial states (input chaining values) P_i and the final states (output chaining values without the feedforward) C_i, i.e. $P_i = f_0^{-1}(S_i), C_i = f_1(S_i), i = 1, \ldots, 4$. Then with probability at least $p^2 q^2$ we have:

$$P_1 \oplus P_2 = \Delta, \qquad\qquad P_3 \oplus P_4 = \Delta$$
$$C_1 \oplus C_3 = \nabla^*, \qquad\qquad C_2 \oplus C_4 = \nabla^*.$$

Extending the following attack to the whole compression function F is trivial – we just have to take into account that $C_i = f(P_i) = F(P_i) \oplus P_i$. For the boomerang quartet (P_1, P_2, P_3, P_4) we get:

$$P_1 \oplus P_2 = \Delta, \qquad\qquad\qquad P_3 \oplus P_4 = \Delta \tag{2}$$
$$[F(P_1) \oplus P_1] \oplus [F(P_3) \oplus P_3] = \nabla^*, \quad [F(P_2) \oplus P_2] \oplus [F(P_4) \oplus P_4] = \nabla^* \tag{3}$$

For a random n-bit compression function F, the complexity of finding the quartet (P_1, P_2, P_3, P_4) with the above relations (2),(3), is around[1] 2^n . Hence when $p^2 q^2 > 2^{-n}$ one can launch a boomerang attack and thus obtain a distinguisher for F. The distinguisher becomes even more powerful if the attacker finds several boomerang quartets with the same differences Δ, ∇^*.

A zero-sum distinguisher, can be obtained based on the boomerangs. If in (3), we XOR the two equations, we get:

$$
\begin{aligned}
0 =& [F(P_1) \oplus P_1] \oplus [F(P_3) \oplus P_3] \oplus \nabla^* \oplus [F(P_2) \oplus P_2] \oplus [F(P_4) \oplus P_4] \oplus \nabla^* = \\
=& F(P_1) \oplus F(P_2) \oplus F(P_3) \oplus F(P_4) \oplus (P_1 \oplus P_2) \oplus (P_3 \oplus P_4) = \\
=& F(P_1) \oplus F(P_2) \oplus F(P_3) \oplus F(P_4) \oplus \Delta \oplus \Delta = \\
=& F(P_1) \oplus F(P_2) \oplus F(P_3) \oplus F(P_4)
\end{aligned}
$$

Finding a zero-sum distinguisher for a random permutation requires $2^{n/4}$ encryptions. However, since we have the additional conditions on the plaintexts (the XORs of the pairs are fixed), the complexity rises to $2^{n/2}$.

It is important to notice that to produce the quartet (for the boomerang or the zero-sum boomerang) one has to start not necessarily from the middle states (S_1, S_2, S_3, S_4). For example, one can start from two input chaining values $(P_1, P_2) = (P_1, P_1 \oplus \Delta)$, produce the values $(S_1, S_2) = (f_0(P_1), f_0(P_2))$, then obtain the values for the two other middle states $(S_3, S_4) = (S_1 \oplus \nabla, S_2 \oplus \nabla)$, and finally get the two input chaining values $(P_3, P_4) = (f_0^{-1}(S_3), f_0^{-1}(S_4))$ and the four output chaining values $(f_1(S_1) \oplus P_1, f_1(S_2) \oplus P_2, f_1(S_3) \oplus P_3, f_1(S_4) \oplus P_4)$. Clearly, the probability of the boomerang stays the same. Starting from the beginning (or from some other particular state before the feedforward) can be beneficial in the cases when one wants to use message modification or wants to have some specific values in one of the four states (as shown further in the case of BLAKE-32).

4   Round-Reduced Differential Trails in BLAKE-32

In order to obtain good differential trails in BLAKE we exploit the structure of the message word permutation. In fact we can easily obtain good 2-round differential trail. The idea is to choose a message word m_j such that

- It appears at Step 1(*Case1*) or at Step 5(*Case2*) in $G_i (0 \le i \le 3)$ at round-r and
- Also appears at Step 5 in $G_i (4 \le i \le 7)$ at round-$(r + 1)$.

If we choose the message word with the above mentioned strategy then with a suitable input difference we may pass 1.5 rounds for free[2] (i.e. with probability 1).

---

[1] This holds only when the difference between the messages is fixed as well. Otherwise, the complexity is only $2^{n/2}$.

[2] A similar technique was used in the analysis presented in [7,1].

Observation 1. *A 2-round differential trail can be obtained in BLAKE-32 with probability* 2^{-1}.

Proof. Choose two rounds with a message word m_j as described previously. In

- *Case1*, we choose $\Delta m_j = \Delta a = $ 0x80000000
- *Case2*, we choose $\Delta m_j = \Delta a = \Delta d = $ 0x80000000

in the corresponding G function (see Fig. 2). After 1.5 rounds we get $\Delta v_k = 0, \forall k \in \{0, \ldots, 15\}$ with probability 1. In the next half of the second round because of our choice of message word and suitable difference, we get one active bit only at step 7 in the corresponding G function (see Fig. 3). Hence we get a differential trail with probability 2^{-1}.

Remark 1. In *Case1* if Δm_j and Δa have any active bits other than MSB then at round-r, probability of the trail is 2^{-t}(where t is the number of active bits in $\Delta m_j (= \Delta a)$ at round-r) and at round-$(r+1)$ the probability is 2^{-s}, where $s = 2t - 1, 2t, 2t + 1$(depending on the position of active bits). So in this case the probability for two rounds will be $1/2^{s+t}$. Also if m_j appears at Step 1 in $G_i(4 \leq i \leq 7)$ at round-$(r+1)$ then probability of a 2-round differential trail decreases further.

Remark 2. In *Case1* if $\Delta m_j = \Delta a = \Delta$, such that Δ has two active bits at ith and $(i+16)$th position and m_j appears at step 1 in $G_i(4 \leq i \leq 7)$ at round-$(r+1)$ then we have 2-round differential trail with probability $2^{-8-1}(= 2^{-9})$ when ith bit is the MSB and $\geq 2^{-12-2}(= 2^{-14})$ otherwise.

In order to construct 3-round trails from these 2-round differential trails we may simply add one more round at the beginning. The occurrence of the chosen message word in this one round does not affect much in terms of probability of the difference propagation.

Observation 2. *A 3-round differential trail may be obtained from the above described two round differential trail with probability* 2^{-s}, *where* $s = 6, 7$ *or* 8

Proof. After obtaining 2-round differential trail with probability 2^{-1}(*Case1*), we add one more round(say, round-$(r-1)$) at the beginning. The probability of this one round differential trail may vary depending on the position of the message word m_j. Suppose the message word occurs in G_l (for some index l) at round r. Then at round $r - 1$:

- If the message word is in $G_i(0 \leq i \leq 3)$ or at step 1 of $G_i(4 \leq i \leq 7)$, probability of this one round trail is 2^{-6}.
- If the message word occurs at step 5 of G_{l+4}, we get differential trail with probability 2^{-5} for this one round.

For all other cases the probability of this one round differential trail is 2^{-7}. Hence we get a 3-round differential trail with probability $2^{-7}, 2^{-6}$ and 2^{-8} respectively.

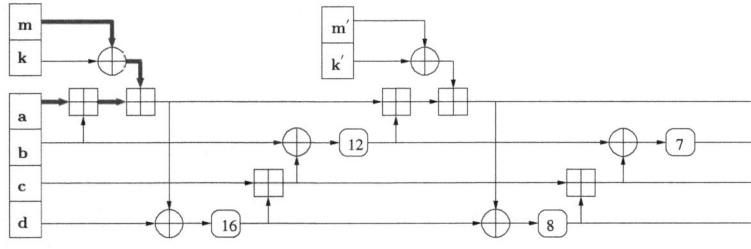

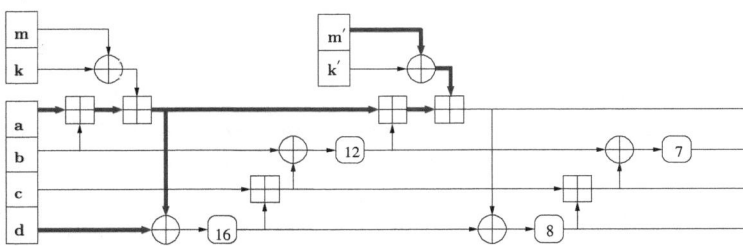

Fig. 2. Two possible differential trails for G at the beginning of 2-round trail. The top trail is *Case1*, while the bottom is *Case2*.

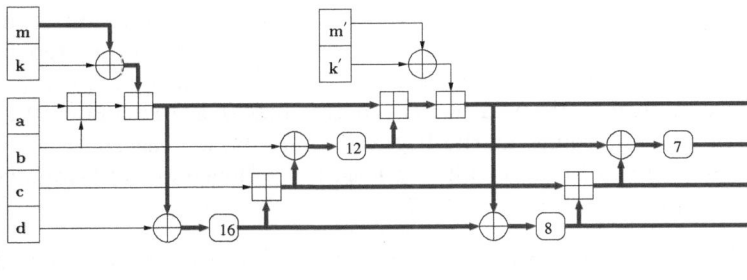

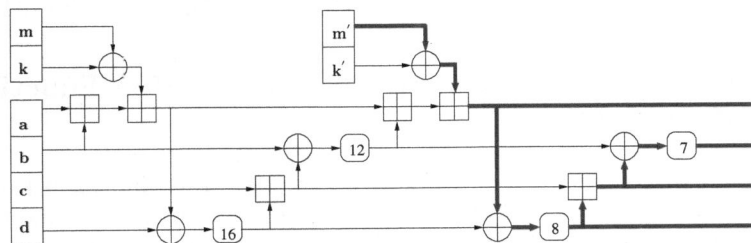

Fig. 3. Two possible differential trails for G at the end of 2-round trail. The top trail is when the message with the difference appears at Step 1, and the bottom at Step 5.

Remark 3. This 3-round differential trail can be extended for half more round in the forward direction. If we add half round at the end of this three rounds and if the chosen message word does not occur there then we can get 3.5-round trail with probability $\geq 2^{-24-8}(= 2^{-32})$.

For this three round differential trail we have to inject two distinct input differences at v_{12} and v_{13} which correspond to the same counter t_0. In order to obtain a 3-round differential trail with consistent input differences at the states corresponding to the counters t_0 and t_1 we use a 2-round trail with lower probability.

Observation 3. *Let $\Delta a = \Delta c = \Delta$ such that Δ has only ith and $(i+16)$th bits active. For a G function if there is no difference in the message words then the differential trail $(\Delta, 0, \Delta, 0) \rightarrow (\Delta, 0, 0, 0)$ occurs with probability 2^{-3} if ith bit is the MSB and with probability 2^{-6} otherwise.*

Observation 4. *A 3-round differential trail with input difference consistent with counters(t_0, t_1) may be obtained with probability 2^{-21} or at least 2^{-36}.*

Proof. Starting with $\Delta m_j = \Delta a = \Delta = $ 0x80008000 we obtain a 2-round differential trail with probability 2^{-9}(as described in *Remark 2*). Then we add one more round at the beginning. The position of the message word m_j in this one round determines which three rounds we should consider in order to obtain the 3-round trail. Such three rounds may be found if we start with round-4. Now in this one round(added at the beginning) we have two G functions with differences as described in *Observation 3* and one G function with difference $(\Delta_1, \Delta_2, \Delta, 0) \rightarrow (0, 0, \Delta, 0)$(with the message difference at step 5 in it). So probability for this one round is $2^{-6-6} = 2^{-12}$. Hence we get a 3-round trail with probability 2^{-21}. If Δ has two active bits (e.g. 0x00080008) then probability of this one round at the beginning may be at least $2^{-12-10} = 2^{-22}$ and probability of the 2-round trail is at least 2^{-14}. Hence we get 3-round differential trail with probability at least 2^{-36}.

The choice of message word for the 3-round differential trail specified in *Observation 4* is available if we start with round-4 and the input differences for the states corresponding to the counters are $\Delta v_{12} = \Delta v_{13} = \Delta v_{14} = \Delta v_{15} = 0$. A similar 2-round and 3-round differential trails exist for BLAKE-64.

5   Boomerang Attacks on the Compression Function of BLAKE-32

The high probability round-reduced differential trails in the permutation of BLAKE-32 can be used to attack the compression function and find boomerang distinguishers. However, due to the *Initialization* procedure, there are a few requirements on the trails. First, since the block index is copied twice, the initial differences in v_{12} and v_{13}, as well as the differences in v_{14} and v_{15}, have to be the same. Second, even in the case when the attacker has a trail with initial differences consistent to the above requirement, if he uses message modification techniques in the higher rounds of the trail, he might end up with inconsistent initial states. For example, if the attacker uses some k-round trail and starts fixing the values of the state and the messages at round k, and then goes backward, he can obtain two states with some predefined difference (as the one predicted by the trail). However, the probability that these two states are consistent with the *Initialization* procedure is 2^{-64} (if $v_{12} \oplus v_{13} = c_4 \oplus c_5$ and $v_{14} \oplus v_{15} = c_6 \oplus c_7$). Note that if one of the states is consistent, then the other one is consistent as well (if the attacker used trails with appropriate initial difference). Therefore, using message modification techniques in later steps of the trail is not trivial (without increasing the complexity of the attack). On the other hand, the modification can still be used at the beginning because the attacker starts with two states consistent with the *Initialization* procedure.

For the boomerang attack on 4 rounds of the compression function of BLAKE-32 we can use two trails each on 2 rounds (see Table 2). Since the probability of these trails is only 2^{-1}, the probability of the boomerang is 2^{-4}. To create a quartet of states, consistent with the *Initialization* procedure, we start with a pair of states (P_1, P_2) that have a difference Δ (note that Δ does not have a difference in the "block index" words) and consistent with the *Initialization* words $v_{12}, v_{13}, v_{14}, v_{15}$ in both of the states, then go two rounds forward and obtain the pair (S_1, S_2). Then we produce the pair $(S_3, S_4) = (S_1 \oplus \nabla, S_2 \oplus \nabla)$ and go backwards two rounds to get the pair of initial states (P_3, P_4). The probability that P_3 (and therefore P_4) is consistent with the *Initialization* is 2^{-64}. Also, from S_1, S_2, S_3, S_4 we go forward two rounds, produce the outputs and apply the *Finalization* to get the new chaining values. Note that *Finalization* is linear, hence the differential trail (with XOR difference) holds with probability 1. Therefore, we can produce the boomerang quartet with a complexity of $4 \cdot 2^{4+64} = 2^{70}$ calls to the 4-round reduced compression function of BLAKE-32.

The boomerang attack on 5 rounds is rather similar. We only need one of the trails to be on 3 rounds, instead of 2 (see Table 3). Such a trail has a probability of 2^{-7}, and we use two round trail with 2^{-3}, hence the boomerang has a probability of $2^{-2 \cdot 3 - 2 \cdot 7} = 2^{-20}$ and the whole attack (taking into account the *Initialization*) has a complexity of around $4 \cdot 2^{20+64} = 2^{86}$ compression function calls.

For the boomerang attack on 6 rounds we will use two 3-round trails (see Table 4). However, we cannot use the optimal trails (the ones that hold with around 2^{-7}) because the starting difference in each such trail is inconsistent with the *Initialization* procedure. Therefore, for the top trail of the boomerang we will use a trail which has lower probability 2^{-34} but has no differences in any of the "block index" words ($v_{12}, v_{13}, v_{14}, v_{15}$). For the bottom trail we can use an optimal trail. The complexity of this boomerang distinguisher on 6 rounds becomes $4 \cdot 2^{2 \cdot 34 + 2 \cdot 7 + 64} = 2^{148}$ calls.

Note, for the top trails for 5 and 6 round boomerangs (see Table 3,4), we did not use the best trails with probability 2^{-1}, 2^{-21}, but instead used trails with lower probability ($2^{-3}, 2^{-34}$). We found that if we use the best trails, then the boomerang does not work, most likely because of the slow diffusion. We cannot get four states in the middle (after the third round), that have pairwise Δ^* and ∇ difference (Δ^* is the end difference of the top trail). However, if we take other trails, as the ones we have taken, the boomerang quartet can be obtained – we confirmed this experimentally, by producing a boomerang quartet.

Each of the above attacks can be improved if we take into account the amplified probabilities for the boomerang attack and if we use message modification. We can obtain the amplified probabilities (and the total probabilities) of the boomerang experimentally: we start with a number of plaintext pairs with the required difference Δ, and then check how many of the returned (by the boomerang) differences are Δ. Also, in the first round, for one side of the boomerang we use message modification, i.e. we pass this round with probability 1. Using these two approaches, we got the following results: the boomerang on 4 rounds has a probability 2^{-1}, on 5 rounds $2^{-5.2}$, and on 6 rounds 2^{-36}. Hence, the attack complexity for 4 rounds drops to $4 \cdot 2^{1+64} = 2^{67}$, for 5 rounds to $4 \cdot 2^{5.2+64} = 2^{71.2}$, and for 6 rounds to $4 \cdot 2^{36+64} = 2^{102}$ compression function calls. An example of boomerang quartet for 6 rounds, with the first pair of plaintext consistent to the *Initialization*, while only the difference in the second is consistent, and therefore obtained with around $4 \cdot 2^{36}$ compression function calls, is given in Table 9. The complexities of the boomerang distinguishers for 4,5, and 6 round are bellow 2^{128}, therefore they can be used as zero-sum boomerang distinguishers, i.e. $P_1 \oplus P_2 = P_3 \oplus P_4 = \Delta$ and $F(P_1) \oplus F(P_2) \oplus F(P_3) \oplus F(P_4) = 0$.

For the boomerang on 6.5 rounds, we use a top trail on 3 rounds (from 0.5 to 3.5) with 2^{-40}, and a bottom trail on 3.5 rounds (from 3.5 to 7), with 2^{-48} (see Table 5). The complexity of producing the boomerang quartet is $4 \cdot 2^{2 \cdot 40+2 \cdot 48+64} = 2^{242}$ compression function calls. The probability of the first round in the top trail is 2^{-3}, hence using message modification does not lower significantly the attack complexity. However, computing the amplified probabilities can improve the attack. Obviously, we cannot do this experimentally, as the probability of the boomerang is too low – $2^{-2 \cdot 40 - 2 \cdot 48} = 2^{-176}$. Therefore, we cannot test for the whole 6.5 rounds, but we can do it for a reduced number of rounds. We tested for only half round at the end of the first trail (round 3 to round 3.5). We start with a pair of states with a difference specified by the top trail at round 3 and go half round forward to obtain a new pair of states. Then, to each element of the pair, we XOR the same difference (the one specified by the bottom trail at round 3.5), and produce a new pair states. Finally, we go backwards a half round, and check if the difference in the pair is at the one we have started with. Note that the half round can be split into four G functions, and for each of them the amplified probabilities can be found independently. By doing so, we found that the amplified probability for this half round of the boomerang is 2^{-26} instead of twice 2^{-33}, i.e. $2^{-2 \cdot 33} = 2^{-66}$. Another low probability part of the boomerang is the top half round of the second trail – round 3.5 to round 4 holds with 2^{-41}. In this part we can use message modification. We start at round 3.5 with four states that have pairwise differences Δ^* and ∇. We go half round forward and obtain four states with pairwise differences as specified by the bottom trail at round 4. To obtain such states we need $4 \cdot 2^{2 \cdot 41} = 2^{84}$.

Once we have this half round boomerang, we can freely change the message words that are not taken as inputs in this half round without altering the input and the output values of the half round. Hence, we have $2^{8 \cdot 32} = 2^{256}$ degrees of freedom. From the middle states we can obtain the initial and final states (and the chaining values). Therefore, the total complexity of the boomerang on 6.5 rounds becomes $2^{84} + 4 \cdot 2^{2 \cdot (3+1+3)+26+2 \cdot (6+1)+128} = 2^{184}$ calls. Note that unlike as in the case of the boomerangs on 4 and 5 rounds, now the probability that the initial states are consistent to the *Initialization* is 2^{-128} because we use message modification in the middle rather than in the beginning. The bottom trail can easily be extended for additional half round (see Table 5) with probability 2^{-24}. Therefore, the boomerang on 7 rounds requires around $2^{184+2 \cdot 24} = 2^{232}$ compression function calls.

6   Boomerang Attacks on the Keyed Permutation of BLAKE-32

Further we present boomerang attacks on the keyed permutation of BLAKE-32, assuming that the key is unknown to the attacker. These attacks can be seen as distinguishers for the internal cipher of BLAKE-32. The cipher takes 512-bit plaintexts and 512-bit key, and after 10 rounds, outputs 512-bit ciphertext (we discard the *Initialization* and *Finalization* procedures).

Switching from the boomerangs for the compression function to the boomerangs for the keyed permutation has advantages and disadvantages for the attacker. On one hand, the attacker is not concern any more about the *Initialization* procedure, and he can use any trails for the boomerang. On the other hand, since the key is unknown, he cannot use message modification techniques to improve the probability of the boomerang.

The boomerangs on 4 and 5 rounds of the keyed permutation of BLAKE-32 have the same probability as in the case of compression function: 2^{-4} for 4 rounds, and 2^{-20} for 5 rounds. For 6 rounds, we can use two high probability trails $(2^{-7}, 2^{-7}$, see Table 6), and therefore, the probability of the boomerang is 2^{-28}. If we take into account the amplified probabilities, and fix the returning difference only in 128 bits (the words v_1, v_5, v_9, v_{13}) instead of in 512 bits, for the total complexity of the boomerang attack we get 2^3 encryptions for 4 rounds, $2^{7.2}$ for 5 rounds, and $2^{11.75}$ for 6 rounds. These results were confirmed on a PC and a boomerang quartet for 6 rounds is presented in Table 8.

The boomerangs for 7 and 8 rounds, are rather similar: for 7 rounds we use two trails on 3.5 rounds (the first from round 2 to round 5.5, and the second from round 5.5 to round 9), and for 8 rounds, we just extend these trails for additional half round (see Table 7). The complexity of the boomerangs is $4 \cdot 2^{2 \cdot 31+2 \cdot 52} = 2^{168}$ for 7 rounds and $4 \cdot 2^{2 \cdot 73+2 \cdot 82} = 2^{312}$ for 8 rounds. Again, as in the case of 6.5-round boomerang on the compression function, we can compute experimentally the lower bounds on the amplified probabilities, by testing only the probability of the first half round of the bottom trail. We get 2^{-48} instead of $2^{-2.44}$. Also, we can fix the returning difference only in 256 bits, instead of 512 bits, and thus increase the probability in the first half round of the top trail by a factor of 2^{-6} for 7 rounds, and 2^{-30} for 8 rounds. Hence, the boomerang on 7 rounds requires at most 2^{122}, and on 8 rounds at most 2^{242} encryptions.

7   Conclusions

In this paper we have shown how to apply the concept of boomerang distinguisher to compression functions, and presented such distinguishers for the compression function of BLAKE-32, as well as classical boomerang distinguishers for the keyed permutation of BLAKE-32. Our attacks work on up to 2/3 of the total number of rounds of the compression function, and on up to 4/5 (the attacks on up to 3/5 have practical complexity) of the total number of rounds of the keyed permutation of BLAKE-32. The attacks can be equally well applied to the other versions of BLAKE. Our attacks do not contradict the security claims of BLAKE.

Interestingly, tweaking the message permutation in BLAKE can reduce the number of attacked rounds only by one. Therefore, either tweaks in the function G or more advanced message expansion is required in order to significantly reduce the number of attacked rounds.

References

1. Aumasson, J.-P., Guo, J., Knellwolf, S., Matusiewicz, K., Meier, W.: Differential and invertibility properties of BLAKE. In: Hong, S., Iwata, T. (eds.) FSE 2010. LNCS, vol. 6147, pp. 318–332. Springer, Heidelberg (2010)
2. Aumasson, J.-P., Henzen, L., Meier, W., Phan, R.C.-W.: SHA-3 proposal BLAKE. Submission to NIST (2008)
3. Ji, L., Liangyu, X.: Attacks on round-reduced BLAKE. Cryptology ePrint Archive, Report 2009/238 (2009), http://eprint.iacr.org/2009/238.pdf
4. Joux, A., Peyrin, T.: Hash functions and the (amplified) boomerang attack. In: Menezes, A. (ed.) CRYPTO 2007. LNCS, vol. 4622, pp. 244–263. Springer, Heidelberg (2007)
5. Kelsey, J., Kohno, T., Schneier, B.: Amplified boomerang attacks against reduced-round MARS and serpent. In: Schneier, B. (ed.) FSE 2000. LNCS, vol. 1978, pp. 75–93. Springer, Heidelberg (2001)
6. National Institute of Standards and Technology. Cryptographic hash algorithm competition, http://csrc.nist.gov/groups/ST/hash/sha-3/index.html
7. Su, B., Wu, W., Wu, S., Dong, L.: Near-collisions on the reduced-round compression functions of Skein and BLAKE. Cryptology ePrint Archive, Report 2010/355 (2010), http://eprint.iacr.org/2010/355.pdf
8. Wagner, D.: The boomerang attack. In: Knudsen, L.R. (ed.) FSE 1999. LNCS, vol. 1636, pp. 156–170. Springer, Heidelberg (1999)

A   Differential Trails for the Boomerangs

Table 2. Differential trails used in the Boomerang Attack on 4 rounds of BLAKE-32. On the left is the top trail, while on the right is the bottom trail of the boomerang. ΔM is the message difference, while ΔV_i are the differences in the state. In the left trail (top trail), ΔV_0 is the starting difference of the trail, i.e. $\Delta V_0 = \Delta$, and ΔV_2 is the ending difference, i e. $\Delta V_2 = \Delta^*$. In the right trail (bottom trail), ΔV_2 is the starting difference of the trail, i.e. $\Delta V_2 = \nabla$, and ΔV_4 is the ending difference, i.e. $\Delta V_4 = \nabla^*$. The numbers 0,1,2, and 2,3,4, indicate the rounds covered by the boomerang – the top trail starts at round 0 and ends after round 1, while the bottom trail starts at round 2 and ends after round 3.

	Δm			Δm
	00000000 00000000 80000000 00000000			00000000 00000000 00000000 00000000
	00000000 00000000 00000000 00000000			00000000 00000000 00000000 00000000
	00000000 00000000 00000000 00000000			80000000 00000000 00000000 00000000
	00000000 00000000 00000000 00000000			00000000 00000000 00000000 00000000

R.	ΔV_i	R.	ΔV_i
0	00000000 80000000 00000000 00000000	2	80000000 00000000 00000000 00000000
	00000000 00000000 00000000 00000000		00000000 00000000 00000000 00000000
	00000000 00000000 00000000 00000000		00000000 00000000 00000000 00000000
	00000000 00000000 00000000 00000000		80000000 00000000 00000000 00000000
	1		1
1	00000000 00000000 00000000 00000000	3	00000000 00000000 00000000 00000000
	00000000 00000000 00000000 00000000		00000000 00000000 00000000 00000000
	00000000 00000000 00000000 00000000		00000000 00000000 00000000 00000000
	00000000 00000000 00000000 00000000		00000000 00000000 00000000 00000000
	2^{-1}		2^{-1}
2	00000000 80000000 00000000 00000000	4	00000000 00000000 00000000 80000000
	00000000 00000000 00010000 00000000		00010000 00000000 00000000 00000000
	00000000 00000000 00000000 00800000		00000000 00800000 00000000 00000000
	00800000 00000000 00000000 00000000		00000000 00000000 00800000 00000000

Table 3. Differential trails used in the Boomerang Attack on 5 rounds of BLAKE-32

	Δm		Δm
	00000000 00000000 40000000 00000000 00000000 00000000 00000000 00000000 00000000 00000000 00000000 00000000 00000000 00000000 00000000 00000000		00000000 00000000 00000000 00000000 00000000 00000000 00000000 00000000 00000000 00000000 00000000 00000000 00000000 80000000 00000000 00000000
R.	ΔV_i	R.	ΔV_i
0	00000000 40000000 00000000 00000000 00000000 00000000 00000000 00000000 00000000 00000000 00000000 00000000 00000000 00000000 00000000 00000000	2	00000800 80008000 80000000 80000000 80000800 80008000 00000000 00000000 80000000 80808080 80000000 00000000 80000000 00800080 80008000 80000000
	2^{-1}		2^{-6}
1	00000000 00000000 00000000 00000000 00000000 00000000 00000000 00000000 00000000 00000000 00000000 00000000 00000000 00000000 00000000 00000000	3	00000000 00000000 80000000 00000000 00000000 00000000 00000000 00000000 00000000 00000000 00000000 00000000 00000000 00000000 00000000 00000000
	2^{-2}		1
2	00000000 40000000 00000000 00000000 00000000 00000000 00008000 00000000 00000000 00000000 00000000 00400000 00400000 00000000 00000000 00000000	4	00000000 00000000 00000000 00000000 00000000 00000000 00000000 00000000 00000000 00000000 00000000 00000000 00000000 00000000 00000000 00000000
			2^{-1}
		5	00000000 00000000 00000000 80000000 00010000 00000000 00000000 00000000 00000000 00800000 00000000 00000000 00000000 00000000 00800000 00000000

Table 4. Differential trails used in the Boomerang Attack on 6 rounds of CF of BLAKE-32

	Δm		Δm
	00080008 00000000 00000000 00000000 00000000 00000000 00000000 00000000 00000000 00000000 00000000 00000000 00000000 00000000 00000000 00000000		00000000 00000000 00000000 00000000 00000000 00000000 00000000 00000000 00000000 00000000 00000000 00000000 00000000 00000000 80000000 00000000
R.	ΔV_i	R.	ΔV_i
4	80088008 00000000 00080008 00000000 80088008 00000000 00000000 00000000 00080008 00000000 00080008 00000000 00000000 00000000 00000000 00000000	7	80008000 00000000 00000000 00000800 80008000 00000000 00000000 80000800 80808080 80000000 00000000 80000000 00800080 00008000 00000000 80000000
	2^{-21}		2^{-6}
5	00000000 00000000 00080008 00000000 00000000 00000000 00000000 00000000 00000000 00000000 00000000 00000000 00000000 00000000 00000000 00000000	8	00000000 80000000 00000000 00000000 00000000 00000000 00000000 00000000 00000000 00000000 00000000 00000000 00000000 00000000 00000000 00000000
	2^{-2}		1
6	00000000 00000000 00000000 00000000 00000000 00000000 00000000 00000000 00000000 00000000 00000000 00000000 00000000 00000000 00000000 00000000	9	00000000 00000000 00000000 00000000 00000000 00000000 00000000 00000000 00000000 00000000 00000000 00000000 00000000 00000000 00000000 00000000
	2^{-11}		2^{-1}
7	00880088 00000000 00000000 00000000 00000000 11011101 00000000 00000000 00000000 00000000 80088008 00000000 00000000 00000000 00000000 80008000		00000000 80000000 00000000 00000000 00000000 00000000 00010000 00000000 00000000 00000000 00000000 00800000 00800000 00000000 00000000 00000000

Table 5. Differential trails used in the Boomerang Attack on 6.5 and 7 rounds of CF of BLAKE-32

	Δm			Δm	
	00000000 0c000000 00000000 00000000			00000000 00000000 80000000 00000000	
	80008000 0c000000 00000000 00000000			00000000 00000000 00000000 00000000	
	00000000 0c000000 00000000 00000000			00000000 00000000 00000000 00000000	
	00000000 0c000000 00000000 00000000			00000000 00000000 00000000 00000000	
R.	ΔV_i		R.	ΔV_i	
0.5	00000000 8c008000 00000000 00000000		3.5	00800880 c8088848 80440044 00008000	
	00000000 0c000000 00000000 00000000			80000000 80800880 488c0888 80040804	
	00000000 0c000000 00000000 80008000			00000800 00008080 80808080 00000000	
	00000000 0c000000 00000000 00000000			80048040 08408840 00800000 80000000	
	2^{-3}			2^{-41}	
1	00000000 8c008000 00000000 00000000		4	80000000 00000000 80000800 80008000	
	00000000 0c000000 00000000 00000000			00000000 00000000 80000800 80008000	
	00000000 0c000000 00000000 00000000			80000000 00000000 80000000 80808080	
	00000000 0c000000 00000000 00000000			80008000 00000000 80000000 00800080	
	2^{-1}			2^{-6}	
2	00000000 0c000000 00000000 00000000		5	80000000 00000000 00000000 00000000	
	00000000 0c000000 00000000 00000000			00000000 00000000 00000000 00000000	
	00000000 0c000000 00000000 00000000			00000000 00000000 00000000 00000000	
	00000000 0c000000 00000000 00000000			00000000 00000000 00000000 00000000	
	2^{-3}			1	
3	00000000 0c000000 00000000 80008000		6	00000000 00000000 00000000 00000000	
	00010001 0c000000 00000000 00000000			00000000 00000000 00000000 00000000	
	00000000 0c800080 00000000 00000000			00000000 00000000 00000000 00000000	
	00000000 0c000000 00800080 00000000			00000000 00000000 00000000 00000000	
	2^{-33}			2^{-1}	
3.5	00010001 08000800 08000800 80088008		7	00000000 00000000 80000000 00000000	
	02000200 10111011 11111111 11101110			00000000 00000000 00000000 00010000	
	00010001 00880088 80888088 88008800			00800000 00000000 00000000 00000000	
	00000000 0c080008 80088008 08000800			00000000 00800000 00000000 00000000	
				2^{-24}	
			7.5	00000800 08000000 80000008 00110010	
				10010010 01101001 10110101 22222022	
				00800008 80080080 08808080 11001101	
				00000008 80080000 08800080 11001100	

Table 6. Differential trails used in the Boomerang Attack on 6 rounds of KP of BLAKE-32

	Δm		Δm
	00000000 00000000 00000000 00000000		00000000 00000000 00000000 00000000
	80000000 00000000 00000000 00000000		00000000 80000000 00000000 00000000
	00000000 00000000 00000000 00000000		00000000 00000000 00000000 00000000
	00000000 00000000 00000000 00000000		00000000 00000000 00000000 00000000
R.	ΔV_i	R.	ΔV_i
0	80008000 80000000 80000000 00000800	3	80008000 00000000 00000000 00000800
	80008000 00000000 00000000 80000800		80008000 00000000 00000000 80000800
	80808080 80000000 00000000 80000000		80808080 00000000 00000000 80000000
	00800080 80008000 00000000 80000000		00800080 00000000 00000000 80000000
	2^{-6}		2^{-6}
1	00000000 80000000 00000000 00000000	4	00000000 80000000 00000000 00000000
	00000000 00000000 00000000 00000000		00000000 00000000 00000000 00000000
	00000000 00000000 00000000 00000000		00000000 00000000 00000000 00000000
	00000000 00000000 00000000 00000000		00000000 00000000 00000000 00000000
	1		1
2	00000000 00000000 00000000 00000000	5	00000000 00000000 00000000 00000000
	00000000 00000000 00000000 00000000		00000000 00000000 00000000 00000000
	00000000 00000000 00000000 00000000		00000000 00000000 00000000 00000000
	00000000 00000000 00000000 00000000		00000000 00000000 00000000 00000000
	2^{-1}		2^{-1}
3	00000000 00000000 00000000 80000000	6	00000000 80000000 00000000 00000000
	00010000 00000000 00000000 00000000		00000000 00000000 00001000 00000000
	00000000 00800000 00000000 00000000		00000000 00000000 00000000 00800000
	00000000 00000000 00800000 00000000		00800000 00000000 00000000 00000000

Table 7. Differential trails used in the Boomerang Attack on 7 and 8 rounds of KP of BLAKE-32

	Δm				Δm		
	00000000	00000000	00000000 00000000		00000000	00000000	00000000 00000000
	00000000	00000000	00000000 00000000		00000000	00000000	00000000 80000000
	00000000	00000000	00000000 00000000		00000000	00000000	00000000 00000000
	00000000	80000000	00000000 00000000		00000000	00000000	00000000 00000000
R.	ΔV_i			R.	ΔV_i		
1.5	80440044	00008000	80800880 48088848	5.5	80808000	80888080	c80c8008 80440044
	488c0888	00040804	80000000 80800880		80040804	80800000	80888000 c8880088
	80808080	80000000	00000880 00008080		80000000	00000800	00808080 80000080
	00800000	80000000	00040040 88c00840		00800000	00048000	08408840 80800000
	2^{-42}				2^{-44}		
2	00000800	80008000	80000000 80000000	6	00008000	80000000	00000000 80000800
	80000800	80008000	00000000 00000000		80008000	00000000	00000000 00000800
	80000000	80808080	80000000 00000000		00808080	80000000	00000000 00000080
	80000000	00800080	80008000 80000000		80808080	80008000	00000000 80800000
	2^{-6}				2^{-7}		
3	00000000	0c000000	80000000 00000000	7	00000000	80000000	00000000 00000000
	00000000	0c000000	00000000 00000000		00000000	00000000	00000000 00000000
	00000000	0c000000	00000000 00000000		00000000	00000000	00000000 00000000
	00000000	0c000000	00000000 00000000		00000000	00000000	00000000 00000000
	1				1		
4	00000000	0c000000	00000000 00000000	8	00000000	00000000	00000000 00000000
	00000000	0c000000	00000000 00000000		00000000	00000000	00000000 00000000
	00000000	0c000000	00000000 00000000		00000000	00000000	00000000 00000000
	00000000	0c000000	00000000 00000000		00000000	00000000	00000000 00000000
	2^{-1}				2^{-1}		
5	00000000	0c000000	00000000 80000000	9	00000000	80000000	00000000 00000000
	00010000	0c000000	00000000 00000000		00000000	00000000	00010000 00000000
	00000000	0c800000	00000000 00000000		00000000	00000000	00000000 00800000
	00000000	0c000000	00800000 00000000		00800000	00000000	00000000 00000008
	2^{-24}				2^{-30}		
5.5	00110010	00000800	08000000 80000008	9.5	08000000	80000008	80110018 00000800
	22222022	10010010	01101001 10110101		01101001	10110101	32332123 10010010
	11001101	00800008	80080080 08808080		80080080	08808080	19809181 00800008
	11001100	00000008	80080000 08800080		80080000	08800080	19801180 00000008

B   Examples of Boomerang quartets

Table 8. Example of a boomerang quartet for 6 round-reduced keyed permutation of BLAKE-32

P_1	7d8a1f02	206849ad	42413a50	d702fa14	facc9c67	11306e7c	eba852eb	4f31f62f
	993e3958	bc426fcc	55033261	b2ac26a9	6dfc2edd	32163c44	ef989577	2d6d6bb4
P_2	fd8a9f02	a06849ad	c2413a50	d702f214	7acc1c67	11306e7c	eba852eb	cf31fe2f
	19beb9d8	3c426fcc	55033261	32ac26a9	6d7c2e5d	b216bc44	ef989577	ad6d6bb4
P_3	de971194	ae012c6a	4422f8ea	fff2d41b	80a79b50	b1d61b36	fe8c23fe	a883faf9
	e1dab487	e4971af1	51dbf40b	6e32fb27	7c797796	19b156e9	16e0ac52	a12eefcb
P_4	5e979194	2e012c6a	c422f8ea	fff2dc1b	00a71b50	b1d61b36	fe8c23fe	2883f2f9
	615a3407	64971af1	51dbf40b	ee32fb27	7cf97716	99b1d6e9	16e0ac52	212eefcb
$P_1 \oplus P_2$	80008000	80000000	80000000	00000800	80008000	00000000	00000000	80000800
	80808080	80000000	00000000	80000000	00800080	80008000	00000000	80000000
$P_3 \oplus P_4$	80008000	80000000	80000000	00000800	80008000	00000000	00000000	80000800
	80808080	80000000	00000000	80000000	00800080	80008000	00000000	80000000
M_1	a0a28e67	1fd77849	83d86d19	4a72bc82	3704f04d	bb57c994	37612239	0f7ad68a
	df14386d	4e2e05c7	55d1a87f	187d8225	fcc527c5	96071c3e	4ae251d8	52de23f2
M_2	a0a28e67	1fd77849	83d86d19	4a72bc82	b704f04d	bb57c994	37612239	0f7ad68a
	df14386d	4e2e05c7	55d1a87f	187d8225	fcc527c5	96071c3e	4ae251d8	52de23f2
M_3	a0a28e67	1fd77849	83d86d19	4a72bc82	3704f04d	3b57c994	37612239	0f7ad68a
	df14386d	4e2e05c7	55d1a87f	187d8225	fcc527c5	96071c3e	4ae251d8	52de23f2
M_4	a0a28e67	1fd77849	83d86d19	4a72bc82	b704f04d	3b57c994	37612239	0f7ad68a
	df14386d	4e2e05c7	55d1a87f	187d8225	fcc527c5	96071c3e	4ae251d8	52de23f2
$M_1 \oplus M_2$	00000000	00000000	00000000	00000000	80000000	00000000	00000000	00000000
	00000000	00000000	00000000	00000000	00000000	00000000	00000000	00000000
$M_1 \oplus M_3$	00000000	00000000	00000000	00000000	00000000	80000000	00000000	00000000
	00000000	00000000	00000000	00000000	00000000	00000000	00000000	00000000
$M_2 \oplus M_4$	00000000	00000000	00000000	00000000	00000000	80000000	00000000	00000000
	00000000	00000000	00000000	00000000	00000000	00000000	00000000	00000000
C_1	928c1f77	3aa097f2	4d5589bb	f307e618	c8ea4ebc	c63769df	64e2b7ba	f2c76b2b
	c909808a	672bcdf3	260608d6	7de7ba36	749c4e7d	aef2defd	b7d3318a	5080389e
C_2	9948791c	21c19a0f	8804efac	d56588e4	c6f6b101	32456224	20c423d5	df0105fe
	33ee8883	23bde21d	bedb2451	2c673c2f	bf7d194d	cfc78321	5ec259f9	a9c8786b
C_3	928c1f77	baa097f2	4d5589bb	f307e618	c8ea4ebc	c63769df	64e3b7ba	f2c76b2b
	c909808a	672bcdf3	260608d6	7d67ba36	741c4e7d	aef2defd	b7d3318a	5080389e
C_4	9948791c	a1c19a0f	8804efac	d56588e4	c6f6b101	32456224	20c523d5	df0105fe
	33ee8883	23bde21d	bedb2451	2ce73c2f	bffd194d	cfc78321	5ec259f9	a9c8786b
$C_1 \oplus C_3$	00000000	80000000	00000000	00000000	00000000	00000000	00010000	00000000
	00000000	00000000	00000000	00800000	00800000	00000000	00000000	00000000
$C_2 \oplus C_4$	00000000	80000000	00000000	00000000	00000000	00000000	00010000	00000000
	00000000	00000000	00000000	00800000	00800000	00000000	00000000	00000000

Table 9. Example of a boomerang quartet for 6 round-reduced compression function of BLAKE-32. Note that the initial states P_1, P_2 are consistent with the *Initialization*.

P_1	30841585	41abc330	447466d0	17ae8472	b94fc56d	e9cb678a	1d9d6e9e	eb558123
	66d322c2	23cbae19	52e9bb2a	dd6b8f2b	ea1cd197	678ad865	6594bdd4	81f42bc5
P_2	b08c958d	41abc330	447c66d8	17ae8472	39474565	e9cb678a	1d9d6e9e	eb558123
	66db22ca	23cbae19	52e1bb22	dd6b8f2b	ea1cd197	678ad865	6594bdd4	81f42bc5
P_3	f3383666	710fc071	1990f347	34475dd7	7d41ddc9	68e231ed	ea9bba79	a4990860
	d7ede8b5	f1c0b054	1c754989	a0e95ceb	3d259f5f	878bffae	f511b0fd	def26a26
P_4	7330b66e	710fc071	1998f34f	34475dd7	fd495dc1	68e231ed	ea9bba79	a4990860
	d7e5e8bd	f1c0b054	1c7d4981	a0e95ceb	3d259f5f	878bffae	f511b0fd	def26a26
$P_1 \oplus P_2$	80088008	00000000	00080008	00000000	80088008	00000000	00000000	00000000
	00080008	00000000	00080008	00000000	00000000	00000000	00000000	00000000
$P_3 \oplus P_4$	80088008	00000000	00080008	00000000	80088008	00000000	00000000	00000000
	00080008	00000000	00080008	00000000	00000000	00000000	00000000	00000000
M_1	7670ae70	c6539713	373c66b6	3d4522c3	b66689d0	37ee4f5d	467de620	9aabd357
	b6b3b13c	c6d41a4c	cb994b4c	b79e16fa	8a9d8079	9914ccb1	9c68b051	86d41e1e
M_2	7678ae78	c6539713	373c66b6	3d4522c3	b66689d0	37ee4f5d	467de620	9aabd357
	b6b3b13c	c6d41a4c	cb994b4c	b79e16fa	8a9d8079	9914ccb1	9c68b051	86d41e1e
M_3	7670ae70	c6539713	373c66b6	3d4522c3	b66689d0	37ee4f5d	467de620	9aabd357
	b6b3b13c	c6d41a4c	cb994b4c	b79e16fa	8a9d8079	9914ccb1	1c68b051	86d41e1e
M_4	7678ae78	c6539713	373c66b6	3d4522c3	b66689d0	37ee4f5d	467de620	9aabd357
	b6b3b13c	c6d41a4c	cb994b4c	b79e16fa	8a9d8079	9914ccb1	1c68b051	86d41e1e
$M_1 \oplus M_2$	00080008	00000000	00000000	00000000	00000000	00000000	00000000	00000000
	00000000	00000000	00000000	00000000	00000000	00000000	00000000	00000000
$M_1 \oplus M_3$	00000000	00000000	00000000	00000000	00000000	00000000	00000000	00000000
	00000000	00000000	00000000	00000000	00000000	00000000	80000000	00000000
$M_2 \oplus M_4$	00000000	00000000	00000000	00000000	00000000	00000000	00000000	00000000
	00000000	00000000	00000000	00000000	00000000	00000000	80000000	00000000
C_1	3f432ef6	5f89fb80	7283d8cf	13731945	344d16f8	2203b3b5	74b3637e	52ed9169
	efcea8db	32b84ffc	57cfa772	2258156c	22696ef4	53cb7ac6	3ab6294a	ce58038c
C_2	f284e034	f866e60d	1e52775f	f6f764cb	ef09e2e8	da83b2d1	a4a869d1	f22eefb0
	821c38c2	6da245e0	7b52665c	0f8ce3ba	7ed4c20c	ef76217d	77835c6d	184a17e3
C_3	3f432ef6	df89fb80	7283d8cf	13731945	344d16f8	2203b3b5	74b2637e	52ed9169
	efcea8db	32b84ffc	57cfa772	22d8156c	22e96ef4	53cb7ac6	3ab6294a	ce58038c
C_4	f284e034	7866e60d	1e52775f	f6f764cb	ef09e2e8	da83b2d1	a4a969d1	f22eefb0
	821c38c2	6da245e0	7b52665c	0f0ce3ba	7e54c20c	ef76217d	77835c6d	184a17e3
$C_1 \oplus C_3$	00000000	80000000	00000000	00000000	00000000	00000000	00010000	00000000
	00000000	00000000	00000000	00800000	00800000	00000000	00000000	00000000
$C_2 \oplus C_4$	00000000	80000000	00000000	00000000	00000000	00000000	00010000	00000000
	00000000	00000000	00000000	00800000	00800000	00000000	00000000	00000000

Practical Near-Collisions
on the Compression Function of BMW

Gaëtan Leurent[1] and Søren S. Thomsen[2],[⋆]

[1] University of Luxembourg
gaetan.leurent@uni.lu
[2] Technical University of Denmark
s.thomsen@mat.dtu.dk

Abstract. Blue Midnight Wish (BMW) is one of the fastest SHA-3 candidates in the second round of the competition. In this paper we study the compression function of BMW and we obtain practical partial collisions in the case of BMW-256: we show a pair of inputs so that 300 pre-specified bits of the outputs collide (out of 512 bits). Our attack requires about 2^{32} evaluations of the compression function. The attack can also be considered as a near-collision attack: we give an input pair with only 122 active bits in the output, while generic algorithm would require 2^{55} operations for the same result. A similar attack can be developed for BMW-512, which will gives message pairs with around 600 colliding bits for a cost of 2^{64}. This analysis does not affect the security of the iterated hash function, but it shows that the compression function is far from ideal.

We also describe some tools for the analysis of systems of additions and rotations, which are used in our attack, and which can be useful for the analysis of other systems.

1 Introduction

Blue Midnight Wish (BMW) is a candidate in the SHA-3 hash function competition [7] which made it to the second round of the competition, but was not selected as a finalist. It is one of the fastest second round candidates in software, and belongs to the ARX family, using only additions, rotations, and xors.

BMW is built by iterating a compression function, similarly to the ubiquitous Merkle-Damgård paradigm [5, 9]. More precisely, BMW uses a chaining value twice as large as the output of the hash function (this is known as wide-pipe, or Chop-MD), and uses a final transformation similar to the HMAC construction. There are several security proofs for this mode of operation and similar modes [2–4], which essentially show that if the compression function behaves like a random function, then the hash function will behave like a random function (up to some level determined by the width of the chaining variable).

In this paper we explain how to find partial-collisions in the BMW-256 compression function. The same technique could be used to find partial-collisions in

---

[⋆] Supported by a grant from the Villum Kann Rasmussen Foundation.

A. Joux (Ed.): FSE 2011, LNCS 6733, pp. 238–251, 2011.

the BMW-512 compression function, but the complexity would be too high to carry out the attack in a reasonable amount of time, and so we have not implemented this attack. The attacks are not affected by the value of the security parameter of BMW.

1.1 Compression Function Attacks

A natural step in the analysis of iterated hash functions is to study the compression function. Most attacks on the compression function do not lead to attacks on the iterated hash function, but they can invalidate the assumptions of the security proofs. This does not weaken the hash function in itself, but it can undermine the confidence in the design, because the security of the hash function is no longer a consequence of a simple assumption (namely the security of the compression function).

Recently, new results have shown that some attacks on the compression function can be integrated inside the security proof of the mode of operation [2]. This shows that the security of the hash function does not need a truly perfect compression function: some classes of weaknesses of the compression function cannot be used to attack the iterated hash function. As a general rule, it seems that most attacks that require control over the chaining value can be covered by this kind of proofs. However, those attacks usually reveal some unwanted properties of the function, and might be extended to attacks on the full hash function using more advanced techniques.

To put such attacks into perspective, one might look at the attacks on MD5. The first attack on the compression function was found in 1993 by den Boer and Bosselaers [6], using a very simple differential path. This attack did not threaten the iterated hash function, but the path used in the attack is a core element of the successful attack of Wang et al. in 2005 [12].

1.2 Description of BMW

BMW comes in four variants BMW-n, with $n \in \{224, 256, 384, 512\}$, returning output size n. There are two variants of the BMW compression functions; The BMW-256 compression function is used in both BMW-224 and BMW-256, and the BMW-512 compression function is used in both BMW-384 and BMW-512.

The compression function of BMW-256 takes two inputs, H and M, of 16 32-bit words each. The general structure of the compression function is shown in Figure 1. It consists of three functions named f_0, f_1, f_2. The function f_0 applies an invertible linear transformation P to $H \oplus M$ and adds H wordwise modulo 2^{32}. We denote by '$\boxplus$' modular addition, by '$\boxminus$' modular subtraction, and by '$\oplus$' the exclusive or. The output of f_0 is a 16-word vector Q. P consists of a matrix multiplication over $\mathbb{Z}_{2^{32}}$, followed by linear functions $s_{i \bmod 5}$ (see Appendix A) applied to each word W_i individually and by a wordwise rotation by 1 position ($W_{i+1} \leftarrow W_i$).

The function f_1 is a feedback shift register. To begin with, the vector Q contains 16 elements; in each one of 16 rounds of f_1, one more element is added

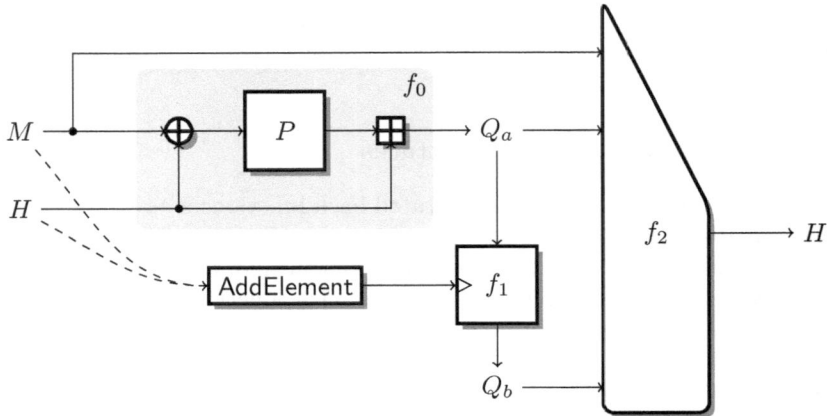

Fig. 1. Compression function of BMW

to Q. This element is computed from the previous 16 elements of Q, and from a value called AddElement(i) (where i is the round number $+ 16$), which is the following function of three words of M and one word of H:

AddElement(i) =
$$(M_i^{\lll 1+(i \bmod 16)} \boxplus M_{i+3}^{\lll 1+(i+3 \bmod 16)} \boxminus M_{i+10}^{\lll 1+(i+10 \bmod 16)} \boxplus K_i) \oplus H_{i+7}$$

(all indices are to be taken modulo 16, and K_i is a round constant). We note that if there is a collision in the output of f_0 and also in the first, say, j instances of AddElement(i), then there is a collision in the first $16 + j$ elements of Q. We denote by Q_a the output of f_0, and by Q_b the 16 elements computed in f_1.

The function f_2 performs some final mixing of the elements in Q with M, and produces the 16-word output of the compression function.

Further details on the compression function of BMW can be found in [7].

1.3   Previous Results

During the first round of the SHA-3 competition, the best attacks on BMW have been pseudo-attacks due to Thomsen [11]. However, BMW was quite heavily tweaked at the end of the first round, and those attacks do not apply to the current version of BMW. In this paper we only consider the second-round version of BMW.

For the current version of BMW, the best results are differential properties of the compression function, due to Aumasson and Guo and Thomsen [1, 8]. These papers essentially show that for some particular differences in the input of the compression function, a few output bits will be biased.

1.4 Our Results

In this paper we describe a partial-collision attack on the compression function of BMW. Our attack is based on differential techniques, and we try to control the propagation of differences inside the compression function. The general idea is to control the differences in M, in Q_a, and in the first instances of AddElement. This means that we also control the difference in the first elements of Q_b, and since the final function f_2 only has limited diffusion, we will control the differences in several output words. We managed to cancel all the differences in $Q_0, \ldots Q_{26}$, and to get small differences in Q_{27}, Q_{28}, Q_{29}. This gives a pair of inputs such that 300 pre-specified output bits collide, for a cost similar to 2^{32} evaluations of the compression function, using negligible memory. We note that for a random function, it is expected to take 2^{150} evaluations before finding such a pair of inputs. Moreover, we expect a difference in only half of the uncontrolled bits, and this gives a near-collision attack better than generic algorithms.

Before describing our new attack, we present some useful tools for the analysis of ARX systems in Section 2. In Section 3, we show how to obtain collisions in f_0 without any message difference, which leads to collisions in Q_a, the first half of the Q register. We then show how to find such collisions with some words of H inactive, which leads to a collision in Q_0 to Q_{22}. In Section 4, we extend this result by introducing some differences in the message, and we use the message differences to cancel the chaining value differences in the AddElement function up to Q_{26}. Finally in Section 5 we use near collisions in AddElement instead of full collisions, and we can control the differences up to Q_{29}.

2 Solving a System of Additions and Xor

An important step in our attack requires to solve a system of equations involving only xors and modular additions. In particular, we will often have to solve $x \oplus \Delta = x \boxplus \delta$, where x is a variable, and Δ and δ are given parameters representing respectively the xor-difference and the modular-difference in x. It is well-known that those systems are T-functions, and can be solved from the least significant bit to the most significant bit. However, the naive approach to solve such a system uses backtracking, and can lead to an exponential complexity in the worst case.[1] A more efficient strategy is to use an approach based on automata: any system of such equations can be represented by an automaton, and solving a particular instance take time proportional to the word length. This kind of approach has been used to study differential properties of S-functions in [10]. Here we use this technique to decide whether a system is solvable, and to compute a solution efficiently.

We consider a system of additions and xors, which involves v variables and p parameters. Our goal is twofold: first determine for which values of the parameters the system is compatible, and second, when the system is compatible, determine the set of solutions.

---

[1] e.g., to solve the system $x \oplus \texttt{0x80000000} = x$, the backtracking algorithm will try all possible values for the 31 lower bits of x before concluding that there is no solution.

The first step in applying this technique is to build an automaton corresponding to the system of equations. The states of this automaton correspond to the possible values of the carry bits: a system with s modular additions gives an automaton with 2^s states. The alphabet is $\{0,1\}^{v+p}$, and each transition reads one bit from each parameter and each variable, starting from the least significant bit. Figure 2 shows an example of such automaton.

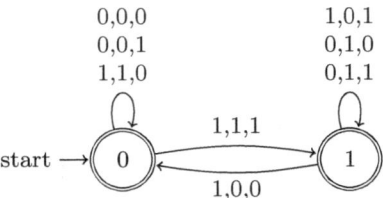

Fig. 2. Carry transitions for $x \oplus \Delta = x \boxplus \delta$. The edges are indexed by Δ, δ, x

Then we remove the variables from the edges, and this gives a non-deterministic automaton which can decide whether a system is solvable or not. We can then build an equivalent deterministic automaton using the powerset construction, as shown in Figure 3.

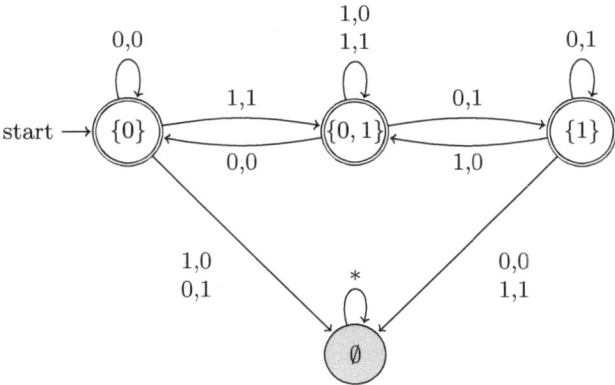

Fig. 3. Decision automaton for $x \oplus \Delta = x \boxplus \delta$. The edges are indexed by Δ, δ

This automaton reveals a lot of information about the system of equation. For instance, one can see that if the state $\{0,1\}$ is reached, then setting $\Delta_i = 1$ assures that the system will have a solution.

In the case of the simple system $x \oplus \Delta = x \boxplus \delta$, we can find an extremely efficient way to check the satisfiability of the system for given parameters, and to find the actual solutions. By looking at Figure 3, we see that the state $\{0\}$ can only be reached as the initial state or after reading $0, 0$, and that the state $\{1\}$ can only be reached after reading $0, 1$. Moreover, reading $0, 0$ can only lead to state $\{0\}$ or $\emptyset$ and reading $0, 1$ can only lead to state $\{1\}$ or $\emptyset$. This allows a very simple description of the parameters that lead to an inconsistent system, i.e., that reach state $\emptyset$:

- $\Delta_0 \neq \delta_0$, or
- one of the following patterns is seen: $(0, 0), (1, 0)$; $(0, 0), (0, 1)$; $(0, 1), (0, 0)$; $(0, 1), (1, 1)$.

The second condition can be expressed as:

$$\exists i : \Delta_i = 0 \quad \text{and} \quad \delta_i \oplus \Delta_{i+1} \oplus \delta_{i+1} = 1$$

Since those conditions are local they can be tested in parallel using bitwise operations. The following C expression evaluates to one if the system is incompatible, and to zero if it is compatible:

```
((⊃^d)&1) || ((((D^d)>>1)^d) & (~D)) << 1
```

Note that the rotation to the left is just used to ignore the MSB of the second expression.

Given a compatible pair (Δ, δ), we can use the automaton in Figure 2 to compute a solution x to the equation $x \oplus \Delta = x \boxplus \delta$. First we can remark that if we are in state 0, the next inputs have to satisfy $\delta_i = \Delta_i$, while if we are in state 1, the next inputs have to satisfy $\delta_i \neq \Delta_i$. We can now express the possibles values for x depending on δ and Δ, by looking at the possible transitions in the automata:

$$\begin{cases} \text{if } (\Delta_i, \delta_i) = (0, 0) & \text{then } x_i \text{ is arbitrary:} & x_i \in \{0, 1\} \\ \text{if } (\Delta_i, \delta_i) = (0, 1) & \text{then } x_i \text{ is arbitrary:} & x_i \in \{0, 1\} \\ \text{if } (\Delta_i, \delta_i) = (1, 0) & \text{then } x_i \text{ is given by the next state:} & x_i = \delta_{i+1} \oplus \Delta_{i+1} \\ \text{if } (\Delta_i, \delta_i) = (1, 1) & \text{then } x_i \text{ is given by the next state:} & x_i = \delta_{i+1} \oplus \Delta_{i+1} \end{cases}$$

This can be expressed by the following C expression, where r is a random value:

```
(D^d)>>1 ^ (r&(~D|0x8000000))
```

3 Using Collisions in f_0

The first step of our attack is to build collisions in f_0 without any message difference. In the following we denote $x = H \oplus M$ and $y = P(H \oplus M)$. We have $f_0(H, M) = y \boxplus H$ (see Figure 4).

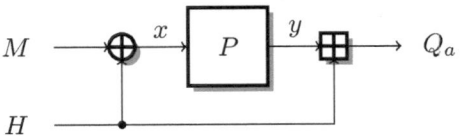

Fig. 4. BMW f_0 function

We propose the following algorithm to find such collisions:

1. Pick a random x, x'.
2. Compute $y = P(x), y' = P(x')$.
3. We have $H \oplus H' = x \oplus x'$ and $H \boxminus H' = y' \boxminus y$. We can solve this and find H using the tools of Section 2.
4. Compute M from x and H: $M = x \oplus H$.

On average, for a random x, x', we expect one solution. However, there is a high probability that there will be no solution for a given x, x', because the xor-difference and the mod-difference for H will not be compatible. (Experiments suggests there is a probability around $2^{-13.9}$ for random differences to be compatible).

To find collisions in practice, we use the degrees of freedom in x to set an xor-difference that has a better probability than a random difference. The best choice is $x' = \neg x$, which works with probability 2^{-1} for each word. However, due to the structure of P, this leads to incompatible systems (the differences in y are constrained by the difference in x). Therefore we use differences of high weight, but we leave some low order bits inactive. This allows to find a compatible system after a few choices of x, x'.

3.1   Collisions in f_0 with Some Words of H Inactive

The next step is to find collisions where some of the words of H are inactive. This will lead to some instances of AddElement being inactive, and some words of Q_b being inactive.

To achieve this, we need an x, x' with some inactive words, but we also require that the same words are inactive in y, y'. Since the inter-word mixing of P is achieved by a linear transformation over $\mathbb{Z}_{2^{32}}$, we can easily find a suitable mod-difference in x. Then we can build the pair x, x' by extending the carries, so that the xor-difference in x is of high Hamming weight.

We use the following algorithm:

1. Pick a random mod-difference in the kernel of the linear transformation P.
2. Build x, x' by extending the carries as much as possible.
3. Compute y, y'.
4. Solve for H.

We can have up to 7 inactive words in H. We use $H_7 \ldots H_{13}$ because they are used in the first 7 AddElement rounds. This gives a collision in $Q_0 \ldots Q_{22}$.

Once we have a solution, we can modify the values of $H_7 \ldots H_{13}$ to generate new solutions by adjusting $M_7 \ldots M_{13}$ (we have to keep the value of x). This can be used to get a small difference in Q_{23} as well.

Each choice of x, x' gives a new value for the xor-difference and the mod-difference in H, and we use the tools of Section 2 to check very efficiently whether those values are compatible. The cost of finding a compatible system is negligible before the cost of 2^{32} that we require in order to put a small difference in Q_{23}. This gives some restrictions on the output of the compression function because the f_2 function has 16 outputs and only 9 active inputs, but we do not have any colliding outputs yet.

4   Using Partial-Collisions in f_0

In order to improve this result, and to have stronger properties on the output, we have to make more values of Q_b collide. To achieve this, we now put some differences in M, so that differences in M and H can cancel each other in the AddElement function.

More precisely, our best path allows to find collisions in $Q_0 \ldots Q_{26}$ using:

- differences in M_{13}, M_{14}, M_{15};
- differences in $H_1 \ldots H_6, H_{10}, H_{11}$ and H_{12}.

The first step of the attack is to choose a pair x, x' such that $x_{0,7,8,9}$ are inactive, and $y_{0,7,3,9,13,14,15}$ are inactive. Moreover, we fix three more differences: $\delta^{\boxplus}x_{13} = 2^{18}$, $\delta^{\boxplus}y_1 = \text{0x04010c43}$, and $\delta^{\boxplus}x_1 = 1$. This is used in order to have $\delta^{\oplus}x_{13} = 2^{18}$, $\delta^{\oplus}y_1 = 1$ (note that $s_0(\text{0x04010c43}) = 1$), and $\delta^{\oplus}x_1 = 1$. This gives 14 constraints so we have a solution space of dimension 2.

After fixing the modular difference in x and y, we choose the values of x, x' by extending the carries as much as possible, in order to have a dense xor-difference in M_{14}, M_{15} and $H_2 \ldots H_6, H_{10}, H_{11}$ and H_{12}. On the other hand, we keep the difference in M_{13} and H_1 sparse so as to have $\delta^{\oplus}x_{13} = 2^{18}$ and $\delta^{\oplus}y_1 = 1$.

When we find a pair x, x' with compatible differences for H, this fixes the values of:

- all active H's: $H_1 \ldots H_6, H_{10}, H_{11}, H_{12}$.
- all M's whose corresponding H is active: $M_1 \ldots M_6, M_{10}, M_{11}, M_{12}$.

The remaining degrees of freedom are:

- H_0, H_7, H_8, H_9 and M_0, M_7, M_8, M_9, but the values of $H_i \oplus M_i = x_i$ are fixed (4 degrees of freedom).
- H_{13}, H_{14}, H_{15}; M_{13}, M_{14}, M_{15}; and $M'_{13}, M'_{14}, M'_{15}$, but the values of $H_i \oplus M_i = x_i$ and $M_i \oplus M'_i = x_i \oplus x'_i$ are fixed (3 degrees of freedom).

In order to achieve a collision in $Q_0 \ldots Q_{26}$, we need to cancel differences in AddElement 19, 20, 21 and 26.

AddElement(19) $(M_3^{\lll 4} \boxplus M_6^{\lll 7} \boxminus M_{13}^{\lll 14} \boxplus K_{19}) \oplus H_{10}$
We use the freedom of M_{13} to extend carries in $(M_3^{\lll 4} \boxplus M_6^{\lll 7} \boxminus M_{13}^{\lll 14})$.

AddElement(20) $(M_4^{\lll 5} \boxplus M_7^{\lll 8} \boxminus M_{14}^{\lll 15} \boxplus K_{20}) \oplus H_{11}$

We use the freedom of M_{14} and M_7 to extend carries in $(M_4^{\lll 5} \boxplus M_7^{\lll 8} \boxminus M_{14}^{\lll 15})$.

AddElement(21) $(M_5^{\lll 6} \boxplus M_8^{\lll 9} \boxminus M_{15}^{\lll 16} \boxplus K_{21}) \oplus H_{12}$

We use the freedom of M_{15} and M_8 to extend carries in $(M_5^{\lll 6} \boxplus M_8^{\lll 9} \boxminus M_{15}^{\lll 16})$.

AddElement(26) $(M_{10}^{\lll 11} \boxplus M_{13}^{\lll 14} \boxminus M_4^{\lll 5} \boxplus K_{26}) \oplus H_1$

We don't have degrees of freedom available to make this collide, but the differences have been selected so that it happens with high probability: the differences in $M_{13}^{\lll 14}$ and H_1 are only in the least significant bit.

Finally, when we have one solution which collides in $Q_0 \ldots Q_{26}$, we use the freedom in M_0 and M_9 to generate many solutions, until we have a collision in $XH = \bigoplus_{i=16}^{31} Q_i$. This gives a collision in the first three output words for a cost of 2^{32} (see Appendix B for details of the f_2 function). Here is an example of an input pair showing this property is given in Table 1

Table 1. Partial-collision example with 96 controlled bits

Chaining Value							
6ae0a10c	4f14abca	57e66e71	6075a601	6ae0a10c	4f14abcb	a819918f	9f8a59fe
bba141a1	46fb0506	e001fffd	e89b2ebf	445ebe5f	8934faf9	9ffe0002	e89b2ebf
cb1e82d3	ae2d53d6	cb55b67f	e6b080a1	cb1e82d3	ae2d53d6	34aa4980	194f7f5e
8b8c0a70	98d0080b	adaacc99	88f0cf2d	7473f58f	98d0080b	adaacc99	88f0cf2d
Message							
4f5381d3	f96e7f0a	72879df2	e8150fa2	4f5381d3	f96e7f0a	72879df2	e8150fa2
476caf9f	fbacf685	d1c47cb8	73a7bf61	476caf9f	fbacf685	d1c47cb8	73a7bf61
445261cf	a4c0f69f	a2316fdd	12dbc43a	445261cf	a4c0f69f	a2316fdd	12dbc43a
e5197bf4	af952392	c2966021	46cab397	e5197bf4	af992392	3d699fde	39354c6b
Output							
fe57177e	d1e1157d	ccf82758	6aecc4d0	fe57177e	d1e1157d	ccf82758	80b0c87d
cf3d27ab	590788dc	eafe31d9	0e95fe74	0f1b49b9	e0b92229	cf1c1fb4	1fd1f3ab
5b069cc1	b1039e9e	a5049da0	c38e8490	174ab741	7768d4bc	947374c1	74ddf4f9
cb6f569c	96fff629	ee5d89a4	71e405a4	8b4d7466	d075a056	0f8d8b0c	d987e0cb

5   Using Near Collisions in AddElement

In order to extend the attack with more colliding bits in the output of the compression function, we use near-collisions in the next instances of AddElement. Since we do not have enough remaining degrees of freedom with a given pair x, x', we use freedom in the choice of the x, x' pair in order to go further.

In a practical implementation of the attack, one computes some words of H as a solution to the equation $x \oplus \Delta = x \boxplus \delta$ as described in the previous sections. As an example, in order to find H_5 and H_5' such that $H_5' \oplus H_5 = \delta^\oplus x_5$ and $H_5' \boxminus H_5 = \delta^\boxplus y_5$, one may compute H_5 as $(\delta^\oplus x_5 \oplus \delta^\boxplus y_5)^{\ggg 1}$ (assuming

Table 2. Partial-collision example with 300 controlled bits

Chaining Value							
59dfd94b	30b036e3	44ad8a65	47461712	59dfd94b	30b036e2	bb52759b	b8b9e8ed
6f56e9b4	425e2d65	40000003	94e62f58	90a9164c	bda1d29a	bffffffc	94e62f58
12c4bf76	17b18302	4f74ffd3	3ec30f93	12c4bf76	17b18302	b08b002c	c13cf06c
8b0f9f9b	7071a4a5	28becf17	6954724f	74f06064	7071a4a5	28becf17	6954724f
Message							
bd050fb4	c6925351	991aa15f	60327d4b	bd050fb4	c6925351	991aa15f	60327d4b
0212e457	9feb065e	d6ab8dac	7b52f8ca	0212e457	9feb065e	d6ab8dac	7b52f8ca
2f8a9774	1f189302	2043dc85	7b0eac19	2f8a9774	1f189302	2043dc85	7b0eac19
08fe0408	01c2f910	19abe45b	00000000	08fe0408	01c6f910	e6541ba4	fffffffe0
Output							
70588aa3	62e38880	4b32cd23	7da56fd2	70588aa3	62e38880	4b32cd23	7da56fd1
54827a61	d78e6b5f	17cce172	0ae88e5a	54827a62	d78e6b5e	f6942bb0	35a96499
232a8830	7f31780e	f0865b01	28cb4150	232a8a30	7f31740e	2ad851f7	362f33fb
39ba3bd2	277e9d52	316a7411	c8dbc618	39ba3bd3	27829d53	d239cc6e	29aa1db7

that the pair $(\delta^\oplus x_5, \delta^\boxplus y_5)$ is compatible). The value y_5 is computed as $s_4(x_1 \boxplus x_2 \boxplus x_9 \boxminus x_{11} \boxminus x_{14})$. Thus, the freedom in the inactive word x_9 can be used to somewhat control H_5 without affecting other conditions. Since M_5 is computed as $H_5 \oplus x_5$, this leads to some freedom in AddElement$(27) = (M_{11}^{\lll 12} \boxplus M_{14}^{\lll 15} \boxminus M_5^{\lll 6} \boxplus K_{27}) \oplus H_2$, and one may use this freedom to search for a collision in AddElement(27). However, the differences on H_2 and $M_{14}^{\lll 15}$ turn out to be incompatible, so one can only hope for a near-collision in AddElement(27). Still, this will lead to a near-collision in Q_{27}, which will lead to a near-collision in output word 3 of the compression function.

In a similar manner, one can use the freedom in x_0 (through H_{12} and thereby M_{12}) to find a full collision in AddElement(28), which (due to the small difference in Q_{27}) will lead to a near-collision in Q_{28}. Since Q_{28} is the only active word affecting output words 4, 8, and 12 of the compression function, all these three words will contain a near-collision.

Finally, we can use the freedom of M_0 to extend carries in AddElement(29). However, we cannot reach a full collision because the differences in $M_{13}^{\lll 14}$ and H_4 are incompatible.

To summarize, we use the following techniques to extend our attack:

AddElement(27) $(M_{11}^{\lll 12} \boxplus M_{14}^{\lll 15} \boxminus M_5^{\lll 6} \boxplus K_{27}) \oplus H_2$
We use the freedom in x_9 (through H_{15} and thereby M_5) to find a near-collision.

AddElement(28) $(M_{12}^{\lll 13} \boxplus M_{15}^{\lll 16} \boxminus M_6^{\lll 7} \boxplus K_{28}) \oplus H_3$
We use the freedom in x_0 (through H_{12} and thereby M_{12}) to find a full collision.

AddElement(29) $(M_{13}^{\lll 14} \boxplus M_0^{\lll 1} \boxminus M_7^{\lll 8} \boxplus K_{29}) \oplus H_4$
We use the freedom of M_0 to extend carries and find a near-collision.

We stress that these (near-)collisions can be found before searching for a collision in XH, and therefore, since the complexity is still below 2^{32}, the full cost of the attack is still around 2^{32}. Due to carries, however, it cannot be said beforehand how many bits will collide, unless one introduces a few additional bit conditions that will slightly increase the complexity. The search of a collision in XH is done using the freedom in M_9.

Table 2 gives an example of an input pair with an output colliding in 300 pre-specified bits (the search for a (near-)collision in $Q_0, \ldots, Q_{28}$ required the equivalent of about $2^{29.5}$ compression function evaluations). This example can be also be considered as a near-collision with 122 active bits.

6   Conclusion

In this paper we describe a technique to build partial-collisions in the compression function of BMW. We managed to build pairs of input which lead to a collision in 300 pre-specified bits, with complexity 2^{32}. Although it does not weaken the security of the iterated hash function, it is a strong distinguisher of the compression function. We also note that if the compression function is truncated like in the final transformation of BMW, we can still build pairs of message which collide in more than 110 bits with complexity 2^{32}. This is the first distinguisher on the truncated compression function of BMW.

A similar attack can be mounted on BMW-512 with complexity 2^{64}. It will give pairs of input of the compression function with about 600 colliding bits, including about 220 bits in the second part of the output.

We believe that the techniques developed for this attacks can be useful for further analysis of BMW, and other ARX based SHA-3 candidates.

Acknowledgments

We would like to thank the anonymous reviewers for helpful comments and suggestions.

References

[1] Aumasson, J.P.: Practical distinguisher for the compression function of Blue Midnight Wish (2010), http://131002.net/data/papers/Aum10.pdf (accessed January 07, 2011)

[2] Bresson, E., Canteaut, A., Chevallier-Mames, B., Clavier, C., Fuhr, T., Gouget, A., Icart, T., Misarsky, J.F., Naya-Plasencia, M., Paillier, P., Pornin, T., Reinhard, J.R., Thuillet, C., Videau, M.: Indifferentiability with Distinguishers: Why Shabal Does Not Require Ideal Ciphers. Cryptology ePrint Archive, Report 2009/199 (2009), http://eprint.iacr.org/

[3] Chang, D., Nandi, M.: Improved Indifferentiability Security Analysis of chopMD Hash Function. In: Nyberg, K. (ed.) FSE 2008. LNCS, vol. 5086, pp. 429–443. Springer, Heidelberg (2008)

[4] Coron, J.-S., Dodis, Y., Malinaud, C., Puniya, P.: Merkle-Damgård Revisited: How to Construct a Hash Function. In: Shoup, V. (ed.) CRYPTO 2005. LNCS, vol. 3621, pp. 430–448. Springer, Heidelberg (2005)

[5] Damgård, I.B.: A Design Principle for Hash Functions. In: Brassard, G. (ed.) CRYPTO 1989. LNCS, vol. 435, pp. 416–427. Springer, Heidelberg (1990)

[6] den Boer, B., Bosselaers, A.: Collisions for the Compression Function of MD-5. In: Helleseth, T. (ed.) EUROCRYPT 1993. LNCS, vol. 765, pp. 293–304. Springer, Heidelberg (1994)

[7] Gligoroski, D., Klíma, V., Knapskog, S.J., El-Hadedy, M., Amundsen, J., Mjøl-snes, S.F.: Cryptographic hash function BLUE MIDNIGHT WISH. Submission to NIST (Round 2) (September 2009),
http://people.item.ntnu.no/~danilog/Hash/BMW-SecondRound/Supporting_Documentation/BlueMidnightWishDocumentation.pdf (March 22, 2010)

[8] Guo, J., Thomsen, S.S.: Deterministic Differential Properties of the Compression Function of BMW. In: Biryukov, A., Gong, G., Stinson, D.R. (eds.) SAC 2010. LNCS, vol. 6544, pp. 338–350. Springer, Heidelberg (2011)

[9] Merkle, R.C.: One Way Hash Functions and DES. In: Brassard, G. (ed.) CRYPTO 1989. LNCS, vol. 435, pp. 428–446. Springer, Heidelberg (1990)

[10] Mouha, N., Velichkov, V., De Cannière, C., Preneel, B.: The Differential Analysis of S-Functions. In: Biryukov, A., Gong, G., Stinson, D.R. (eds.) SAC 2010. LNCS, vol. 6544, pp. 36–56. Springer, Heidelberg (2011)

[11] Thomsen, S.S.: Pseudo-cryptanalysis of the Original Blue Midnight Wish. In: Hong, S., Iwata, T. (eds.) FSE 2010. LNCS, vol. 6147, pp. 304–317. Springer, Heidelberg (2010)

[12] Wang, X., Yu, H.: How to Break MD5 and Other Hash Functions. In: Cramer, R. (ed.) EUROCRYPT 2005. LNCS, vol. 3494, pp. 19–35. Springer, Heidelberg (2005)

A   Details on the Permutation P Used in f_0

The matrix multiplication taking place in f_0 can be described as follows. Let $x = H \oplus M$. Let C denote the matrix. If $z = C \cdot x$, where x is considered a 16-element vector over $\mathbb{Z}_{2^{32}}$, then

$$z_0 = x_5 \boxminus x_7 \boxplus x_{10} \boxplus x_{13} \boxplus x_{14}$$
$$z_1 = x_6 \boxminus x_8 \boxplus x_{11} \boxplus x_{14} \boxminus x_{15}$$
$$z_2 = x_0 \boxplus x_7 \boxplus x_9 \boxminus x_{12} \boxplus x_{15}$$
$$z_3 = x_0 \boxminus x_1 \boxplus x_8 \boxminus x_{10} \boxplus x_{13}$$
$$z_4 = x_1 \boxplus x_2 \boxplus x_9 \boxminus x_{11} \boxminus x_{14}$$
$$z_5 = x_3 \boxminus x_2 \boxplus x_{10} \boxminus x_{12} \boxplus x_{15}$$
$$z_6 = x_4 \boxminus x_0 \boxminus x_3 \boxminus x_{11} \boxplus x_{13}$$
$$z_7 = x_1 \boxminus x_4 \boxminus x_5 \boxminus x_{12} \boxminus x_{14}$$
$$z_8 = x_2 \boxminus x_5 \boxminus x_6 \boxplus x_{13} \boxminus x_{15}$$

$$z_9 = x_0 \boxminus x_3 \boxplus x_6 \boxminus x_7 \boxplus x_{14}$$
$$z_{10} = x_8 \boxminus x_1 \boxminus x_4 \boxminus x_7 \boxplus x_{15}$$
$$z_{11} = x_8 \boxminus x_0 \boxminus x_2 \boxminus x_5 \boxplus x_9$$
$$z_{12} = x_1 \boxplus x_3 \boxminus x_6 \boxminus x_9 \boxplus x_{10}$$
$$z_{13} = x_2 \boxplus x_4 \boxplus x_7 \boxplus x_{10} \boxplus x_{11}$$
$$z_{14} = x_3 \boxminus x_5 \boxplus x_8 \boxminus x_{11} \boxminus x_{12}$$
$$z_{15} = x_{12} \boxminus x_4 \boxminus x_6 \boxminus x_9 \boxplus x_{13}$$

The subfunctions s_i, $0 \le i \le 4$, used in f_0 are defined as follows.

$$s_0(x) = x^{\ggg 1} \oplus x^{\lll 3} \oplus x^{\lll 4} \oplus x^{\lll 19}$$
$$s_1(x) = x^{\ggg 1} \oplus x^{\lll 2} \oplus x^{\lll 8} \oplus x^{\lll 23}$$
$$s_2(x) = x^{\ggg 2} \oplus x^{\lll 1} \oplus x^{\lll 12} \oplus x^{\lll 25}$$
$$s_3(x) = x^{\ggg 2} \oplus x^{\lll 2} \oplus x^{\lll 15} \oplus x^{\lll 29}$$
$$s_4(x) = x^{\ggg 1} \oplus x$$

B   Description of the f_2 Function

The f_2 function performs the following computations:

$$XL = \bigoplus_{i=16}^{23} Q_i \qquad\qquad XH = \bigoplus_{i=16}^{31} Q_i$$

$$
\begin{aligned}
HH_0 &= & (XH^{\ggg 5} &\oplus Q_{16}^{\ggg 5} \oplus M_0) \boxplus (XL &\oplus Q_{24} \oplus Q_0)\\
HH_1 &= & (XH^{\lll 7} &\oplus Q_{17}^{\lll 8} \oplus M_1) \boxplus (XL &\oplus Q_{25} \oplus Q_1)\\
HH_2 &= & (XH^{\ggg 5} &\oplus Q_{18}^{\lll 5} \oplus M_2) \boxplus (XL &\oplus Q_{26} \oplus Q_2)\\
HH_3 &= & (XH^{\ggg 1} &\oplus Q_{19}^{\lll 5} \oplus M_3) \boxplus (XL &\oplus Q_{27} \oplus Q_3)\\
HH_4 &= & (XH^{\ggg 3} &\oplus Q_{20} \oplus M_4) \boxplus (XL &\oplus Q_{28} \oplus Q_4)\\
HH_5 &= & (XH^{\lll 6} &\oplus Q_{21}^{\ggg 6} \oplus M_5) \boxplus (XL &\oplus Q_{29} \oplus Q_5)\\
HH_6 &= & (XH^{\ggg 4} &\oplus Q_{22}^{\lll 6} \oplus M_6) \boxplus (XL &\oplus Q_{30} \oplus Q_6)\\
HH_7 &= & (XH^{\ggg 11} &\oplus Q_{22}^{\lll 2} \oplus M_7) \boxplus (XL &\oplus Q_{31} \oplus Q_7)\\
HH_8 &= HH_4^{\lll 9} \boxplus & (XH &\oplus Q_{24} \oplus M_8) \boxplus (XL^{\lll 8} &\oplus Q_{23} \oplus Q_8)\\
HH_9 &= HH_5^{\lll 10} \boxplus & (XH &\oplus Q_{25} \oplus M_9) \boxplus (XL^{\ggg 6} &\oplus Q_{16} \oplus Q_9)\\
HH_{10} &= HH_6^{\lll 11} \boxplus & (XH &\oplus Q_{26} \oplus M_{10}) \boxplus (XL^{\lll 6} &\oplus Q_{17} \oplus Q_{10})\\
HH_{11} &= HH_7^{\lll 12} \boxplus & (XH &\oplus Q_{27} \oplus M_{11}) \boxplus (XL^{\lll 4} &\oplus Q_{18} \oplus Q_{11})\\
HH_{12} &= HH_0^{\lll 13} \boxplus & (XH &\oplus Q_{28} \oplus M_{12}) \boxplus (XL^{\ggg 3} &\oplus Q_{19} \oplus Q_{12})\\
HH_{13} &= HH_1^{\lll 14} \boxplus & (XH &\oplus Q_{29} \oplus M_{13}) \boxplus (XL^{\ggg 4} &\oplus Q_{20} \oplus Q_{13})\\
HH_{14} &= HH_2^{\lll 15} \boxplus & (XH &\oplus Q_{30} \oplus M_{14}) \boxplus (XL^{\ggg 7} &\oplus Q_{21} \oplus Q_{14})\\
HH_{15} &= HH_3^{\lll 16} \boxplus & (XH &\oplus Q_{31} \oplus M_{15}) \boxplus (XL^{\ggg 2} &\oplus Q_{22} \oplus Q_{15})
\end{aligned}
$$

In the attack of Section 4, we have differences in M_{13}, M_{14}, M_{15}, and $Q_{27}, \ldots Q_{31}$, with no difference in XL and XH. In the first part of f_2, this results in differences in $HH_3, \ldots HH_7$. In the second part, outputs $HH_8, \ldots HH_{15}$ are active.

In the attack of Section 5, we have dense differences in M_{14}, M_{15}, Q_{30}, Q_{31}, and small differences in M_{13}, Q_{27}, Q_{28} and Q_{29}, with no difference in XL and XH. In the first part of f_2, this results in small differences in HH_3, HH_4, HH_5, and dense differences in HH_6 and HH_7. In the second part, there are dense differences in $HH_{10}, HH_{11}, HH_{14}, HH_{15}$, and small differences in HH_8, HH_9, HH_{12} and HH_{13}.

Higher-Order Differential Properties of KECCAK and *Luffa*[*]

Christina Boura[1,2], Anne Canteaut[1], and Christophe De Cannière[3]

[1] SECRET Project-Team - INRIA Paris-Rocquencourt - B.P. 105
78153 Le Chesnay Cedex - France
[2] Gemalto - 6, rue de la Verrerie - 92447 Meudon sur Seine - France
[3] Department of Electrical Engineering ESAT/SCD-COSIC
Katholieke Universiteit Leuven - Kasteelpark Arenberg 10
B-3001 Heverlee - Belgium
{Christina.Boura,Anne.Canteaut}@inria.fr,
Christophe.DeCanniere@esat.kuleuven.be

Abstract. In this paper, we identify higher-order differential and zero-sum properties in the full KECCAK-f permutation, in the *Luffa* v1 hash function and in components of the *Luffa* v2 algorithm. These structural properties rely on a new bound on the degree of iterated permutations with a nonlinear layer composed of parallel applications of a number of balanced Sboxes. These techniques yield zero-sum partitions of size 2^{1575} for the full KECCAK-f permutation and several observations on the *Luffa* hash family. We first show that *Luffa* v1 applied to one-block messages is a function of 255 variables with degree at most 251. This observation leads to the construction of a higher-order differential distinguisher for the full *Luffa* v1 hash function, similar to the one presented by Watanabe *et al.* on a reduced version. We show that similar techniques can be used to find all-zero higher-order differentials in the *Luffa* v2 compression function, but the additional blank round destroys this property in the hash function.

Keywords: Hash functions, degree, higher-order differentials, zero-sums, SHA-3.

1 Introduction

The algebraic degrees of some hash function proposals and of their building blocks have been studied for analyzing their security. In particular, the fact that some inner primitive in a hash function has a relatively low degree can often be used to construct higher-order differential distinguishers, or zero-sum structures. This direction has been investigated in [1,13,3] for three SHA-3 candidates, Luffa, Hamsi and KECCAK. Here, we show how to deduce a new bound for the degree of iterated permutations for a special category of SP-networks. This category

---

[*] Partially supported by the French Agence Nationale de la Recherche through the SAPHIR2 project under Contract ANR-08-VERS-014.

A. Joux (Ed.): FSE 2011, LNCS 6733, pp. 252–269, 2011.

includes functions that have for non-linear layer, a number of smaller balanced Sboxes. This class of functions is though quite general: it includes functions with a large number of small Sboxes (e.g. Sboxes operating on 3 or 4 bits), but it also includes any nonlinear permutation which can be decomposed as several independent Sboxes, even of large size. Our new bound shows in particular that, when it is iterated, the degree of the function grows in a much smoother way than expected when it approaches the number of variables.

For instance, this new bound enables us to find zero-sum partitions for the full inner permutations of the hash functions KECCAK [2] and for the *Luffa* v1 hash function [5]. Furthermore, by applying a technique similar to that used in [13], and by combining it with the results given by the new bound, we show that the degree of the *Luffa* v2 compression function [6] is slightly lower than expected. This also enables us to find distinguishers for the Q_j permutations and for the compression function of *Luffa* v2. These results do not seem to affect the security of *Luffa* v2, but are another confirmation of the fact that the internal components of *Luffa* do not behave as ideal random functions.

The rest of the paper is organized as follows. In Section 2, a new bound on the degree of iterated permutations is presented when the nonlinear layer consists of several parallel applications of smaller balanced Sboxes. Section 3 recalls how a low algebraic degree can be exploited for mounting higher-order differential distinguishers and zero-sum distinguishers. An application to the full KECCAK-*f* permutation is presented in Section 4, while applications to the *Luffa* hash family are described in Section 5.

2   A New Bound on the Degree of Some Iterated Permutations

In the whole paper, the addition in $\mathbb{F}_2^n$, *i.e.* the bitwise exclusive-or will be denoted by $+$, while $\oplus$ will be used for denoting the direct sum of subspaces of $\mathbb{F}_2^n$.

A *Boolean function* f *of* n *variables* is a function from $\mathbb{F}_2^n$ into $\mathbb{F}_2$. It can be expressed as a polynomial, called *algebraic normal form*. The *degree* of f, denoted by $deg(f)$, is the degree of its algebraic normal form. Moreover, the *degree* of a vectorial function F from $\mathbb{F}_2^n$ into $\mathbb{F}_2^m$ is defined as the highest degree of its coordinates. The Hamming weight of a Boolean function, f, is denoted by $\mathrm{wt}(f)$. It corresponds to the number of x such that $f(x) = 1$. Any function F from $\mathbb{F}_2^n$ into $\mathbb{F}_2^m$ is said to be *balanced* if each element in $\mathbb{F}_2^m$ has exactly 2^{n-m} preimages under F.

In this paper, we are interested in estimating the degree of a composed function $G \circ F$. Obviously, we can bound the degree of the composition $G \circ F$ by $deg(G \circ F) \leq deg(G)deg(F)$. Though, this trivial bound is often very little representative of the true degree of the permutation, in particular if we are trying to estimate the degree after a high number of rounds. A first improvement of the trivial bound was provided by Canteaut and Videau [7] when the values occurring in the Walsh spectrum of F are divisible by a high power of 2, *i.e.* if the values $\mathrm{wt}(\varphi_b \circ F + \varphi_a)$ for all $a \in \mathbb{F}_2^n$ and $b \in \mathbb{F}_2^m$ are divisible by a high power of 2, where φ_a denotes the linear function $x \mapsto a \cdot x$.

Theorem 1. [7] *Let F be a function from $\mathbb{F}_2^n$ into $\mathbb{F}_2^n$ such that all values*

$$\mathrm{wt}(\varphi_b \circ F + \varphi_a), \quad a, b \in \mathbb{F}_2^n, b \neq 0$$

are divisible by 2^ℓ, for some integer ℓ. Then, for any $G : \mathbb{F}_2^n \to \mathbb{F}_2^n$, we have

$$\deg(G \circ F) \leq n - 1 - \ell + \deg(G).$$

In particular, this result applies to the functions composed of a nonlinear layer followed by a linear permutation, where the nonlinear layer is defined by the concatenation of m smaller balanced Sboxes $S_1, \ldots, S_m$, defined over $\mathbb{F}_2^{n_0}$, $n_0 \geq 2$. Indeed, since all elements $\mathrm{wt}(\varphi_b \circ S_i + \varphi_a)$ for all smaller functions $S_1, \ldots, S_m$ are divisible by 2, then we deduce that, for the whole permutation, $\mathrm{wt}(\varphi_b \circ F + \varphi_a)$ is divisible by 2^{2m-1}. We will show here how this bound can be further improved in this particular case. The result mainly comes from the following observation.

Proposition 1. *Let F be a balanced function from $\mathbb{F}_2^n$ into $\mathbb{F}_2^m$, and let k be an integer with $1 \leq k \leq m$. Then, all products of k coordinates of F have the Hamming weight 2^{n-k}.*
In particular, if $k < n$, the product of any k coordinates of F has degree at most $(n-1)$.

Proof. Let $(f_1, \ldots, f_m)$ denote the coordinates of F. Let I be any subset of $\{1, \ldots, m\}$ of size k, and let F_I be the function from $\mathbb{F}_2^n$ into $\mathbb{F}_2^k$ whose coordinates are the $f_i, i \in I$. Since F_I is balanced, the multiset $\{F_I(x), x \in \mathbb{F}_2^n\}$ consists of all elements in $\mathbb{F}_2^k$, each one with multiplicity 2^{n-k}. Therefore, there exist exactly 2^{n-k} values of x such that $F_I(x)$ is the all-one vector, or equivalently such that $\prod_{i \in I} f_i = 1$. ◇

From the last part of Proposition 1, we deduce the following theorem.

Theorem 2. *Let F be a function from $\mathbb{F}_2^n$ into $\mathbb{F}_2^n$ corresponding to the concatenation of m smaller Sboxes, $S_1, \ldots, S_m$, defined over $\mathbb{F}_2^{n_0}$. Let δ_k be the maximal degree of the product of any k coordinates of anyone of these smaller Sboxes. Then, for any function G from $\mathbb{F}_2^n$ into $\mathbb{F}_2^\ell$, we have*

$$\deg(G \circ F) \leq n - \frac{n - \deg(G)}{\gamma}, \tag{1}$$

where

$$\gamma = \max_{1 \leq i \leq n_0 - 1} \frac{n_0 - i}{n_0 - \delta_i}.$$

Most notably, if all Sboxes are balanced, we have

$$\deg(G \circ F) \leq n - \frac{n - \deg(G)}{n_0 - 1}.$$

Moreover, if $n_0 \geq 3$ and all Sboxes are balanced functions of degree at most $n_0 - 2$, we have

$$\deg(G \circ F) \leq n - \frac{n - \deg(G)}{n_0 - 2} \,,$$

Proof. Let us denote by π the product of d output coordinates of F. Some of the coordinates involved in π may belong to the same Sbox. Then, for any i, $1 \leq i \leq n_0$, we denote by x_i the integer corresponding to the number of Sboxes for which exactly i coordinates are involved in π. Obviously, we have

$$\deg(\pi) \leq \max_{(x_1,\ldots,x_{n_0})} \sum_{i=1}^{n_0} \delta_i x_i$$

where the maximum is taken over all vectors $(x_1, \ldots, x_{n_0})$ satisfying

$$\sum_{i=1}^{n_0} i x_i = d \text{ and } \sum_{i=1}^{n_0} x_i \leq m \,.$$

Then, we have

$$\gamma \deg(\pi) - d \leq \gamma \sum_{i=1}^{n_0} \delta_i x_i - \sum_{i=1}^{n_0} i x_i$$

$$\leq (\gamma - 1)n_0 x_{n_0} + \sum_{i=1}^{n_0-1} (\gamma \delta_i - i) x_i$$

$$\leq (\gamma - 1)n_0 \sum_{i=1}^{n_0} x_i - \sum_{i=1}^{n_0-1} ((\gamma - 1)n_0 - \gamma \delta_i + i) x_i$$

$$\leq (\gamma - 1)n - \sum_{i=1}^{n_0-1} ((\gamma - 1)n_0 - \gamma \delta_i + i) x_i \leq (\gamma - 1)n \,,$$

where the last inequality comes from the fact that all coefficients in the sum are positive. Actually, we have

$$(\gamma - 1)n_0 - \gamma \delta_i + i = \gamma(n_0 - \delta_i) - (n_0 - i) \geq 0$$

by definition of γ. Thus, since $\gamma \deg(\pi) - d \leq (\gamma - 1)n$, we deduce that

$$\gamma (n - \deg(\pi)) \geq n - d \,.$$

Now, we first show that, if all Sboxes are balanced, then $\gamma \leq n_0 - 1$. Indeed, for any $1 \leq i \leq n_0 - 1$, we have

$$\frac{n_0 - i}{n_0 - \delta_i} \leq \frac{n_0 - 1}{1} \,,$$

since we know from Proposition 1 that the degree of the product of $(n_0 - 1)$ coordinates of a balanced $n_0 \times n_0$ Sbox cannot be equal to n_0, and thus $\delta_i \leq n_0 - 1$. Also, we can prove that, if the degrees of all Sboxes satisfy $\deg S < n_0 - 1$, then $\gamma \leq n_0 - 2$. Indeed, for $i = 1$, we have

$$\frac{n_0 - i}{n_0 - \delta_i} = \frac{n_0 - 1}{n_0 - \delta_1} \leq \frac{n_0 - 1}{2} \leq n_0 - 2$$

since $n_0 \geq 3$. Similarly, for any i, $2 \leq i < n_0$, we have $\delta_i \leq n_0 - 1$, implying that

$$\frac{n_0 - i}{n_0 - \delta_i} \leq n_0 - i \leq n_0 - 2 . \qquad \diamond$$

It is worth noticing that Bound (1) and the trivial bound are in some sense symetric. Indeed, we have

$$\frac{\deg(G \circ F)}{\deg G} \leq \max_{1 \leq i < n_0} \frac{\delta_i}{i} \quad \text{and} \quad \frac{n - \deg(G \circ F)}{n - \deg G} \geq \left(\max_{1 \leq i < n_0} \frac{n_0 - i}{n_0 - \delta_i} \right)^{-1} .$$

In other words, when representing $\deg(G \circ F)$ as a function of $\deg G$, the trivial bound states that the degree of $G \circ F$ is upper-bounded by a line through the origin with coefficient $\deg F$. When representing the "degree deficiency" $(n - \deg(G \circ F))$ as a function of $(n - \deg G)$, (1) states that the degree deficiency of $G \circ F$ is lower-bounded by a line through the origin with coefficient γ^{-1}. This can be observed on Figure 1 where the parameters correspond to the inverse of the Keccak permutation.

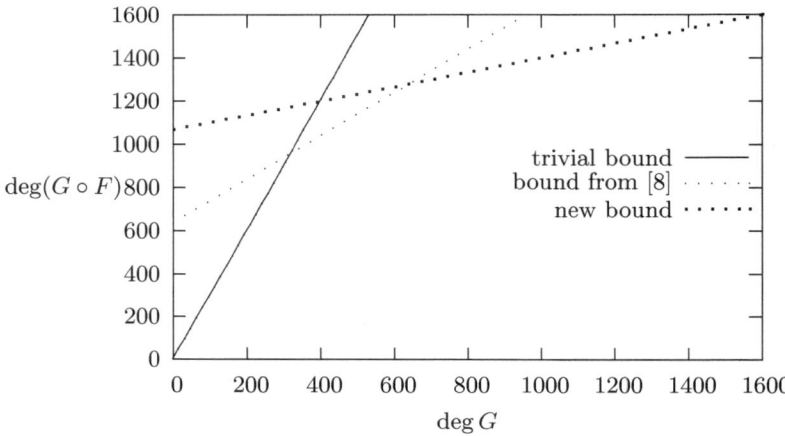

Fig. 1. Evolution of the degree of $G \circ F$ where F is a 1600-variable function composed of 320 cubic permutations over $\mathbb{F}_2^5$ corresponding to the inverse of Keccak χ function

3   Distinguishing Properties Related to the Algebraic Degree

3.1   Higher-Order Derivatives

The algebraic degree of a permutation F provides some particular distinguishers, which correspond to the values of any derivative of F with respect to a subspace of $\mathbb{F}_2^n$ with dimension $(\deg(F)+1)$. This result comes from the following property of higher-order derivatives of a function.

Definition 1. *[12] Let F be a function from $\mathbb{F}_2^n$ into $\mathbb{F}_2^m$. For any $a \in \mathbb{F}_2^n$ the derivative of F with respect to a is the function $D_a F(x) = F(x+a) + F(x)$. For any k-dimensional subspace V of $\mathbb{F}_2^n$ and for any basis of V, $\{a_1, \ldots, a_k\}$, the k-th order derivative of F with respect to V is the function defined by*

$$D_V F(x) = D_{a_1} D_{a_2} \ldots D_{a_k} F(x) = \sum_{v \in V} F(x+v), \forall x \in \mathbb{F}_2^n .$$

It is well-known that the degree of any first-order derivative of a function is strictly less than the degree of the function. This simple remark, which is exploited in higher-order differential attacks [10], implies that for every subspace V of dimension $(\deg F + 1)$,

$$D_V F(x) = \sum_{v \in V} F(x+v) = 0, \quad \text{for every } x \in \mathbb{F}_2^n.$$

3.2   Zero-Sum Structures

The existence of zero-sum structures is a distinguishing property which has been recently investigated by Aumasson and Meier [1], Knudsen and Rijmen [11] and by Boura and Canteaut [3].

Definition 2. *Let F be a function from $\mathbb{F}_2^n$ into $\mathbb{F}_2^m$. A zero-sum for F of size K is a subset $\{x_1, \ldots, x_K\} \subset \mathbb{F}_2^n$ of elements which sum to zero and for which the corresponding images by F also sum to zero, i.e.,*

$$\sum_{i=1}^{K} x_i = \sum_{i=1}^{K} F(x_i) = 0 .$$

It has been shown in [3] that any function from $\mathbb{F}_2^n$ into $\mathbb{F}_2^m$ has zero-sums of size less than or equal to 5. However, when F is a permutation over $\mathbb{F}_2^n$, a much stronger property, named *zero-sum partition*, can be investigated.

Definition 3. *Let P be a permutation from $\mathbb{F}_2^n$ into $\mathbb{F}_2^n$. A zero-sum partition for P of size $K = 2^k$ is a collection of 2^{n-k} disjoint zero-sums $X_i = \{x_{i,1}, \ldots, x_{i,2^k}\} \subset \mathbb{F}_2^n$ i.e.,*

$$\bigcup_{i=1}^{2^{n-k}} X_i = \mathbb{F}_2^n \text{ and } \sum_{j=1}^{2^k} x_{i,j} = \sum_{j=1}^{2^k} P(x_{i,j}) = 0, \forall 1 \leq i \leq 2^{n-k} .$$

Here, we focus on the search for zero-sum partitions coming from structural properties of the permutation P, when P is an iterated permutation of the form

$$P = R_r \circ \ldots \circ R_1,$$

where all R_i are simpler permutations over $\mathbb{F}_2^n$, named the *round permutations*. The fact that the permutation used in a hash function does not depend on any secret parameter allows to exploit the previous property starting from the middle, *i.e.*, from an intermediate internal state. This property was used by Aumasson and Meier [1] and also by Knudsen and Rijmen in the case of a known-key property of a block cipher [11]. The only information needed for finding such zero-sums on the iterated permutation using this first approach is an upper bound on the algebraic degrees of both the round transformation and its inverse.

More precisely, we consider $P = R_r \circ \ldots \circ R_1$, and we choose some integer t, $1 \le t \le r$. We define the following functions involved in the decomposition of P: F_{r-t} consists of the last $(r-t)$ round transformations, *i.e.*, $F_{r-t} = R_r \circ \ldots \circ R_{t+1}$ and G_t consists of the inverse of the first t round transformations, *i.e.*, $G_t = R_1^{-1} \circ \ldots \circ R_t^{-1}$. Then, as detailed in [1] and in [3], we can find many zero-sum partitions for P of size 2^{d+1} where $d = \max(\deg(F_{r-t}), \deg(G_t))$.

Besides the degree of the round transformation, it has been shown in [3] that some properties of the linear layer in the round transformation may also be exploited for constructing zero-sum partitions, in particular when the nonlinear layer of the round transformation consists of parallel applications of smaller functions defined over $\mathbb{F}_2^{n_0}$. In the following, we denote by B_i, $0 \le i < m$, the n_0-dimensional subspaces corresponding to the inputs of these smaller Sboxes, *i.e.*,

$$B_i = \langle e_{n_0 i}, \ldots, e_{n_0 i + n_0 - 1} \rangle$$

where $e_0, \ldots, e_{n-1}$ denotes the canonical basis of $\mathbb{F}_2^n$ and where the positions of the n bits in the internal state are numbered such that the n_0-bit Sboxes apply on n_0 consecutive input variables. Then, it was shown in [3] that it is possible to extend a number of zero-sum partitions that have been found for t rounds, to $t+1$ rounds, without increasing the complexity.

Proposition 2. *[3] Let d_1 and d_2 be such that $\deg(F_{r-t-1}) \le d_1$ and $\deg(G_t) \le d_2$. Let us decompose the round transformation after t rounds into $R_{t+1} = A_2 \circ \chi \circ A_1$ where both A_1 and A_2 have degree 1 and χ corresponds to the concatenation of m smaller permutations defined over $\mathbb{F}_2^{n_0}$. Let $\mathcal{I}$ be any subset of $\{0, \ldots, m-1\}$ of size $\lceil (d+1)/n_0 \rceil$, let*

$$V = \bigoplus_{i \in \mathcal{I}} B_i$$

and W be its complement. Then, the sets

$$X_a = \{(G_t \circ A_1^{-1})(a + z), \ z \in V\}, \ a \in W$$

form a zero-sum partition of $\mathbb{F}_2^n$ of size 2^k, with $k = n_0 \lceil \frac{d+1}{n_0} \rceil$, for the r-round permutation P.

It is worth noticing that the zero-sum partitions deduced from this proposition correspond to a structural property which can be described by means of some close formula. This implies for instance that they can be used for proving that some given permutations do not satisfy the expected property, and this may only require the evaluation of the permutation on a few sets X_i. In this sense, they differ from the zero-sum partitions found by a generic algorithm since all generic algorithms known so far require the evaluation of the permutation at almost all points (see [3] for a discussion on generic algorithms for finding zero-sums and zero-sum partitions).

4 Application to the KECCAK-f Permutation

4.1 The KECCAK-f Permutation

KECCAK [2] is one of the fourteen hash functions selected for the second round of the SHA-3 competition. Its mode of operation is the sponge construction. The inner primitive in KECCAK is a permutation, composed of several iterations of very similar round transformations. Within the KECCAK-family, the SHA-3 candidate operates on a 1600-bit state, which is represented by a 3-dimensional binary matrix of size $5 \times 5 \times 64$. Then, the state can be seen as 64 parallel slices, each one containing 5 rows and 5 columns. The permutation in KECCAK is denoted by KECCAK-$f[b]$, where b is the size of the state. So, for the SHA-3 candidate, $b = 1600$.

The number of rounds in KECCAK-$f[1600]$ was 18 in the original submission, and it has been updated to 24 for the second round. Every round R consists of a sequence of 5 permutations modifying the state:

$$R = \iota \circ \chi \circ \pi \circ \rho \circ \theta.$$

The functions θ, ρ, π, ι are transformations of degree 1 providing diffusion in all directions of the 3-dimensional state. Then, keeping the same notation as in the previous section, we have $A_1 = \pi \circ \rho \circ \theta$, which is linear and $A_2 = \iota$, which corresponds to the addition of a constant value. Therefore, the linear part of $A = A_1 \circ A_2$ corresponds to $L = \pi \circ \rho \circ \theta$. The nonlinear layer, χ, is a quadratic permutation which is applied to each row of the 1600-bit state. In other words, 320 parallel applications of χ_0 are implemented in order to provide confusion. The inverse permutation, denoted by χ^{-1}, is a permutation of degree 3.

4.2 Zero-Sum Partitions for the Full KECCAK-f Permutation

We apply here Theorem 2 to the KECCAK-f round permutation, which is denoted by R. For any F,

$$\deg(F \circ R) = \deg(F \circ \chi) \leq n - \frac{n - \deg(F)}{3}$$

and

$$\deg(F \circ R^{-1}) = \deg((F \circ L^{-1}) \circ \chi^{-1}) \le n - \frac{n - \deg(F)}{3}$$

by using that the inverse of χ has degree 3. By combining this bound with the trivial bound, we get the bound presented in Table 1 on the degree of several iterations of the round permutation of KECCAK-f and of its inverse. With this

Table 1. Upper bounds on the degree of several rounds of KECCAK-f and of its inverse (the results in bold are obtainéd with the new bound, while the other ones correspond to the trivial bound)

forward		backward		
# rounds	bound on $\deg(R^r)$	# rounds	bound on $\deg(R^{-r})$	bound on $\deg(R^{-r})$ using [8]
1	2	1	3	3
2	4	2	9	9
3	8	3	27	27
4	16	4	81	81
5	32	5	243	243
6	64	6	729	729
7	128	7	**1309**	**1164**
8	256	8	**1503**	**1382**
9	512	9	**1567**	**1491**
10	1024	10	**1589**	**1545**
11	**1408**	11	**1596**	**1572**
12	**1536**	12	**1598**	**1586**
13	**1578**	13	**1599**	**1593**
14	**1592**	14	**1599**	**1596**
15	**1597**	15	**1599**	**1598**
16	**1599**	16	**1599**	**1599**

new bound, we can use the technique presented in Proposition 2 for finding zero-sum partitions for the full KECCAK-f permutation. Namely, we consider the intermediate states after the linear layer $L = \pi \circ \rho \circ \theta$ in the 11-th round. Let us choose any subspace V in $\mathbb{F}_2^{1600}$ corresponding to a collection of 318 rows (out of the 320), implying $\dim V = 1590$. Then, the sets

$$X_a = \{(G_{10} \circ L^{-1})(a + z), \ z \in V\}, \quad a \in \mathbb{F}_2^{1600},$$

where G_{10} denotes the inverse of the first 10 rounds, form a zero-sum partition of size 2^{1590} for the full KECCAK-f permutation. This comes directly from Proposition 2 and from the fact that the inverse of the first 10 rounds of the permutation has degree at most $1589 < \dim V$, and that the last 13 rounds have degree at most $1578 < \dim V$.

Recently, Duan and Lai [8] have observed that the inverse of χ has the following remarkable property: the product of any 2 components of χ^{-1} has degree at most 3 (instead of 4 which is the bound obtained with Proposition 1). Then, we deduce that the value of the coefficient γ for χ^{-1} involved in Theorem 2 is $\gamma = 2$ since $\delta_2 = 3$. By using this particular property of χ^{-1}, the previous result can be improved as follows: for any F,

$$\deg(F \circ R^{-1}) = \deg((F \circ L^{-1}) \circ \chi^{-1}) \leq n - \frac{n - \deg(F)}{2}.$$

This leads to the new bound on several iterations of R^{-1} as presented in the last column of Table 1. Now, by choosing the intermediate states after the linear layer on the 12-th round of KECCAK-f in any subspace V corresponding to a collection of 315 rows, we obtain a zero-sum partition for the full 24-round KECCAK-f permutation of size 2^{1575}. This comes from the fact that the inverse of the first 11 rounds have degree at most $1572 < \dim V$ and that the last 12 rounds have degree at most $1536 < \dim V$.

5 Application to the Hash Function *Luffa*

5.1 The Luffa Hash Function

The *Luffa* hash function [5,6] is also a Round-2 candidate of the NIST SHA-3 competition. Its mode of operation is based on a variant of the sponge design. The internal state in *Luffa* consists of w 256-bit words where w equals 3, 4 and 5 for the output lengths 256, 384 and 512 bits respectively. At each iteration, a 256-bit message block is processed by applying a linear message injection function *MI*. Then, a permutation is applied to the output as follows: the state is split into w 256-bit words and w parallel 256-bit permutations Q_j are applied to each word independently.

The internal state of each permutation Q_j is now divided in 8 words of 32 bits, denoted by $a_0, \ldots, a_7$. Each permutation consists of an input tweak applied only once at the beginning of each permutation and 8 rounds of a round transformation Step. The Step function consists of a nonlinear transformation

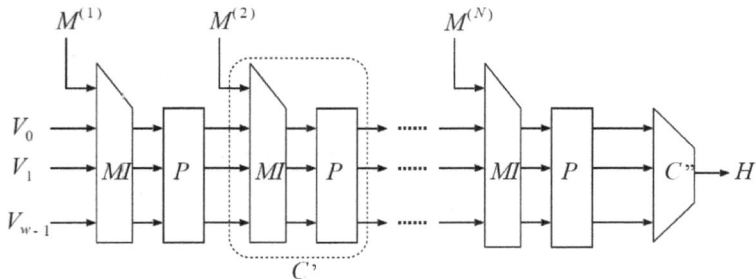

Fig. 2. The *Luffa* construction

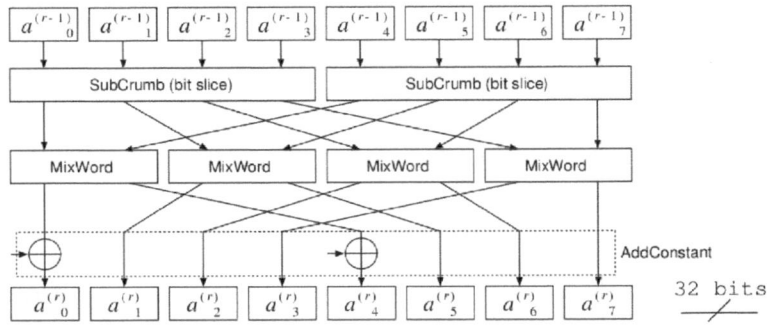

Fig. 3. The Step function

called SubCrumb, a linear transformation MixWord and an addition of constants AddConstant. The nonlinear part SubCrumb consists of 64 parallel applications of a 4×4 cubic permutation.

Finally, a finalization step is applied. It consists of several iterations of a blank round with fixed message 0x0...00 followed by a linear output function OF. In *Luffa* v1, a blank round with message block 0x0...00 is applied at the beginning of the finalization, only if the number of padded message blocks is strictly greater than one. In *Luffa* v2 such a blank round is always applied, in order to prevent higher-order differential attacks.

The SubCrumb Permutation. The input of every Sbox has four bits, every one coming from a different word a_k: S substitutes the ℓ-th bits of a_0, a_1, a_2, a_3 (or a_4, a_5, a_6, a_7) by a 4×4 Sbox of degree 3. The Sbox used in the original submission, *Luffa* v1, was

$$S_1[16] = \{7, 13, 11, 10, 12, 4, 8, 3, 5, 15, 6, 0, 9, 1, 2, 14\},$$

but the terms of degree 3 in the first three coordinates of this Sbox are similar. This property has been exploited in [13] for showing that the degree of Q_j does not grow as expected. In particular, Q_j reduced to 5 rounds out of 8 has degree at most 130, and the sum of the first two coordinates of Q_j after 6 rounds has degree at most 214. In order to avoid these unsuitable properties, the designers have modified the Sbox according to the strategy detailed in [4]. The new Sbox used in *Luffa* v2, is then

$$S_2[16] = \{13, 14, 0, 1, 5, 10, 7, 6, 11, 3, 9, 12, 15, 8, 2, 4\},$$

and the algebraic normal forms of its outputs are

$$y_0 = 1 + x_0 + x_1 + x_1x_2 + x_0x_3 + x_1x_3 + x_0x_1x_3 + x_0x_2x_3$$
$$y_1 = x_0 + x_3 + x_0x_1 + x_1x_2 + x_0x_3 + x_1x_3 + x_0x_1x_3 + x_0x_2x_3$$
$$y_2 = 1 + x_1 + x_3 + x_0x_2 + x_1x_2 + x_1x_3 + x_2x_3 + x_0x_1x_2 + x_0x_1x_3$$
$$y_3 = 1 + x_1 + x_2 + x_0x_3 + x_0x_2 + x_1x_2 + x_1x_3 + x_2x_3 + x_0x_1x_2 + x_0x_1x_3$$

Then the substitution by S is given by

$$b_{3,\ell}||b_{2,\ell}||b_{1,\ell}||b_{0,\ell} = S[a_{3,\ell}||a_{2,\ell}||a_{1,\ell}||a_{0,\ell}], \quad 0 \le \ell < 32,$$
$$b_{7,\ell}||b_{6,\ell}||b_{5,\ell}||b_{4,\ell} = S[a_{7,\ell}||a_{6,\ell}||a_{5,\ell}||a_{4,\ell}], \quad 0 \le \ell < 32.$$

in *Luffa* v1. In *Luffa* v2, the order of the last four input words is modified when entering the Sbox in order to break the symmetries exploited in [13]:

$$b_{3,\ell}||b_{2,\ell}||b_{1,\ell}||b_{0,\ell} = S[a_{3,\ell}||a_{2,\ell}||a_{1,\ell}||a_{0,\ell}], \quad 0 \le \ell < 32,$$
$$b_{4,\ell}||b_{7,\ell}||b_{6,\ell}||b_{5,\ell} = S[a_{4,\ell}||a_{7,\ell}||a_{6,\ell}||a_{5,\ell}], \quad 0 \le \ell < 32.$$

The MixWord Permutation. MixWord is a linear permutation of two words. If $z_0, \ldots, z_7$ are the 8 words of the state after the application of Step we have that

$$(z_0, z_4) = \text{MixWord}(b_0, b_4),$$
$$(z_1, z_5) = \text{MixWord}(b_1, b_5),$$
$$(z_2, z_6) = \text{MixWord}(b_2, b_6),$$
$$(z_3, z_7) = \text{MixWord}(b_3, b_7).$$

5.2 Algebraic Degree of the Q_j Permutation and Its Inverse

We now show that the approach used in [13] still applies to some extent to the *Luffa* v2 nonlinear function, and that this approach can be combined with Theorem 2 in order to find a new upper bound on the degree of several iterations of the Step function.

The remarkable property comes from the fact that the sum of the four coordinates of S_2 has degree 2 only: indeed, we deduce from the algebraic normal forms of the coordinates of S_2 that

$$d = y_0 + y_1 + y_2 + y_3 = 1 + x_1 + x_2 + x_0 x_1 + x_0 x_3 .$$

Let $x_i^r = \left(x_{i,\ell}^r \right)_{0 \le \ell < 32}$ denote the output words of r rounds of Step, and let $d_{0,\ell}^r$ (resp. $d_{4,\ell}^r$) denote the sum $x_{0,\ell}^r + x_{1,\ell}^r + x_{2,\ell}^r + x_{3,\ell}^r$ (resp. $x_{4,\ell}^r + x_{5,\ell}^r + x_{6,\ell}^r + x_{7,\ell}^r$). Now, let us consider the sum of any two distinct monomials of degree 3 in 4 variables. Any such two monomials share two variables. Then, if we denote by d the sum of all four variables, we obtain that

$$x_i x_j x_k + x_i x_j x_{k'} = x_i x_j x_k + x_i x_j (x_i + x_j + x_k + d)$$
$$= x_i x_j x_k + x_i x_j + x_i x_j + x_i x_j x_k + x_i x_j d$$
$$= x_i x_j d .$$

It follows that, since all coordinates of the Sboxes S_2 contain an even number of distinct monomials of degree 3, the degrees of their outputs (and then the degree of the output of $(r+1)$ rounds) satisfy

$$\deg x_{i,\ell}^{r+1} \leq 2 \max_{0 \leq j \leq 3} \deg x_{j,\ell}^{r} + \deg d_{0,\ell}^{r} \ \forall 0 \leq i \leq 3 . \tag{2}$$

Moreover, this property holds for any ordering of the inputs and outputs of the Sbox, implying

$$\deg x_{i,\ell}^{r+1} \leq 2 \max_{4 \leq j \leq 7} \deg x_{j,\ell}^{r} + \deg d_{4,\ell}^{r} \ \forall 4 \leq i \leq 7 .$$

Now, since the linear layer consists of the same function applied to all pairs of words (b_k, b_{k+4}) for $0 \leq k \leq 3$ separately, we deduce that

$$d_{0,\ell}^{r+1} = x_{0,\ell}^{r+1} + x_{1,\ell}^{r+1} + x_{2,\ell}^{r+1} + x_{3,\ell}^{r+1}$$

$$= \sum_{i=0}^{3} \texttt{MixWord}_{0,\ell}(b_i, b_{i+4}) = \texttt{MixWord}_{0,\ell}\left(\sum_{i=0}^{3} b_i, \sum_{i=0}^{3} b_{i+4}\right)$$

and

$$d_{4,\ell}^{r+1} = \texttt{MixWord}_{1,\ell}\left(\sum_{i=0}^{3} b_i, \sum_{i=0}^{3} b_{i+4}\right) .$$

Therefore, the degrees of $d_{0,\ell}^{r+1}$ and of $d_{4,\ell}^{r+1}$ correspond to the degrees of the sum of the four coordinates of the Sboxes, implying

$$\deg d_{i,\ell}^{r+1} \leq 2 \max_{i \leq j \leq i+3} \deg x_{j,\ell}^{r} , i \in \{0, 4\} . \tag{3}$$

Both recurrence relations (2) and (3) lead to the bounds presented in Table 2 on the degrees of several iterations of Step for the new nonlinear layer (i.e. for the new Sbox S_2 and the ordering of the input variables).

Table 2. Upper bounds on the algebraic degree of the output of r iterations of Step for *Luffa* v2 (and comparison with the results obtained in [13] for *Luffa* v1)

r	Luffa v2		Luffa v1	
	$\deg x_{i,\ell}^{r}$	$\deg d_{i,\ell}^{r}$	$\deg x_{i,\ell}^{r}$	$\deg d_{i,\ell}^{r}$
1	3	2	3	2
2	8	6	8	5
3	22	16	20	13
4	60	44	51	33
5	164	120	130	84
6		-		214

Now, for $r \geq 6$, we apply Theorem 2, exploiting the fact that Step is the composition of a linear layer and of several parallel applications of a smaller balanced Sbox of degree 3 defined over $\mathbb{F}_2^4$. Then, for any G, we have

$$\deg(G \circ \texttt{Step}) \leq \frac{512 + \deg(G)}{3} \; ,$$

implying

$$\max_{i,\ell} \deg(x_{i,\ell}^r) \leq \frac{512 + \max_{i,\ell} \deg(x_{i,\ell}^{r-1})}{3}$$

$$\max_{i,\ell} \deg(d_{i,\ell}^r) \leq \frac{512 + \max_{i,\ell} \deg(d_{i,\ell}^{r-1})}{3} \; .$$

These new bounds are given in Table 3.

Table 3. Upper bounds on the algebraic degree of the output of r iterations of $\texttt{Step}$ for *Luffa* v1 and *Luffa* v2

r	*Luffa* v2		*Luffa* v1	
	$\deg x_{i,\ell}^r$	$\deg d_{i,\ell}^r$	$\deg x_{i,\ell}^r$	$\deg d_{i,\ell}^r$
1	3	2	3	2
2	8	6	8	5
3	22	16	20	13
4	60	44	51	33
5	164	120	130	84
6	225	210	214	198
7	245	240	242	236
8	252	250	251	249

It is worth noticing that the same upper bounds hold for the degree of r iterations of the inverse of $\texttt{Step}$ in *Luffa* v2 since the algebraic normal form of the inverse of S_2 is

$$y_0 = x_0 + x_2 + x_3 + x_2 x_3$$
$$y_1 = 1 + x_3 + x_0 x_1 + x_0 x_2 + x_0 x_3 + x_2 x_3 + x_0 x_2 x_3 + x_1 x_2 x_3$$
$$y_2 = x_1 + x_2 + x_3 + x_0 x_1 + x_1 x_2 + x_0 x_3 + x_1 x_3 + x_0 x_1 x_3 + x_0 x_1 x_2 + x_0 x_2 x_3$$
$$\qquad + x_1 x_2 x_3$$
$$y_3 = x_1 + x_2 + x_3 + x_0 x_2 + x_2 x_3 + x_0 x_1 x_2 + x_0 x_1 x_3 \; .$$

Then, the sum of the four coordinates of S_2^{-1} is equal to

$$1 + x_0 + x_2 + x_1 x_2 + x_1 x_3 + x_2 x_3$$

and has degree 2 only. Moreover, all four coordinates of S_2^{-1} have an even number of monomials of degree 3. Then, the previously described technique for upper-bounding the degree of several iterations of the round function applies similarly when computing the inverse.

5.3   Higher-Order Differentials for the Compression Function of *Luffa* v2

The compression function in *Luffa* v2 takes as input a $256w$-bit chaining value and a 256-bit message block and it outputs a new $256w$-bit chaining value, where w equals $3, 4$ and 5 when the output length is $256, 384$ and 512. Then, we have proved that this function has degree at most 252, while it is expected from its construction to have degree 255.

A first consequence is the existence of all-zero higher-order differentials for the full compression function of *Luffa* v2, similar to those found in [13] for *Luffa* v1 reduced to 7 steps. Let us choose a position ℓ_0 among the 32 possible positions in a word, $0 \le \ell_0 < 32$, and let us consider any coset of the linear subspace V of the set of all possible message blocks defined as

$$V = \langle e_{i,\ell}, 0 \le i \le 7, \ell \ne \ell_0 \rangle ,$$

where $e_{i,\ell}$ denotes the 256-bit word of Hamming weight 1 having a one at position ℓ in word i. Then, V has dimension 248. For any fixed chaining value, the message injection function MI stabilizes the subspaces $\langle e_{i,\ell}, 0 \le i \le 7 \rangle$, implying that the input of each Q_j is a coset of V. Now, the tweak function at the beginning of each Q_j rotates the least significant four words by a number of bits depending on j. Its output then corresponds to a coset of a subspace V', which is the direct sum of 4-dimensional subspaces of the form $\langle e_{i,\ell}, 0 \le i \le 3 \rangle$ or $\langle e_{i,\ell}, 4 \le i \le 7 \rangle$. Since the first nonlinear layer applies to those 4-dimensional subspaces separately, it stabilizes the structure of V'. Therefore, the output of the first iteration of Step in each Q_j varies in a coset of a subspace of dimension 248. Then, the outputs of the compression function, *i.e.*, after 8 iterations of Step, sum to zero when the message block varies in any coset of V, since $\dim V = 248 > 246 > \deg(\text{Step}^7)$. This observation holds for any size of the hash value. It should be noted that, by nature, this algebraic property is very different from the properties exploited in previously known distinguishers on the compression function of *Luffa* v2 (e.g. [9]).

5.4   Zero-Sum Partitions for the Q_j Permutations

We consider the subspace V generated by the first 23 bits in a given word, that is

$$V = \langle e_{i_0,0}, e_{i_0,1}, \ldots, e_{i_0,22} \rangle ,$$

for some $0 \le i_0 \le 7$. Then, we can show that the sets

$$X_a = \{\text{Tweak}_j^{-1} \circ \left(\text{Step}^{-1}\right)^4 (a + z), \ z \in V\}, a \in \mathbb{F}_2^{256} .$$

form a zero-sum partition of size 2^{23} for each Q_j.

We first consider any coset of V as input of 4 rounds of Step. Then, the 23 active bits in V correspond to the inputs of 23 different Sboxes and thus the first round of `Step` is a function of degree 1 in these 23 input bits. As 3 iterations of `Step` have degree at most 22, we deduce that

$$\sum_{x \in X_a} Q_j(x) = 0 .$$

We now focus on the backward computation. We first take the image of V under the inverse of the linear application `MixWord`. As all the variables are in the same word a_{i_0}, after the application of the inverse of the linear layer, all words are constant except the words of index i_0 and $(i_0 + 4)$. But the bits of the words a_{i_0} and a_{i_0+4} all go to different Sboxes, implying that the first round backwards is linear. As we have proven that the inverse of 3 iterations of `Step` has degree at most 22, we deduce that

$$\sum_{x \in X_a} x = 0 .$$

There exist $\binom{32}{23} \times 8 = 2^{27.7}$ such zero-sum partitions for each Q_j corresponding to all possible choices for V, *i.e.*, for all possible choices for i_0 and for the 23 positions within the word of index i_0.

5.5   Higher-Order Differentials for the Full *Luffa* v1 Hash Function

It is shown in [13] that, when hashing messages of length at most 256 bits, the reduced version of *Luffa* v1 hash function, with 7 out of 8 steps in each Q_j, does not behave as a random function. Actually, if the message block varies in some particular subspace of dimension 216, then some linear combination of the output words of this reduced version of *Luffa* v1 sums to zero. This property comes from the fact that *Luffa* v1 does not perform any blank round for one-block messages, and that, after 6 rounds of `Step`, some linear combinations of the output words have degree at most 214. Even if the advantage that this property could give to an attacker is unclear, this unsuitable property has led the designers to modify the function for the second round of the SHA-3 competition. In particular, a blank round is performed for any message length in *Luffa* v2.

It turns out that this was probably a prudent decision, as the new upper bound on the degree of Q_j for *Luffa* v1 given in Table 3 now shows that a similar distinguisher can be exhibited for the full *Luffa* v1, since the degree of the two words obtained by

$$(y_0 + y_1 + y_2 + y_3, y_4 + y_5 + y_6 + y_7)$$

after 7 iterations of `Step` is at most 236. We then get a similar distinguisher based on the fact that the corresponding linear combinations of the bits of the hash values sum to zero when the message block varies in some particular subspace of dimension 240.

More interestingly, we have shown that the full *Luffa* v1 hash function, when applied to one-block messages, has degree at most 251 in the 255 bits of the message.

5.6   Degree of the Full *Luffa* v2 Hash Function with Chosen IVs

The previous observation does not hold for *Luffa* v2 since a blank round is performed for any message length. However, if we would consider the $256w$ bits of the initial value of *Luffa* v2 as an additional input which can freely be chosen, then we can still make some theoretical observations for the hash function applied to one-block messages. Recall that w equals 3, 4 and 5 for a message digest of 256, 384 and 512 bits respectively.

In this setting, *Luffa* v2 is a function from $(256(w+1)-1)$ bits to $128(w-1)$ bits, where the input bits correspond to the bits of the initial value and of the message block. But, *Luffa* v2 is composed of a linear message injection function, followed by a function G from $256w$ bits to $128(w-1)$ bits. Therefore, the degree of the $(256(w+1)-1)$-bit function *Luffa* v2 is equal to the degree of G and cannot exceed $256w$. Moreover, we can show that the degree of this function is even smaller than $256w$ due to the particular design of the inner permutation.

This new upper bound on the degree of *Luffa* v2 comes from the fact that G can be decomposed as the inner permutation P, *i.e.*, the parallel applications of w independent nonlinear permutations Q_j of $n_0 = 256$ variables with degree less than $(n_0 - 2)$, followed by some rounds of the finalization function `Final`. Moreover, the first 256 bits of the message digest are extracted after a single application of `Final`. Then, using that the finalization function consists of 8 iterations of `Step` and has then degree at most 252, Theorem 2 implies that

$$\deg(\texttt{Final} \circ P) \leq 256w - \frac{256w - \deg(\texttt{Final})}{254}$$
$$\leq 256w - \frac{256w - 252}{254}$$
$$< 256w - (w-1)\,.$$

For the $(128(w-1))$-bit version of *Luffa* v2, we get that the first 256 output bits of *Luffa* v2 have degree at most $(256w - w)$. This property must be compared to the probability that this property holds for a randomly chosen function from $\mathbb{F}_2^{256(w+1)-1}$ to $\mathbb{F}_2^{256}$ bits. Such a function can be written as a polynomial with coefficients in $\mathbb{F}_{2^{256}}$ and the number of its monomials of degree greater than $(256w - w + 1)$ is

$$\sum_{i=0}^{256+w-1} \binom{256(w+1)-1}{i}\,.$$

Therefore, the probability that a randomly chosen function with the same parameters as *Luffa* v2-256 has degree at most 765 is $2^{-2^{837}}$.

For the 384-bit version (resp. for the 512-bit version), *i.e.*, for $w = 4$ (resp. $w = 5$), we get that *Luffa* v2 has 1280 variables (resp. 1536 variables) and degree at most 1020 (resp. 1275). The probability that this property holds for a randomly chosen function with the same parameters is $2^{-2^{933}}$ (resp. $2^{-2^{1010}}$).

6 Conclusions

We have found a new bound for the degree of iterated permutations. This improved bound has firstly led to zero-sum distinguishers for the full KECCAK-f permutation. Even if the security of the hash function is not affected, our results contradict the so-called hermetic sponge strategy. Additionally, a number of structural properties related to the existence of all-zero higher-order differentials and of zero-sum partitions have been presented for the *Luffa* hash family.

Acknowledgments

We would like to thank Gilles Van Assche for his valuable comments and interesting suggestions.

References

1. Aumasson, J.-P., Meier, W.: Zero-sum distinguishers for reduced KECCAK -f and for the core functions of Luffa and Hamsi. Presented at the Rump Session of Cryptographic Hardware and Embedded Systems - CHES 2009 (2009)
2. Bertoni, G., Daemen, J., Peeters, M., Van Assche, G.: KECCAK sponge function family main document. Submission to NIST, Round 2 (2009)
3. Boura, C., Canteaut, A.: Zero-Sum Distinguishers for Iterated Permutations and Application to KECCAK-f and Hamsi-256. In: Biryukov, A., Gong, G., Stinson, D.R. (eds.) SAC 2010. LNCS, vol. 6544, pp. 1–17. Springer, Heidelberg (2011)
4. De Cannière, C., Sato, H., Watanabe, D.: The reasons for the change of Luffa. Supplied with the Second Round Package
5. De Cannière, C., Sato, H., Watanabe, D.: Hash Function Luffa: Specification. Submission to NIST, Round 1 (2008)
6. De Cannière, C., Sato, H., Watanabe, D.: Hash Function Luffa: Specification. Submission to NIST, Round 2 (2009)
7. Canteaut, A., Videau, M.: Degree of composition of highly nonlinear functions and applications to higher order differential cryptanalysis. In: Knudsen, L.R. (ed.) EUROCRYPT 2002. LNCS, vol. 2332, pp. 518–533. Springer, Heidelberg (2002)
8. Duan, M., Lai, X.: Improved zero-sum distinguisher for full round KECCAK -f permutation. IACR ePrint Report 2011/023 (January 2011), http://eprint.iacr.org/2011/023
9. Khovratovich, D., Naya-Plasencia, M., Röck, A., Schläffer, M.: Cryptanalysis of Luffa v2 components. In: Biryukov, A., Gong, G., Stinson, D.R. (eds.) SAC 2010. LNCS, vol. 6544, pp. 388–409. Springer, Heidelberg (2011)
10. Knudsen, L.R.: Truncated and higher order differentials. In: Preneel, B. (ed.) FSE 1994. LNCS, vol. 1008, pp. 196–211. Springer, Heidelberg (1995)
11. Knudsen, L.R., Rijmen, V.: Known-key distinguishers for some block ciphers. In: Kurosawa, K. (ed.) ASIACRYPT 2007. LNCS, vol. 4833, pp. 315–324. Springer, Heidelberg (2007)
12. Lai, X.: Higher order derivatives and differential cryptanalysis. In: Proc. Symposium on Communication, Coding and Cryptography, in Honor of J. L. Massey on the Occasion of His 60'th Birthday. Kluwer Academic Publishers, Dordrecht (1994)
13. Watanabe, D., Hatano, Y., Yamada, T., Kaneko, T.: Higher Order Differential Attack on Step-Reduced Variants of *Luffa* v1. In: Hong, S., Iwata, T. (eds.) FSE 2010. LNCS, vol. 6147, pp. 270–285. Springer, Heidelberg (2010)

Cryptanalysis of PRESENT-Like Ciphers with Secret S-Boxes

Julia Borghoff[*], Lars R. Knudsen, Gregor Leander, and Søren S. Thomsen[**]

Department of Mathematics, Technical University of Denmark

Abstract. At Eurocrypt 2001, Biryukov and Shamir investigated the security of AES-like ciphers where the substitutions and affine transformations are all key-dependent and successfully cryptanalysed two and a half rounds. This paper considers PRESENT-like ciphers in a similar manner. We focus on the settings where the S-boxes are key dependent, and repeated for every round. We break one particular variant which was proposed in 2009 with practical complexity in a chosen plaintext/chosen ciphertext scenario. Extrapolating these results suggests that up to 28 rounds of such ciphers can be broken. Furthermore, we outline how our attack strategy can be applied to an extreme case where the S-boxes are chosen uniformly at random for each round and where the bit permutation is secret as well.

Keywords: Symmetric key, block cipher, PRESENT, differential cryptanalysis.

1 Introduction

Small computing devices are becoming more and more popular and establish a part of the pervasive communication infrastructure. One example of these tiny computing devices are RFID systems which are used e.g., for identifying and tracking animals or on toll roads. A prediction for the future is that RFID tags will replace bar codes. But this extensive deployment of computing devices is not only useful and convenient, it also carries a wide range of security risks. At the same time we are talking about extremely resource constrained environments. Therefore, the demand for lightweight encryption algorithms increases. The block cipher PRESENT [1] is an important example of a lightweight cipher. It consists of alternate layers of substitutions and permutations.

Important design principles of lightweight ciphers are an efficient hardware implementation, a good performance and a moderate security level. Usually there is a trade-off between the performance and the security level. In order to speed up the algorithm we want as few rounds of encryption as possible but there is a minimum number of rounds required to assure the security level.

PRESENT is a 64-bit iterated block cipher that comes in two variants, one with an 80-bit key and one with a 128-bit key. Both run in 31 rounds, each

[*] Supported by a grant from the Danish research council, grant number 274-07-0246.
[**] Supported by a grant from the Villum Kann Rasmussen Foundation.

A. Joux (Ed.): FSE 2011, LNCS 6733, pp. 270–289, 2011.

round has three layers, a substitution layer consisting of 16 parallel applications of the same 4-bit S-box, a permutation layer consisting of a bit-wise permutation of 64 bits, and a key addition layer, where a subkey is exclusive-ored to the text. PRESENT was designed to allow fast and compact implementation in hardware. The best known cryptanalytic attack on PRESENT is a linear attack on 26 of the 31 rounds [2]. The attack requires all possible 2^{64} texts and has a running time of 2^{72}. Although this attack is hardly practical, it illustrates that the number of rounds used should not be dramatically reduced.

An idea of how to strengthen the cipher in a way that enables one to reduce the number of rounds has been presented by two researchers from Princeton University. The cipher Maya [3] is a 16-round SP-network similar to PRESENT. The main difference is that the substitution layer of Maya consists of 16 different S-boxes which are key dependent and therefore kept secret. The bit permutation between the S-box layers is fixed and public. In each round a round key is xored to the text. It is argued that this cipher can be implemented efficiently in practice and also that "differential cryptanalysis is infeasible". In this paper we will investigate the question if such a cipher would be stronger than the original, and if so, how much stronger.

The Maya design is one particular way of designing a PRESENT-like cipher with secret components. In an extreme case one could choose 16 S-boxes uniformly at random and independently for every round. Furthermore one could also make the bit permutation part of the key and chosen uniformly at random from the set of all such permutations and used repeatedly or as another extreme, a bit permutation is chosen for each round uniformly at random and independently for every round.

The idea of having ciphers where the substitutions are not publicly known and part of the secret key is not new. Notable examples are Khufu [4],the Khufu variation Blowfish [5] and GOST [6] as well as other proposals [7,8].

Our results. In this paper we focus on the Maya case. We present a novel differential-style attack which enables us to find the S-boxes in the first round one by one.

The attack was implemented and successfully recovered the secret key in versions up to 16 rounds. The complexity of the attack on the 16-round version is approximately 2^{38} using a similar number of chosen plaintexts/chosen ciphertexts. In particular, the proposed cipher Maya can be broken with practical complexity. In our experiments the correct key was usually found in less than one week on a standard PC.

To better understand the running time of the attack, we establish a simplified, mathematical model for the complexity of this attack and verify by numerous experiments that the model fits the real world. Extrapolation of the experimental data, backed up by our model, indicate that the attack has the potential to break up to 28 rounds with a chosen plaintext complexity less than 2^{64}.

Furthermore, we outline how even the extreme case of PRESENT-like ciphers with secret components, that is the case where all components in all rounds are chosen uniformly and independent at random can be attacked.

Related work. Biryukov and Shamir investigated the security of iterated ciphers where the substitutions and permutations are all key-dependent [9]. In particular they analysed an AES-like cipher with 128-bit blocks using eight-bit S-boxes. An attack was presented on five layers (SASAS, where S stands for substitution and A stands for affine mapping) of this construction which finds all secret components (up to an equivalence) using 2^{16} chosen plaintexts and with a time complexity of 2^{28}. Using the terminology of "rounds" as in the AES, this version consists of two and a half rounds.

The extreme case of our cipher, where the S-boxes and the bit-permutation are chosen at random for each round, is a special instance of the SASAS cipher [9]. In fact the attack of Biryukov and Shamir applies to three rounds of this variant and has a running time of 2^{16} using 2^8 chosen texts. However the complexity of the attack for more than three rounds is unclear, but seems to grow very quickly [9]. The SASAS attack is a multiset attack whereas we use a differential-style attack to recover the S-boxes. Also, the technique to recover the bit permutation is different.

There have been other attempts to cryptanalyse ciphers with secret S-boxes. Gilbert and Chauvaud presented a differential attack on the cipher Khufu [10]. Khufu is an unbalanced Feistel cipher and the attack exploits the relatively slow diffusion in the cipher and bears some resemblance with our work. Also, Vaudenay provided cryptanalysis of reduced-round variants of Blowfish [11]. Moreover, the cipher C2, which has a secret S-box, was cryptanalysed by Borghoff et al. [12].

Organisation. The paper is organised as follows. In Section 2 the cipher is presented. Section 3 explains the approach for recovering the secret S-boxes. In Section 4, practical issues of the attack are discussed. In Section 5 we give experimental results for the attack when applied to the Maya cipher [3]. Section 6 describes our model to back up the extrapolations of the experimental data. We outline the more general case and further improvements in Section 7. Section 8 holds the conclusion.

2   The Cipher

We focus on a PRESENT-like cipher where the secret consists of one round key for each round and 16 secret S-boxes. We assume that the round keys and the S-boxes are randomly chosen. In practice these secret components might be derived from a master key using a key schedule which generates key dependent round keys and S-boxes. These 16 randomly chosen S-boxes form the substitution layer which is used repeatedly throughout all the rounds. The permutation layer consists of a bit permutation which is fixed and publicly known.

One round of encryption works as follows (cf. Algorithm 1). The current text is divided into nibbles of 4 bits which are processed by the 16 S-boxes in parallel. Then the bit permutation is applied to the concatenation of the output of the S-boxes and the output is xored with the round-key.

Require: X is a 64-bit plaintext
Ensure: $C = E_K(X)$ where E_K means the encryption function with key K
1: Derive 16 S-boxes S_i and N round keys K_i from K
2: STATE $\leftarrow X$
3: **for** $i = 1$ to N **do**
4:    Parse STATE as $\text{STATE}_0 \| \cdots \| \text{STATE}_{15}$, where each STATE_j is a four-bit nibble
5:    **for** $j = 0$ to 15 **do** /*Substitution layer*/
6:       $\text{STATE}_j \leftarrow S_j(\text{STATE}_j)$
7:    **end for**
8:    Reassemble STATE
9:    Apply bit permutation to STATE
10:   Add round key K_i to STATE
11: **end for**
12: $C \leftarrow$ STATE

Algorithm 1. Pseudo-code of a PRESENT-like cipher with secret S-boxes. The number of rounds is N.

The cipher Maya, proposed by Gomathisankaran and Lee [3], is an instance of the cipher described in Algorithm 1 with $N = 16$. The authors claim that it is efficient in a hardware implementation.

We attack this cipher by recovering all 16 S-boxes. However, in the general case, we do not know the last-round key, and therefore what we recover is in fact the 16 S-boxes xored with the last round key. Once this is done, we can peel off the first and last layers of encryption, and attack the cipher with two rounds less; this time, the S-boxes are known and a standard differential or linear attack can be mounted to extract the round keys. What we obtain in the end is an equivalent description of the cipher, but not necessarily the key. Still, the equivalent description of the cipher will allow us to encrypt or decrypt any text of our choice.

Furthermore, we shall outline how our attack can be applied to a generalization. Here, the S-boxes are chosen uniformly at random for each round. Additionally, the bit permutation can be chosen randomly for each round and kept secret as part of the key. In this case, the addition of the round keys is not necessary because it can be seen as part of the S-boxes. Furthermore the permutation is omitted in the last round. This extreme variant can be compared with an instance of SASAS [9]. Note that in this variant nothing but the block size and the number of rounds is known. The pseudo-code of this variant is described as Algorithm 2.

3   Principle of the Attack

In this section, we explain the idea of our approach to recover the S-boxes in the basic variant of a PRESENT-like cipher with secret S-boxes. It is a differential-style attack and the complexity is analysed in Section 6.

Recall that in the basic variant of the cipher (cf. Algorithm 1), there are 16 secret S-boxes which are applied in all rounds. We denote these 16 S-boxes S_i,

Require: X is a 64-bit plaintext
Ensure: $C = E_K(X)$ where E_K means the encryption function with key K
 1: Derive $16 \cdot N$ S-boxes $S_{i,j}$, $1 \leq i \leq N$, $0 \leq j \leq 15$ and $N - 1$ bit permutations P_i
    from K
 2: STATE $\leftarrow X$
 3: **for** $i = 1$ to N **do**
 4:     Parse STATE as $\mathsf{STATE}_0 \| \cdots \| \mathsf{STATE}_{15}$, where each STATE_j is a four-bit nibble
 5:     **for** $j = 0$ to 15 **do** /*Substitution layer*/
 6:         $\mathsf{STATE}_j \leftarrow S_{i,j}(\mathsf{STATE}_j)$
 7:     **end for**
 8:     Reassemble STATE
 9:     **if** $i < N$ **then**
10:         Apply bit permutation P_i to STATE
11:     **end if**
12: **end for**
13: $C \leftarrow$ STATE

Algorithm 2. Pseudo-code of a PRESENT-like cipher with secret S-boxes and secret bit permutations, all unique for each of the N rounds.

$0 \leq i < 16$, and we note that all S_i are bijective mappings with the signature $\mathbb{F}_2^4 \rightarrow \mathbb{F}_2^4$. For convenience, we introduce the following notation.

Definition 1. *Given the S-box S and $e \in \mathbb{F}_2^4$, we denote the set of all pairs $\{x, y\}$ such that $S(x) \oplus S(y) = e$ by D_e. Here, we consider the pairs $\{x, y\}$ and $\{y, x\}$ to be identical. A pair $\{x, y\}$ belonging to a set D_e where e has Hamming weight 1 is called a* slender *pair. A set consisting of slender pairs is called a* slender set.

Without loss of generality, we explain how to recover the leftmost S-box S_0. In order to obtain information about S_0, we encrypt a certain number t of structures P_{r_i} of plaintexts of the form

$$P_{r_i} = \{(x \| r_i) \mid x \in \mathbb{F}_2^4\}$$

where each $r_i \in \mathbb{F}_2^{60}$ for $0 \leq i < t$ is chosen uniformly at random. Two different plaintexts $(x \| r_i), (y \| r_i)$ in P_{r_i} have an input difference of the form

$$(x \| r_i) \oplus (y \| r_i) = (? \| 0^{60}),$$

where 0^n denotes the bit string consisting of n zeros.

We shall be looking at the corresponding ciphertexts in order to see if there is an input pair for which only one S-box is active in the ciphertext. For now, let $p(\{x, y\})$ denote the probability that only one S-box is active in the ciphertext difference when the plaintext pair is $\{x \| r, y \| r\}$, taken over all the different choices of $r \in \mathbb{F}_2^{60}$. The attack is based on some assumptions. The first assumption is a standard one in differential cryptanalysis:

Assumption 1. *The probability $p(\{x, y\})$ depends only on the value of $S(x) \oplus S(y)$, not specifically on the pair $\{x, y\}$. Hence, given $e = S(x) \oplus S(y)$, we can denote this probability p_e.*

We shall be particularly interested in identifying slender pairs. In order to do this, we need the following assumption, which has been experimentally verified to hold in most cases.

Assumption 2. *The probability p_e is higher when e has Hamming weight 1, than when e has Hamming weight greater than 1.*

Learning all the probabilities p_e would require encryptions of all 2^{64} possible plaintexts, but we can estimate the probabilities by introducing counters

$$C(\{x,y\}) = \left| \{ r_i \mid \exists j : E(x\|r_i) \oplus E(y\|r_i) = 0^{4j}\|?\|0^{60-4j} \} \right|$$

for all pairs $\{x,y\}$, $x,y \in \mathbb{F}_2^4$. Hence, the counter $C(\{x,y\})$ counts how often only one S-box is active in the ciphertext pair when the input pair to S-box S_0 is $\{x,y\}$.

Assumption 1 says that pairs belonging to the same set D_e should also have similar counter values when sufficiently many plaintexts have been encrypted. Assumption 2 says that the highest counter values will (usually) correspond to slender pairs. In the attack we are going to try to identify the slender sets, and this will be relatively easy if the probabilities p_e and $p_{e'}$, $e \neq e'$, are sufficiently different. Experiments show that this condition is often satisfied.

The counter C consists of 120 values since there are $\binom{16}{2} = 120$ different pairs $\{x,y\}$. After encrypting sufficiently many structures we may sort C in descending order, and thereby hopefully obtain a partitioning of the 120 pairs into a number of sets corresponding to D_e for different values of e. For every $e \neq 0$ it holds $|D_e| = 8$. We shall return to this partitioning method in a moment. Our final goal will be to learn all four slender sets D_e.

Generalizing to all S-boxes and their inverses. In a practical attack we do not only want to eventually recover the S-box S_0, but all S-boxes. The above observations can clearly be generalized to all S-boxes by introducing additional types of structures and additional counters.

Moreover, the symmetry between encryption and decryption in the cipher we are considering here means that one may obtain the same type of information about the inverse S-boxes as one obtains about the S-boxes themselves. This can even be done in a chosen-plaintext setting, although it may require more texts than in a chosen-ciphertext setting.

Assume now that we have identified u slender sets for some S-box S, and v slender sets for its inverse S^{-1}. The following table shows the average number of S-boxes that would give rise to the same $u + v$ sets; these averages are based on 100,000 randomly generated S-boxes.

$u \backslash v$	1	2	3	4
1	207	3.52	1.44	1.19
2	3.52	1.16	1.03	1.01
3	1.44	1.03	1.01	1.01
4	1.19	1.01	1.01	1.01

Evidently, if $u + v \geq 6$, the S-box is usually uniquely determined from the $u + v$ sets, and in many cases, fewer sets are sufficient. However, there exist S-boxes S which are not uniquely determined even if all four slender sets are known for both S and S^{-1}.

On a side note: if D_e and $D_{e'}$ are known for some S-box S, then $D_{e \oplus e'}$ does not give any new information about S, since $D_{e \oplus e'}$ can be derived from D_e and $D_{e'}$. Clearly, if $\{x, y\} \in D_e$ and $\{x, z\} \in D_{e'}$, then $\{y, z\} \in D_{e \oplus e'}$. This observation generalizes to more than two sets. In general, given sets D_{e_i} one can construct all sets D_e where e can be written as a linear combination of the vectors e_i, see Lemma 2 in Appendix A. Therefore, we shall generally only be interested in the four slender sets, since all other sets give no additional information about the S-box.

We now describe a number of ways to partition the pairs into sets and to check that this partitioning is correct.

Partitioning pairs into sets. Assume again that we are trying to recover S-box S_0. Our starting point for partitioning pairs (in particular the slender pairs) into sets is the counter C.

The straightforward partitioning method simply sorts C in descending order, and takes the first eight pairs as the first set, the next eight pairs as a second set, etc. Using this method obviously means that we shall often make the wrong partitioning into sets, but the partitioning can be checked using the very strong *filtering methods* described in the following subsection.

Filtering methods. Given u sets for some S-box S and v sets for its inverse S^{-1}, the most indicative method to check whether these sets may be correct is to see how many S-boxes would give rise to the same sets. If no S-box gives rise to these sets, then clearly the sets must be wrong. However, counting the number of S-boxes that give rise to these sets is somewhat inefficient (see, however, Section 4), and as we have seen, if we only know a few sets, there are usually several S-boxes that give rise to the same sets, and so the probability of a false positive is high in this case. We call this filter the *existence filter*.

A much more efficient method is based on the trivial observation that for any valid set D_e, we have that $\{x, y : \{x, y\} \in D_e\} = \mathbb{F}_2^4$. In other words, a valid set "covers" all values in $\mathbb{F}_2^4$. Hence, if we have identified a candidate set D containing two pairs $\{x, y\}$ and $\{x, z\}$, then D cannot be a valid set. Although this method is very simple, it is in fact a very strong filter; the probability that eight randomly chosen pairs among the 120 pairs cover all values in $\mathbb{F}_2^4$ is only

$$\prod_{i=1}^{7} \frac{\binom{2i}{2}}{\binom{16}{2} - i} \approx 2^{-18.7},$$

and therefore in practice, many wrong candidate sets are discovered by this method. We call this filter the *cover filter*.

It should be noted that one can prove that the cover filter is not only necessary, but also sufficient; see Appendix A.

The final filtering method that we describe here is based on the observation that if $\{x_1, y_1\}$ and $\{x_2, y_2\}$ belong to the same set D_e, then $\{x_1, y_2\}$ and $\{x_2, y_1\}$ will also belong to the same set $D_{e'}$ for some $e' \neq e$, and likewise, $\{x_1, x_2\}$ and $\{y_1, y_2\}$ will belong to the same set $D_{e''}$ for some $e'' \notin \{e, e'\}$. To see this, note that if $\{x_1, y_1\}$ and $\{x_2, y_2\}$ belong to the same set D_e, then (by definition) $S(x_1) \oplus S(y_1) = S(x_2) \oplus S(y_2) = e$, and therefore $S(x_1) \oplus S(y_2) = S(x_2) \oplus S(y_1) = e \oplus S(y_1) \oplus S(y_2) \neq e$, etc. Hence, assume that we know two sets D' and D'' (both already known to cover $\mathbb{F}_2^4$), and that $\{a, b\} \in D'$ and $\{a, c\} \in D''$. Now, if $\{c, d\} \in D'$, then for these two sets to both be valid, it must hold that $\{b, d\} \in D''$. We call this filter the *bowtie filter*; if one follows the "partner" b of a in the set D' and jumps to the next set D'' to find the partner d of b there and so forth, then one should come back to the pair $\{a, b\}$ in D' after two jumps back and forth between the two sets, hence forming a bowtie-shaped cycle:

$$D' = \{ \{a, b\}, \{c, d\} \cdots$$

$$D'' = \{ \{a, c\}, \{b, d\} \cdots$$

3.1 Relaxed Truncated Differentials

The method considered so far increments a counter only when there is a single active S-box in the ciphertext pair. The probability of this event is relatively low, so many plaintext pairs are needed before it is possible to partition pairs into sets.

It is much more likely that the weight one difference spreads moderately through the cipher resulting in a few active S-boxes in the ciphertext. Hence, we might find slender pair candidates more efficiently by looking at ciphertext pairs with more than one active S-box. The more active S-boxes we allow, the more noise we will get, and so there is a tradeoff between the signal-to-noise ratio, and the strength of the signal.

It turns out that allowing even a relatively large number of active S-boxes does not introduce too much noise. This can be used to make the attack more efficient. For each input S-box S_i and for each pair $\{x, y\}$ we introduce counters $C_{i,j}(\{x, y\})$. We increment the counter $C_{i,j}(\{x, y\})$ every time the input pair $\{x, y\}$ to S-box S_i (with a random but fixed input to the other S-boxes) leads to exactly j S-boxes being active, where j ranges from 1 to 15. When we have done a number of encryptions we may sort the counters $C_{i,j}$ for some pair i, j. If the cover filter identifies sets based on this sorting, we assume that these are correct slender sets. When we have several sets, we use the bowtie filter to check the validity of the sets. We do this for increasing j from 1 to 15. Since the cover filter is a very strong filter, the risk of errors is low, both in the cases where the signal is weak (small values of j), and also in the cases where there is a lot of noise (large values of j).

4   The Attack in Practice

We now describe how the attack is carried out in practice. The attack consists of a data collection phase followed by an S-box recovery phase, and those two phases are repeated until all or almost all S-boxes have been recovered.

4.1   Data Collection Phase

In the data collecting phase we simply encrypt structures and increment counters when applicable. Each structure consists of 16 plaintexts differing in only a single input S-box. Which S-box is active is a random choice among the S-boxes that have not already been recovered.

After encryption, we check all 120 pairs of ciphertexts to see if any of them are active in less than 16 S-boxes. If so, we increment the corresponding counter for the input pair to the S-box that was active in the plaintext.

We also carry out decryptions in order to obtain information about the inverse S-boxes.

4.2   S-Box Recovery Phase

Every once in a while, we stop collecting data and try identifying sets for each S-box. This is done by first sorting the counters for each number of active output S-boxes. We start with the lowest number of active output S-boxes. We check if the top eight counter values in the sorted list passes the cover filter. If so, we consider these eight pairs a slender set and add it to a collection of identified sets, unless the set is already present in the collection. When there are multiple sets in the collection, we check that they pass the bowtie filter. We then look at the next eight pairs and so forth. We stop adding sets when we have identified four sets, or we run into an inconsistency such as a failing bowtie test or non-disjoint sets. In case of an inconsistency, we give up identifying sets for this S-box.

The bowtie filter can also be used to filter out candidate sets that can be derived from existing sets. Consider as an example a situation where the following two candidate sets D_e and $D_{e'}$ (passing the bowtie test) have been identified:

$$D_e = \{\{0,1\}, \{2,3\}, \{4,5\}, \{6,7\}, \{8,9\}, \{a,b\}, \{c,d\}, \{e,f\}\}$$
$$D_{e'} = \{\{0,2\}, \{1,3\}, \{4,6\}, \{5,7\}, \{8,a\}, \{9,b\}, \{c,e\}, \{d,f\}\}.$$

From these two sets we can derive the set $D_{e \oplus e'}$ directly as

$$D_{e \oplus e'} = \{\{0,3\}, \{1,2\}, \{4,7\}, \{5,6\}, \{8,b\}, \{9,a\}, \{c,f\}, \{d,e\}\}$$

As an example, $S(0) \oplus S(3) = (S(0) \oplus S(1)) \oplus (S(1) \oplus S(3)) = e \oplus e'$. Hence, if we identify a set which can be derived from two sets already identified, then we should not add the third set to our collection (on the assumption that the first two sets are slender, which means the third is not).

We note that if one swaps two "bowtie pairs" in two valid sets (e.g., the pairs $\{0,1\}$ and $\{2,3\}$ could be swapped with $\{0,2\}$ and $\{1,3\}$ in D_e and $D_{e'}$ above), then the resulting sets will still pass both the cover and the bowtie test. This is a potential cause for errors; if two sets have roughly the same probability of causing a single active S-box in the ciphertext, and the distribution of the probabilities for each output S-box is similar for the two sets, then we are likely to generate wrong sets that pass both the cover and the bowtie test. This error may be caught by the existence filter (cf. the following), but if not, then we'll be recovering the wrong S-box. This does happen in practice, although it is rather rare.

We repeat the above method of identifying sets for the inverse S-boxes as well, maintaining separate counters for these.

Once we have identified as many sets as possible using this method (for both the S-box and its inverse), we can apply the existence filter to check if these sets can possibly be valid; if there is no S-box generating these sets, then the sets are obviously not valid. As mentioned in Section 3, applying the existence filter is not terribly efficient; on the other hand, it is not terribly slow either. A reasonably efficient way to implement it is by making guesses for values of $S(0)$ and the exact values e for the identified sets D_e until one runs into an inconsistency with the candidate sets. Note that once these guesses have been made, we may find the "partner" of 0 in all candidate sets. For instance, if the two sets D_e and $D_{e'}$ in the example above are our candidate sets, and we guess that $S(0) = 0$, then we would know that $S(1) = 2^i$ and $S(2) = 2^j$ for some (guessed) i, j, $i \neq j$ and $0 \leq i, j < 4$. We would obtain similar information about the inverse S-box from the candidate sets for the inverse S-box. This method is able to find all candidate S-boxes in a fraction of a second given at least one set for the S-box and one set for its inverse.

If an S-box (or a candidate for it) has been recovered, we stop considering this S-box both in the data collection and the S-box recovery phase. If not all S-boxes have been recovered, we continue the data collection phase. In some cases, we have to give up recovering one or more S-boxes because we are unable to identify sufficiently many sets, or because we consistently get no candidates for the S-box based on the identified sets. In the latter case, there is obviously an error in the partitioning into sets. If we consistently obtain multiple candidates for an S-box, we may also accept this and consider the S-box recovered, keeping a record of all candidates.

5   Case Study: The Block Cipher Maya

Maya is a block cipher proposed at WCC 2009 [3]. It is a PRESENT-like cipher with key dependent S-boxes (repeated in every round) and a fixed, known bit permutation (see Fig. 1). Each round also contains an addition of a round key. The round keys and the S-boxes are derived from the 1024-bit master key.

Since the S-boxes are the same in every round, using the differential-style attack described above, we are able to get information on the S-boxes and their inverses. We get information on both directions for every encrypted pair and can choose to also do decryptions to obtain information about the inverse of a specific

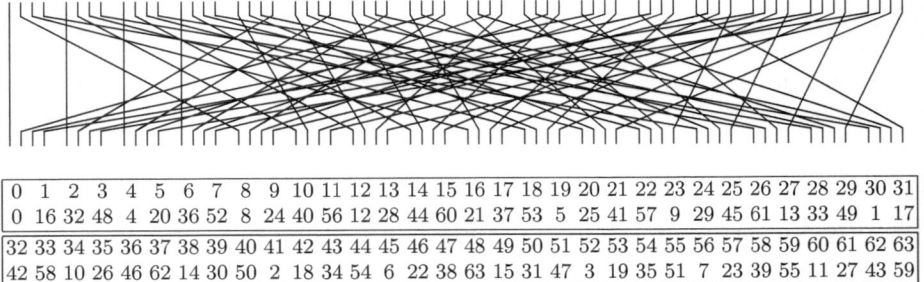

0	1	2	3	4	5	6	7	8	9	10	11	12	13	14	15	16	17	18	19	20	21	22	23	24	25	26	27	28	29	30	31
0	16	32	48	4	20	36	52	8	24	40	56	12	28	44	60	21	37	53	5	25	41	57	9	29	45	61	13	33	49	1	17

32	33	34	35	36	37	38	39	40	41	42	43	44	45	46	47	48	49	50	51	52	53	54	55	56	57	58	59	60	61	62	63
42	58	10	26	46	62	14	30	50	2	18	34	54	6	22	38	63	15	31	47	3	19	35	51	7	23	39	55	11	27	43	59

Fig. 1. The Maya bit permutation.

S-box. In this way we often recover at least two sets in each direction, which usually means all the S-boxes can be determined uniquely. The key addition, however, means that we only obtain the correct S-boxes up to an xor by the last round key, which is unknown. However, this still enables us to peel off the first and the last round of encryption, after which the attack can be repeated on this reduced cipher. Moreover, we expect that once the S-boxes are known, a dedicated differential or linear attack is more efficient than our general attack. In the end, we obtain a description of an equivalent cipher.

The standard number of rounds in Maya is 16 and below the log of the complexity to recover the secret S-boxes for a number of different randomly chosen example keys is given. Complexities in *italics* are extrapolated values from running the attack on fewer rounds.

Case	1	2	3	4	5	6	7	8	9	10	11	12	13	14	15	16
Complexity	*45.5*	36.0	35.9	36.9	35.7	*39.3*	37.4	37.1	*40.6*	38.5	*39.4*	*39.5*	36.0	36.7	*38.3*	37.4

Moreover, Table 1 (Appendix B) shows the log of the complexity (number of texts) as a function of the number of rounds for the same example keys. See Fig. 2 for a graphical representation. The complexities refer to obtaining all 16 S-boxes (whenever possible, see discussion below), so that the first and the last round can be peeled off, and the cipher with two round less can then be attacked.

In this implementation of the attack, an S-box was considered correctly recovered if only one S-box gave rise to the given partitioning into sets (or the given top 32 pairs). However, if a substantial amount of time had been spent on an S-box, the conditions were relaxed such that even if there were more than one candidate S-box, work on this S-box was still discontinued and all candidates were printed. In extreme cases, where there were no candidate S-boxes after a lot of time had been spent trying to recover the S-box, that S-box was given up. The choice of when to accept multiple candidates, or when to give up an S-box, obviously affects the complexity of the attack. A more sophisticated implementation might adapt better to these situations. As an example, if the

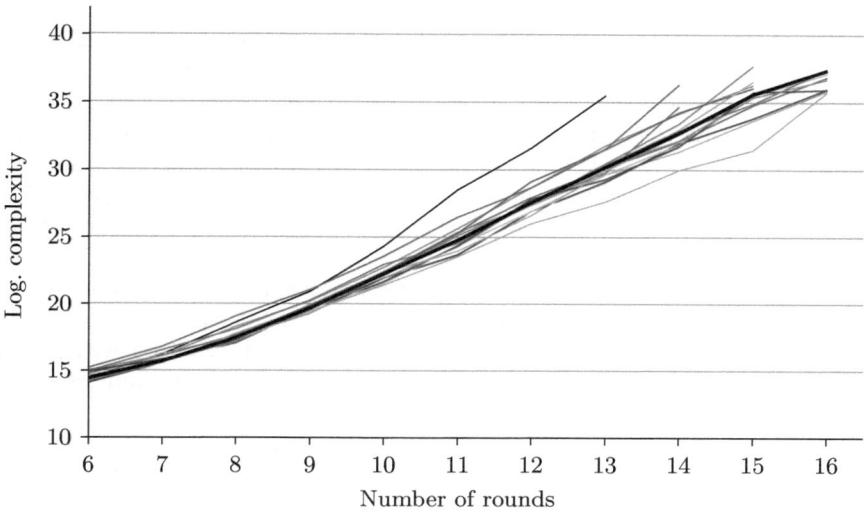

Fig. 2. A graphical representation of the data in Table 1. The thick line represents the median computed for each number of rounds.

program consistently gives rise to the same partitioning into sets, and there are no candidates for this partitioning, one might try swapping elements between sets in such a way that the bowtie condition still holds.

The error rate of the attack is very low. If we consider the highest number of rounds broken in each of the 16 test cases, then the total number of S-boxes that had to be recovered was $16 \cdot 16 = 256$. Of these, 245 were correctly recovered with only a single S-box candidate. For seven S-boxes, there were multiple candidates, and the correct S-box was always one of these. The number of candidates ranged from two to four. Three out of 256 S-boxes were incorrectly recovered with only a single S-box candidate. One S-box was given up due to too much time spent trying to recover it.

In a real attack, the fact that some S-boxes were incorrectly recovered would be discovered after attempting to break the cipher reduced by the first and the last rounds. By making sure that a large amount of information about the identified sets and the counter values is recorded, it is likely that one would be able to locate the S-box causing the problem. For instance, there may be 16 counters that are all similar, meaning that it is likely that two sets have been mixed up.

6   Model for the Complexity of Recovering Sets D_e

For a small number of rounds the attack to recover one or more sets D_e has small complexity and it is possible to get sufficient experimental data. However, to be able to extrapolate the attack complexity we describe a theoretical model below.

We again focus on recovering a single S-box e.g., S_0. In the attack we are faced with the problem to group 120 counters $C(\{x, y\})$, each belonging to an input pair to an S-box of the first round, into 15 distinct groups. All pairs within a group should yield the same output difference, i.e., belong to a set D_e for some e.

Interpreting the counters $C(\{x, y\})$ as random variables, a counter $C(\{x, y\})$, with $S(x) \oplus S(y) = e$ is binomially distributed with parameters n and p_e . Here p_e is the probability that the difference $(e||0^{60})$ after the first layer of S-boxes yields to only one active S-box in the output and n is the number of text pairs.

Assumption 2 states that counters $C(\{x, y\})$ such that $S(x) \oplus S(y)$ has a weight greater than one are significantly smaller than others and we therefore focus only on the 32 counters corresponding to slender pairs. Thus, we consider 8 counters distributed with parameters (n, p_1), 8 distributed with parameters (n, p_2), 8 distributed with parameters (n, p_4) and finally 8 counters distributed with parameters (n, p_8) (here we identified $e = (0, 0, 0, 1)$ with 1, $e = (0, 0, 1, 0)$ with 2 etc.). Without loss of generality we assume $p_1 \geq p_2 \geq p_4 \geq p_8$ and that holds $p_1 \neq p_2$. The attack works by looking at the 8 highest counters and is successful if those counters correspond to the same output difference, e.g., $e = 1$, of the S-box. The attack fails whenever there exists a pair $\{x_1, y_1\}$ with output difference '1' and a pair $\{x_2, y_2\}$ with $S(x_2) \oplus S(y_2) \neq 1$ such that $C(\{x_1, y_1\}) \leq C(\{x_2, y_2\})$. In the following we estimate this failure probability depending on the number of samples n.

To simplify the problem for now, we consider only two pairs $\{x_1, y_1\}$ and $\{x_2, y_2\}$ and their corresponding counters where $C(\{x_1, y_1\})$ is distributed with parameters (n, q) and $C(\{x_2, y_2\})$ is distributed with parameters (n, p) for $q > p$. The attack fails if $C(\{x_1, y_1\}) \leq C(\{x_2, y_2\})$ and thus we denote $Z = C(\{x_2, y_2\}) - C(\{x_1, y_1\})$ and

$$\text{err} = \Pr(C(\{x_1, y_1\}) \leq C(\{x_2, y_2\})) = \Pr(Z \geq 0).$$

To investigate this error further consider the usual approximation of the binomial distribution by the normal distribution, $C(\{x_1, y_1\}) \sim N(nq, nq(1 - q))$ and $C(\{x_2, y_2\}) \sim N(np, np(1 - p))$. With this approximation, the distribution of Z can be approximated by $Z \sim N(\mu, \sigma^2)$, where $\mu = n(p - q)$ and $\sigma = n(p(1 - p) + q(1 - q))$.

The density function for the normal distribution with mean μ and variance σ^2 is given by the following formula: $f(x) = \frac{1}{\sqrt{2\pi}\sigma} e^{-\frac{(x-\mu)^2}{2\sigma^2}}$. The integral of the normal density function is the normal distribution function

$$N(t) = \frac{1}{\sqrt{2\pi}} \int_{-\infty}^{t} e^{-\frac{1}{2}x^2} dx.$$

The error we make is thus described by

$$\text{err} \approx 1 - \frac{1}{\sqrt{2\pi}\sigma} \int_{-\infty}^{0} e^{-\frac{(x-\mu)^2}{2\sigma^2}} = 1 - \frac{1}{\sqrt{2\pi}} \int_{-\infty}^{\frac{-\mu}{\sigma}} e^{-\frac{x^2}{2}} = 1 - N\left(\frac{-\mu}{\sigma}\right).$$

The following lemma gives an estimate of the 'tail' $1 - N(x)$ which is useful to approximate the error.

Lemma 1 ([13]). *Let* $\phi(x) = \frac{1}{\sqrt{2\pi}} e^{-\frac{x^2}{2}}$ *be the normal distribution. As* $x \to \infty$

$$1 - N(x) \approx x^{-1}\phi(x).$$

Using the approximation of Lemma 1 yields

$$\text{err} \approx 1 - N(-\frac{\mu}{\sigma}) \approx -\frac{\sigma}{\mu}\frac{1}{\sqrt{2\pi}}e^{-\frac{1}{2}(\frac{\mu}{\sigma})^2}. \tag{1}$$

From (1) it follows that for a given failure probability err the sample must be of size

$$n > \frac{-c(p^2 - p + q^2 - q)}{(p - q)^2}, \tag{2}$$

where $c = \text{LambertW}\left(\frac{1}{2\,\text{err}^2\,\pi}\right)$ [14] is a small constant depending on the error. As example we can assume that $q = 2p$ then in order for the attack to be successful we need a sample of size $\frac{3c}{p}$.

After having estimated the failure probability for 2 counters, assuming independence, the total error probability err_t, that is, the probability of the event that one of the 8 counters with parameter (n, p_1) being smaller than one of the 24 counters with parameters $(n, p_2), (n, p_4), (n, p_8)$ can be bounded as

$$\text{err}_t \leq 1 - (1 - \text{err})^{8 \cdot 24}.$$

If we allow an error probability of $\text{err}_t \leq 0.5$, which in light of the strong cover filter is clearly sufficient, we need $\text{err} \leq 1 - 0.5^{1/(8 \cdot 24)} \approx 0.0036$. For this $c = 8$ is sufficient.

The next step is to find a way to estimate the probabilities p_e. Assuming the cipher is a Markov cipher we can model the propagation of differences through the cipher as a matrix multiplication of the difference distribution matrices and the permutation matrices. Considering the difference distribution table for the whole layer of S-boxes would yield a $2^{64} \times 2^{64}$ matrix. Therefore we determine the difference distribution matrix which contains only the probabilities for 1 to 1 bit differences, which as it turns out when comparing to experimental data, is a good approximation. This matrix is of size only 64×64. This enables us to simulate the propagation of 1 to 1 bit differences through a number of rounds using matrix multiplications. For the resulting matrix an entry (i, j) contains the probability that given the single, active input bit i after the first layer of S-boxes, a single output bit j in the second last round will be active. This matrix can therefore be used to get an estimate for the parameters of the counters. We determine the probability that given a fixed 1 bit difference after the first round exactly one S-box is active in the last round (analogously for the inverse). This can be done by summing over the corresponding matrix entries. Then we use formula (2) to calculate the number of plaintexts needed to recover at least two sets D_e in each direction. Note that in the original attack we do not restrict ourselves to having a single active S-box in the last round but a limited number

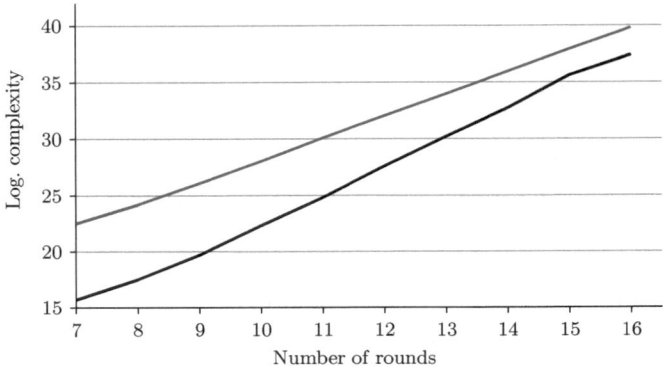

Fig. 3. Comparison between the medians of the experimental data and the model for recovering two sets D_e in each direction. The black line shows the experimental data while the red (gray) line shows the data from the model. The complexity unit is one plaintext.

of active S-boxes. Furthermore, we can expect that a single active S-box will on average not lead to 16 active S-box after two rounds of encryption. Thus we believe that in practice we can break at least two more rounds of encryption with the sample size determined by the model, meaning the model yields an upper bound for the complexity.

The comparison between the experimental data and the modeled data support this assumption.

To justify the introduced model we implemented the attack for a small number of rounds (see Section 5). For each number of rounds we sampled 1000 ciphers in our model to determine the sample size needed to distinguish between the two distributions. Fig. 3 gives a comparison of the experimental data with that of the model for the case that we want to recover at least four sets D_e for all 16 S-boxes. The black line shows the experimental data and the red line shows the model for an error of around 0.3% which corresponds to $c = 8$. The complexity denotes the logarithm of the number of plaintexts used. As seen, the model seems to give an upper bound on the complexity of the attack. In some rare cases the difference between p and q is close to zero, which leads to a very high attack complexity. These rare cases have a strong influence on the average complexity, hence we considered the median instead of the mean to estimate the complexity of the attack.

The modeled data suggest that we are able to break up to 28 rounds before we reach the bound of 2^{64} available plaintexts.

7 Extensions

In this section we outline some possible extensions of our attack. This includes some further improvements (cf. Section 7.1) as well as attacks on the more general variant of the cipher where all components in all rounds are chosen independently and uniformly at random (cf. Section 7.2).

7.1 Linear Cryptanalysis

In the differential-style attack one hypothesis is that the probability of a characteristic with a single-bit difference at the output of the S-box layer in the first round is correlated to a single-bit difference at the input to the S-box layer in the last round or to the number of active S-boxes in the last round. Using a similar hypothesis for linear characteristics one can mount a linear attack to extract information about the secret S-boxes. In the differential-style attack one tries to identify sets of eight pairs of values related to a certain differential. In a linear-style attack one tries to identify pairs of eight values related to a certain linear characteristic. It was confirmed in a small number of experiments on ciphers with a small number of rounds that this approach can be used to derive information about the S-boxes. One natural future direction of research is to combine the differential-style attack outline in this paper with a similar linear-style attack.

7.2 Fully Random PRESENT-Like Ciphers

In this section we consider PRESENT-like ciphers where the S-boxes and the bit permutations of all rounds are chosen independently and uniformly at random, that is ciphers given by Algorithm 2.

For such a cipher one would not get information about the inverse S-boxes like in the case of Maya. Moreover, the S-boxes are not uniquely determined, cf. Appendix A for more details. One needs to recover all four slender sets D_e for each S-boxes. We implemented a series of attacks on such ciphers and the results show that recovering four sets is indeed possible, but not for all S-boxes. The following table shows the results of our tests to fully recover one S-box in the first round. The complexity is the number of chosen plaintexts needed and is given as the median of 500 tests.

Rounds	Complexity	Probability
4	$2^{12.5}$	73%
5	$2^{15.5}$	82%
8	$2^{24.5}$	81%

In each test the computation was stopped if not all 4 slender sets where obtained with 2^{30} structures. The tests are very time-consuming which is why results for 6 and 7 rounds were not implemented.

Summing up, the attack does not seem to be able to fully recover all S-boxes of the first (or last) round, merely about 80%. However in the remaining cases, the attack identifies one, two or three sets S_e, which means that only a limited number of choices for these S-boxes remain. Depending on exactly how many choices of the S-boxes are left, one possible way to proceed is to simply make a guess, and repeat the attack on a reduced number of rounds. If S-boxes in other rounds cannot be successfully recovered, the guess might have been wrong. This is a topic for further research.

Recovering the Bit Permutations. Once the first S-box layer has been recovered, one can start recovering the first bit permutation layer. Here we outline the technique.

The idea is similar to the method of recovering S-boxes; one encrypts plaintext pairs differing in (e.g.) two bit positions. Whenever the output difference is small (e.g., one active S-box), one increments a counter for the pair of positions differing in the plaintext. This is repeated a number of times for all pairs of bit positions. One may now assume that the highest counter values correspond to pairs of bit positions that are mapped to the same S-box input.

This leads to information about which bit positions are mapped to the same S-box input in the next round. One can also vary three or four bit positions in order to obtain more information. The complexity of this method has not been thoroughly investigated, but preliminary results indicate that it is similar to (if not lower than) the complexity of recovering S-boxes.

8   Conclusion

In this paper a novel differential-style attack was presented and applied to 64-bit PRESENT-like ciphers with secret components. A variant with 16 secret S-boxes can be attacked for up to 28 rounds with a data complexity of less than 2^{64}. It is interesting to note that the best known attack on PRESENT, a linear attack, can be used to cryptanalyse up to 26 rounds of PRESENT (which has publicly known but carefully chosen S-boxes and bit permutation).

Also, the variant where the S-boxes and bit permutations are chosen at random for every round can also be attacked with a data complexity of less than 2^{64} for up to 16 rounds.

It is clear that our attacks exploit that there are weak differential properties for some randomly chosen four-bit S-boxes, and they do not apply to ciphers where the S-boxes are chosen as in PRESENT. However, a restriction to strong S-boxes (w.r.t to differential cryptanalysis) would also limit the size of the key.

References

1. Bogdanov, A., Knudsen, L.R., Leander, G., Paar, C., Poschmann, A., Robshaw, M.J.B., Seurin, Y., Vikkelsø, C.: PRESENT: An Ultra-Lightweight Block Cipher. In: Paillier, P., Verbauwhede, I. (eds.) CHES 2007. LNCS, vol. 4727, pp. 450–466. Springer, Heidelberg (2007)
2. Cho, J.: Linear Cryptanalysis of Reduced-Round PRESENT. In: Pieprzyk, J. (ed.) CT-RSA 2010. LNCS, vol. 5985, pp. 302–317. Springer, Heidelberg (2010)
3. Gomathisankaran, M., Lee, R.B.: Maya: A Novel Block Encryption Function. In: Proceedings of International Workshop on Coding and Cryptography (2009), http://palms.princeton.edu/system/files/maya.pdf (February 14, 2010)
4. Merkle, R.C.: Fast Software Encryption Functions. In: Menezes, A., Vanstone, S.A. (eds.) CRYPTO 1990. LNCS, vol. 537, pp. 476–500. Springer, Heidelberg (1991)

5. Schneier, B.: Description of a New Variable-Length Key, 64-bit Block Cipher (Blow-fish). In: Anderson, R.J. (ed.) FSE 1993. LNCS, vol. 809, pp. 191–204. Springer, Heidelberg (1994)
6. GOST: Gosudarstvennyi standard 28147-89, cryptographic protection for data pro-cessing systems. Government Committee of the USSR for Standards (1989) (in Russian)
7. Biham, E., Biryukov, A.: How to Strengthen DES Using Existing Hardware. In: Pieprzyk, J., Safavi-Naini, R. (eds.) ASIACRYPT 1994. LNCS, vol. 917, pp. 398–412. Springer, Heidelberg (1995)
8. Schneier, B., Kelsey, J., Whiting, D., Wagner, D., Hall, C., Ferguson, N.: Twofish: A 128-Bit Block Cipher Submitted as candidate for AES (February 05, 2010), http://www.schneier.com/paper-twofish-paper.pdf
9. Biryukov, A., Shamir, A.: Structural Cryptanalysis of SASAS. In: Pfitzmann, B. (ed.) EUROCRYPT 2001. LNCS, vol. 2045, pp. 394–405. Springer, Heidelberg (2001)
10. Gilbert, H., Chauvaud, P.: A Chosen Plaintext Attack of the 16-round Khufu Cryptosystem. In: Desmedt, Y.G. (ed.) CRYPTO 1994. LNCS, vol. 839, pp. 359–368. Springer, Heidelberg (1994)
11. Vaudenay, S.: On the Weak Keys of Blowfish. In: Gollmann, D. (ed.) FSE 1996. LNCS, vol. 1039, pp. 27–32. Springer, Heidelberg (1996)
12. Borghoff, J., Knudsen, L.R., Leander, G., Matusiewicz, K.: Cryptanalysis of C2. In: Halevi, S. (ed.) CRYPTO 2009. LNCS, vol. 5677, pp. 250–266. Springer, Heidelberg (2009)
13. Feller, W.: An Introduction to Probability Theory and Its Applications, 3rd edn. John Wiley and Sons, Chichester (1968)
14. Corless, R.M., Gonnet, G.H., Hare, D., Jeffery, D.J.: Lambert's W function in Maple. Maple Technical Newsletter 9 (1993)

A   What We Learn about the S-Boxes from the Sets

In this section, we discuss in detail how much we actually learned about an S-box after recovering one or more sets D_e. Here we focus on sets for the S-box itself and not on sets for its inverse. Before doing so, we remark that it is not possible to recover the S-boxes uniquely when no set for the inverse S-box is given. In particular, when two S-boxes S and S' differ by a permutation of the output bits and by adding a constant after the S-box, in other words, there exists a bit permutation P and a constant c such that

$$S'(x) = P(S(x)) + c,$$

then those S-boxes cannot be distinguished. We therefore call two S-boxes ful-filling the above relation *equivalent*.

Lemma 2. *Given r sets $D_{e_1}, \ldots, D_{e_r}$ for $1 \leq r \leq 4$, and $e_i \in \mathbb{F}_2^4$ we can construct all sets D_y where $y \in \text{span}(e_1, \ldots, e_r)$.*

Proof. If $y \in \text{span}(e_1, \ldots, e_r)$ then there exists a (not unique) chain of values

$$y_0 = e_{j_0}, y_1, \ldots, y_s = y$$

such that $y_i \oplus y_{i+1} = e_{j_i}$ for $j_i \in \{1, \ldots, r\}$. We can inductively construct the sets D_{y_i}. First note that we already know the set $D_{y_0} = D_{e_{j_0}}$ and we can construct $D_{y_{i+1}}$ using the set D_{y_i} and $D_{e_{j_i}}$ given that

$$\{a, b\} \in D_{y_i \oplus e_{j_i}} \Leftrightarrow \exists c \in \mathbb{F}_2^4 \text{ such that } \{a, c\} \in D_{y_i} \text{ and } \{c, b\} \in D_{e_{j_i}} \qquad \square$$

Having this technical lemma in place, we can prove the following theorem.

Theorem 3. *Let* $S : \mathbb{F}_2^4 \to \mathbb{F}_2^4$ *be a (bijective) S-box and for* $e \in \mathbb{F}_2^4$ *with* $\mathrm{wt}(e) = 1$,

$$D_e = \{\{x, y\} \mid S(x) \oplus S(y) = e\}.$$

Given r *sets* $D_{e_1}, \ldots, D_{e_r}$ *for* $1 \le r \le 4$, *up to equivalence, there are*

$$\prod_{i=1}^{2^{4-r}-1} 2^4 - i 2^r$$

possibilities for S. *More concretely,*

1. *given 4 sets the S-box is determined uniquely,*
2. *given 3 sets there are 8 possible S-boxes,*
3. *given 2 sets there are 384 possible S-boxes, and*
4. *given 1 set there are 645120 possible S-boxes.*

Proof. Assume we are given r sets $D_{e_1}, \ldots, D_{e_r}$. First, up to equivalence, we can assume that $S(0) = 0$ and furthermore $e_1 = (1, 0, 0, 0)$, $e_2 = (0, 1, 0, 0)$ and so on. We claim that given this information, S is fixed on the set

$$\{x \mid S(x) \in \mathrm{span}(e_1, \ldots, e_r)\}.$$

For this, let $y \in \mathrm{span}(e_1, \ldots, e_r)$ be given. From Lemma 2 we know that we can construct the set D_y. As D_y passes the cover filter, there exists a pair $\{0, x\} \in D_y$ for some $x \in \mathbb{F}_2^4$. It follows that we found an $x \in \mathbb{F}_2^4$ such that

$$S(0) + S(x) = S(x) = y.$$

More generally, the same argument shows that, given $D_{e_1}, \ldots, D_{e_r}$, fixing $S(x') = y'$ the values of S are fixed for all x such that $S(x)$ is in the coset $y' \oplus \mathrm{span}(e_1, \ldots, e_r)$. Noting there are 2^{4-r} cosets of $\mathrm{span}(e_1, \ldots, e_r)$ and taking into account the bijectivity of the S-box, the theorem follows. $\qquad \square$

In particular, the proof of Theorem 3 implies the following.

Corollary 1. *The cover filter is necessary and sufficient. That is to say that given a number of sets* D_e *where* e *runs through a subspace of* $\mathbb{F}_2^4$, *there exists an S-box corresponding to these sets if and only if each of the sets* D_e *passes the cover filter.*

B   Example Complexities for Maya

Table 1. The log of the complexity (number of texts encrypted or decrypted) of 16 test runs of the attack on Maya as a function of the number of rounds. The complexities in italics are extrapolations based on the assumption of a linear relationship between the number of rounds and the log complexity. The median was computed on the assumption that non-existent complexities are infinite.

Case	Rounds										
	6	7	8	9	10	11	12	13	14	15	16
1	14.4	16.2	18.6	21.0	24.3	28.5	31.6	35.5	40.5		*46.8*
2	14.1	15.6	17.3	19.7	22.0	23.7	26.9	29.1	32.0	33.8	36.0
3	14.3	16.3	17.4	19.5	22.2	24.7	27.4	29.7	31.3	33.6	35.9
4	14.8	16.1	17.6	19.8	22.3	25.3	27.9	30.1	32.1	34.8	36.9
5	14.6	15.7	17.4	19.4	21.4	23.5	26.0	27.6	30.0	31.4	35.7
6	15.0	16.1	18.3	20.2	22.7	25.6	28.7	31.8	34.2	36.3	*39.3*
7	14.2	15.6	17.7	19.7	22.4	25.4	27.4	29.9	32.6	35.4	37.4
8	14.5	15.7	17.5	19.4	21.5	24.4	26.9	29.6	31.9	35.5	37.1
9	15.2	16.8	19.1	21.1	23.6	26.5	28.7	31.5	36.3	39.0	*41.2*
10	14.9	16.5	18.1	20.2	23.0	24.5	27.6	29.8	34.7	38.6	38.5
11	14.4	15.6	17.5	19.8	22.1	25.1	27.5	30.5	33.4	37.7	*39.4*
12	15.0	15.7	17.5	19.9	22.4	25.3	29.1	31.5	34.2	36.1	*39.5*
13	14.9	15.9	17.1	19.6	21.7	24.4	27.9	29.3	31.8	35.8	36.0
14	14.4	15.6	17.5	19.3	21.9	24.3	27.7	30.3	32.1	35.4	36.7
15	14.4	15.6	17.2	19.5	22.3	24.0	26.6	29.9	33.0	36.5	40.5
16	14.2	15.7	17.4	19.7	22.4	24.9	27.6	30.4	32.9	34.9	37.4
Median	14.4	15.7	17.5	19.7	22.3	24.8	27.6	30.2	32.5	35.6	37.4

A Single-Key Attack
on the Full GOST Block Cipher

Takanori Isobe

Sony Corporation
1-7-1 Konan, Minato-ku, Tokyo 108-0075, Japan
Takanori.Isobe@jp.sony.com

Abstract. The GOST block cipher is the Russian encryption standard published in 1989. In spite of considerable cryptanalytic efforts over the past 20 years, a key recovery attack on the full GOST block cipher without any key conditions (*e.g.*, weak keys and related keys) has not been published yet. In this paper, we show a first single-key attack, which works for all key classes, on the full GOST block cipher. To construct the attack, we develop a new attack framework called *Reflection-Meet-in-the-Middle Attack*. This approach combines techniques of the reflection attack and the meet-in-the-middle attack. We apply it to the GOST block cipher with further novel techniques which are the effective MITM techniques using equivalent keys on short rounds. As a result, a key can be recovered with 2^{225} computations and 2^{32} known plaintexts.

Keywords: block cipher, GOST, single-key attack, reflection attack, meet-in-the-middle attack, equivalent keys.

1 Introduction

The GOST block cipher [22] is known as the former Soviet encryption standard GOST 28147-89 which was standardized as the Russian encryption standard in 1989. It is based on a 32-round Feistel structure with 64-bit block and 256-bit key size. The round function consists of a key addition, eight 4 × 4-bit S-boxes and a rotation. Since values of S-boxes are not specified in the GOST standard [22], each industry uses a different set of S-boxes. For example, one of the S-boxes used in the Central Bank of the Russian Federation is known as in [27].

The GOST block cipher is well-suited for compact hardware implementations due to its simple structure. Poschmann *et al.* showed the most compact implementation requiring only 651 GE [24]. Therefore, the GOST block cipher is considered as one of ultra lightweight block ciphers such as PRESENT [6] and KATAN family [8], which are suitable for the constrained environments including RFID tags and sensor nodes. Note that for the remainder of this paper we refer to the GOST block cipher as GOST.

Over the past 20 years, several attacks on GOST have been published. A differential attack on 13-round GOST was proposed by Seki and Kaneko [28]. In the related-key setting, an attack is improved up to 21 rounds. Ko *et al.* showed

A. Joux (Ed.): FSE 2011, LNCS 6733, pp. 290–305, 2011.

Table 1. Key recovery attack on GOST

Key setting	Type of attack	Round	Complexity	Data	Paper
Single key	Differential	13	Not given	2^{51} CP	[28]
	Slide	24	2^{63}	2^{63} ACP	[2]
	Slide	30	$2^{253.7}$	2^{63} ACP	[2]
	Reflection	30	2^{224}	2^{32} KP	[17]
	Reflection-Meet-in-the-Middle	32	2^{225}	2^{32} KP	**This paper**
Single key	Slide (2^{128} weak keys)	32	2^{63}	2^{63} ACP	[2]
(Weak key)	Reflection (2^{224} weak keys)	32	2^{192}	2^{32} CP	[17]
Related key	Differential	21	Not given	2^{56} CP	[28]
	Differential[†]	32	2^{244}	2^{35} CP	[19]
	Boomerang [‡]	32	2^{248}	$2^{7.5}$ CP	[15]

CP : Chosen plaintext, ACP : Adaptive chosen plaintext, KP : Known plaintext.

† The attack can recover 12 bits of the key with 2^{36} computations and 2^{35} CP.

‡ The attack can recover 8 bits of the key with $2^{7.5}$ computations and $2^{7.5}$ CP.

a related-key differential attack on the full GOST [19]. These results work on only the GOST that employs the S-boxes of the Central Bank of the Russian Federation [27]. Fleischmann *et al.* presented a related-key boomerang attack on the full GOST which works for any S-boxes [15]. As other types of attacks, Biham *et al.* showed slide attacks on the reduced GOST [2]. Their attack utilizes self similarities among round functions of the encryption process, and does not also depend on used values of S-boxes. Even if an attacker does not know the values of S-boxes, the 24-round GOST can be attacked by this approach. If the values are known, this attack can be improved up to 30 rounds. In addition, for a class of 2^{128} weak keys, the full GOST can be attacked by this approach. After that, Kara proposed a reflection attack on 30-round GOST [17]. This attack also uses self similarities among round functions, and works for any bijective S-boxes. The difference from the slide attack proposed by Biham *et al.* [2] is to use similarities of both encryption and decryption processes. The reflection attack utilizes these similarities in order to construct fixed points of some round functions. Moreover, for a class of 2^{224} weak keys, the full GOST can be attacked by using the reflection technique.

In spite of considerable cryptanalytic efforts, a key recovery attack on the full GOST without any key assumptions (*e.g.*, weak keys and related keys) has not been published so far. Furthermore, a weak-key attack and a related-key attack are arguable in the practical sense, because of their strong assumptions. A weak-key attack is generally applicable to very few keys, *e.g.*, in the attack of [17], the rate of weak keys is $2^{-32}(= 2^{224}/2^{256})$. Hence, almost all keys, $(2^{256} - 2^{224}) \approx 2^{256}$ keys, can not be attacked by [17]. Besides, the attacker can not even know whether a target key is included in a weak key class or not. A related-key attack assumes that the attacker can access to the encryption/decryption under multiple unknown keys such that the relation between them is known to the attacker. Though this type of attack is meaningful during the design and certification of ciphers, it does not lead to a realistic threat in practical security protocols which use the block cipher in a standard way as stated in [13]. Therefore, the security under the single-key setting is the most important issue from the aspect of the practical security. In particular, an ultra lightweight block cipher does not need a security against related-key attacks in many cases. For example, in low-end devices such as a passive RFID tag, the key may not be changed in its life cycle as mentioned in [6, 8]. Indeed, KTANTAN supports only a fixed key [8] and the compact implementation of GOST proposed by Poschmann *et al.* also uses a hard-wired fixed key [24]. Therefore, it can be said that GOST has not been theoretically broken.

Recently, Bogdanov and Rechberger showed a new variant of the Meet-in-the-Middle (MITM) attack on block ciphers called 3-subset MITM attack [7]; it was applied to KTANTAN [8]. This attack is based on the techniques of the recent MITM preimage attacks on hash functions [1, 25]. It seems to be effective for the block cipher whose key schedule is simple, *e.g.*, a bit or a word permutation. In fact, the key schedule function of KTANTAN consists of a bit permutation. Since GOST also has a simple key schedule function, which is a word permutation, the 3-subset MITM attack seems applicable to it. However, it does not work well on the full GOST, because the key dependency of the full GOST is stronger than that of KTANTAN due to the iterative use of key words during many round functions.

Our Contributions. In this paper, we first introduce a new attack framework called *Reflection-Meet-in-the-Middle (R-MITM) Attack* ; it is a combination of the reflection attack and the 3-subset MITM attack. The core idea of this combination is to make use of fixed points of the reflection attack to enhance the 3-subset MITM attack. If some round functions have fixed points, we can probabilistically remove these rounds from the whole cipher. Since this skip using fixed points allows us to disregard the key bits involved in the removed rounds, the key dependency is consequently weakened. Thus, our attack is applicable to more rounds compared to the original 3-subset MITM attack if fixed points can be constructed with high probability. Then, we apply it to the full GOST block cipher with further novel techniques which make the MITM approach more efficient by using equivalent keys on short rounds. As a result, we succeed in constructing a first key recovery attack on the full GOST block cipher in the single key setting. It can recover a key with 2^{225} computations and 2^{32} known plaintext/ciphertext pairs. An important point to emphasize is that our attack does not require any assumptions for a key unlike the previous attacks. In addition, our attack can be applied to any S-boxes as long as they are bijective. These results are summarized in Table 1.

Table 2. Key schedule of GOST

Round	1	2	3	4	5	6	7	8	9	10	11	12	13	14	15	16
Key	k_1	k_2	k_3	k_4	k_5	k_6	k_7	k_8	k_1	k_2	k_3	k_4	k_5	k_6	k_7	k_8
Round	17	18	19	20	21	22	23	24	25	26	27	28	29	30	31	32
Key	k_1	k_2	k_3	k_4	k_5	k_6	k_7	k_8	k_8	k_7	k_6	k_5	k_4	k_3	k_2	k_1

Outline of the Paper. This paper is organized as follows. A brief description of GOST, a 3-subset MITM attack and a reflection attack are given in Section 2. The R-MITM attack is introduced in Section 3. In Section 4, we present a R-MITM attack on the full GOST. Finally, we present conclusions in Section 5.

2 Preliminaries

In this section, we give a brief description of GOST, a 3-subset MITM attack and a reflection attack.

2.1 Description of GOST

GOST is a block cipher based on a 32-round Feistel structure with 64-bit block and 256-bit key size. The F-function consists of a key addition, eight 4×4-bit S-boxes S_j ($1 \leq j \leq 8$) and a 11-bit left rotation (See Fig.1).

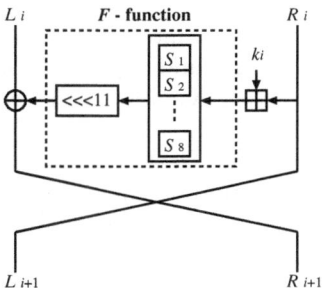

Fig. 1. One round of the GOST block cipher

The 256-bit master key K is divided into eight 32-bit words, *i.e.*, $K = (k_1, k_2, \ldots, k_8)$, $k_i \in \{0, 1\}^{32}$. Each k_i is used as a round key in each round function as shown in Table 2.

In the GOST standard [22], the S-boxes are not specified. Each industry uses a different set of S-boxes. In this paper, we do not care about specific values of the S-boxes as long as they are bijective.

2.2   3-Subset MITM Attack

The basic concept of the MITM attack was proposed by Diffie and Helman [12]. So far, this attack has been applied to several block ciphers [9–11, 13, 14, 16]. Furthermore, over the past few years, this attack has been improved in a line of preimage attacks on hash functions, and several novel techniques are introduced, *e.g.*, a partial matching [1] and an initial structure [25]. Recently, by using these novel techniques, Bogdanov and Rechberger showed a new variant of MITM attack on block ciphers called 3-subset MITM attack [7]; it was applied to KTANTAN [8].

This attack consists of two stages: a MITM stage and a key testing stage. First, the MITM stage filters out part of wrong keys from key candidates by using MITM techniques. Then, the key testing stage finds a correct key from the surviving key candidates in a brute force manner.

Let $E_K : \{0,1\}^b \to \{0,1\}^b$ be a block cipher with an l-bit key K and a b-bit block. Assume that E_K is a composition of round functions as follows;

$$E_K(x) = F_{k_r} \circ F_{k_{r-1}} \circ \cdots \circ F_{k_1}(x), \ x \in \{0,1\}^b,$$

where r is the number of rounds, $k_1,\ldots,k_r$ are round keys and F_{k_i} is the i-th round function, $F_{k_i} : \{0,1\}^b \to \{0,1\}^b$. The composition of $j - i + 1$ functions starting from i is denoted by $F_K[i,j]$ defined as

$$F_K[i,j](x) = F_{k_j} \circ \cdots \circ F_{k_i}(x), \ 1 \le i < j \le r.$$

In the following, we give details of each stage of the 3-subset MITM attack.

MITM stage : $E_k(X)$ is divided into two functions as $E_K(X) = F_K[a+1,r] \circ F_K[1,a], 1 < a < r-1$ [1]. Let K_1 and K_2 be sets of key bits used in $F_K[1,a]$ and $F_K[a+1,r]$, respectively. $A_0 = K_1 \cap K_2$ is the common set of key bits used in both $F_K[1,a]$ and $F_K[a+1,r]$. $A_1 = K_1 \setminus K_1 \cap K_2$ and $A_2 = K_2 \setminus K_1 \cap K_2$ are the sets of key bits used in only $F_K[1,a]$ and only $F_K[a+1,r]$, respectively. In this stage, we use only one plaintext/ciphertext pair (P,C).

The procedure of the MITM stage is as follows. Fig. 2 shows the overview of the MITM stage.

1. Guess a value of A_0.
2. Compute $v = F_K[1,a](P)$ for all values of A_1 and make a table of (v, A_1) pairs. In this step, $2^{|A_1|}$ pairs are generated, where $|A_i|$ is the bit length of A_i and $2^{|A_i|}$ is the number of elements of A_i.
3. Compute $u = F_K^{-1}[a+1,r](C)$ for all values of A_2. In this step, $2^{|A_2|}$ pairs are generated.

---

[1] As in the attack of KTANTAN [7], by using the partial matching technique, E_K is divided into $F_K[1,a]$ and $F_K[a+t,r]$, $t > 1$. However, in this paper, we consider only the case of $t = 1$, because we do not use the partial matching.

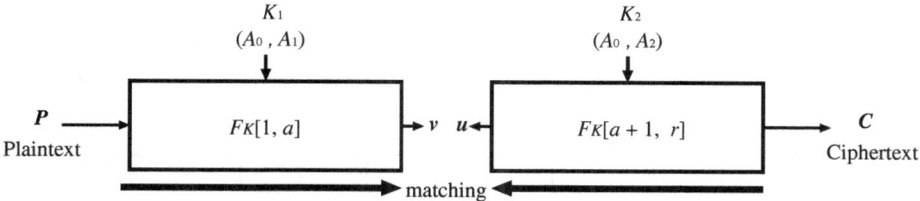

Fig. 2. Meet-in-the-middle stage

4. Add key candidates for which the equation $v = u$ is satisfied to the list of surviving keys.
   The number of surviving keys is $2^{|A_1|+|A_2|}/2^b$.
5. Repeat 2-4 for each different value of A_0. ($2^{|A_0|}$ times)

In this stage, 2^{l-b} key candidates survive, because $2^{|A_1|+|A_2|}/2^b \times 2^{|A_0|} = 2^l/2^b$.

Key testing stage : We test surviving keys in a brute force manner by using additional plaintext/ciphertext pairs.

We evaluate the cost of this attack. The whole attack complexity C_{comp} is estimated as

$$C_{comp} = \underbrace{2^{|A_0|}(2^{|A_1|} + 2^{|A_2|})}_{\text{MITM stage}} + \underbrace{(2^{l-b} + 2^{l-2b} + \ldots)}_{\text{Key testing stage}}.$$

The number of required plaintext/ciphertext pair is $\lceil \frac{l}{b} \rceil$. The required memory is $\max(2^{|A_1|}, 2^{|A_2|})$, which is the cost of the table used in the MITM stage. When $\min(|A_1|, |A_2|) > 1$ the attack is more effective than an exhaustive search. Therefore, the point of the 3-subset MITM attack is to find independent sets of master key bits such as A_1 and A_2.

2.3   Reflection Attack

The reflection attack was first introduced by Kara and Manap [18]; it was applied to Blowfish [26]. After that, the attack was generalized by Kara [17]. In this section, we introduce a basic principle of the reflection attack used in our attack. See [17, 18] for details about the reflection attack.

The reflection attack is a kind of a self-similarity attack such as the slide attack [4, 5]. Though the reflection attack utilizes similarities of some round functions of both encryption and decryption processes, the slide attack exploits similarities among the round functions of only the encryption process. In the reflection attack, by using these similarities, fixed points of some round functions are constructed.

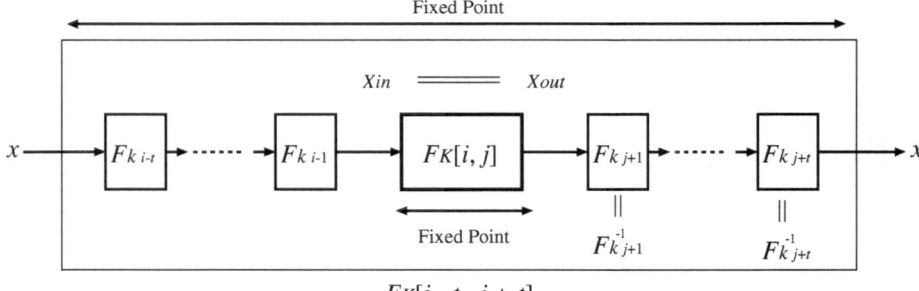

$$F_K[i - t, \ j + t]$$

Fig. 3. Basic principle of the reflection attack

Let $U_K(i, j)$ be the set of fixed points of the function $F_K[i, j]$ defined as follows;

$$U_K(i, j) = \{x \in \{0, 1\}^n \mid F_K[i, j](x) = x\}.$$

The basic principle of the reflection attack is given by the following Lemma.

Lemma 1. [17] *Let i and j be integers such that $0 \leq j - i < i + j < r$. Assume that $F_{k_{i-t}} = F_{k_{j+t}}^{-1}$ for all $t : 1 \leq t < i$. If $F_K[i - t, i - 1](x) \in U_k(i, j)$, then $x \in U_k(i - t, j + t)$ for all $t : 1 < t < i$. In addition, if $x \in U_K(i - t, j + t)$ for certain $t : 1 < t < i$, then $F_K[i - t, i - 1](x) \in U_K(i, j)$.*

From Lemma 1, if the round functions hold the conditions, a local fixed point is expanded to previous and next rounds as shown in Fig. 3. Roughly speaking, fixed points of some round functions can be constructed easily in the certain setting. These fixed points enable us to probabilistically skip the round functions from a whole cipher.

We give an example to explain this skip in detail. Let i and j be integers such that $0 < j - i < i + j < r$. Assume that $F_{k_{i-t}} = F_{k_{j+t}}^{-1}$ for all $t : 1 < t < i$, and $E_K(x)$ is expressed as follows;

$$E_K(x) = F_K[j + i, r] \circ F_K[j + 1, j + i - 1] \circ F_K[i, j] \circ F_K[1, i - 1](x),$$
$$= F_K[j + i, r] \circ F_K^{-1}[1, i - 1] \circ F_K[i, j] \circ F_K[1, i - 1](x).$$

Besides, assuming $F_K[1, i-1](x) \in U_K(i, j)$, then $F_K[1, j+i-1](x) = x$ (Lemma 1). Thus $E_K(x)$ is expressed as

$$E_K(x) = F[j + i, r](x).$$

In this case, the round functions $F_K[1, j + i - 1]$ can be skipped from E_K. The probability P_{ref} of that above skip occurs for arbitrary x is $|U_K(i, j)|/2^b$. If $P_{ref} > 2^{-b}$ (*i.e.*, $|U_K[i, j]| > 1$), this skip occurs at $F_K[1, j + i - 1]$ with higher probability than a random function.

3   Reflection-Meet-in-the-Middle Attack

We propose a new attack framework called *reflection-meet-in-the-middle (R-MITM) attack*, which is a combination of the reflection attack and the 3-subset MITM attack. As mentioned in Section 2.2, the point of the 3-subset MITM attack is to construct independent sets of master key bits. In general, if the master key bits are used iteratively in each round and the use of key bits is not biased among rounds[2], it seems to be difficult to find the independent sets of master key bits, because such cipher have the strong key dependency on even small number of rounds.

To overcome this problem, we utilize the technique of the reflection attack. In the reflection attack, some rounds satisfying certain conditions can be skipped from the whole cipher with the probability P_{ref}. From now on, we call this skip a *reflection skip*. Since key bits used in skipped round functions can be omitted, it becomes easier to construct independent sets of master key bits. This is the concept of the R-MITM attack. In the following, we give the detailed explanation of the attack.

3.1   Details of the R-MITM Attack

Suppose that E_K is expressed as follows;

$$E_K(x) = F_K[a_3 + 1, r] \circ F_K[a_2 + 1, a_3] \circ F_K[a_1 + 1, a_2] \circ F_K[1, a_1](x),$$

where $2 < a_1 + 1 < a_2 < a_3 - 1 < r - 2$ and the reflection skip occurs at $F_K[a_2 + 1, a_3]$ with the probability P_{ref}. Then, E_K can be redescribed as follows and denoted by $E'_K(x)$,

$$E'_K(x) = F_K[a_3 + 1, r] \circ F_K[a_1 + 1, a_2] \circ F_K[1, a_1](x).$$

The R-MITM attack consists of three stages; a data collection stage, a R-MITM stage and a key testing stage. In the following, we explain each stage.

Data collection stage : We collect plaintext/ciphertext pairs to obtain a pair in which the reflection skip occurs at $F_K[a_2 + 1, a_3]$. Since the probability of this event is P_{ref}, the number of required plaintext/ciphertext pairs is P_{ref}^{-1}.

After that, the R-MITM stage and the key testing stage are executed for all plaintext/ciphertext pairs obtained in the data collection stage.

R-MITM stage :   We divide E_K into two functions: $F_K[1, a_1]$ and $F_K[a_1 + 1, r]$[3]. In this stage, we ignore $F_K[a_2 + 1, a_3]$ as follows;

$$F'_K[a_1 + 1, r] = F_K[a_3 + 1, r] \circ F_K[a_1 + 1, a_2],$$

---

[2] In KTANTAN [8], 6 bits of master key are not used in the first 111 rounds and other 6 bits of master key are also not used in the last 131 rounds. The attack of [7] utilizes this bias of used key bits among rounds.

[3] Though there are many choices of divisions, we use it as an example.

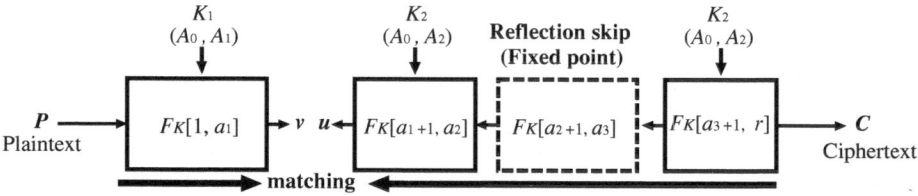

Fig. 4. Reflection-meet-in-the-middle stage

assuming that the reflection skip occurs. Let K_1 and K_2 be sets of key bits used in $F_K[1, a_1]$ and $F'_K[a_1 + 1, r]$, respectively. $A_0 = K_1 \cap K_2$ is the set of key bits used in both $F_K[1, a_1]$ and $F'_K[a_1 + 1, r]$. $A_1 = K_1 \setminus K_1 \cap K_2$ and $A_2 = K_2 \setminus K_1 \cap K_2$ are the sets of key bits used in only $F_K[1, a_1]$ and only $F'_K[a_1 + 1, r]$, respectively. Figure 4 illustrates the R-MITM stage.

The procedure of the R-MITM stage is almost same as the MITM stage of Section 2.2. The difference is that in the R-MITM stage, we assume that reflection skip occurs, *i.e.*, $F_K[a_2 + 1, a_3]$ is ignored. After this stage, 2^{l-b} key candidates survive.

Key testing stage : We test surviving keys in a brute force manner by using plaintext/ciphertext pairs.

3.2   Evaluation of the R-MITM Attack

We evaluate the cost of the R-MITM attack. The whole attack complexity C_{comp} is estimated as

$$C_{comp} = (\underbrace{(2^{|A_0|}(2^{|A_1|} + 2^{|A_2|}))}_{\text{R-MITM stage}} + \underbrace{(2^{l-b} + 2^{l-2b} + \ldots)}_{\text{Key testing stage}}) \times R_{ref}^{-1}.$$

The number of required plaintext/ciphertext pair is $\max(\lceil l/b \rceil, R_{ref}^{-1})$. The required memory is $\max(2^{|A_1|}, 2^{|A_2|})$, which is the cost of the table in the R-MITM stage. When $\min(2^{|A_1|}, 2^{|A_2|}, 2^b) > (R_{ref}^{-1})$, the attack is more effective than an exhaustive search.

Compared with the basic 3-subset MITM attack in Section 2.2, the number of required plaintext/ciphertext pairs increases, because the R-MITM attack utilizes the probabilistic event, *i.e.*, reflection skip. In addition, more independent key bits are needed for the successful attack. However, this attack has a distinct advantage, which is to be able to skip some round functions by the reflection skip. Recall that the most important point of the 3-subset MITM attack is to find independent sets of master key bits. Since the reflection skip enables us to disregard key bits involved in some round, it obviously becomes easier to construct such independent sets. Thus, this attack seem to be applicable to more rounds than the 3-subset MITM attack when the reflection skip occurs with high probability.

4   R-MITM Attack on the Full GOST Block Cipher

In this section, we apply the R-MITM attack to the full GOST block cipher [22]. From Table. 2, in full 32 rounds, the master key is iteratively used four times and all master key bits are involved in every 8 rounds. The basic 3-subset MITM attack in Section 2.2 is not applicable to the full GOST, because independent sets of master key bits can not be constructed in any divisions of 32 rounds. However, by using the R-MITM attack, we can construct independent sets and mount a key recovery attack on the full GOST.

We first introduce the reflection property of GOST proposed by Kara [17] to construct the reflection skip. Next, we present effective MITM techniques to enhance the R-MITM stage. These techniques make use of the equivalent keys of short round functions. Finally, we evaluate our attack.

4.1   Reflection Property of GOST

The reflection attack on GOST has been proposed by Kara [17][4]. The GOST block cipher $E_K : \{0, 1\}^{64} \rightarrow \{0, 1\}^{64}$ is expressed as

$$E_K = S \circ F_K[25, 32] \circ F_K[17, 24] \circ F_K[9, 16] \circ F_K[1, 8],$$
$$= F_K^{-1}[1, 8] \circ S \circ F_K[1, 8] \circ F_K[1, 8] \circ F_K[1, 8],$$

where S is the swap of the Feistel structure.

S has 2^{32} fixed points, because the probability of that the right halves equal to the left halves is 2^{-32}. From Lemma 1, $F_K^{-1}[1, 8] \circ S \circ F_K[1, 8]$ also has 2^{32} fixed points, i.e., $|U_K(17, 32)| = 2^{32}$. Thus, with the probability $P_{ref} = 2^{-32}$ ($= (2^{32}/2^{64})$), $F_K[17, 32]$ can be ignored. E_K is redescribed as follows and denoted by E'_K

$$E'_K = F_K[1, 8] \circ F_K[1, 8].$$

Figure 5 shows this reflection skip of GOST.

Therefore, in the data collection stage, we need to collect $P_{ref}^{-1} = 2^{32}$ plaintext/ciphertext pairs. In 2^{32} collected pairs, there is a pair in which the reflection skip occurs, i.e., last 16 rounds can be removed as E'_K.

4.2   Effective MITM Technique Using Equivalent Keys on Short Rounds

In the R-MITM stage, we mount the MITM approach on only $E'_K = F_K[1, 8] \circ F_K[1, 8]$ for all 2^{32} collected pairs.

As mentioned in Section 3.2, we need to construct independent sets A_1 and A_2 which hold the condition, $\min(2^{|A_1|}, 2^{|A_2|}) > 2^{32}$. However, despite the reduction

---

[4] The similar technique for constructing a fixed point is also used in the attacks on the GOST hash function [20, 21].

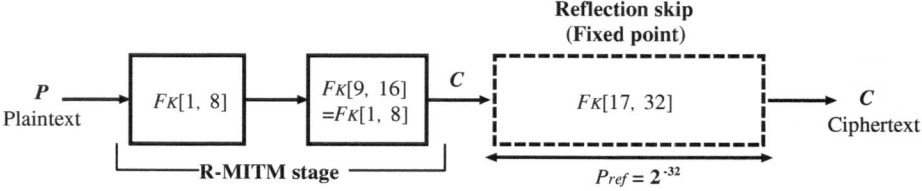

Fig. 5. Reflection skip of GOST

of rounds by the reflection skip, in the straightforward method, we can not find such sets in any divisions of 16 rounds, due to the strict condition of independent sets.

We introduce effective MITM techniques which make use of equivalent keys of short round functions (*i.e.*, 4 round). The aim of these techniques is to ignore the first and the last 4 rounds and to mount the MITM approach in only intermediate 8 rounds. These techniques enable us to construct independent sets enough for the successful attack.

We treat E'_K as following four-round units;

$$E'_K = F_K[5, 8] \circ F_K[1, 4] \circ F_K[5, 8] \circ F_K[1, 4].$$

In the following, we first explain equivalent keys used in our attack. Then, we present detail of the R-MITM stage using the equivalent keys.

Equivalent Keys on Short Rounds. Define a set of equivalent keys on $F_K[i, j]$ as $Z(F_K[i, j], x, y)$ as follows:

$$Z(F_K[i, j], x, y) = \{ek \in \{0, 1\}^{256} \mid F_{ek}[i, j](x) = y\},$$

where $(x, y) \in \{0, 1\}^{64}$. Note that the class of keys defined above is the equivalent keys with respect to only one input/output pair. To put it more concretely, if equivalent keys $ek \in Z(F_K[i, j], x, y)$ are used, input x is always transformed to y in $F_K[i, j]$. For other input/output pairs, these relations do not hold even if the same equivalent keys are used.

GOST has an interesting property regarding the equivalent keys on short rounds as described in the following observation.

Observation 1: *Given any x and y, $Z(F_K[1, 4], x, y)$ and $Z(F_K^{-1}[5, 8], x, y)$ can be easily obtained, and the number of each equivalent keys is 2^{64}.*

For $F_K[1, 4]$, k_1, k_2, k_3 and k_4 are added in each round. Given the values of k_1 and k_2, the other values of k_3 and k_4 are determined from $F_K[1, 2](x)$ and y as follows:

$$k_3 = F^{-1}(z_L + y_L) - z_R, \tag{1}$$
$$k_4 = F^{-1}(z_R + y_R) - y_R, \tag{2}$$

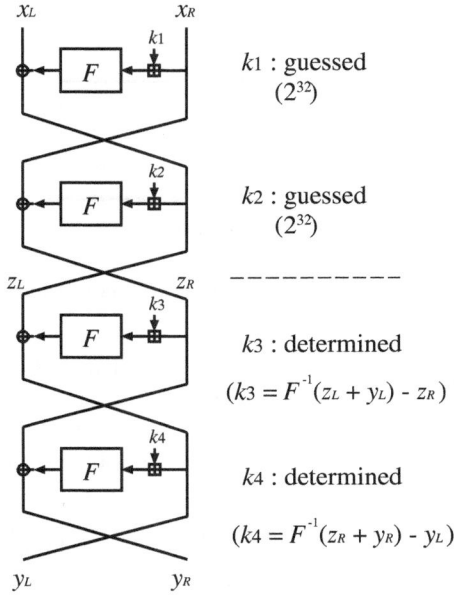

k_1 : guessed
(2^{32})

k_2 : guessed
(2^{32})

----

k_3 : determined

($k_3 = F^{-1}(z_L + y_L) - z_R$)

k_4 : determined

($k_4 = F^{-1}(z_R + y_R) - y_L$)

Fig. 6. Equivalent keys of 4 rounds

where F^{-1} is the inverse of F function, y_L and y_R are left and right halves of y, and z_L and z_R are those of $F_K[1,2](x)$. Since values of (k_1, k_2) are 64 bits, the number of $Z(F_K[1,4], x, y)$ is 2^{64}. Figure 6 shows this procedure. A similar property holds for $F_K^{-1}[5,8]$.

From Observation 1, we can easily obtain 2^{64} equivalent keys of the first and the last 4 rounds for any inputs and outputs. Moreover, $F_K[1,4]$ and $F_K^{-1}[5,8]$ use different master key bits each other, $K_a = (k_1||k_2||k_3||k_4)$ and $K_b = (k_5||k_6||k_7||k_8)$, respectively. Thus, $Z(F_K[1,4], x, y)$ and $Z(F_K^{-1}[5,8], x, y)$ are expressed by sets of only K_a and K_b as follows;

$$Z_{Ka}(F_K[1,4], x, y) = \{ek_a \in \{0,1\}^{128} \mid F_{ek_a}[1,4](x) = y\},$$
$$Z_{Kb}(F_K^{-1}[5,8], x, y) = \{ek_b \in \{0,1\}^{128} \mid F_{ek_b}^{-1}[5,8](x) = y\}.$$

Since K_a and K_b are independent sets of mater key, $Z_{Ka}(F_K[1,4], x, y)$ and $Z_{Kb}(F_K^{-1}[5,8], x, y)$ are also independent sets.

Detail of the R-MITM Stage using Equivalent Keys. Let S and T be $F_K[1,4](P)$ and $F^{-1}[5,8](C)$, which are input and output values of 8 intermediate rounds, *i.e.*, $F_K[5,12] = F_K[1,4] \circ F_K[5,8]$.

From Observation 1, given values of P, C, S and T, two sets of 2^{64} equivalent keys, $Z_{K_a}(F_K[1,4], P, S)$ and $Z_{K_b}(F_K^{-1}[5,8], C, T)$, can be easily obtained.

When $Z_{K_a}(F_K[1,4], P, S)$ and $Z_{K_b}(F_K^{-1}[5,8], C, T)$ are used, S and T are not changed. Thus by using these equivalent keys, the first and the last 4 round

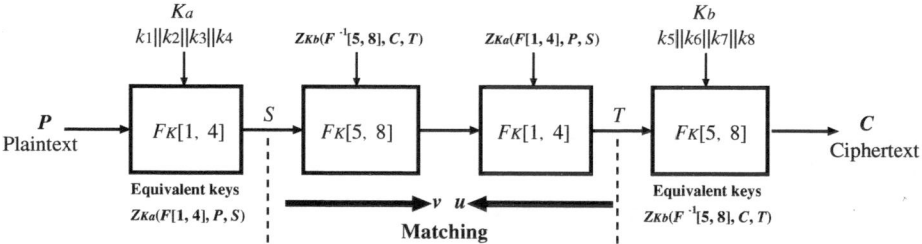

Fig. 7. R-MITM stage using equivalent keys

can be ignored, and we can mount the MITM attack between $F_K[5, 8](S)$ and $F_K^{-1}[1, 4](T)$. The number of elements in each independent set is 2^{64}, which is enough for the successful attack.

The procedure of the R-MITM stage is as follows and illustrated in Fig. 7.

1. Guess the values S and T.
2. Compute $v = F_K[5, 8](S)$ with 2^{64} K_b in $Z_{K_b}(F_K^{-1}[5, 8], C, T)$ and make a table of (v, K_b) pairs.
3. Compute $u = F_K^{-1}[1, 4](T)$ with 2^{64} K_a in $Z_{K_a}(F_K[1, 4], P, S)$.
4. Add key candidates for which the equation $v = u$ is satisfied to the list of surviving keys. The number of surviving keys is $2^{64+64}/2^{64} = 2^{64}$.
5. Repeat 2-4 with the different values of S and T. (2^{128} times).

After this procedure, 2^{192} ($=2^{64} \times 2^{128}$) key candidates survive. These key candidates are evaluated in the key testing stage.

The R-MITM stage utilizes equivalent-key sets of $Z_{K_a}(F_K[1, 4], P, S)$ and $Z_{K_b}(F_K^{-1}[5, 8], C, T)$, $0 \le S$, $T < 2^{64}$, where each set includes 2^{64} elements. For $Z_{K_a}(F_K[1, 4], P, S)$, $0 \le S < 2^{64}$, all elements of every set are different, because if the values of S are different, equivalent keys of $F_K[1, 4]$ are surely different from Eq. (1) and (2) as long as S-boxes are bijective. Thus, $Z_{K_a}(F_K[1, 4], P, S)$, $0 \le S < 2^{64}$ covers all 2^{128} ($= 2^{64} \times 2^{64}$) values of K_a. A similar property holds for K_b. Therefore, all possible values for the master key are tested and the set of surviving key candidates surely contain the correct key if the reflection skip occurs.

4.3 Evaluation

The whole attack complexity C_{comp} is estimated as

$$C_{comp} = (\underbrace{(2^{128}(2^{64} + 2^{64}))}_{\text{R-MITM stage}} + (\underbrace{2^{256-64} + 2^{256-128} + \ldots)}_{\text{Key testing stage}}) \times 2^{32},$$
$$= 2^{225}.$$

The number of required known plaintext/ciphertext pairs is $\max(\lceil l/b \rceil, R_{ref}^{-1}) = \max(\lceil 256/64 \rceil, 2^{32}) = 2^{32}$. The required memory is $\max(2^{64}, 2^{64}) = 2^{64}$, which is the cost of the table used in the R-MITM stage. Therefore, this attack can recover a key with 2^{225} computations, 2^{32} known plaintext/ciphertext pairs and 2^{64} memory. It is more effective than an exhaustive attack.

5 Conclusion

This paper has presented a first single-key attack on the full GOST block cipher without relying on weak key classes. To build the attack, we introduced a new attack framework called *Reflection-Meet-in-the-Middle Attack*, which is the combination of the reflection and the 3-subset MITM attacks. The advantage of this attack over the basic 3-subset MITM attack, some rounds can be probabilistically removed from the whole cipher. Since this allows us to disregard the key bits involved in the removed rounds, it becomes easier to construct the independent sets of the key bits. Thus, our attack seems to be applicable to more rounds when the reflection skip occurs with high probability. Then we applied it to the full GOST block cipher with further novel techniques which make use of equivalent keys of short round functions (*i.e.*, 4 rounds). These techniques enable us to mount the effective MITM approach. As a result, we succeeded in constructing a first key recovery attack on the full GOST without any key conditions, which works for any bijective S-boxes. Our result shows that GOST does not have the 256-bit security for all key classes, even if a fixed key is used such as [24].

The idea of the R-MITM attack seems applicable to other block ciphers in which the fixed point can be constructed with high probability and its key schedule is simple in the sense that the key dependency is not strong. Furthermore, the basic principle of the attack does not constrain the reflection property and fixed points. Other non-random properties of round functions may also be able to be utilized as the skip techniques, *e.g.*, the strong correlations among round functions.

Acknowledgments. We would like to thank to Taizo Shirai, Kyoji Shibutani, Özgül Küçük, and anonymous referees for their insightful comments and suggestions.

References

1. Aoki, K., Sasaki, Y.: Preimage Attacks on One-Block MD4, 63-Step MD5 and More. In: Avanzi, R.M., Keliher, L., Sica, F. (eds.) SAC 2008. LNCS, vol. 5381, pp. 103–119. Springer, Heidelberg (2009)
2. Biham, E., Dunkelman, O., Keller, N.: Improved Slide Attacks. In: Biryukov, A. (ed.) [3], pp. 153–166
3. Biryukov, A. (ed.): FSE 2007. LNCS, vol. 4593. Springer, Heidelberg (2007)
4. Biryukov, A., Wagner, D.: Slide attacks. In: Knudsen, L.R. (ed.) FSE 1999. LNCS, vol. 1636, pp. 245–259. Springer, Heidelberg (1999)
5. Biryukov, A., Wagner, D.: Advanced Slide Attacks. In: Preneel, B. (ed.) EUROCRYPT 2000. LNCS, vol. 1807, pp. 589–606. Springer, Heidelberg (2000)

6. Bogdanov, A.A., Knudsen, L.R., Leander, G., Paar, C., Poschmann, A., Robshaw, M.J.B., Seurin, Y., Vikkelsoe, C.: PRESENT: An Ultra-Lightweight Block Cipher. In: Paillier, P., Verbauwhede, I. (eds.) CHES 2007. LNCS, vol. 4727, pp. 450–466. Springer, Heidelberg (2007)

7. Bogdanov, A., Rechberger, C.: A 3-Subset Meet-in-the-Middle Attack: Cryptanalysis of the Lightweight Block Cipher KTANTAN. In: Biryukov, A., Gong, G., Stinson, D.R. (eds.) SAC 2010. LNCS, vol. 6544, pp. 229–240. Springer, Heidelberg (2011)

8. De Cannière, C., Dunkelman, O., Knežević, M.: KATAN and KTANTAN — A Family of Small and Efficient Hardware-Oriented Block Ciphers. In: Clavier, C., Gaj, K. (eds.) CHES 2009. LNCS, vol. 5747, pp. 272–288. Springer, Heidelberg (2009)

9. Chaum, D., Evertse, J.-H.: Cryptanalysis of DES with a Reduced Number of Rounds Sequences of Linear Factors in Block Cipher. In: Williams, H.C. (ed.) CRYPTO 1985. LNCS, vol. 218, pp. 192–211. Springer, Heidelberg (1986)

10. Demirci, H., Selçuk, A.A.: A Meet-in-the-Middle Attack on 8-Round AES. In: Nyberg, K. (ed.) [23], pp. 116–126

11. Demirci, H., Taşkın, İ., Çoban, M., Baysal, A.: Improved Meet-in-the-Middle Attacks on AES. In: Roy, B.K., Sendrier, N. (eds.) INDOCRYPT 2009. LNCS, vol. 5922, pp. 144–156. Springer, Heidelberg (2009)

12. Diffie, W., Hellman, M.E.: Exhaustive Cryptanalysis of the NBS Data Encryption Standard. Computer 10, 74–84 (1977)

13. Dunkelman, O., Keller, N., Shamir, A.: Improved Single-Key Attacks on 8-Round AES-192 and AES-256. In: Abe, M. (ed.) ASIACRYPT 2010. LNCS, vol. 6477, pp. 158–176. Springer, Heidelberg (2010)

14. Dunkelman, O., Sekar, G., Preneel, B.: Improved Meet-in-the-Middle Attacks on Reduced-Round DES. In: Srinathan, K., Rangan, C.P., Yung, M. (eds.) INDOCRYPT 2007. LNCS, vol. 4859, pp. 86–100. Springer, Heidelberg (2007)

15. Fleischmann, E., Gorski, M., Hüehne, J., Lucks, S.: Key Recovery Attack on full GOST. Block Cipher with Negligible Time and Memory. In: Western European Workshop on Research in Cryptology (WEWoRC). LNCS, vol. 6429. Springer, Heidelberg (2009)

16. Indesteege, S., Keller, N., Dunkelman, O., Biham, E., Preneel, B.: A Practical Attack on KeeLoq. In: Smart, N.P. (ed.) EUROCRYPT 2008. LNCS, vol. 4965, pp. 1–18. Springer, Heidelberg (2008)

17. Kara, O.: Reflection Cryptanalysis of Some Ciphers. In: Chowdhury, D.R., Rijmen, V., Das, A. (eds.) INDOCRYPT 2008. LNCS, vol. 5365, pp. 294–307. Springer, Heidelberg (2008)

18. Kara, O., Manap, C.: A New Class of Weak Keys for Blowfish. In: Biryukov, A. (ed.) [3], pp. 167–180

19. Ko, Y., Hong, S.H., Lee, W.I., Lee, S.-J., Kang, J.-S.: Related Key Differential Attacks on 27 Rounds of XTEA and Full-Round GOST. In: Roy, B.K., Meier, W. (eds.) FSE 2004. LNCS, vol. 3017, pp. 299–316. Springer, Heidelberg (2004)

20. Mendel, F., Pramstaller, N., Rechberger, C.: A (Second) Preimage Attack on the GOST Hash Function. In: Nyberg, K. (ed.) [23], pp. 224–234.

21. Mendel, F., Pramstaller, N., Rechberger, C., Kontak, M., Szmidt, J.: Cryptanalysis of the GOST Hash Function. In: Wagner, D. (ed.) CRYPTO 2008. LNCS, vol. 5157, pp. 162–178. Springer, Heidelberg (2008)

22. National Soviet Bureau of Standards. Information Processing System - Cryptographic Protection - Cryptographic Algorithm GOST 28147-89 (1989)

23. Nyberg, K. (ed.): FSE 2008. LNCS, vol. 5086. Springer, Heidelberg (2008)

24. Poschmann, A., Ling, S., Wang, H.: 256 Bit Standardized Crypto for 650 GE – GOST Revisited. In: Mangard, S., Standaert, F.-X. (eds.) CHES 2010. LNCS, vol. 6225, pp. 219–233. Springer, Heidelberg (2010)
25. Sasaki, Y., Aoki, K.: Finding Preimages in Full MD5 Faster Than Exhaustive Search. In: Joux, A. (ed.) EUROCRYPT 2009. LNCS, vol. 5479, pp. 134–152. Springer, Heidelberg (2009)
26. Schneier, B.: Description of a New Variable-Length Key, 64-bit Block Cipher (Blowfish). In: Anderson, R.J. (ed.) FSE 1993. LNCS, vol. 809, pp. 191–204. Springer, Heidelberg (1994)
27. Schneier, B.: Applied Cryptography, 2nd edn. Protocols, Algorithms, and Source Code in C. John Wiley & Sons, Inc., New York (1995)
28. Seki, H., Kaneko, T.: Differential Cryptanalysis of Reduced Rounds of GOST. In: Stinson, D.R., Tavares, S.E. (eds.) SAC 2000. LNCS, vol. 2012, pp. 315–323. Springer, Heidelberg (2001)

The Software Performance of Authenticated-Encryption Modes

Ted Krovetz[1] and Phillip Rogaway[2]

[1] Computer Science, California State University, Sacramento, CA 95819 USA
[2] Computer Science, University of California, Davis, CA 95616 USA

Abstract. We study software performance of authenticated-encryption modes CCM, GCM, and OCB. Across a variety of platforms, we find OCB to be substantially faster than either alternative. For example, on an Intel i5 ("Clarkdale") processor, good implementations of CCM, GCM, and OCB encrypt at around 4.2 cpb, 3.7 cpb, and 1.5 cpb, while CTR mode requires about 1.3 cpb. Still we find room for algorithmic improvements to OCB, showing how to trim one blockcipher call (most of the time, assuming a counter-based nonce) and reduce latency. Our findings contrast with those of McGrew and Viega (2004), who claimed similar performance for GCM and OCB.

Keywords: authenticated encryption, cryptographic standards, encryption speed, modes of operation, CCM, GCM, OCB.

1   Introduction

BACKGROUND. Over the past few years, considerable effort has been spent constructing schemes for *authenticated encryption* (AE). One reason is recognition of the fact that a scheme that delivers both privacy and authenticity may be more efficient than the straightforward amalgamation of separate privacy and authenticity techniques. A second reason is the realization that an AE scheme is less likely to be incorrectly used than a privacy-only encryption scheme.

While other possibilities exist, it is natural to build AE schemes from blockciphers, employing some *mode of operation*. There are two approaches. In a *composed* ("two-pass") AE scheme one conjoins essentially separate privacy and authenticity modes. For example, one might apply CTR-mode encryption and then compute some version of the CBC MAC. Alternatively, in an *integrated* ("one-pass") AE scheme the parts of the mechanism responsible for privacy and for authenticity are tightly coupled. Such schemes emerged around a decade ago, with the work of Jutla [21], Katz and Yung [23], and Gligor and Donescu [11].

Integrated AE schemes were invented to improve performance of composed ones, but it has not been clear if they do. In the only comparative study to date [32], McGrew and Viega found that their composed scheme, GCM, was about as fast as, and sometimes faster than, the integrated scheme OCB [36] (hereinafter OCB1, to distinguish it from a subsequent variant we'll call OCB2 [35]). After McGrew and Viega's 2004 paper, no subsequent performance study

A. Joux (Ed.): FSE 2011, LNCS 6733, pp. 306–327, 2011.
© International Association for Cryptologic Research 2011

scheme	ref	date	ty	high-level description
EtM	[1]	2000	C	Encrypt-then-MAC (and other) generic comp. schemes
RPC	[23]	2000	I	Insert counters and sentinels in blocks, then ECB
IAPM	[21]	2001	I	Seminal integrated scheme. Also IACBC
XCBC	[11]	2001	I	Concurrent with Jutla's work. Also XECB
✓ OCB1	[36]	2001	I	Optimized design similar to IAPM
TAE	[29]	2002	I	Recasts OCB1 using a tweakable blockcipher
✓ CCM	[40]	2002	C	CTR encryption + CBC MAC
CWC	[24]	2004	C	CTR encryption + $GF(2^{127}-1)$-based CW MAC
✓ GCM	[32]	2004	C	CTR encryption + $GF(2^{128})$-based CW MAC
EAX	[2]	2004	C	CTR encryption + CMAC, a cleaned-up CCM
✓ OCB2	[35]	2004	I	OCB1 with AD and alleged speed improvements
CCFB	[30]	2005	I	Similar to RPC [23], but with chaining
CHM	[18]	2006	C	Beyond-birthday-bound security
SIV	[37]	2006	C	Deterministic/misuse-resistant AE
CIP	[17]	2008	C	Beyond-birthday-bound security
HBS	[20]	2009	C	Deterministic AE. Single key
BTM	[19]	2009	C	Deterministic AE. Single key, no blockcipher inverse
✓ OCB3	new	2010	I	Refines the prior versions of OCB

Fig. 1. Authenticated-encryption schemes built from a blockcipher. Checks indicate schemes included in our performance study. Column **ty** (type) specifies if the scheme is integrated (I) or composed (C). Schemes EtM, CCM, GCM, EAX, OCB2, and SIV are in various standards (ISO 19772, NIST 800-38C and 800-38D, RFC 5297).

was ever published. This is unfortunate, as there seems to have been a major problem with their work: reference implementations were compared against optimized ones, and none of the results are repeatable due to the use of proprietary code. In the meantime, CCM and GCM have become quite important to cryptographic practice. For example, CCM underlies modern WiFi (802.11i) security, while GCM is supported in IPsec and TLS.

McGrew and Viega identified two performance issues in the design of OCB1. First, the mode uses $m + 2$ blockcipher calls to encrypt a message of $m = \lceil |M|/128 \rceil$ blocks. In contrast, GCM makes do with $m + 1$ blockcipher calls. Second, OCB1 twice needs one AES result before another AES computation can proceed. Both in hardware and in software, this can degrade performance. Beyond these facts, existing integrated modes cannot exploit the "locality" of counters in CTR mode—that high-order bits of successive input blocks are usually unchanged, an observation first exploited, for software speed, by Hongjun Wu [4].

Given all of these concerns, maybe GCM really is faster than OCB—and, more generally, maybe composed schemes are the fastest way to go. The existence of extremely high-speed MACs supports this possibility [3,5,25].

CONTRIBUTIONS. We begin by refining the definition of OCB to address the performance concerns just described. When the provided nonce is a counter, the mode that we call OCB3 shaves off one AES encipherment per message encrypted about 98% of the time. In saying that the nonce is a counter we mean that, in a given session, its top portion stays fixed, while, with each successive message, the bottom portion gets bumped by one. This is the approach recommended in RFC 5116 [31, Section 3.2] and, we believe, the customary way to use an AE scheme. We do not introduce something like a $GF(2^{128})$ multiply to compensate for the usually-eliminated blockcipher call, and no significant penalty is paid, compared to OCB1, if the provided nonce is not a counter (one just fails to save the blockcipher call). We go on to eliminate the latency that used to occur when computing the "checksum" and processing the AD (associated data).

Next we study the relative software performance of CCM, GCM, and the different versions of OCB. We employ the fastest publicly available code for Intel x86, both with and without Intel's new instructions for accelerating AES and GCM. For other platforms—ARM, PowerPC, and SPARC—we use a refined and popular library, OpenSSL. We test the encryption speed on messages of every byte length from 1 byte to 1 Kbyte, plus selected lengths beyond. The OCB code is entirely in C, except for a few lines of inline assembly on ARM and compiler intrinsics to access byteswap, trailing-zero count, and SSE/AltiVec functionality.

We find that, across message lengths and platforms, OCB, in any variant, is well faster than CCM and GCM. While the performance improvements from our refining OCB are certainly measurable, those differences are comparatively small. Contrary to McGrew and Viega's findings, the speed differences we observe between GCM and OCB1 are large and favor OCB1.

As an example of our experimental findings, for 4 KB messages on an Intel i5 ("Clarkdale") processor, we clock CCM at 4.17 CPU cycles per byte (cpb), GCM at 3.73 cpb, OCB1 at 1.48 cpb, OCB2 at 1.80 cpb, and OCB3 at 1.48 cpb. As a baseline, CTR mode runs at 1.27 cpb. See Fig. 2. These implementations exploit the processor's AES New Instructions (AES-NI), including "carryless multiplication" for GCM. The OCB3 *authentication overhead*—the time the mode spends in excess of the time to encrypt with CTR—is about 0.2 cpb, and the difference between OCB and GCM overhead is about a factor of 10. Even written in C, our OCB implementations provide, on this platform, the fastest reported times for AE.

The means for refining OCB are not complex, but it took much work to understand what optimization would and would not help. First we wanted to arrange that nonces agreeing on all but their last few bits be processed using the same blockcipher call. To accomplish this in a way that minimizes runtime state and key-setup costs, we introduce a new hash-function family, a *stretch-then-shift* xor-universal hash. The latency reductions are achieved quite differently, by changes in how the mode defines and operates on the Checksum.

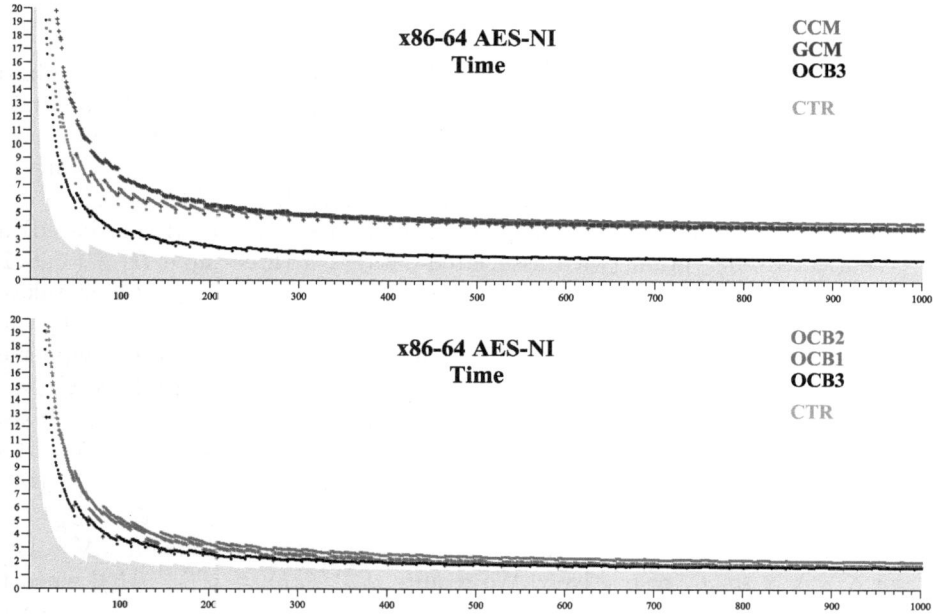

Fig. 2. Top: Performance of CCM, GCM, and OCB3 on an x86 with AES-NI. The x-coordinate is the message length, in bytes; the y-coordinate is the measured number of cycles per byte. From top-to-bottom on the right-hand side, the curves are for CCM, GCM, and OCB3. The shaded region shows the time for CTR mode. This and subsequent graphs are best viewed in color. **Bottom:** Performance of OCB variants on an x86 with AES-NI. From top-to-bottom, the curves are for OCB2, OCB1, and OCB3. The shaded region shows the time for CTR mode.

One surprising finding is that, on almost all platforms, OCB2 is slightly slower than OCB1. To explain, recall that most integrated schemes involve computing an *offset* for each blockcipher call. With OCB1, each offset is computed by xoring a key-dependent value, an approach going back to Jutla [21]; with OCB2, each offset is computed by a "doubling" in $GF(2^{128})$. The former approach turns out to be faster.

During our work we investigated novel ways to realize a maximal period, software-efficient, 128-bit LFSR; such constructions can also be used to make the needed offsets. A computer-aided search identified constructions like sending $A \parallel B \parallel C \parallel D$ to $C \parallel D \parallel B \parallel ((A \ll 1) \oplus (A \gg 1) \oplus (D \ll 15))$ (where A, B, C, D are 32 bits); see Appendix A. While very fast, such maps are still slower than xoring a precomputed value. Our findings thus concretize Chakraborty and Sarkar's suggestion [6] to improve OCB using a fast, 128-bit, word-oriented LFSR—but, in the end, we conclude that the idea doesn't really help. Of course software-optimized 128-bit LFSRs may have other applications.

All code and data used in this paper, plus a collection of clickable tables and graphs, are available from the second author's webpage.

2   The Mode OCB3

PRELIMINARIES. We begin with a few basics. A *blockcipher* is a deterministic algorithm $E \colon \mathcal{K} \times \{0,1\}^n \to \{0,1\}^n$ where $\mathcal{K}$ is a finite set and $n \geq 1$ is a number, the *key space* and *blocklength*. We require $E_K(\cdot) = E(K, \cdot)$ be a permutation for all $K \in \mathcal{K}$. Let $D = E^{-1}$ be the map from $\mathcal{K} \times \{0,1\}^n$ to $\{0,1\}^n$ defined by $D_K(Y) = D(K, Y)$ being the unique point X such that $E_K(X) = Y$.

Following recent formalizations [1,23,34,36], a scheme for (nonce-based) authenticated encryption (with associated-data) is a three-tuple $\Pi = (\mathcal{K}, \mathcal{E}, \mathcal{D})$. The *key space* $\mathcal{K}$ is a finite, nonempty set. The *encryption algorithm* $\mathcal{E}$ takes in a *key* $K \in \mathcal{K}$, a *nonce* $N \in \mathcal{N} \subseteq \{0,1\}^*$, a *plaintext* $M \in \mathcal{M} \subseteq \{0,1\}^*$, and *associated data* $A \in \mathcal{A} \subseteq \{0,1\}^*$. It returns, deterministically, either a ciphertext $C = \mathcal{E}_K^{N, A}(M) \in \mathcal{C} \subseteq \{0,1\}^*$ or the distinguished value INVALID. Sets $\mathcal{N}, \mathcal{M}, \mathcal{C}$, and $\mathcal{A}$ are called the *nonce space*, *message space*, *ciphertext space*, and *AD space* of Π. The *decryption algorithm* $\mathcal{D}$ takes a tuple $(K, N, A, C) \in \mathcal{K} \times \mathcal{N} \times \mathcal{A} \times \mathcal{C}$ and returns, deterministically, either INVALID or a string $M = \mathcal{D}_K^{N, A}(C) \in \mathcal{M} \subseteq \{0,1\}^*$. We require that $\mathcal{D}_K^{N, A}(C) = M$ for any string $C = \mathcal{E}_K^{N, A}(M)$ and that $\mathcal{E}$ and $\mathcal{D}$ return INVALID if provided an input outside of $\mathcal{K} \times \mathcal{N} \times \mathcal{A} \times \mathcal{M}$ or $\mathcal{K} \times \mathcal{N} \times \mathcal{A} \times \mathcal{C}$, respectively. We require $|\mathcal{E}_K^{N, A}(M)| = |\mathcal{E}_K^{N, A}(M')|$ when the encryptions are strings and $|M| = |M'|$. If this value is always $|M| + \tau$ we call τ the *tag length* of the scheme.

DEFINITION OF OCB3. Fix a blockcipher $E \colon \mathcal{K} \times \{0,1\}^{128} \to \{0,1\}^{128}$ and a tag length $\tau \in [0\mathbin{..}128]$. In Fig. 3 we define the AE scheme $\Pi = \text{OCB3}[E, \tau] = (\mathcal{K}, \mathcal{E}, \mathcal{D})$. The nonce space $\mathcal{N}$ is the set of all binary strings with fewer than 128 bits.[1] The message space $\mathcal{M}$ and AD-space $\mathcal{A}$ are all binary strings. The ciphertext space $\mathcal{C}$ is the set of all strings whose length is at least τ bits. Fig. 3's procedure Setup is implicitly run on or before the first call to $\mathcal{E}$ or $\mathcal{D}$. The variables it defines are understood to be global. In the protocol definition we write $\text{ntz}(i)$ for the number of trailing zeros in the binary representation of positive integer i (eg, $\text{ntz}(1) = \text{ntz}(3) = 0$, $\text{ntz}(4) = 2$), we write $\text{msb}(X)$ for the first (most significant) bit of X, we write $A \land B$ for the bitwise-and of A and B, and we write $A \ll i$ for the shift of A by i positions to the left (maintaining string length, leftmost bits falling off, zero-bits entering at the right). At lines 111 and 311 we regard Bottom as a number instead of a string.

DESIGN RATIONALE. We now explain some of the design choices made for OCB3. While not a large departure from OCB1 or OCB2, the refinements do help.

Trimming a blockcipher call. OCB1 and OCB2 took $m + 2$ blockcipher calls to encrypt an m-block string M: one to map the nonce N into an initial offset Δ; one for each block of M; one to encipher the final Checksum. The first of these is easy to eliminate if one is willing to replace the $E_K(N)$ computation by, say,

---

[1] In practice one would either restrict nonces to byte strings of 1–15 bytes, or else demand that nonces have a fixed length, say exactly 12-bytes. Under RFC 5116, a conforming AE scheme *should* use a 12-byte nonce.

101 **algorithm** $\mathcal{E}_K^{N\,A}(M)$	301 **algorithm** $\mathcal{D}_K^{N\,A}(\mathcal{C})$
102 **if** $\lvert N\rvert \geq 128$ **return** INVALID	302 **if** $\lvert N\rvert \geq 128$ **or** $\lvert\mathcal{C}\rvert < \tau$ **return** INVALID
103 $M_1 \cdots M_m\, M_* \leftarrow M$ where each	303 $C_1 \cdots C_m\, C_*\, T \leftarrow \mathcal{C}$ where each
104 $\lvert M_i\rvert = 128$ and $\lvert M_*\rvert < 128$	304 $\lvert C_i\rvert = 128$ and $\lvert C_*\rvert < 128$ and $\lvert T\rvert = \tau$
105 Checksum $\leftarrow 0^{128}$; $\quad C \leftarrow \varepsilon$	305 Checksum $\leftarrow 0^{128}$; $\quad M \leftarrow \varepsilon$
106 Nonce $\leftarrow 0^{127-\lvert N\rvert}\, 1\, N$	306 Nonce $\leftarrow 0^{127-\lvert N\rvert}\, 1\, N$
107 Top $\leftarrow$ Nonce $\wedge\, 1^{122}\, 0^6$	307 Top $\leftarrow$ Nonce $\wedge\, 1^{122}\, 0^6$
108 Bottom $\leftarrow$ Nonce $\wedge\, 0^{122}\, 1^6$	308 Bottom $\leftarrow$ Nonce $\wedge\, 0^{122}\, 1^6$
109 Ktop $\leftarrow E_K(\text{Top})$	309 Ktop $\leftarrow E_K(\text{Top})$
110 Stretch $\leftarrow$ Ktop $\parallel$ (Ktop $\oplus$ (Ktop$\ll 8$))	310 Stretch $\leftarrow$ Ktop $\parallel$ (Ktop $\oplus$ (Ktop$\ll 8$))
111 $\Delta \leftarrow$ (Stretch $\ll$ Bottom)[1..128]	311 $\Delta \leftarrow$ (Stretch $\ll$ Bottom)[1..128]
112 **for** $i \leftarrow 1$ **to** m **do**	312 **for** $i \leftarrow 1$ **to** m **do**
113 $\Delta \leftarrow \Delta \oplus L[\text{ntz}(i)]$	313 $\Delta \leftarrow \Delta \oplus L[\text{ntz}(i)]$
114 $C \leftarrow E_K(M_i \oplus \Delta) \oplus \Delta$	314 $M \leftarrow D_K(C_i \oplus \Delta) \oplus \Delta$
115 Checksum $\leftarrow$ Checksum $\oplus M_i$	315 Checksum $\leftarrow$ Checksum $\oplus M_i$
116 **if** $M_* \neq \varepsilon$ **then**	316 **if** $C_* \neq \varepsilon$ **then**
117 $\Delta \leftarrow \Delta \oplus L_*$	317 $\Delta \leftarrow \Delta \oplus L_*$
118 Pad $\leftarrow E_K(\Delta)$	318 Pad $\leftarrow E_K(\Delta)$
119 $C \leftarrow M_* \oplus \text{Pad}[1..\lvert M_*\rvert]$	319 $M \leftarrow M_* \leftarrow C_* \oplus \text{Pad}[1..\lvert C_*\rvert])$
120 Checksum $\leftarrow$ Checksum $\oplus M_*\, 10^*$	320 Checksum $\leftarrow$ Checksum $\oplus M_*\, 10^*$
121 $\Delta \leftarrow \Delta \oplus L_\$$	321 $\Delta \leftarrow \Delta \oplus L_\$$
122 Final $\leftarrow E_K(\text{Checksum} \oplus \Delta)$	322 Final $\leftarrow E_K(\text{Checksum} \oplus \Delta)$
123 Auth $\leftarrow \text{Hash}_K(A)$	323 Auth $\leftarrow \text{Hash}_K(A)$
124 Tag $\leftarrow$ Final $\oplus$ Auth	324 Tag $\leftarrow$ Final $\oplus$ Auth
125 $T \leftarrow \text{Tag}[1..\tau]$	325 $T' \leftarrow \text{Tag}[1..\tau]$
126 **return** $C \parallel T$	326 **if** $T = T'$ **then return** M
	327 **else return** INVALID
201 **algorithm** Setup(K)	401 **algorithm** Hash$_K(A)$
202 $L_* \leftarrow E_K(0^{128})$	402 $A_1 \cdots A_m\, A_* \leftarrow A$ where each
203 $L_\$ \leftarrow \text{double}(L_*)$	403 $\lvert A_i\rvert = 128$ and $\lvert A_*\rvert < 128$
204 $L[0] \leftarrow \text{double}(L_\$)$	404 Sum $\leftarrow 0^{128}$
205 **for** $i \leftarrow 1, 2, \cdots$ **do** $L[i] \leftarrow \text{double}(L[i-1])$	405 $\Delta \leftarrow 0^{128}$
206 **return**	406 **for** $i \leftarrow 1$ **to** m **do**
	407 $\Delta \leftarrow \Delta \oplus L[\text{ntz}(i)]$
	408 Sum $\leftarrow$ Sum $\oplus E_K(A_i \oplus \Delta)$
	409 **if** $A_* \neq \varepsilon$ **then**
	410 $\Delta \leftarrow \Delta \oplus L_*$
211 **algorithm** double(X)	411 Sum $\leftarrow$ Sum $\oplus E_K(A_*\, 10^* \oplus \Delta)$
212 **return** $(X\ll 1) \oplus (\text{msb}(X) \cdot 135)$	412 **return** Sum

Fig. 3. Definition of OCB3$[E,\tau]$. Here $E\colon \mathcal{K} \times \{0,1\}^{128} \to \{0,1\}^n$ is a blockcipher and $\tau \in [0..128]$ is the tag length. Algorithms $\mathcal{E}$ and $\mathcal{D}$ are called with arguments $K \in \mathcal{K}$, $N \in \{0,1\}^{\leq 127}$, and $M, C \in \{0,1\}^*$.

$K_1 \cdot N$, the product in $\text{GF}(2^{128})$ of nonce N and a variant K_1 of K. The idea has been known since Halevi [14]. But such a change would necessitate implementing a $\text{GF}(2^{128})$ multiply for just this one step. Absent hardware support, one would need substantial precomputation and enlarged internal state to see any savings; not a net win. We therefore compute the initial offset Δ using a different xor-universal hash function: $\Delta = H_K(N) = (\text{Stretch}\ll \text{Bottom})[1..128]$ where Bottom is the last six bits of N and the $(128+64)$-bit string Stretch is made by a process involving enciphering N with its last six bits zeroed out. This *stretch-then-shift* hash will be proven xor-universal in Section 4.1. Its use ensures that, when the nonce N is a counter, the initial offset Δ can be computed without a new blockcipher call $63/64 \approx 98\%$ of the time. In this way we reduce cost from

$m + 2$ blockcipher calls to an amortized $m + 1.016$ blockcipher calls, plus tiny added time for the hash.

Reduced latency. Assume the message being encrypted is not a multiple of 128 bits; there is a final block M_* having 1–127 bits. In prior versions of OCB one would need to wait on the penultimate blockcipher call to compute the Checksum and, from it, the final blockcipher call. Not only might this result in pipeline stalls [32], but if the blockcipher's efficient implementation needs a long string to ECB, then the lack of parallelizability translates to extra work. For example, Käsper and Schwabe's bit-sliced AES [22] ECB-encrypts eight AES blocks in a single shot. Using this in OCB1 or OCB2 would result in enciphering 24 blocks to encrypt a 100-byte string—three times more than what "ought" to be needed—since twice one must wait on AES output to form the next AES input. In OCB3 we restructure the algorithm so that the Checksum never depends on any ciphertext. Concretely, Checksum $= M_1 \oplus M_2 \oplus M_{m-1} \oplus M_m 10^*$ for a short final block, and Checksum $= M_1 \oplus M_2 \oplus M_{m-1} \oplus M_m$ for a full final block. The fact that you can get the same Checksum for distinct final blocks is addressed by using different offsets in these two cases.

Incrementing offsets. In OCB1, each noninitial offset is computed from the prior one by xoring some key-derived value; the ith offset is constructed by $\Delta \leftarrow \Delta \oplus L[\mathrm{ntz}(i)]$. In OCB2, each noninitial offset is computed from the prior one by multiplying it, again in $\mathrm{GF}(2^{128})$, by a constant: $\Delta \leftarrow (\Delta \ll 1) \oplus (\mathrm{msb}(\Delta) \cdot 135)$, an operation that has been called *doubling*. Not having to go to memory or attend to the index i, doubling was thought to be faster than the first method. In our experiments, it is not. While doubling can be coded in five Intel x86-64 assembly instructions, it still runs more slowly. In some settings, doubling loses big: it is expensive on 32-bit machines, and some compilers do poorly at turning C/C++ code for doubling into machine code that exploits the available instructions. On Intel x86, the 128-bit SSE registers lack the ability to be efficiently shifted one position to the left. Finally, the doubling operation is not endian neutral: if we must create a bit pattern in memory to match the sequence generated by doubling (and AES implementations generally do expect their inputs to live in memory) we will effectively favor big-endian architectures. We can trade this bias for a little-endian one by redefining double() to include a byteswap. But one is still favoring one endian convention over the other, and not just at key-setup time. See Appendix A for some of the alternatives to repeated doubling that we considered.

Further design issues. Unlike OCB1 and OCB2, each 128-bit block of plaintext is now processed in the same way whether or not it is the final 128 bits. This change facilitates implementing a clean incremental API, since one is able to output each 128-bit chunk of ciphertext after receiving the corresponding chunk of plaintext, even if it is not yet known if the plaintext is complete.

All AD blocks can now be processed concurrently; in OCB2, the penultimate block's output was needed to compute the final block's input, potentially creating pipeline stalls or inefficient use of a blockcipher's multi-block ECB interface. Also, each 128-bit block of AD is treated the same way if it is or isn't the message's end, simplifying the incremental provisioning of AD.

We expect the vast majority of processors running OCB3 will be little-endian; still, the mode's definition does nothing to favor this convention. The issue arises each time "register oriented" and "memory oriented" values interact. These are the same on big-endian machines, but are opposite on little-endian ones. One could, therefore, favor little-endian machines by building into the algorithm byte swaps that mimic those that would occur naturally each time memory and register oriented data interact. We experimentally adapted our implementation to do this but found that it made very little performance difference. This is due, first, to good byte reversal facilities on most modern processors (eg, `pshufb` can reverse 16 bytes on our x86 in a single cycle). It is further due to the fact that OCB3's table-based approach for incrementing offsets allows for the table to be endian-adjusted at key setup, removing most endian-dependency on subsequent encryption or decryption calls. Since it makes little difference to performance, and big-endian specifications are conceptually easier, OCB3 does not make any gestures toward little-endian orientation.

A low-level choice where OCB and GCM part ways is in the representation of field points. In GCM the polynomial $a_{127}\mathrm{x}^{127} + \cdots a_1\mathrm{x} + a_0$ corresponds to string $a_0 \ldots a_{127}$ rather than $a_{127} \ldots a_0$. McGrew and Viega call this the little-endian representation, but, in fact, this choice has nothing to do with endianness. The usual convention on machines of all kinds is that the msb is the left-most bit of any register. Because of this, GCM's "reflected-bit" convention can result in extra work to be performed even on Intel chips having instructions specifically intended for accelerating GCM [12,13]. Among the advantages of following the msb-first convention is that a left shift by one can be implemented by adding a register to itself, an operation often faster than a logical shift.

SECURITY OF OCB3. First we provide our definitions. Let $\Pi = (\mathcal{K}, \mathcal{E}, \mathcal{D})$ be an AE scheme. Given an adversary (algorithm) $\mathcal{A}$, we let $\mathbf{Adv}_{\Pi}^{\mathrm{priv}}(\mathcal{A}) = \Pr[K \xleftarrow{\$} \mathcal{K} : \mathcal{A}^{\mathcal{E}_K(\cdot,\cdot,\cdot)} \Rightarrow 1] - \Pr[\mathcal{A}^{\$(\cdot,\cdot,\cdot)} \Rightarrow 1]$ where queries of $\$(N, A, M)$ return a uniformly random string of length $|\mathcal{E}_K^{N,A}(M)|$. We demand that $\mathcal{A}$ never asks two queries with the same first component (the N-value), that it never ask a query outside of $\mathcal{N} \times \mathcal{A} \times \mathcal{M}$, and that it never repeats a query. Next we define authenticity. For that, let $\mathbf{Adv}_{\Pi}^{\mathrm{auth}}(\mathcal{A}) = \Pr[K \xleftarrow{\$} \mathcal{K} : \mathcal{A}^{\mathcal{E}_K(\cdot,\cdot,\cdot)}$ forges] where we say that the adversary *forges* if it outputs a value $(N, A, C) \in \mathcal{N} \times \mathcal{A} \times \mathcal{C}$ such that $\mathcal{D}_K^{N,A}(C) \neq$ INVALID yet there was no prior query (N, A, M') that returned C. We demand that $\mathcal{A}$ never asks two queries with the same first component (the N-value), never asks a query outside of $\mathcal{N} \times \mathcal{A} \times \mathcal{M}$, and never repeats a query.

When $E : \mathcal{K} \times \{0,1\}^n \to \{0,1\}^n$ is a blockcipher define $\mathbf{Adv}_E^{\pm\mathrm{prp}}(\mathcal{A}) = \Pr[\mathcal{A}^{E_K(\cdot), E_K^{-1}(\cdot)} \Rightarrow 1] - \Pr[\mathcal{A}^{\pi(\cdot), \pi^{-1}(\cdot)} \Rightarrow 1]$ where K is chosen uniform from $\mathcal{K}$ and $\pi(\cdot)$ is a uniform permutation on $\{0,1\}^n$. Define $\mathbf{Adv}_E^{\mathrm{prp}}(\mathcal{A}) = \Pr[\mathcal{A}^{E_K(\cdot)} \Rightarrow 1] - \Pr[\mathcal{A}^{\pi(\cdot)} \Rightarrow 1]$ by removing the decryption oracle. The *ideal* blockcipher of blocksize n is the blockcipher $\text{Bloc}[n]\colon \mathcal{K} \times \{0,1\}^n \to \{0,1\}^n$ where each key K names a distinct permutation.

The security of OCB3 is given by the following theorem. We give the result in its information-theoretic form. Passing to the complexity-theoretic setting, where the idealized blockcipher $\text{Bloc}[n]$ is replaced by a conventional blockcipher secure as a strong-PRP, is standard.

Theorem 1. *Fix* $n = 128$, $\tau \in [0\,..\,n]$, *and let* $\Pi = \text{OCB3}[E, \tau]$ *where* $E = \text{Bloc}[n]$ *is the ideal blockcipher on* n *bits. If* $\mathcal{A}$ *asks encryption queries that entail* σ *total blockcipher calls, then* $\mathbf{Adv}_{\Pi}^{\text{priv}}(\mathcal{A}) \leq 6\,\sigma^2/2^n$. *Alternatively, if* $\mathcal{A}$ *asks encryption queries then makes a forgery attempt that together entail* σ *total blockcipher calls, then* $\mathbf{Adv}_{\Pi}^{\text{auth}}(\mathcal{A}) \leq 6\sigma^2/2^n + (2^{n-\tau})/(2^n - 1)$. $\qquad\square$

When we speak of the number of blockcipher calls entailed we are adding up the (rounded-up) blocklength for all the different strings output by the adversary and adding in $q + 2$ ($q =$number of queries), to upper-bound blockcipher calls for computing L_* and the initial Δ values. Main elements of the proof are described in Section 4; see the full paper for further details [26].

3   Experimental Results

SCOPE AND CODEBASE. We empirically study the software performance of OCB3, and compare this with state-of-the-art implementations of GCM, which delivers the fastest previously reported AE times. Both modes are further compared against CTR, the fastest privacy-only mode, which makes a good baseline for answering how much extra one pays for authentication. Finally, we consider CCM, the first NIST-approved AE scheme, and also OCB1 and OCB2, which are benchmarked to show how the evolution of OCB has affected performance.

Intensively optimized implementations of CTR and GCM are publicly available for the x86. Käsper and Schwabe hold the speed record for 64-bit code with no AES-NI, reporting peak rates of 7.6 and 10.7 CPU cycles per byte (cpb) for CTR and GCM [22]. With AES-NI, developmental versions of OpenSSL achieve 1.3 cpb for CTR [33] and 3.3 cpb for GCM.[2] These various results use different x86 chips and timing mechanisms. Here we use the Käsper-Schwabe AES, CTR, and GCM, the OpenSSL CTR, CCM, and GCM, augment the collection with new code for OCB, and compare performance on a single x86 and use a common timing mechanism, giving the fairest comparison to date.

The only non-proprietary, architecture-specific non-x86 implementations for AES and GCM that we could find are those in OpenSSL. Although these implementations are hand-tuned assembly, they are designed to be timing-attack resistant, and are therefore somewhat slow. This does not make comparisons with them irrelevant. OCB is timing-attack resistant too (assuming the underlying blockcipher is), making the playing field level. We adopt the OpenSSL

---

[2] Andy Polyakov, personal communication, August 27, 2010. The fastest published AES-NI time for GCM is 3.5 cpb on 8KB messages, from Gueron and Kounavis [13].

implementations for non-x86 comparisons and emphasize that timing-resistant implementations are being compared, not versions written for ultimate speed.

The OCB1 and OCB2 implementations are modifications of our OCB3 implementation, and therefore are similarly optimized. These implementations are in C, calling out to AES. No doubt further performance improvements can be obtained by rewriting the OCB code in assembly.

HARDWARE AND SOFTWARE ENVIRONMENTS. We selected five representative instruction-set architectures: (1) 32-bit x86, (2) 64-bit x86, (3) 32-bit ARM, (4) 64-bit PowerPC, and (5) 64-bit SPARC. Collectively, these architectures dominate the workstation, server, and portable computing marketplace. The x86 processor used for both 32- and 64-bit tests is an Intel Core i5-650 "Clarkdale" supporting the AES-NI instructions. The ARM is a Cortex-A8. The PowerPC is a 970fx. The SPARC is an UltraSPARC IIIcu. Each runs Debian Linux 6.0 with kernel 2.6.35 and GCC 4.5.1. Compilation is done with -O3 optimization, -mcpu or -march set according to the host processor, and -m64 to force 64-bit compilation when needed.

TESTING METHODOLOGY. The number of CPU cycles needed to encrypt a message is divided by the length of the message to arrive at the cost per byte to encrypt messages of that length. This is done for every message length from 1 to 1024 bytes, as well as 1500 and 4096 bytes. So as not to have performance results overly influenced by the memory subsystem of a host computer, we arrange for all code and data to be in level-1 cache before timing begins. Two timing strategies are used: C clock and x86 time-stamp counter. In the clock version, the ANSI C clock() function is called before and after repeatedly encrypting the same message, on sequential nonces, for a little more than one second. The clock difference determines how many CPU cycles were spent on average per processed byte. This method is highly portable, but it is time-consuming when collecting an entire dataset. On x86 machines there is a "time-stamp counter" (TSC) that increments once per CPU cycle. To capture the average cost of encryption—including the more expensive OCB3 encryptions that happen once every 64 calls—the TSC is used to time encryption of the same message 64 times on successive counter-based nonces. The TSC method is not portable, working only on x86, but is fast. Both methods have their potential drawbacks. The clock method depends on the hardware having a high-resolution timer and the OS doing a good job of returning the time used only by the targeted process. The TSC read instruction might be executed out of order, in some cases it has high latency, and it continues counting when other processes run.[3] In the end, we found that both timing methods give similar results. For example, in the

---

[3] To lessen these problems we read the TSC once before and after encrypting the same message 65 times, then read the TSC once before and after encrypting the same message once more. Subtracting the second timing from the first gives us the cost for encrypting the message 64 times, and mitigates the out-of-order and latency problems. To avoid including context-switches, we run experiments multiple times and keep only the median timing.

x86-64 AES-NI				
Mode	T_{4K}	T_{IPI}	Size	Init
CCM	4.17	4.57	512	265
GCM	3.73	4.53	656	337
OCB1	1.48	2.08	544	251
OCB2	1.80	2.41	448	185
OCB3	1.48	1.87	624	253
CTR	1.27	1.37	244	115

x86-32 AES-NI				
Mode	T_{4K}	T_{IPI}	Size	Init
CCM	4.18	4.70	512	274
GCM	3.88	4.79	656	365
OCB1	1.60	2.22	544	276
OCB2	1.79	2.42	448	197
OCB3	1.59	2.04	624	270
CTR	1.39	1.52	244	130

x86-64 Käsper-Schwabe				
Mode	T_{4K}	T_{IPI}	Size	Init
GCM	22.4	26.7	1456	3780
GCM-8K	10.9	15.2	9648	2560
OCB1	8.28	13.4	3008	3390
OCB2	8.55	13.6	2912	3350
OCB3	8.05	9.24	3088	3480
CTR	7.74	8.98	1424	1180

ARM Cortex-A8				
Mode	T_{4K}	T_{IPI}	Size	Init
CCM	51.3	53.7	512	1390
GCM	50.8	53.9	656	1180
OCB1	29.3	31.5	672	1920
OCB2	28.5	31.8	576	1810
OCB3	28.9	30.9	784	1890
CTR	25.4	25.9	244	236

PowerPC 970				
Mode	T_{4K}	T_{IPI}	Size	Init
CCM	75.7	77.8	512	1510
GCM	53.5	56.2	656	1030
OCB1	38.2	41.0	672	2180
OCB2	38.1	41.1	576	2110
OCB3	37.5	39.6	784	2240
CTR	37.5	37.8	244	309

UltraSPARC III				
Mode	T_{4K}	T_{IPI}	Size	Init
CCM	49.4	51.7	512	1280
GCM	39.3	41.5	656	904
OCB1	25.5	27.7	672	1720
OCB2	24.8	27.0	576	1700
OCB3	25.0	26.5	784	1730
CTR	24.1	24.4	244	213

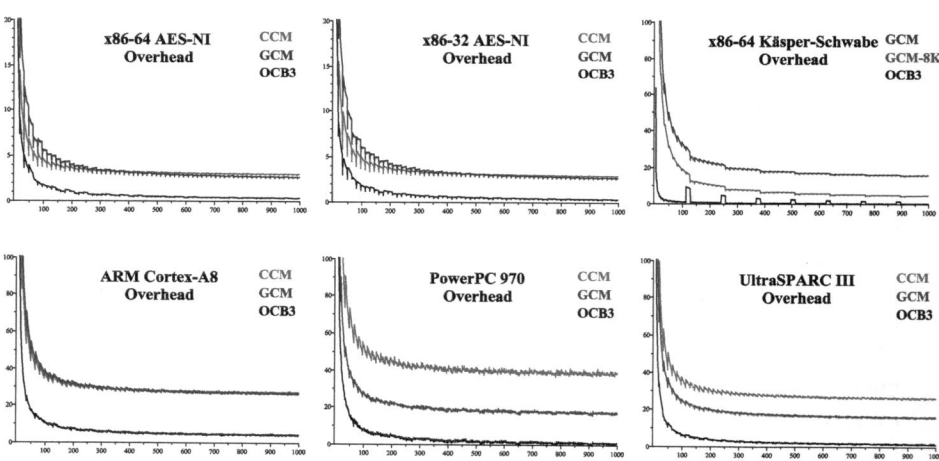

Fig. 4. Empirical performance of AE modes. For each architecture we give time to encrypt 4KB messages (in CPU cycles per byte), time to encrypt a weighted basket of message lengths (IPI, also in cpb), size of the implementation's context (in bytes), and time to initialize key-dependent values (in CPU cycles). Next we graph the same data, subtracting the CTR time and dropping the curves for OCB1 and OCB2, which may be visually close to that of OCB3. The CCM and GCM curves are visually hard to distinguish in the x86-64 AES NI, x86-32 AES NI, and ARM Cortex-A8 graphs.

eighteen x86 test runs done for this paper, the Internet Performance Index values computed by the two methods varied by no more than 0.05 cpb 10 times, no more than 0.10 cpb 15 times, and no more than 0.20 cpb all 18 times.

RESULTS. Summary findings are presented in Figs. 2 and 4. On all architectures and message lengths, OCB3 is significantly faster than GCM and CCM. Except on very short messages, it is nearly as fast as CTR. On x86, GCM's most competitive platform, OCB3's authentication overhead (its cost beyond CTR encryption) is 4–16%, with or without AES-NI, on both an Internet Performance Index (IPI)[4] and 4KB message length basis. In all our tests, CCM never has IPI or 4KB rates better than GCM, coming close only when small registers make GCM's multiplications expensive, or AES-NI instructions speed CCM's block encipherments. Results are similar on other architectures. The overhead of OCB3 does not exceed 12% that of GCM or CCM on PowerPC or SPARC, or 18% on ARM, when looking at either IPI or 4KB message encryption rates.

To see why OCB3 does so well, consider that there are four phases in OCB3 encryption: initial offset generation, encryption of full blocks, encryption of a partial final block (if there is one), and tag generation. On all but the shortest messages, full-block processing dominates overall cost per byte. Here OCB3, and OCB1, are particularly efficient. An unrolled implementation of, say, four blocks per iteration, will have, as overhead on top of the four blockcipher calls and the reads and writes associated to them: 16 xor operations (each on 16-byte words), 1 ntz computation, and 1 table lookup of a 16-byte value. On x86, summing the latencies of these 18 operations—which ignores the potential for instruction-level parallelism (ILP)—the operations require 23 cycles, or 0.36 cpb. In reality, on 64-bit x64 using AES-NI, we see CTR taking 1.27 cpb on 4KB messages while OCB3 uses 1.48, an overhead of 0.21 cpb, the savings coming from the ILP.

Short messages are optimized for too. When there is little or no full-block processing, it is the other three phases of encryption that determine performance. One gets a sense of the cost of these by looking at the cost to encrypt a single byte. On x86, OpenSSL's AES-NI based CTR implementation does this in 86 cycles, while CCM, GCM, and OCB3 use 257, 354, and 249 cycles, respectively. CCM remains competitive with OCB3 only for very short strings. On 64-bit x86 without AES-NI, using Käsper-Schwabe's bit-sliced AES that processes eight blocks at once, OCB3's performance lead is much greater, as its two blockcipher calls can be computed concurrently, unlike CCM and GCM. In this scenario, single-byte encryption rates for CCM, GCM, OCB3, CTR are 2600, 2230, 1080, 1010 cycles. On the other three architectures we see the following single-byte encryption times for (CCM, GCM, OCB3; CTR): ARM (1770, 1950, 1190; 460), PowerPC (2520, 1860, 1450; 309), and SPARC (1730, 1520, 1770; 467).

With hardware support making AES very cheap, authentication overhead becomes more prominent. AES-NI instructions enable AES-128 throughput of around 20 cycles per block. VIA's xcrypt assembly instruction is capable of 10

---

[4] The IPI is a weighted average of timings for messages of 44 bytes (5%), 552 bytes (15%), 576 bytes (20%), and 1500 bytes (60%) [32]. It is based on Internet backbone studies from 1998. We do not suggest that the IPI reflects a contemporary, real-world distribution of message lengths, only that it is useful to have *some* metric that attends to shorter messages and those that are not a multiple of 16 bytes. Any metric of this sort will be somewhat arbitrary in its definition.

cycles per block on long ECB sequences [39]. Speeds like these can make authentication overhead more expensive than encryption. With the Käsper-Schwabe code (no AES-NI), for example, on an IPI basis, OCB3 overhead is only 3% of encryption cost, but under AES-NI it rises to 27%. Likewise, GCM overhead rises from 41% to 70%. One might think CCM would do well using AES-NI since its overhead is mostly blockcipher calls, but its use of (serial) CBC for authentication reduces AES throughput to around 60 cycles per block, causing authentication overhead of about 70%.[5]

As expected, OCB1 and OCB3 long-message performance is the same due to having identical full-block processing. OCB2 is slower on long messages on all tested platforms but SPARC (computing ntz is slow on SPARC). With a counter-based nonce, OCB3 computes its initial encryption offset using a few bitwise shifts of a cached value rather than generating it with a blockcipher as both OCB1 and OCB2 do. This results in significantly improved average performance for encryption of short messages. The overall effect is that on an IPI basis on, say, 64-bit x86 using AES-NI, OCB3's authentication overhead is only 65% of that for OCB1 and only 40% of that for OCB2. When the provided nonce is *not* a counter, OCB3 performance is, in most of our test environments, indistinguishable from that of OCB1.

4   Proof of Security for OCB3

We describe three elements in the proof of OCB3's security: (1) the new xor-universal hash function it employs; (2) the definition and proof for a simple TBC (tweakable blockcipher) based generalization of OCB3; and (3) the proof that the particular TBC used by OCB3 is good.

4.1   Stretch-then-Shift Universal Hash

A new hash function H underlies the mapping of the low-order bits of the nonce to a 128-bit string (lines 108, 110, and 111 of Fig. 3). While an off-the-shelf hash would have worked alright, we were able to do better for this step. We start with the needed definitions.

DEFINITION. Let $\mathcal{K}$ be a finite set and let $H : \mathcal{K} \times \mathcal{X} \to \{0,1\}^n$ be a function. We say that H is *strongly xor-universal* if for all distinct $x, x' \in \mathcal{X}$ we have that $H_K(x) \oplus H_K(x')$ is uniformly distributed in $\{0,1\}^n$ and, also, $H_K(x)$ is uniformly distributed in $\{0,1\}^n$ for all $x \in \mathcal{X}$. The first requirement is the usual definition for H being *xor-universal*; the second we call *universal-1*.

---

[5] Intel released their Sandy Bridge microarchitecture January 2011, too late for a thorough update of this paper. Sandy Bridge increases both AES throughput and latency. Under Sandy Bridge, OCB and CTR will be substantially faster (likely under 1.0 cpb on long messages) because their work is dominated by parallel AES invocations. GCM will be just a little faster because most of its time is spent in authentication, which does not benefit from Sandy Bridge. CCM will be slower because longer latencies negatively affect CBC authentication.

THE TECHNIQUE. We aim to construct strongly xor-universal hash-functions $H: \mathcal{K} \times \mathcal{X} \rightarrow \{0,1\}^n$ where $\mathcal{K} = \{0,1\}^{128}$, $\mathcal{X} = [0 \,.. \,\text{domSize} - 1]$, and $n = 128$. We want domSize to be at least some modest-size number, say domSize ≥ 64, and intend that computing $H_K(x)$ be almost as fast as doing a table lookup. Fast computation of H should not require any large table, nor the preprocessing of K. Our desire for extreme speed in the absence of preprocessing and big tables rules out methods based on $\text{GF}(2^{128})$ multiplication, the obvious first attempt.

The method we propose is to stretch the key K into a longer string $stretch(K)$, and then extract its bits $x+1$ to $x+128$. Symbolically, $H_K(x) = (stretch(K))[x+1 \,.. \,x + 128]$ where $S[a \,.. \,b]$ denotes bits a through b of S, indexing beginning with 1. Equivalently, $H_K(x) = (stretch(K) \ll x)[1 \,.. \,128]$. We call this a *stretch-then-shift* hash.

How to stretch K? It seems natural to have $stretch(K)$ begin with K, so let's assume that $stretch(K) = K \parallel s(K)$ for some function s. It's easy to see that $s(K) = K$ and $s(K) \ll c$ won't work, but $s(K) = K \oplus (K \ll c)$, for some constant c, looks plausible for accommodating modest-sized domain. We now demonstrate that, for well-chosen c, this function does the job.

ANALYSIS. To review, we are considering $H_K^c(x) = (\text{Stretch} \ll x)[1 \,.. \,128]$ where $\text{Stretch} = stretch(K) = K \parallel (K \oplus (K \ll c))$ and $c \in [0 \,.. \,127]$. We'd like to know the maximal value of domSize for which $H_K(x)$ is xor-universal on the domain $\mathcal{X} = [0 \,.. \,\text{domSize}(c) - 1]$. This can be calculated by a computer program, as we now explain. Fix c and consider the 256×128 entry matrix $A = \begin{pmatrix} I \\ J \end{pmatrix}$ where I is the 128×128 identity matrix and J is the 128×128-bit matrix for which $J_{ij} = 1$ iff $j = i$ or $j = i + c$. Let A_i denote the 128×128 submatrix of A that includes only A's rows i to $i + 127$. Then $H_K^c(x) = A_{x+1}K$, the product in $\text{GF}(2)$ of the matrix A_{i+1} and the column vector K. Let $B_{i,j} = A_i + A_j$ be the indicated 128×128 matrix, the matrix sum over $\text{GF}(2)$. We would like to ensure that, for arbitrary $0 \leq i < j < \text{domSize}(c)$ and a uniform $K \in \{0,1\}^{128}$ that the 128-bit string $H_K^c(i) + H_K^c(j)$ is uniform—which is to say that $A_{i+1}K + A_{j+1}K = (A_{i+1} + A_{j+1})K = B_{i+1,j+1}K$ is uniform. This will be true if and only if $B_{i,j}$ is invertible in $\text{GF}(2)$ for all $1 \leq i < j \leq \text{domSize}(c)$. Thus domSize$(c)$ can be computed as the largest number domSize(j) such that $B_{i,j}$ is full rank, over $\text{GF}(2)$, for all $1 \leq i < j \leq \text{domSize}(j)$. Recalling the universal-1 property we also demand that A_i have full rank for all $1 \leq i \leq \text{domSize}(c)$. Now for any c, the number of matrices $A_{i,j}$ to consider is at most 2^{13}, and finding the rank in $\text{GF}(2)$ of that many 128×128 matrices is a feasible calculation.

Our results are tabulated in Fig. 5. The most interesting cases are H^5 and H^8, which are strongly xor-universal on $\mathcal{X} = [0 \,.. \,123]$ and $\mathcal{X} = [0 \,.. \,84]$, respectively. We offer no explanation for why these functions do well and various other H^c do not. As both H^5 and H^8 work on $[0 \,.. \,63]$ we select the latter map for use in OCB3 and single out the following result:

Lemma 1. *Let $H_K(x)$ be the first 128 bits of* $\text{Stretch} \ll x$ *where* $\text{Stretch} = K \parallel (K \oplus (K \ll 8))$, $|K| = 128$, $x \in [0 \,.. \,63]$. *Then H is strongly xor-universal.* $\square$

c	1	2	3	4	5	6	7	8	9	10	11	12	13	14	15	16	17	18	19	20
domSize(c)	3	15	7	3	124	7	3	85	120	3	118	63	3	31	63	3	7	31	3	7

Fig. 5. Stretch-then-shift hash. Largest $\mathcal{X} = [0 .. \text{domSize}(c) - 1]$ such that $H_K^c(x) = (\text{Stretch}(K) \ll x)[1 .. 128]$ is strongly xor-universal when $c \in [1 .. 16]$, $K \in \{0,1\}^{128}$, $x \in \mathcal{X}$, and $\text{Stretch}(K) = K \parallel (K \oplus (K \ll c))$.

EFFICIENCY. On 64-bit computers, assuming $K \parallel (K \oplus (K \ll 8))$ is precomputed and in memory, the value of $H_K(x)$ can be computed by three memory loads and two multiprecision shifts, requiring fewer than ten cycles on most architectures. If only K is in memory then the first 64 bits of $K \oplus (K \ll 8)$ can be computed with three additional assembly instructions. In the absence of a preprocessed table or special hardware-support, a method based on $\text{GF}(2^{128})$ multiplies would not fare nearly as well.

Computing successive H_K^c values requires a single extended-precision shift, making stretch-then-shift a reasonable approach for incrementing offsets. Unfortunately, it is not endian-neutral.

4.2   The TBC-Based Generalization of OCB3

Following the insight of Liskov, Rivest, and Wagner [29], OCB3 can be understood as an instantiation of an AE scheme that depends on a *tweakable blockcipher* (TBC). This is a deterministic algorithm $\widetilde{E}$ having signature $\widetilde{E} : \mathcal{K} \times \mathcal{T} \times \{0,1\}^n \to \{0,1\}^n$ where $\mathcal{K}$ and $\mathcal{T}$ are sets and $n \geq 1$ is a number—the *key space, tweak space,* and *blocklength,* respectively. We require $\widetilde{E}_K^T(\cdot) = \widetilde{E}(K, T, \cdot)$ be a permutation for all $K \in \mathcal{K}$ and $T \in \mathcal{T}$. Write $\widetilde{D} = \widetilde{E}^{-1}$ for the map from $\mathcal{K} \times \mathcal{T} \times \{0,1\}^n$ to $\{0,1\}^n$ defined by $\widetilde{D}_K^T(Y) = \widetilde{D}(K, T, Y)$ being the unique X such that $\widetilde{E}_K^T(X) = Y$. The *ideal* TBC for a tweak set $\mathcal{T}$ and blocksize n is the blockcipher $\text{Bloc}[\mathcal{T}, n] : \mathcal{K} \times \mathcal{T} \times \{0,1\}^n \to \{0,1\}^n$ where the keys name distinct permutations for each tweak T. For $\mathcal{T} = \mathcal{T}^{\pm} \cup \mathcal{T}^+$, $\mathcal{T}^{\pm} \cap \mathcal{T}^+ = \emptyset$, let $\mathbf{Adv}_{\widetilde{E}}^{\text{prp}[\mathcal{T}^{\pm}]}(\mathcal{A}) = \Pr[K \xleftarrow{\$} \mathcal{K} : \mathcal{A}^{\widetilde{E}_K(\cdot,\cdot), \widetilde{D}_K(\cdot,\cdot)} \Rightarrow 1] - \Pr[\mathcal{A}^{\pi(\cdot,\cdot), \pi^{-1}(\cdot,\cdot)} \Rightarrow 1]$ where π is chosen uniformly from $\text{Bloc}[\mathcal{T}, n]$ and adversary $\mathcal{A}$ is only allowed to ask decryption queries (T, Y) with $T \in \mathcal{T}^{\pm}$. Write $\mathbf{Adv}_{\widetilde{E}}^{\pm\text{prp}}(\mathcal{A})$ for $\mathbf{Adv}_{\widetilde{E}}^{\text{prp}[\mathcal{T}]}(\mathcal{A})$ and $\mathbf{Adv}_{\widetilde{E}}^{\text{prp}}(\mathcal{A})$ for $\mathbf{Adv}_{\widetilde{E}}^{\text{prp}[\emptyset]}(\mathcal{A})$. Our definition unifies PRP and strong-PRP security, allowing forward queries for all tweaks and backwards queries for those in $\mathcal{T}^{\pm}$. A conventional blockcipher can be regarded as a TBC with a singleton tweak space.

THE ΘCB3 SCHEME. Fix an arbitrary set of nonces $\mathcal{N}$; for concreteness, say $\mathcal{N} = \{0,1\}^{<128}$. Define from this set the corresponding tweak space $\mathcal{T}$ by

$$\mathcal{T} = \mathcal{N} \times \mathbb{N}_1 \cup \mathcal{N} \times \mathbb{N}_0 \times \{*\} \cup \mathcal{N} \times \mathbb{N}_0 \times \{\$\} \cup \mathcal{N} \times \mathbb{N}_0 \times \{*\$\} \cup \mathbb{N}_1 \cup \mathbb{N}_0 \times \{*\}$$

where $\mathbb{N}_1$ and $\mathbb{N}_0$ are the positive and nonnegative integers, respectively. Tweaks, it can be seen, are of six mutually exclusive "types." Tweaks of the first type are

```
101 algorithm E_K^{N,A}(M)                              201 algorithm D_K^{N,A}(C)
102 if N ∉ N then return INVALID                        202 if N ∉ N or |C| < τ then return INVALID
103 M_1 ··· M_m M_* ← M where each                      203 C_1 ··· C_m C_* T ← C where each
104     |M_i| = n and |M_*| < n                         204     |C_i| = n, |C_*| < n, and |T| = τ
105 Checksum ← 0^n,  C_* ← ε                            205 Checksum ← 0^n,  M_* ← ε
106 for i ← 1 to m do                                   206 for i ← 1 to m do
107     C_i ← Ẽ_K^{N i}(M_i)                            207     M_i ← D̃_K^{N i}(C_i)
108     Checksum ← Checksum ⊕ M_i                       208     Checksum ← Checksum ⊕ M_i
109 if M_* = ε then Final ← Ẽ_K^{N m $}(Checksum)       209 if C_* = ε then Final ← Ẽ_K^{N m $}(Checksum)
111 else  Pad ← Ẽ_K^{N m *}(0^n)                        211 else  Pad ← Ẽ_K^{N m *}(0^n)
111     C_* ← M_* ⊕ Pad[1 .. |M_*|]                     211     M_* ← C_* ⊕ Pad[1 .. |C_*|]
112     Checksum ← Checksum ⊕ M_* 10^*                  212     Checksum ← Checksum ⊕ M_* 10^*
113     Final ← Ẽ_K^{N m * $}(Checksum)                 213     Final ← Ẽ_K^{N m * $}(Checksum)
114 Auth ← Hash_K(A)                                    214 Auth ← Hash_K(A)
115 Tag ← Final ⊕ Auth                                  215 Tag ← Final ⊕ Auth
116 T ← Tag[1 .. τ]                                      216 T' ← Tag[1 .. τ]
117 return C_1 ··· C_m C_* ‖ T                          217 if T = T' then return M_1 ··· M_m M_*
                                                        218             else return INVALID

301 algorithm Hash_K(A)
302 Sum ← 0^n
303 A_1 ··· A_m A_* ← A for |A_i| = n, |A_*| < n
304 for i ← 1 to m do
305     Sum ← Sum ⊕ Ẽ_K^i(A_i)
306 if A_* ≠ ε then
307     Sum ← Sum ⊕ Ẽ_K^{m *}(A_* 10^*)
308 return Sum
```

Fig. 6. Definition of ΘCB3$[\widetilde{E}, \tau]$. Here $\widetilde{E} \colon \mathcal{N} \times \mathcal{T} \times \{0,1\}^n \to \{0,1\}^n$ is a tweakable blockcipher and $\tau \in [0 .. n]$ is the tag length. We have that OCB3$[E, \tau] = \Theta$CB3$[\widetilde{E}, \tau]$ for an appropriately chosen $\widetilde{E}$.

in the set $\mathcal{T}^{\pm} = \mathcal{N} \times \mathbb{N}_1$. Omitting parenthesis and commas when writing tweaks, TBC calls will look like $\widetilde{E}_K^{N\,i}(X)$, $\widetilde{E}_K^{N\,i\,*}(X)$, $\widetilde{E}_K^{N\,i\,\$}(X)$, $\widetilde{E}_K^{N\,i\,*\,\$}(X)$, $\widetilde{E}_K^{i}(X)$, or $\widetilde{E}_K^{i\,*}(X)$. Now given such a TBC $\widetilde{E} \colon \mathcal{K} \times \mathcal{T} \times \{0,1\}^n \to \{0,1\}^n$ and given a tag length $\tau \in [0 .. n]$, we construct the AE scheme $\Pi = \Theta$CB3$[\widetilde{E}, \tau] = (\mathcal{K}, \mathcal{E}, \mathcal{D})$ as defined in Fig. 6. The scheme's nonce space is $\mathcal{N}$, the message space is $\mathcal{M} = \{0,1\}^*$, the AD space is $\mathcal{A} = \{0,1\}^*$, and the ciphertext space is $\mathcal{C} = \{0,1\}^{\geq \tau}$. The scheme is illustrated in Fig. 7.

We now describe the security of ΘCB3 when using an ideal TBC. The proof is in the full paper [26].

Lemma 2. *Let* $\Pi = \Theta$CB3$[\widetilde{E}, \tau]$ *where* $\widetilde{E} = \mathrm{Bloc}[\mathcal{T}, n] \colon \mathcal{K} \times \mathcal{T} \times \{0,1\}^n \to \{0,1\}^n$ *is ideal. Let* $\mathcal{A}$ *be an adversary. Then* $\mathbf{Adv}_{\Pi}^{\mathrm{priv}}(\mathcal{A}) = 0$ *and* $\mathbf{Adv}_{\Pi}^{\mathrm{auth}}(\mathcal{A}) \leq (2^{n-\tau})/(2^n - 1)$. □

4.3 Instantiating the TBC

Continuing to assume that $n = 128$ and $\mathcal{N} = \{0,1\}^{<n}$, map each blockcipher $E \colon \mathcal{K} \times \{0,1\}^n \to \{0,1\}^n$ to the TBC $\widetilde{E} = \mathrm{Tw}[E]$, $\widetilde{E} \colon \mathcal{K} \times \mathcal{T} \times \{0,1\}^n \to \{0,1\}^n$, where $\mathcal{T} = \mathcal{N} \times \mathbb{N}_1 \cup \mathcal{N} \times \mathbb{N}_0 \times \{*\} \cup \mathcal{N} \times \mathbb{N}_0 \times \{\$\} \cup \mathcal{N} \times \mathbb{N}_0 \times \{*\$\} \cup \mathbb{N}_1 \cup \mathbb{N}_0 \times \{*\}$ by the construction of Fig. 8. There, multiplication is in GF(2^{128}) using the irreducible polynomial $\mathrm{x}^{128} + \mathrm{x}^7 + \mathrm{x}^7 + \mathrm{x}^2 + \mathrm{x} + 1$. We use the standard facts on the Gray code sequence $a \colon \mathbb{N}_0 \to \mathbb{N}_0$ that it is a permutation and $0 \leq a(i) \leq 2i$. It follows that coefficients $\Lambda = \{\lambda_i, \lambda_j^*, \lambda_j^\$, \lambda_j^{*\$} \colon 1 \leq i \leq 2^{120}$,

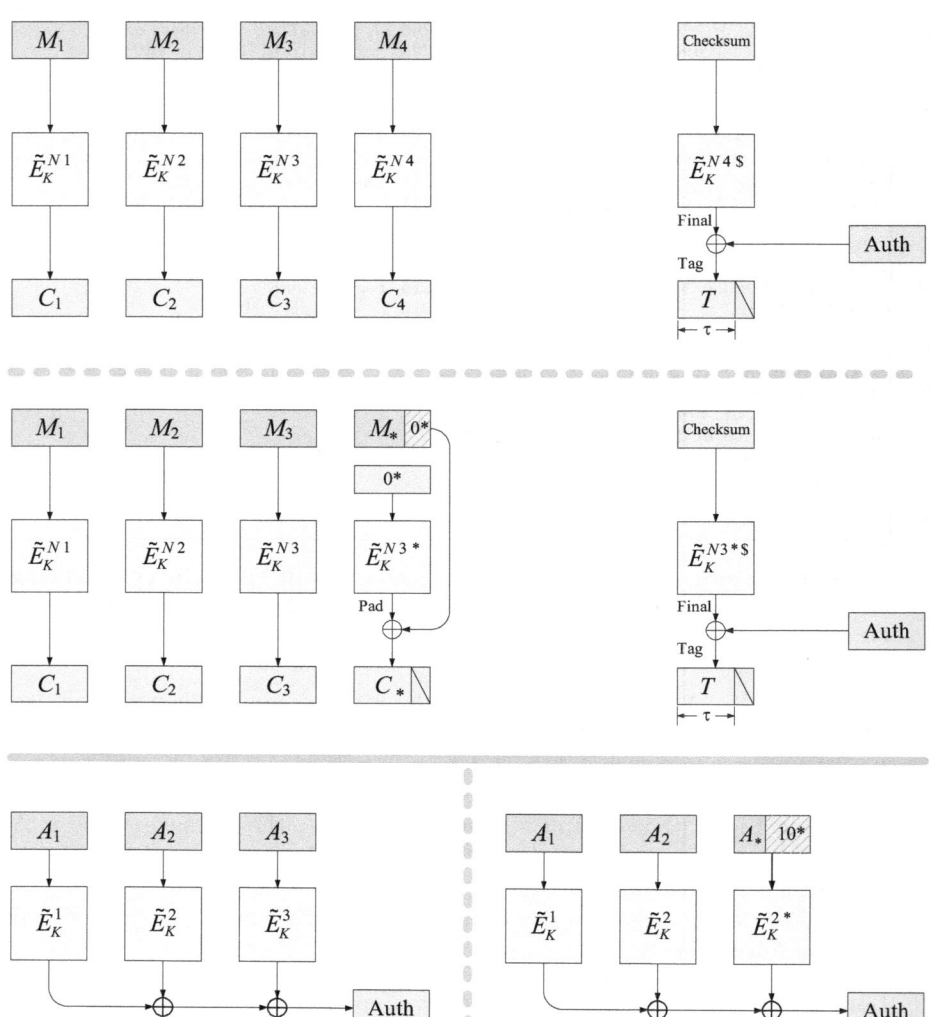

Fig. 7. Illustration of ΘCB3. The scheme depends on tweakable blockcipher $\widetilde{E}\colon \mathcal{N}\times \mathcal{T}\times\{0,1\}^n \to \{0,1\}^n$ and tag length $\tau \in [0\mathrel{..}n]$. The top figure shows the treatment of a message M having a full final block ($|M_4|=n$) (Checksum $= M_1 \oplus M_2 \oplus M_3 \oplus M_4$) while the middle picture shows the treatment of a message M having a short final block ($1 \le |M_*| < n$) (Checksum $= M_1 \oplus M_2 \oplus M_3 \oplus M_*10^*$). The bottom-left picture shows the processing of a three-block AD; on bottom-right, an AD with two full blocks and a short one. Algorithm OCB3$[E,\tau]$ coincides with ΘCB3$[\widetilde{E},\tau]$ for a particular TBC $\widetilde{E} = \mathrm{Tw}[E]$ constructed from E.

$$\widetilde{E}_K^{Ni}(X) = E_K(X \oplus \Delta) \oplus \Delta \ \text{with} \ \ \Delta = \text{Initial} \oplus \lambda_i\,L \quad \text{for } i \geq 1$$
$$\widetilde{E}_K^{Ni*}(X) = E_K(X \oplus \Delta) \qquad \text{with} \ \ \Delta = \text{Initial} \oplus \lambda_i^*\,L \quad \text{for } i \geq 0$$
$$\widetilde{E}_K^{Ni\$}(X) = E_K(X \oplus \Delta) \qquad \text{with} \ \ \Delta = \text{Initial} \oplus \lambda_i^\$\,L \quad \text{for } i \geq 0$$
$$\widetilde{E}_K^{Ni*\$}(X) = E_K(X \oplus \Delta) \qquad \text{with} \ \ \Delta = \text{Initial} \oplus \lambda_i^{*\$}\,L \quad \text{for } i \geq 0$$
$$\widetilde{E}_K^{i}(X) = E_K(X \oplus \Delta) \qquad \text{with} \ \ \Delta = \lambda_i\,L \qquad\qquad \text{for } i \geq 1$$
$$\widetilde{E}_K^{i*}(X) = E_K(X \oplus \Delta) \qquad \text{with} \ \ \Delta = \lambda_i^*\,L \qquad\qquad \text{for } i \geq 0$$

where

$$\text{Nonce} = 0^{127-|N|}\, 1\, N$$
$$\text{Top} = \text{Nonce} \wedge 1^{122}\, 0^6$$
$$\text{Bottom} = \text{Nonce} \wedge 0^{122}\, 1^6$$
$$\text{Ktop} = E_K(\text{Top})$$
$$\text{Stretch} = \text{Ktop} \parallel (\text{Ktop} \oplus (\text{Ktop}{\ll}8))$$
$$\text{Initial} = (\text{Stretch} \ll \text{Bottom})[1..128]$$

$$L = E_K(0^{128})$$
$$\lambda_i = 4\,a(i)$$
$$\lambda_i^* = 4\,a(i) + 1$$
$$\lambda_i^\$ = 4\,a(i) + 2$$
$$\lambda_i^{*\$} = 4\,a(i) + 3$$
$$a(0) = 0 \quad //\text{Grey code seq } 0, 1, 3, 2, 6, \ldots$$
$$a(i) = a(i-1) \oplus 2^{\text{ntz}(i)} \ \text{ if } i \geq 1$$

Fig. 8. Definition of $\widetilde{E} = \text{Tw}[E]$**, the tweakable blockcipher built from** E

$0 \leq j \leq 2^{120}\}$ are distinct and nonzero points of $\text{GF}(2^{128})$. The reader can check that $\text{OCB3}[E, \tau] = \Theta\text{CB3}[\text{Tw}[E], \tau]$.

SECURITY OF THE CONSTRUCTED TBC. We show that $\widetilde{E} = \text{Tw}[E]$ is a good TBC if E is a good blockcipher. In formalizing this, forward queries may be asked throughout $\mathcal{T}$, but backwards queries must be of the form $\widetilde{E}_K^{Ni}$.

Lemma 3. *Let* $n = 128$ *and let* $E = \text{Bloc}[n]$ *be the ideal blockcipher on* n *bits. Let* $\widetilde{E} = \text{Tw}[E]$*, the tweak space being* $\mathcal{T}$*, and let* $\mathcal{T}^{\pm} = \mathcal{N} \times \mathbb{N}_1$*. Let* $\mathcal{A}$ *be an adversary that asks at most* q *queries, non employing an* i*-value in excess of* 2^{120}*. Then* $\mathbf{Adv}_{\widetilde{E}}^{\text{prp}[\mathcal{T}^{\pm}]}(\mathcal{A}) \leq 6q^2/2^n$. $\qquad\qquad\square$

The proof is in the full version [26]. Combining it with Lemma 2 gives Theorem 1.

Acknowledgments

Phil Rogaway had interesting discussions with Tariq Ahmad (University of Massachusetts) on hardware aspects of GCM and OCB3.

The authors appreciate the support of NSF CNS 0904380.

References

1. Bellare, M., Namprempre, C.: Authenticated encryption: relations among notions and analysis of the generic composition paradigm. J. Cryptology 21(4), 469–491 (2008); Earlier version in ASIACRYPT 2000
2. Bellare, M., Rogaway, P., Wagner, D.: The EAX mode of operation. In: Roy, B., Meier, W. (eds.) FSE 2004. LNCS, vol. 3017, pp. 389–407. Springer, Heidelberg (2004)

3. Bernstein, D.: The Poly1305-AES message-authentication code. In: Gilbert, H., Handschuh, H. (eds.) FSE 2005. LNCS, vol. 3557, pp. 32–49. Springer, Heidelberg (2005)

4. Bernstein, D.J., Schwabe, P.: New AES software speed records. In: Chowdhury, D.R., Rijmen, V., Das, A. (eds.) INDOCRYPT 2008. LNCS, vol. 5365, pp. 322–336. Springer, Heidelberg (2008)

5. Black, J., Halevi, S., Krawczyk, H., Krovetz, T., Rogaway, P.: UMAC: fast and secure message authentication. In: Wiener, M. (ed.) CRYPTO 1999. LNCS, vol. 1666, pp. 216–233. Springer, Heidelberg (1999)

6. Chakraborty, D., Sarkar, P.: A general construction of tweakable block ciphers and different modes of operations. IEEE Trans. on Information Theory 54(5) (May 2008)

7. Dworkin, M.: Recommendation for block cipher modes of operation: the CCM mode for authentication and confidentiality. NIST Special Publication 800-38C (May 2004)

8. Dworkin, M.: Recommendation for block cipher modes of operation: Galois/Counter Mode (GCM) and GMAC. NIST Special Publication 800-38D (November 2007)

9. Ekdahl, P., Johansson, T.: A new version of the stream cipher SNOW. In: Nyberg, K., Heys, H.M. (eds.) SAC 2002. LNCS, vol. 2595, pp. 47–61. Springer, Heidelberg (2003)

10. Ferguson, N., Whiting, D., Schneier, B., Kelsey, J., Lucks, S., Kohno, T.: Helix: fast encryption and authentication in a single cryptographic primitive. In: Johansson, T. (ed.) FSE 2003. LNCS, vol. 2887, pp. 330–346. Springer, Heidelberg (2003)

11. Gligor, V., Donescu, P.: Fast encryption and authentication: XCBC encryption and XECB authentication modes. In: Matsui, M. (ed.) FSE 2001. LNCS, vol. 2355, pp. 92–108. Springer, Heidelberg (2002)

12. Gueron, S.: Intel's New AES instructions for enhanced performance and security. In: Dunkelman, O. (ed.) FSE 2009. LNCS, vol. 5665, pp. 51–66. Springer, Heidelberg (2009)

13. Gueron, S., Kounavis, M.: Intel carry-less multiplication instruction and its usage for computing the GCM mode (revision 2) (May 2010), White paper, available from http://www.intel.com

14. Halevi, S.: An observation regarding Jutla's modes of operation. Cryptology ePrint report 2001/015, April 2 (2001)

15. IEEE Standard 802.11i-2004. Part 11: Wireless Medium Access Control (MAC) and Physical Layer (PHY) Specifications: Medium Access Control (MAC) Security Enhancements (2004)

16. ISO/IEC 19772. Information technology – Security techniques – Authenticated encryption, 1st edn. (February 15, 2009)

17. Iwata, T.: Authenticated encryption mode for beyond the birthday bound security. In: Vaudenay, S. (ed.) AFRICACRYPT 2008. LNCS, vol. 5023, pp. 125–142. Springer, Heidelberg (2008)

18. Iwata, T.: New blockcipher modes of operation with beyond the birthday bound security. In: Robshaw, M.J.B. (ed.) FSE 2006. LNCS, vol. 4047, pp. 310–327. Springer, Heidelberg (2006)

19. Iwata, T., Yasuda, K.: BTM: A single-key, inverse-cipher-free mode for deterministic authenticated encryption. In: Jacobson Jr., M.J., Rijmen, V., Safavi-Naini, R. (eds.) SAC 2009. LNCS, vol. 5867, pp. 313–330. Springer, Heidelberg (2009)

20. Iwata, T., Yasuda, K.: HBS: a single-key mode of operation for deterministic authenticated encryption. In: Dunkelman, O. (ed.) FSE 2009. LNCS, vol. 5665, pp. 394–415. Springer, Heidelberg (2009)

21. Jutla, C.: Encryption modes with almost free message integrity. In: Pfitzmann, B. (ed.) EUROCRYPT 2001. LNCS, vol. 2045, pp. 529–544. Springer, Heidelberg (2001)

22. Käsper, E., Schwabe, P.: Faster and timing-attack resistant AES-GCM. In: Clavier, C., Gaj, K. (eds.) CHES 2009. LNCS, vol. 5747, pp. 1–17. Springer, Heidelberg (2009)

23. Katz, J., Yung, M.: Unforgeable encryption and chosen ciphertext secure modes of operation. In: Schneier, B. (ed.) FSE 2000. LNCS, vol. 1978, p. 284. Springer, Heidelberg (2001)

24. Kohno, T., Viega, J., Whiting, D.: CWC: A high-performance conventional authenticated encryption mode. In: Roy, B., Meier, W. (eds.) FSE 2004. LNCS, vol. 3017, pp. 408–426. Springer, Heidelberg (2004)

25. Krovetz, T.: Message authentication on 64-bit architectures. In: Biham, E., Youssef, A.M. (eds.) SAC 2006. LNCS, vol. 4356, pp. 327–341. Springer, Heidelberg (2007)

26. Krovetz, T., Rogaway, P.: The software performance of authenticated-encryption modes. Full version of this paper (January 2011)

27. Leiserson, C., Prokop, H., Randall, K.: Using de Bruijn sequences to index a 1 in a computer word (July 7, 1998) (unpublished manuscript)

28. Lidl, R., Niederreiter, H.: Introduction to finite fields and their applications (Revised Edition). Cambridge University Press, Cambridge (1994)

29. Liskov, M., Rivest, R., Wagner, D.: Tweakable block ciphers. In: Yung, M. (ed.) CRYPTO 2002. LNCS, vol. 2442, pp. 31–46. Springer, Heidelberg (2002)

30. Lucks, S.: Two-pass authenticated encryption faster than generic composition. In: Gilbert, H., Handschuh, H. (eds.) FSE 2005. LNCS, vol. 3557, pp. 284–298. Springer, Heidelberg (2005)

31. McGrew, D.: An interface and algorithms for authenticated encryption. IETF RFC 5116 (January 2008)

32. McGrew, D., Viega, J.: The security and performance of the Galois/Counter Mode (GCM) of operation. In: Canteaut, A., Viswanathan, K. (eds.) INDOCRYPT 2004. LNCS, vol. 3348, pp. 343–355. Springer, Heidelberg (2004); Also Cryptology ePrint report 2004/193, with somewhat different performance results

33. OpenSSL: The Open Source Toolkit for SSL/TLS, http://www.openssl.org/

34. Rogaway, P.: Authenticated-encryption with associated-data. In: CCS 2002. ACM Press, New York (2002)

35. Rogaway, P.: Efficient instantiations of tweakable blockciphers and refinements to modes OCB and PMAC. In: Lee, P.J. (ed.) ASIACRYPT 2004. LNCS, vol. 3329, pp. 16–31. Springer, Heidelberg (2004)

36. Rogaway, P., Bellare, M., Black, J.: OCB: A block-cipher mode of operation for efficient authenticated encryption. ACM Trans. on Information and System Security 6(3), 365–403 (2003); Earlier version, with T. Krovetz, in CCS 2001

37. Rogaway, P., Shrimpton, T.: A provable-security treatment of the key-wrap problem. In: Vaudenay, S. (ed.) EUROCRYPT 2006. LNCS, vol. 4004, pp. 373–390. Springer, Heidelberg (2006)

38. Tsaban, B., Vishne, U.: Efficient linear feedback shift registers with maximal period. Finite Fields and Their Applications 8(2), 256–267 (2002), Also CoRR cs.CR/0304010 (2003)

39. VIA Technologies. VIA Padlock programming guide (2005)

40. Whiting, D., Housley, R., Ferguson, N.: AES encryption & authentication using CTR mode & CBC-MAC. IEEE P802.11 doc 02/001r2 (May 2002)
41. Whiting, D., Housley, R., Ferguson, N.: Counter with CBC-MAC (CCM). IETF RFC 3610 (September 2003)
42. Zeng, G., Han, W., He, K.: High efficiency feedback shift register: σ-LFSR. Cryptology ePrint report 2007/114 (2007)

A   New Word-Oriented LFSRs

Recall that in OCB2 each 128-bit offset is computed from the prior one by multiplying it, in $GF(2^{128})$, by the constant $x = 2 = 0^{126}10$. Concretely, the point $X \in \{0, 1\}^{128}$ is stepped (or "incremented" or "doubled") by applying the map $S(X) = (X \ll 1) \oplus (\mathrm{msb}(X) \cdot 135)$. The constant 135 (decimal) represents (without the x^{128} term) the primitive polynomial $g(x) = x^{128} + x^7 + x^2 + x + 1$.

Chakraborty and Sarkar suggested [6] that there might be an incrementing function more efficient than S; they suspected that one might achieve efficiency gains with a *word-oriented* LFSR [38], as exemplified by the blockcipher SNOW [9]. After all, multiplication by x and reducing mod $g(x)$ is just the "Galois configuration" of a particular 128-bit LFSR [28], and one that has not been optimized for software performance. Some other 128-bit LFSRs might run faster.

To develop this idea, let S be an $n \times n$ binary matrix that is invertible over $GF(2)$. Then we may regard S as the feedback matrix of an LFSR that transforms the row vector $X \in \{0, 1\}^n$ into the row vector $X \cdot S$, a process we refer to as *stepping* the string X under S. The t-fold stepping of X by S is realized by matrix S^t. If the characteristic polynomial of S is primitive (over $GF(2)$) then the order of S in the general linear group $GL(n, GF(2))$ will be $2^n - 1$ and the map $X \mapsto X \cdot S$ will have two cycles: the length-1 cycle from 0^n to itself and the cycle of length $2^n - 1$ passing through all remaining n-bit strings [28]. The matrices $\langle S \rangle = \{S^i : 1 \leq i \leq 2^{n-1} - 1\}$, along with the matrix $n \times n$ zero matrix, can be regarded as a representation of $GF(2^n)$ under the operations of matrix multiplication and matrix addition, both mod 2.

Based on the paragraph above, the following is a simple way to obtain maximal and fast-to-compute 128-bit LFSRs. Generate candidate LFSRs by randomly combining a small number of shifts, ands, xors, using small or random constants. Represent each scheme by its feedback matrix. For each candidate matrix, check if it has a primitive characteristic polynomial. This is roughly the same approach taken by Zeng, Han, and He [42] to devise some software-efficient maximal-period shift registers intended for stream-cipher use. Using it, we generated and tested thousands of 128-bit stepping functions. Some efficient-to-compute schemes giving rise to maximal LFSRs are as follows:

$$X\,Y \mapsto Y \quad ((X \ll 1) \oplus (\mathrm{msb}(X) \cdot 10^{120}1010001) \oplus Y) \tag{1}$$
$$X\,Y \mapsto Y \quad ((X \ll 1) \oplus (X \gg 1) \oplus (Y \wedge 148)) \tag{2}$$
$$A\,B\,C\,D \mapsto C \quad D \quad B \quad ((A \ll 1) \oplus (\mathrm{msb}(A) \cdot 831) \oplus B \oplus D) \tag{3}$$
$$A\,B\,C\,D \mapsto C \quad D \quad B \quad ((A \ll 1) \oplus (A \gg 1) \oplus (D \wedge 107)) \tag{4}$$
$$A\,B\,C\,D \mapsto C \quad D \quad B \quad ((A \ll 1) \oplus (A \gg 1) \oplus (D \ll 15)) \tag{5}$$

Here $|X| = |Y| = 64$ and $|A| = |B| = |C| = |D| = 32$. Our experience searching for such maximal LFSRs suggests that they are rather finicky and sparse.

Implementing the candidate LFSRs on a variety of platforms revealed no clear winner. Beyond this, we found that none of the stepping functions were competitive with xoring in a pre-computed 128-bit value. All of the candidate stepping function introduce endian favoritism. In the end, then, we decided against using an LFSR stepping function to update offsets, going back to the OCB1 approach, instead.

Cryptanalysis of Hummingbird-1

Markku-Juhani O. Saarinen

Revere Security
4500 Westgrove Drive, Suite 335, Addison, TX 75001, USA
mjos@reveresecurity.com

Abstract. Hummingbird-1 is a lightweight encryption and message authentication primitive published in RISC '09 and WLC '10. Hummingbird-1 utilizes a
256-bit secret key and a 64-bit IV. We report a chosen-IV, chosen-message attack
that can recover the full secret key with a few million chosen messages processed
under two related IVs. The attack requires at most 2^{64} off-line computational
effort. The attack has been implemented and demonstrated to work against a real-
life implementation of Hummingbird-1. By attacking the differentially weak E
component, the overall attack complexity can be reduced by a significant factor. Our cryptanalysis is based on a differential divide-and-conquer method with
some novel techniques that are uniquely applicable to ciphers of this type.

Keywords: Hummingbird cipher, constrained devices, lightweight cryptography, stream cipher cryptanalysis.

1 Introduction

The advent of small-form wireless control and communication devices, sensors and
authentication tags is affecting commercial, military and domestic security engineering
in ways which were almost unimaginable only 10–20 years ago.

An important selection criterion when choosing cryptographic security components
for such extremely constrained devices is obviously *cost*, which directly relates to the
complexity of hardware and software implementation of the component and its computational efficiency. These lightweight cryptographic solutions must also meet stringent
security requirements as they are often critical links in the overall "chain of security" –
user authentication with a RFID token, a private conversation using a wireless hands-
free set and encryption of key presses on a wireless keyboard are some examples.

Hummingbird-1 [2,5] is a recent cryptographic algorithm proposal for RFID tags and
other constrained devices. It is covered by several pending patents and is being commer-
cially marketed by the Revere Security [7]. Revere has invested into Hummingbird's
cryptographic security assurance before its publication by contracting ISSI, a private
consultancy employing some ex-NSA staff [6] and members of U. Waterloo CACR [4].
After this work was originally done, an improved version, Hummingbird-2, has been
developed.

In the present report we show that the published version of Hummingbird-1 is sus-
pectible to a chosen-IV, chosen message attack that has an attack complexity of significantly less than 2^{64} operations and data complexity of only few megabytes, the entire

A. Joux (Ed.): FSE 2011, LNCS 6733, pp. 328–341, 2011.

256-bit secret key can be recovered. The attack has been implemented and demonstrated to work against a validated implementation of Hummingbird-1.

This paper is structured as follows. In Section 2 we give a description of Hummingbird-1 and make a key observations about its initialization procedure. In Section 3 we build an attack, step by step, that breaks Hummingbird-1. Section 4 contains a discussion about the implementation and implications of the attack, followed by conclusions in Section 5.

2 Description of Hummingbird-1

Hummingbird-1 [2,4,5] is an encryption and message authentication primitive that has a 256-bit secret key, uses a 64-bit IV (nonce) and optionally produces a 64-bit authenticator for the message. Hummingbird-1 is similar to ciphers such as Helix [3] and Phelix [10] in that it is a word-based stream cipher that can also be used for authentication. We have not analyzed the security of the proposed authentication functionality and it will not be discussed in this paper.

2.1 Notation and Parameters

The 256-bit secret key K is indexed as a vector of four 64-bit subkeys $K^{(i)}$. Each one of the 64-bit subkeys further consists of 16-bit words $K_j^{(i)}$ as follows:

$$K = (K^{(1)}, K^{(2)}, K^{(3)}, K^{(4)})$$
$$K^{(1)} = (K_1^{(1)}, K_2^{(1)}, K_3^{(1)}, K_4^{(1)})$$
$$K^{(2)} = (K_1^{(2)}, K_2^{(2)}, K_3^{(2)}, K_4^{(2)})$$
$$K^{(3)} = (K_1^{(3)}, K_2^{(3)}, K_3^{(3)}, K_4^{(3)})$$
$$K^{(4)} = (K_1^{(4)}, K_2^{(4)}, K_3^{(4)}, K_4^{(4)}).$$

The 80-bit internal state of Hummingbird-1 at round t consists of four 16-bit registers $RS1_t$, $RS2_t$, $RS3_t$, $RS4_t$ and the independent shift register $LFSR_t$.

When considering differential attacks, we denote by Δ the additive difference between two values. In our differential analysis we will be working on pairs of related instances of Hummingbird-1 which share the same secret key K. The state of the first and second instance at round t is written as

$$(RS1_t, RS2_t, RS3_t, RS4_t, LFSR_t)$$
$$and$$
$$(RS1'_t, RS2'_t, RS3'_t, RS4'_t, LFSR'_t).$$

The additive state difference $\Delta(RS1_t, RS2_t, RS3_t, RS4_t, LFSR_t)$ is

$$(RS1_t \boxminus RS1'_t, RS2_t \boxminus RS2'_t, RS3_t \boxminus RS3'_t, RS4_t \boxminus RS4'_t, LFSR_t \boxminus LFSR'_t).$$

Here $\boxminus$ denotes two's complement subtraction modulo 2^{16}. We will also write $\Delta P_i = P_i \boxminus P_i'$ and $\Delta C_i = C_i \boxminus C_i'$ to denote plaintext and ciphertext difference at message word i. Numerical values for differentials are in hexadecimal notation.

2.2   The 16-Bit Permutation E

The 16-bit permutation component $E(x, K^{(i)})$ consists of five invocations of four S-Boxes, interleaved with a mixing of a 16-bit subkey and a linear transform L. Figure 1 illustrates the operation of the the block cipher E.

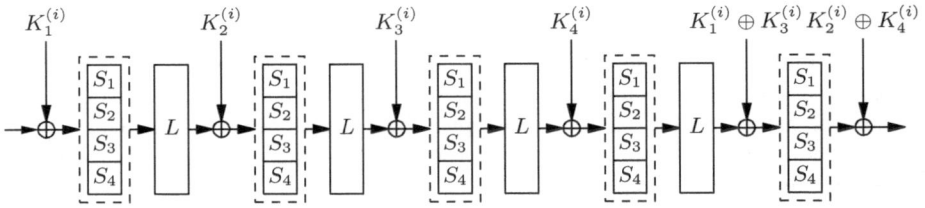

Fig. 1. The "E box" is a 16-bit permutation with a 64-bit key. L is a 16-bit linear transform $L(x) = x \oplus (x \lll 6) \oplus (x \lll 10)$.

Four permutations of values $0..15$ are used as the four-bit S-boxes $S_1(x)$, $S_2(x)$, $S_3(x)$ and $S_4(x)$. We have discovered that at least two variants of the four S-Boxes exist, one set being described in [5] and an another set in ISSI's analysis [6]. The second set of S-Boxes is equivalent to S4-S7 of Serpent-1 [1] and is compatible with test vectors provided by Revere Security [9]. Tables 1 and 2 give both S-Boxes in full.

Table 1. Hummingbird S-Boxes as reported in [5]

x	0	1	2	3	4	5	6	7	8	9	10	11	12	13	14	15
$S_1(x)$	8	6	5	15	1	12	10	9	14	11	2	4	7	0	13	3
$S_2(x)$	0	7	14	1	5	11	8	2	3	10	13	6	15	12	4	9
$S_3(x)$	2	14	15	5	12	1	9	10	11	4	6	8	0	7	3	13
$S_4(x)$	0	7	3	4	12	1	10	15	13	14	6	11	2	8	9	5

Table 2. The actual Hummingbird S-Boxes in an implementation obtained from its authors [9]

x	0	1	2	3	4	5	6	7	8	9	10	11	12	13	14	15
$S_1(x)$	1	15	8	3	12	0	11	6	2	5	4	10	9	14	7	13
$S_2(x)$	15	5	2	11	4	10	9	12	0	3	14	8	13	6	7	1
$S_3(x)$	7	2	12	5	8	4	6	11	14	9	1	15	13	3	10	0
$S_4(x)$	1	13	15	0	14	8	2	11	7	4	12	10	9	3	5	6

Any particular choice of S-Boxes does not affect the main cryptanalysis presented in this paper. In fact, the attack is applicable regardless of what type of E function is used as long as it is keyed with only 64 bits. Hence the particular choice of the number of rounds, S-Boxes and the linear transformation has little effect to the overall security of the cipher.

We define the linear transform $L(x)$ as

$$L(x) = x \oplus (x \lll 6) \oplus (x \lll 10), \tag{1}$$

where $\lll$ is a left circular shift operator. By $S(x)$ we denote the application of the four S-boxes in parallel on the four nibbles of $x = x_0 \mid x_1 \mid x_2 \mid x_3$:

$$S(x) = S_1(x_0) \mid S_2(x_1) \mid S_3(x_2) \mid S_4(x_3). \tag{2}$$

The complete 16-bit keyed permutation $E(x, K^{(i)})$ is described by:

$$u_0 = x \oplus K_1^{(i)}$$
$$u_1 = L(S(u_0)) \oplus K_2^{(i)}$$
$$u_2 = L(S(u_1)) \oplus K_3^{(i)}$$
$$u_3 = L(S(u_2)) \oplus K_4^{(i)}$$
$$u_4 = L(S(u_3)) \oplus K_1^{(i)} \oplus K_3^{(i)}$$
$$E(x, K^{(i)}) = S(u_4) \oplus K_2^{(i)} \oplus K_4^{(i)}.$$

2.3 Initialization

To set up Hummingbird-1, we first load the 64-bit IV value to the state registers:

$$(\text{RS1}_{-4}, \text{RS2}_{-4}, \text{RS3}_{-4}, \text{RS4}_{-4}) = (\text{IV}_1, \text{IV}_2, \text{IV}_3, \text{IV}_4). \tag{3}$$

After this, four rounds of special stepping is performed for $t = -4, -3, -2, -1$:

$$v12_t = E((\text{RS1}_t \boxplus \text{RS3}_t) \boxplus \text{RS1}_t, \ K^{(1)})$$
$$v23_t = E(v12_t \boxplus \text{RS2}_t, \ K^{(2)})$$
$$v34_t = E(v23_t \boxplus \text{RS3}_t, \ K^{(3)})$$
$$tv_t = E(v34_t \boxplus \text{RS4}_t, \ K^{(4)})$$
$$\text{RS1}_{t+1} = \text{RS1}_t \boxplus tv_t$$
$$\text{RS2}_{t+1} = \text{RS2}_t \boxplus v12_t$$
$$\text{RS3}_{t+1} = \text{RS3}_t \boxplus v23_t$$
$$\text{RS4}_{t+1} = \text{RS4}_t \boxplus v34_t.$$

Here the $\boxplus$ operator denotes addition modulo 2^{16}. After the final round, we set the bit 12 (or the 13th bit as it is expressed in the specification) in the tv temporary variable and assign that as the LFSR value:

$$\text{LFSR}_0 = tv_3 \ \vee \ 1000. \tag{4}$$

Therefore the 80-bit state after the initialization phase consists of the five words

$$(\text{RS1}_0 \ \text{RS2}_0 \ \text{RS3}_0 \ \text{RS4}_0 \ \text{LFSR}_0). \tag{5}$$

Observation 1. *The Hummingbird-1 initialization function has a high-bit XOR differential that holds with probability 1:*

$$\Delta(\text{IV}_1, \text{IV}_2, \text{IV}_3, \text{IV}_4) = (8000, 0000, 0000, 0000)$$

$$\Downarrow$$

$$\Delta(\text{RS1}_0, \text{RS2}_0, \text{RS3}_0, \text{RS4}_0, \text{LFSR}_0) = (8000, 0000, 0000, 0000, 0000).$$

2.4   The Encryption Function

Each Hummingbird-1 encryption round accepts a 16-bit plaintext word P_i to produce a ciphertext word C_i. Figure 2 illustrates one round of Hummingbird encryption.

For $t \geq 0$ (after initialization) we have

$$v_{12_t} = E(P_t \boxplus \text{RS1}_t, \ K^{(1)})$$

$$v_{23_t} = E(v_{12_t} \boxplus \text{RS2}_t, \ K^{(2)})$$

$$v_{34_t} = E(v_{23_t} \boxplus \text{RS3}_t, \ K^{(3)})$$

$$C_t = E(v_{34_t} \boxplus \text{RS4}_t, \ K^{(4)})$$

$$\text{LFSR}_{t+1} = \text{STEP}(\text{LFSR}_t)$$

$$\text{RS1}_{t+1} = \text{RS1}_t \boxplus v_{34_t}$$

$$\text{RS4}_{t+1} = \text{RS4}_t \boxplus v_{12_t} \boxplus \text{RS1}_{t+1}$$

$$\text{RS2}_{t+1} = \text{RS2}_t \boxplus v_{12_t} \boxplus \text{RS4}_{t+1}$$

$$\text{RS3}_{t+1} = \text{RS3}_t \boxplus v_{23_t} \boxplus \text{LFSR}_{t+1}.$$

The Hummingbird LFSR has been implemented in a slightly unusual right-cyclical fashion, which is best desribed in the C language:

```
lfsr = (lfsr >> 1) ^ (-(lfsr & 1) & 0xCA44);
```

THe LFSR operates independently from the other registers as there is no feedback from them or the plaintext to it. The particular LFSR selection or its operation does not affect on our attack in any way.

In this paper we will denote by $\text{HB}(\text{IV}, v) = z$ a query for encryption of vector v with the given IV value. Conversely, $\text{HB}^{-1}(\text{IV}, z) = v$ is a decryption query. Since Hummingbird is attacked in a "black box" fashion in this chosen-IV, chosen message attack, we don't include the unknown secret key into the notation of encryption/decryption queries.

3   Building an Attack

Our attack proceeds in several stages, first attacking the initialization function and then each 64-bit subkey individually, proceeding from the "outer layer" subkeys $K^{(1)}$ and

$K^{(4)}$ towards the "inner layer" subkeys $K^{(3)}$ and $K^{(2)}$. Each stage of the attack is constructed differently.

The line of attack described in this paper is just one of many. A small modification of the algorithm or adjustment of the usage model may lead to wholly different security properties.

We will first describe a very simple chosen-IV distinguisher for Hummingbird, which will be a part of subsequent stages of the attack. For any two nonces (IVs) that have a difference in the most significant bit (MSB) of the first word, we can simply flip the MSB of the plaintext word and the ciphertext words will match.

Observation 2. *There is a Chosen-IV distinguisher for Hummingbird that works with probability $P = 65535/65536$ and has data complexity of 1 word. One can use the high-bit differential of Observation 1 and the following differential for the first round:*

$$\Delta(P_0, RS1_0, RS2_0, RS3_0, RS4_0, LFSR_0) = (8000, 8000, 0000, 0000, 0000, 0000)$$
$$\Updownarrow$$
$$\Delta(C_0, RS1_1, RS2_1, RS3_1, RS4_1, LFSR_1) = (0000, 8000, 8000, 0000, 8000, 0000)$$

The differential works both ways (chosen plaintext and chosen ciphertext). If we decipher the same word, say, 0000 under the two different nonces that are related by only having a MSB difference in the first word, there will be a high-bit difference in the first word of the corresponding plaintext. This constitutes the distinguisher.

3.1 An Iterative Differential

Observation 3. *There is a one-round iterated differential that works if a collision occurs inside the cipher as follows:*

$$\Delta v_{12_t} = 8000 \, , \; \Delta v_{23_t} = 0000 \, , \; \Delta v_{34_t} = 0000$$
$$\Delta(RS1_t, RS2_t, RS3_t, RS4_t, LFSR_t) = (8000, 8000, 0000, 8000, 0000)$$
$$\Updownarrow$$
$$\Delta(RS1_{t+1}, \cdots RS4_{t+1}, LFSR_{t+1}) = (8000, 8000, 0000, 8000, 0000).$$

The initial condition for $t = 5$ can be satisfied using the initialization and first-round encryption differentials given in Observations 1 and 2.

To verify Observation 3, one may find it useful to trace the high-bit differentials (and their internal cancellation) in Figure 2 with a highlighting pen. We note that each one of the conditions $\Delta v_{12} = 8000$, $\Delta v_{23} = 0000$, $\Delta v_{34} = 0000$ implies the other two if the input (or output) state differential holds.

From the algorithm description we see that the internal value v_{34} satisfies

$$\Delta v_{34} = \Delta E^{-1}(C_i, K^{(4)}) \boxminus \Delta RS4_t. \tag{6}$$

For the condition $\Delta v_{34} = 0000$ to be satisfied and the iterative differential to work it suffices to find a pair of ciphertext words $C_i = a$ and $C_i' = b$ such that

$$E^{-1}(a, K^{(4)}) \boxminus E^{-1}(b, K^{(4)}) = 8000. \tag{7}$$

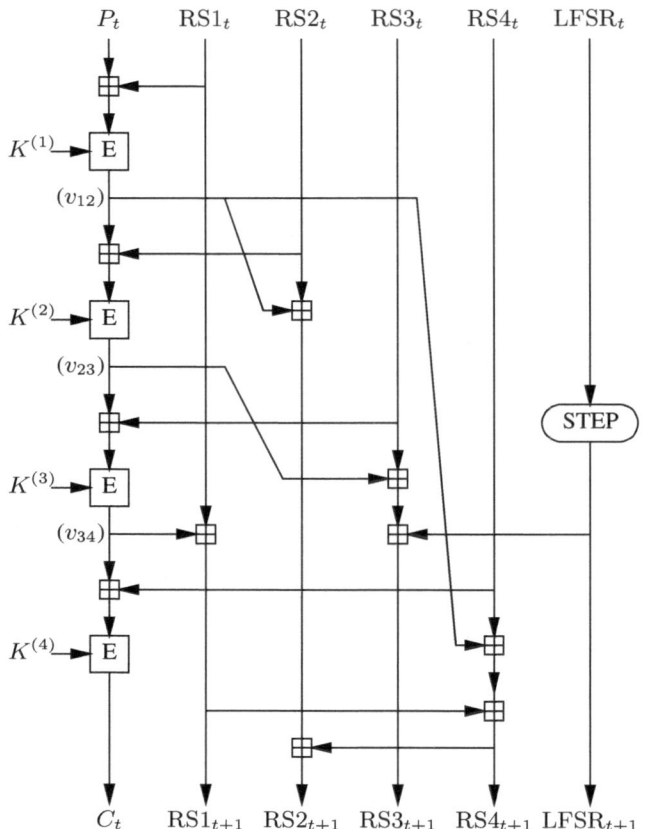

Fig. 2. Encrypting a single 16-bit word P_t to produce a ciphertext word C_t with Hummingbird. After initialization $t \geq 0$.

The first stage of our overall attack is based on chosen-ciphertext queries of the type

$$P = \mathrm{HB}^{-1}\big((0000, 0000, 0000, 0000), (x, a, a, \ldots, a)\big) \qquad (8)$$

$$P' = \mathrm{HB}^{-1}\big((8000, 0000, 0000, 0000), (x, b, b, \ldots, b)\big). \qquad (9)$$

If a and b are related in as in Equation 7, the iterative differential of Observation 3 will hold for all $t \geq 1$ in Equations 8 and 9 above. The initial x word is arbitrary; the differential will work as long as $C_0 = C_0'$. This will result in $\Delta P_0 = 8000$.

For our attack *any* pair (a, b) satisfying Equation 7 will suffice. It is easy to see that there are 2^{16} such pairs. By the birthday paradox, by decrypting about $\sqrt{2^{16}} = 2^8$ vectors of the form given in Equations 8 and 9, we should have found one such pair. How to distinguish it from the other pairs ?

From the algorithm definition we can see that if the iterative differential holds, then $\Delta v_{12} = 8000$, $\Delta RS1_t = 8000$ and the plaintext words satisfy for all $t > 0$

$$\Delta P_t = E^{-1}(v_{12_t}, K^{(1)}) \boxminus \left(E^{-1}(v_{12_t} \boxplus 8000, K^{(1)}) \boxplus 8000\right). \tag{10}$$

To analyze this condition, we may consider a random bijective function F on n-bit values and the behavior of the differential

$$\Delta F(x) = F(x) \boxminus F(x \boxplus c) \tag{11}$$

where c is some nonzero constant and x takes on all values $0 \le x \le 2^n$. It is easy to show that the behavior of $\Delta F(x)$ resembles that of a random function in that its range can be expected to be $2^n(1 - e^{-1}) \approx 0.6321 \times 2^n$ rather than 2^n. For ease of exposition we will be considering the absolute delta value

$$\text{abs}(\Delta x) = x - x' \text{ if } x > x' \text{ and } x' - x \text{ otherwise.} \tag{12}$$

ΔP_i in Equation 10 has similarly limited range if the iterative differential holds. If the differential does not hold, ΔP_i may have any value. We use this feature to test for the right pair; if the iterative differential holds for some ciphertext words x and y, the range of $\text{abs}(\Delta P_i)$ values will be close to $2^{15}(1 - e^{-1}) \approx 20713$ rather than $2^{15} = 32768$. The procedure is given by Algorithm 1. The complexity of Algorithm 1 is less than 2^{30} operations and data complexity is equivalent to decrypting eight megabytes of data. The choice of looping through 2^9 values of i and using 2^{12} words of data in Algorithm 1 may not be optimal, but will be sufficient for actually finding a correct pair with a reasonable probability.

In practice the algorithm finds a right pair in a few seconds. The current implementation also rechecks the pair with longer decryptions and performs a retry if the count of the absolute range is larger than 25000.

Algorithm 1. Probabilistically find a pair (a, b) satisfying Equation 7 as discussed in Section 3.1

```
for i = 1, 2, ..., 2^9 do
    v = (0000, i, i, ..., i),   a vector of 2^12 words.
    x[i][1..2^12] = HB^-1((0000, 0000, 0000, 0000), v).
    y[i][1..2^12] = HB^-1((8000, 0000, 0000, 0000), v).
end for
a = 0, b = 0, m = 2^15.
for i = 0, 1, ..., 2^9 do
    for j = 0, 1, ..., 2^9 do
        Count the number of different words n in the set defined by abs(x[i][k] ⊟ y[j][k]).
        if n < m then
            a = i, b = j, m = n.
        end if
    end for
end for
```

3.2 Attacking $K^{(1)}$

Our first target is to attack the 64-bit subkey $K^{(1)}$. With the (a, b) ciphertext word pair obtained with Algorithm 1, and further chosen-message queries, we will extract the entire range $\mathbf{S_1}$ of the function δ_1 defined by

$$\delta_1(x) = \mathrm{abs}\big(E^{-1}(x, K^{(1)}) \boxminus E^{-1}(x \boxplus 8000, K^{(1)})\big). \tag{13}$$

The expected size of $\mathbf{S_1}$ is $2^{15}(1 - \frac{1}{e}) \approx 20713$ elements. To compute $\mathbf{S_1}$, we decrypt two at least megaword-long vectors consisting of the a and b words:

$$P = \mathrm{HB}^{-1}\big((0000, 0000, 0000, 0000), (0000, a, a, \ldots, a)\big) \tag{14}$$
$$P' = \mathrm{HB}^{-1}\big((8000, 0000, 0000, 0000), (0000, b, b, \ldots, b)\big). \tag{15}$$

Since the iterative differential of Observation 3 holds for all rounds $t > 1$ but the internal state is otherwise evolving and can be modelled as random, each difference in corresponding plaintext words can be simply inserted into the set $\mathbf{S_1}$:

$$\mathrm{abs}(P_i \boxminus P_i' \oplus 8000) \in \mathbf{S_1} \text{ when } i > 0. \tag{16}$$

Note that the completeness of $\mathbf{S_1}$ is highly dependent on the length of the ciphertext vectors; one million words will yield a complete set with high certainty, but one hundred thousand words with very low certainty.

Armed with the set $\mathbf{S_1}$, we can perform an off-line attack on the first subkey. To test a subkey candidate $K^{(1)}$ it suffices to loop through values x doing the membership test $\delta(x) \in \mathbf{S_i}$, as indicated by Equation 13. For a false key candidate the membership test will fail with probability of roughly 63.2%. Most key candidates can be discarded after two trials. Since each membership test (for x) is independent, the certainty that a correct key has not been found after n successful trials $(1 - \frac{1}{e})^n$. $n = 97$ trials gives a 2^{-64} uncertainty. Our implementation performs all $n = 2^{15}$ trials, as the performance penalty is negligible due to the early exit strategy.

3.3 Attacking $K^{(4)}$

The next subkey to be attacked after $K^{(1)}$ is the last to be used during encryption, $K^{(4)}$. There are several ways to do this efficiently. We will describe the one we implemented.

We use our knowledge of $K^{(1)}$ and the differential of Observation 3 to find more ciphertext pairs (C_i, C_i') that have a $\Delta = 8000$ input difference to the last invocation of E. This implies that these ciphertext pairs satisfy the equation

$$E^{-1}(C_i, K^{(4)}) = E^{-1}(C_i', K^{(4)}) \boxplus 8000. \tag{17}$$

If at least four such ciphertext word pairs are available, we may do a conclusive exhaustive search over the entire 64-bit subkey $K^{(4)}$ by using Equation 17 as a test.

We will first obtain a known value for $\mathrm{RS1}_1$. We use the known (a, b) pair from Section 3.1 and Algorithm 1 and decrypt a set of two-word vectors for few running values of initial ciphertext word x:

$$P = \mathrm{HB}^{-1}\big((0000, 0000, 0000, 0000), (\mathsf{x}, \mathsf{a})\big) \tag{18}$$
$$P' = \mathrm{HB}^{-1}\big((8000, 0000, 0000, 0000), (\mathsf{x}, \mathsf{b})\big). \tag{19}$$

For each decryption $\Delta P_0 = 8000$ as indicated by Observation 2. The second plaintext word will satisfy

$$E(P_1 \boxminus \mathrm{RS1}_1, K^{(1)}) = E(P_1' \boxplus \mathrm{RS1}_1 \boxplus 8000, K^{(1)}) \boxplus 8000 \tag{20}$$

since $\Delta \mathrm{RS1}_1 = 8000$. There usually is only one or at most few possible values of $\mathrm{RS1}_1$ that satisfy Equation 20. Such an unique value is found for some x by simply searching through all possible 2^{16} values of $\mathrm{RS1}_1$ using the knowledge of the subkey $K^{(1)}$ (that was obtained in the previous section). This gives us information about the internal state of the cipher after one encryption round.

Let $y = P_0 = P_3' \boxplus 8000$ for some pair of related decryptions described in Equations 18 and 19 such that an unique value for $\mathrm{RS1}_1$ can be established. To create pairs suitable for testing by Equation 17 we again turn into a chosen-plaintext attack and encrypt few vectors for a chosen running value of v_{12_1}:

$$C = \mathrm{HB}\big((0000, 0000, 0000, 0000), (y, \, E^{-1}(v_{12_1}, K^{(1)}) \boxminus \mathrm{RS1}_1)\big)$$
$$C' = \mathrm{HB}\big((8000, 0000, 0000, 0000),$$
$$(y \boxplus 8000, E^{-1}(v_{12_1} \boxplus 8000, K^{(1)}) \boxminus \mathrm{RS1}_1 \boxplus 8000)\big).$$

The ciphertext words C_1 and C_1' can be used for exhaustive search of the 64-bit subkey $K^{(4)}$ using Equation 17.

3.4   Attacking $K^{(3)}$

Thus far we have recovered 128 bits of the secret key K, $K^{(1)}$ and $K^{(4)}$ using MSB differentials only. The next in turn is $K^{(3)}$, which appears to require a slightly more complicated attack also involving second highest bit.

We will be using the two new differentials in addition to the ones given in Observation 1 for initialization rounds $t = -4, \ldots, -1$ and Observation 2 for $t = 0$. For $t = 1$ the differential is:

$$\Delta v_{12_1} = \mathtt{C000} \, , \; \Delta v_{23_1} = d \, , \; \Delta v_{34_1} = 8000$$
$$\Delta(\mathrm{RS1}_1, \mathrm{RS2}_1, \mathrm{RS3}_1, \mathrm{RS4}_1, \mathrm{LFSR}_1) = (8000, 8000, 0000, 8000, 0000)$$
$$\Downarrow$$
$$\Delta(\mathrm{RS1}_2, \mathrm{RS2}_2, \mathrm{RS3}_2, \mathrm{RS4}_2, \mathrm{LFSR}_2) = (0000, 8000, d, 4000, 0000).$$

To make this differential work, we will use the known value for $\mathrm{RS1}_1$ obtained in Section 3.3. Loop through the values $y = v_{12_1} = 0, 1, \ldots, 2^{16} - 1$ and for each one of those make the following two-word encryption queries until $C_1 = C_1'$ condition is reached:

$$C = \mathrm{HB}\big((0000, 0000, 0000, 0000), (x, E^{-1}(y, K^{(1)}) \boxminus \mathrm{RS1}_1)\big)$$
$$C' = \mathrm{HB}\big((8000, 0000, 0000, 0000),$$
$$(x \boxplus 8000, E^{-1}(y \boxplus \mathtt{C000}, K^{(1)}) \boxminus \mathrm{RS1}_1 \boxminus 8000)\big)$$

From the $C_1 = C_1'$ condition we will know that $\Delta v_{34_1} = 8000$ as it cancels out the differential $\Delta RS4_1 = 8000$ before invocation of the last E function. When the condition is met by some x, $d = \Delta v_{23_1} = \Delta RS3_2$ will be a quantity that satisfies

$$E^{-1}(v_{34_1}, K^{(4)}) \boxminus E^{-1}(v_{34_1} \boxplus 8000, K^{(4)}) = d. \tag{21}$$

Now we will extend the chosen-plaintext attack by one more round. We will use the differential:

$$\Delta v_{12_2} = 8000 \,,\, \Delta v_{23_2} = 0000 \,,\, \Delta v_{34_2} = 8000$$
$$\Delta(RS1_2, RS2_2, RS3_2, RS4_2, LFSR_2) = (0000, 8000, d, 4000, 0000)$$
$$\Downarrow$$
$$\Delta(RS1_3, RS2_3, RS3_3, RS4_3, LFSR_3) = (8000, 4000, d, 4000, 0000).$$

We now proceed to deriving the contents of $RS1_2$. We choose the first two plaintext words P_0, P_0', P_1, P_1' as before. For some z and $y = 0, 1, \ldots, 2^{16} - 1$ the third words will be chosen as

$$P_2 = E^{-1}(z, K^{(1)}) \boxminus y \tag{22}$$
$$P_2' = E^{-1}(z \boxplus 8000, K^{(1)}) \boxminus y \tag{23}$$

until the corresponding ciphertext

$$C = HB\big((0000, 0000, 0000, 0000), (P_0, P_1, P_2)\big)$$
$$C' = HB\big((8000, 0000, 0000, 0000), (P_0', P_1', P_2')\big)$$

satisfies the previous conditions and the additional condition

$$E^{-1}(C_2, K^{(4)}) \boxminus E^{-1}(C_2', K^{(4)}) = 4000. \tag{24}$$

This will imply that the second differential works and the conditions $\Delta v_{12_2} = 8000$, $\Delta v_{23_2} = 0000$, and $\Delta v_{34_2} = 8000$ hold. Furthermore we will have the contents of register $RS1_2 = y$ and $v_{12_2} = z$. Note that if the guess for $RS1_2 = y$ is correct, then Equation 24 will hold for any z in Equations 22 and 23.

We now have sufficient information about the internal state of Hummingbird to mount a "quartet" attack on $K^{(3)}$. Additional quantities of the internal state can be derived as follows:

$$v_{34_1} = RS1_2 \boxminus RS1_1 \tag{25}$$
$$RS4_1 = E^{-1}(C_1, K^{(4)}) \boxminus v_{34_1} \tag{26}$$
$$RS4_2 = RS4_1 \boxplus E(P_1 \boxplus RS1_2, K^{(1)}) \boxplus RS1_2 \tag{27}$$
$$v_{34_2} = E^{-1}(C_2, K^{(4)}) \boxminus RS4_2. \tag{28}$$

We can now perform an exhaustive search for $K^{(3)}$ that satisfies

$$E^{-1}(v_{34_1}, K^{(3)}) \boxminus E^{-1}(v_{34_1} \boxplus 8000, K^{(3)}) = d \quad \text{and} \tag{29}$$
$$E^{-1}(v_{34_2}, K^{(3)}) \boxminus E^{-1}(v_{34_2} \boxplus 8000, K^{(3)}) = d \tag{30}$$

for some value d. We call this a "quartet test" as it involves four (inverse) E invocations. To get more quartets (you will need at least four), increase z in Equations 22 and 23 and perform more chosen-plaintext queries.

3.5   Attacking $K^{(2)}$

After the recovery of $K^{(1)}$, $K^{(3)}$ and $K^{(4)}$, there is only 64 bits of unknown keying material left to discover. A simple known-plaintext exhaustive search for $K^{(2)}$ will suffice to recover this last missing piece.

4   Discussion

Hummingbird has some superficial similarities to the Helix [3] and Phelix [10] ciphers – these are stream ciphers where message data is used to modify the internal state of the cipher and an authentication code is produced. An analysis by Muller also used a lack of high-bit propagation in a distinguishing attack [8].

4.1   Implementing the Attack

Our attack on Hummingbird-1 was implemented using the C language on Linux platform. Due to the divide-and-conquer technique that we are using, we may efficiently demonstrate the attack with keys that have limited entropy in each one of the subkeys. Our demonstration code attacks a variant that has four 24-bit subkeys, bringing the total effective key size to 96 bits. Note that this is not a reduced cipher; the subkey entropy has simply been reduced.

The demonstration code first performs a self-test of its Hummingbird-1 implementation against test vectors supplied by Revere Security It then chooses a random key and lets the attack code perform black-box chosen-IV encryption or decryption queries. Typical execution time is 15-20 seconds before the correct 96-bit key is found on an Intel Core 2 Duo clocked at 3.16 GHz.

It seems reasonable to assume that the the E function described in Section 2.2 offers less than 2^{64} security since its diffusion properties are far from perfect. To illustrate this, we note that in Figure 1 it is easy to see that the 16-bit subkey $K_4^{(i)}$ affects two invocations of the S-Box layer and a single bit linear diffusion layer – therefore a single bit change in this subkey won't even necessary affect all ciphertext bits. Since the security of Hummingbird-1 is reduced to the security of the E function by the techniques described in this paper, we feel confident in estimating that Hummingbird-1 offers significantly less than 64 bits of security.

Throughout this paper the any constant pair of IVs can be used as long as

$$\Delta(RS1_0, RS2_0, RS3_0, RS4_0, LFSR_0) = (8000, 0000, 0000, 0000, 0000).$$

This initial condition follows from $\Delta IV = (8000, 0000, 0000, 0000)$ by Observation 1, but if the flaw in the initialization function is fixed, we may find such pairs by the birthday paradox. If the initialization function would be completely random, finding such a pair would require about $\sqrt{2^{80}} = 2^{40}$ queries. Testing for the condition can be done with Observation 2.

4.2 Lessons Learned

Due to its extremely light-weight application target scenario, the security margins used in the design of Hummingbird-1 are very small. In addition to the unfortunate bug in the initialization function (Observation 1), the security of Hummingbird-1 seems to suffer from the fact its state size is very small and that chosen input can directly affect almost all of its internal state bits (apart from the LFSR "counter") in an adaptive attack. We suggest that the number of state bits which run independently from input data should be increased in future encryption algorithm designs of the Hummingbird type.

5 Conclusions

We have described a key-recovery attack against the 256-bit authenticated encryption primitive Hummingbird-1. The attack is based on a divide-and-conquer and differential techniques and has complexity upper bounded by 2^{64} operations. Significant improvements to this bound are possible by attacking the E function. The attack requires processing of few megabytes of chosen messages under two related nonces (IVs).

The attack proceeds in four stages, attacking each one of the 64-bit subkeys individually. The attacks are mainly based on differentials in in the high bits of words. It is noteworthy that the described attacks work regardless of the design of the main nonlinear component, the E keyed permutation. The present line of attack are made effective by a clear design flaw in the Hummingbird-1 initialization function, but similar attacks can be envisioned for many possible straightforward fixes.

We conclude that the published version of Hummingbird-1 may not offer adequate security for some cryptographic applications. The Revere Security team is actively developing an improved version that will remedy the security issues reported in this paper.

References

1. Anderson, R., Biham, E., Knudsen, L.: Serpent: A Proposal for the Advanced Encryption Standard (1999), http://www.cl.cam.ac.uk/~rja14/Papers/serpent.pdf
2. Fan, X., Hu, H., Gong, G., Smith, E.M., Engels, D.: Lightweight Implementation of Hummingbird Cryptographic Algorithm on 4-Bit Microcontroller. In: The 1st International Workshop on RFID Security and Cryptography 2009 (RISC 2009), pp. 838–844 (2009)
3. Ferguson, N., Whiting, D., Schneier, B., Kelsey, J., Lucks, S., Kohno, T.: Helix: Fast encryption and authentication in a single cryptographic primitive. In: Johansson, T. (ed.) FSE 2003. LNCS, vol. 2887, pp. 330–346. Springer, Heidelberg (2003)
4. Engels, D., Fan, X., Gong, G., Hu, H., Smith, E.M.: Ultra-Lightweight Cryptography for Low-Cost RFID Tags: Hummingbird Algorithm and Protocol. Centre for Applied Cryptographic Research (CACR) Technical Reports, CACR-2009-29, http://www.cacr.math.uwaterloo.ca/techreports/2009/cacr2009-29.pdf
5. Engels, D., Fan, X., Gong, G., Hu, H., Smith, E.M.: Hummingbird: Ultra-Lightweight Cryptography for Resource-Constrained Devices. In: 1st International Workshop on Lightweight Cryptography for Resource-Constrained Devices (WLC 2010), Tenerife, Canary Islands, Spain (January 2010)

6. Frazer, R. (ed.): An Analysis of the Hummingbird Cryptographic Algorithm. Commercial security analysis report by Information Security Systems Inc., April 26 (2009), http://www.reveresecurity.com/pdfs/ISSI_Hummingbird.pdf
7. Revere Security. Web page and infomation on the Hummingbird cipher. Fetched November 03 (2010), http://www.reveresecurity.com/
8. Muller, F.: Differential Attacks against the Helix Stream Cipher. In: Roy, B., Meier, W. (eds.) FSE 2004. LNCS, vol. 3017, pp. 94–108. Springer, Heidelberg (2004)
9. Smith, E.M.: Personal Communication, July 7 (2010)
10. Whiting, D., Schneier, B., Lucks, S., Muller, F.: Phelix – Fast Encryption and Authentication in a Single Cryptographic Primitive. ECRYPT Stream Cipher Project Report 2005/027 (2005), http://www.schneier.com/paper-phelix.html

The Additive Differential Probability of ARX[*]

Vesselin Velichkov[1,2,**], Nicky Mouha[1,2,***],
Christophe De Cannière[1,2,†], and Bart Preneel[1,2]

[1] Department of Electrical Engineering ESAT/SCD-COSIC,
Katholieke Universiteit Leuven. Kasteelpark Arenberg 10, B-3001 Heverlee, Belgium
[2] Interdisciplinary Institute for BroadBand Technology (IBBT), Belgium
{Vesselin.Velichkov,Nicky.Mouha,Christophe.DeCanniere}@esat.kuleuven.be

Abstract. We analyze $\mathrm{adp}^{\mathrm{ARX}}$, the probability with which additive differences propagate through the following sequence of operations: modular addition, bit rotation and XOR (ARX). We propose an algorithm to evaluate $\mathrm{adp}^{\mathrm{ARX}}$ with a linear time complexity in the word size. This algorithm is based on the recently proposed concept of S-functions. Because of the bit rotation operation, it was necessary to extend the S-functions framework. We show that $\mathrm{adp}^{\mathrm{ARX}}$ can differ significantly from the multiplication of the differential probability of each component. To the best of our knowledge, this paper is the first to propose an efficient algorithm to calculate $\mathrm{adp}^{\mathrm{ARX}}$. Accurate calculations of differential probabilities are necessary to evaluate the resistance of cryptographic primitives against differential cryptanalysis. Our method can be applied to find more accurate differential characteristics for ARX-based constructions.

Keywords: Additive differential probability, differential cryptanalysis, symmetric-key, ARX.

1 Introduction

Many cryptographic primitives are built using the operations modular addition, bit rotation and XOR (ARX). The advantage of using these operations is that they are very fast when implemented in software. At the same time, they have desirable cryptographic properties. Modular addition provides non-linearity, bit rotation provides diffusion within a single word, and XOR provides diffusion between words and linearity. A disadvantage of using these operations is that the diffusion is typically slow. This is often compensated for by adding more rounds to the designed primitive.

---

[*] This work was supported in part by the Research Council K.U.Leuven: GOA TENSE, and by the IAP Program P6/26 BCRYPT of the Belgian State (Belgian Science Policy), and in part by the European Commission through the ICT program under contract ICT-2007-216676 ECRYPT II.
[**] DBOF Doctoral Fellow, K.U.Leuven, Belgium.
[***] This author is funded by a research grant of the Institute for the Promotion of Innovation through Science and Technology in Flanders (IWT-Vlaanderen).
[†] Postdoctoral Fellow of the Research Foundation – Flanders (FWO).

A. Joux (Ed.): FSE 2011, LNCS 6733, pp. 342–358, 2011.

Examples of cryptographic algorithms that make use of the addition, XOR and rotate operations, are the stream ciphers Salsa20 [2] and HC-128 [16], the block cipher XTEA [13], the MD4-family of hash functions (including MD5 and SHA-1), as well as 6 out of the 14 candidates of NIST's SHA-3 hash function competition [12]: BLAKE [1], Blue Midnight Wish [7], CubeHash [3], Shabal [4], SIMD [8] and Skein [6].

Differential cryptanalysis is one of the main techniques to analyze cryptographic primitives. Therefore, it is essential that the differential properties of ARX are well understood both by designers and attackers. Several important results have been published in this direction. In [15], Meier and Staffelbach present the first analysis of the propagation of the carry bit in modular addition. Later, Lipmaa and Moriai proposed an algorithm to compute the XOR differential probability of modular addition (xdp^+) [9]. Its dual, the additive differential probability of XOR ($\text{adp}^\oplus$), was analyzed by Lipmaa, Wallén and Dumas in [10]. The latter proposed new algorithms for the computation of both xdp^+ and $\text{adp}^\oplus$, based on matrix multiplications. The differential properties of bit rotation have been analyzed by Daum in [5].

In [11], Mouha et al. propose the concept of S-functions. S-functions are a class of functions that can be computed bitwise, so that the i-th output bit is computed using only the i-th input bits and a finite state $S[i]$. Although S-functions have been analyzed before, [11] is the first paper to present a fully generic and efficient framework to determine their differential properties. The methods used in the proposed framework are based on graph theory, and the calculations can be efficiently performed using matrix multiplications.

In this paper, we extend the S-function framework to compute the differential probability adp^{ARX} of the following sequence of operations: addition, bit rotation and XOR. We describe a method to compute adp^{ARX} based on the matrix multiplication technique proposed in [10], and generalized in [11]. The time complexity of our algorithm is linear in the word size. We provide a formal proof of its correctness, and also confirm it experimentally. We performed experiments on all combinations of 4-bit inputs and on a number of random 32-bit inputs.

We observe that adp^{ARX} can differ significantly from the probability obtained by multiplying the differential probabilities of addition, rotation and XOR. This confirms the need for an efficient calculation of the differential probability for the ARX operation. We are unaware of any results in existing literature where adp^{ARX} is calculated efficiently. Accurate and efficient calculations of differential probabilities are required for the efficient search for characteristics used in differential cryptanalysis.

The outline of the paper is as follows. In Sect. 2, we define the additive differential probability of bit rotation ($\text{adp}^{\lll}$). We give an overview of S-functions and we describe how they can be used to compute the additive differential probability of XOR ($\text{adp}^\oplus$) in Sect. 3. The additive differential probability of ARX (adp^{ARX}) is defined in Sect. 4. We show that adp^{ARX} can deviate significantly from the product of the probabilities of rotation and XOR. In Sect. 5, we propose a method for the calculation of adp^{ARX}. The theorem stating its correctness is formulated in

Table 1. Notation

Symbol	Meaning
n	Number of bits in one word
x	n-bit word
$x[i]$	Select the $(i \mod n)$-th bit (or element) of the n-bit word x, $x[0]$ is the least-significant bit (or element)
$+$	Addition modulo 2^n
$-$	Subtraction modulo 2^n
r	Rotation constant, $0 \leq r < n$
$\lll r$	Left bit rotation by r positions
$\ggg r$	Right bit rotation by r positions
$\gg 1$	A signed shift by one position to the right (e.g. $-1 \gg 1 = -1$)
$\oplus$	Exclusive-OR (XOR)
Δx	n-bit additive difference $(x_2 - x_1) \mod 2^n$
$\|$	Concatenation of bit strings
ARX	The sequence of the operations: $+, \lll, \oplus$
$\text{adp}^{\lll}$	The additive differential probability of bit rotation
$\text{adp}^{\oplus}$	The additive differential probability of XOR
adp^{ARX}	The additive differential probability of ARX
x_2	Number x in binary representation
$\Delta\alpha \to \Delta\beta$	Input difference $\Delta\alpha$ propagates to output difference $\Delta\beta$

Sect. 6. In Sect. 7, we confirm the computation of adp^{ARX} experimentally. Section 8 concludes the paper. The matrices used to compute adp^{ARX} are given in Appendix A. Appendix B contains the full proof of correctness of the adp^{ARX} algorithm. Throughout the paper, we use the notation listed in Table 1.

2   Definition of $\text{adp}^{\lll}$

The additive differential probability of bit rotation, denoted by $\text{adp}^{\lll}$, is the probability with which additive differences propagate through bit rotation. This probability was studied by Daum in [5]. We give a brief summary of the results in [5] that are relevant to our work.

Let $\Delta\alpha$ be a fixed additive difference. Let a_1 be an n-bit word chosen uniformly at random and $(a_1, a_1 + \Delta\alpha)$ be a pair of n-bit words input to a left rotation by r positions. Let $\Delta\beta$ be the output additive difference between the rotated inputs:

$$\Delta\beta = ((a_1 + \Delta\alpha) \lll r) - (a_1 \lll r) \ . \tag{1}$$

In [5, Corollary 4.14, Case 2] it is shown that there are four possibilities for $\Delta\beta$:

$$\Delta\beta \in \{\Delta\beta_{u,v} = (\Delta\alpha \lll r) - u2^r + v, \quad u,v \in \{0,1\}\} \ . \tag{2}$$

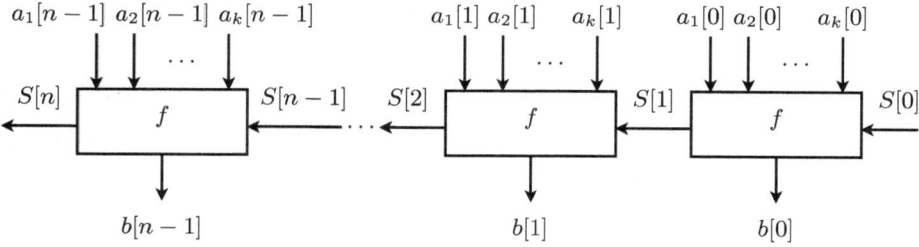

Fig. 1. Representation of an S-function

The probabilities for the output differences $\Delta\beta$ are:

$$P_{0,0} = P(\Delta\alpha \rightarrow \Delta\beta_{0,0}) = 2^{-n}(2^r - \Delta\alpha_L)(2^{n-r} - \Delta\alpha_R) \ , \tag{3}$$

$$P_{0,1} = P(\Delta\alpha \rightarrow \Delta\beta_{0,1}) = 2^{-n}(2^r - \Delta\alpha_L - 1)\Delta\alpha_R \ , \tag{4}$$

$$P_{1,0} = P(\Delta\alpha \rightarrow \Delta\beta_{1,0}) = 2^{-n}\Delta\alpha_L(2^{n-r} - \Delta\alpha_R) \ , \tag{5}$$

$$P_{1,1} = P(\Delta\alpha \rightarrow \Delta\beta_{1,1}) = 2^{-n}(\Delta\alpha_L + 1)\Delta\alpha_R \ . \tag{6}$$

In the above equations, $\Delta\alpha_L$ is the word composed of the r most significant bits of $\Delta\alpha$ and $\Delta\alpha_R$ is the word composed of the $n - r$ least significant bits of $\Delta\alpha$ such that

$$\Delta\alpha = \Delta\alpha_L \parallel \Delta\alpha_R \ . \tag{7}$$

We define the additive differential probability of bit rotation as

$$\mathrm{adp}^{\lll}(\Delta\alpha \xrightarrow{r} \Delta\beta) = \begin{cases} P_{u,v} & , \text{ if } \Delta\beta = \Delta\beta_{u,v} \text{ for some } u,v \in \{0,1\} \ , \\ 0 & , \text{ otherwise } . \end{cases} \tag{8}$$

3   Computation of adp$^{\oplus}$ Using S-Functions

S-functions were introduced by Mouha et al. in [11]. An S-function (short for *state*-function) accepts n-bit words $a_1, a_2, \ldots, a_k$ and a list of states $S[i]$ (for $0 \le i < n$) as input, and produces an n-bit output word b in the following way:

$$(b[i], S[i + 1]) = f(a_1[i], a_2[i], \ldots, a_k[i], S[i]), \qquad 0 \le i < n \ . \tag{9}$$

Initially, we set $S[0] = 0$. A schematic representation of an S-function is given in Fig. 1.

In [11], S-functions were used to compute the additive differential probability of XOR ($\mathrm{adp}^{\oplus}$). This is the probability with which additive differences propagate through the XOR operation. The results of [11] confirm the calculation of $\mathrm{adp}^{\oplus}$ obtained in [10]. They are relevant to the calculation of $\mathrm{adp}^{\mathtt{ARX}}$, and will therefore be briefly described below.

Fix the additive differences $\Delta\alpha, \Delta\beta, \Delta\gamma$. With Δe we designate the additive difference:

$$\Delta e = e_2 - e_1 = ((c_1 + \Delta\alpha) \oplus (d_1 + \Delta\beta)) - (c_1 \oplus d_1) . \tag{10}$$

The probability $\mathrm{adp}^\oplus$ is equal to the number of pairs (c_1, d_1) for which $\Delta e = \Delta\gamma$, divided by the total number of pairs (c_1, d_1):

$$\mathrm{adp}^\oplus(\Delta\alpha, \Delta\beta \to \Delta\gamma) = \frac{|\{(c_1, d_1) : \Delta e = \Delta\gamma\}|}{|\{(c_1, d_1)\}|} . \tag{11}$$

The i-th bit of the output difference $\Delta e[i]$ can be computed from the i-th bits of the input differences $\Delta\alpha[i], \Delta\beta[i]$ and the state $S[i]$. The state $S[i]$ consists of the carries $s_1[i], s_2[i]$ and the borrow $s_3[i]$:

$$s_1[i] = (c_1[i-1] + \Delta\alpha[i-1] + s_1[i-1]) \gg 1 , \tag{12}$$

$$s_2[i] = (d_1[i-1] + \Delta\beta[i-1] + s_2[i-1]) \gg 1 , \tag{13}$$

$$s_3[i] = (e_2[i-1] - e_1[i-1] + s_3[i-1]) \gg 1 , \tag{14}$$

where $s_1[0] = s_2[0] = s_3[0] = 0$. Note that the bit shift by one position to the right in (14) is a signed shift (e.g. $-1 \gg 1 = -1$). The S-function for $\mathrm{adp}^\oplus$ is defined as

$$(\Delta e[i], S[i+1]) = f(c_1[i], d_1[i], \Delta\alpha[i], \Delta\beta[i], S[i]), \quad 0 \le i < n . \tag{15}$$

By definition, the state $S[i]$ of an S-function has the same fixed size for every $0 \le i < n$. In the case of $\mathrm{adp}^\oplus$, this size is 3 bits. Therefore, there are eight distinct states $S[i]$ in total for any bit position $0 \le i < n$. For fixed input differences, the transition between consecutive states $S[i]$ and $S[i+1]$ can be described by an 8×8 adjacency matrix. There are eight such matrices in total – one for each value of the 3-tuple $(\Delta\alpha[i], \Delta\beta[i], \Delta\gamma[i])$. These eight matrices are derived in [11] and are shown to be equal (up to a permutation) to the matrices previously computed in [10].

The probability $\mathrm{adp}^\oplus$ is computed by iterating over all bit positions 0 through $n-1$. At each position i, one of the eight matrices is selected depending on the value of the bits of the differences $\Delta\alpha[i], \Delta\beta[i], \Delta\gamma[i]$. All n matrices that are selected in this way are multiplied. The resulting matrix is right-multiplied by the column vector representing the initial state. We now obtain a column vector. After summing its elements, we end up with the number of pairs (a_1, b_1) for which $\Delta e = \Delta\gamma$. The probability $\mathrm{adp}^\oplus$ is computed by dividing the obtained value by the total number of pairs (a_1, b_1). This whole process is summarized by the following formula:

$$\mathrm{adp}^\oplus(\Delta\alpha, \Delta\beta \to \Delta\gamma) = 2^{-2n} L A_{w[n-1]} \cdots A_{w[1]} A_{w[0]} C . \tag{16}$$

In (16), the factor 2^{2n} corresponds to the total number of n-bit pairs (c_1, d_1). $C = (1\,0\,0\,0\,0\,0\,0\,0)^T$ is a column vector indicating the initial state $S[0] = 0$,

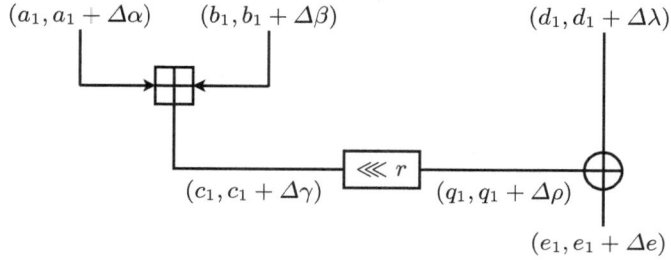

Fig. 2. Additive differences passing through the ARX operation

corresponding to the two initial carries $s_1[0], s_2[0]$ and the initial borrow $s_3[0]$ being equal to zero. Multiplication by the row vector $L = \begin{pmatrix} 1\,1\,1\,1\,1\,1\,1\,1 \end{pmatrix}$ is equivalent to adding the elements of the column vector resulting from the product $A_{w[n-1]} \cdots A_{w[0]} C$. The matrix indices $w[i], 0 \leq i < n$ are in the set $\{0, 1, \ldots, 7\}$. Index $w[i]$ is obtained by concatenating the i-th bits of the differences: $w[i] = \Delta\alpha[i] \parallel \Delta\beta[i] \parallel \Delta\gamma[i]$. At every bit position i, the index $w[i]$ selects one of the eight distinct adjacency matrices $A_{w[i]}$. They are given in Appendix A.

In the next sections, we define the additive differential probability of ARX and we describe a method to compute this probability using S-functions.

4   Definition of adp^{ARX}

The operation ARX is defined as:

$$\text{ARX}(a, b, d, r) \overset{\text{def}}{=} ((a + b) \lll r) \oplus d \ . \tag{17}$$

Let the additive differences $\Delta\alpha, \Delta\beta, \Delta\lambda, \Delta\eta$ be fixed. Let Δe be the difference between two outputs of ARX:

$$\Delta e = e_2 - e_1 = \text{ARX}(a_1 + \Delta\alpha, b_1 + \Delta\beta, d_1 + \Delta\lambda, r) - \text{ARX}(a_1, b_1, d_1, r) \ . \tag{18}$$

Equation (18) is illustrated in Fig. 2. Additive differences pass through modular addition with probability one. Therefore we can directly compute the output difference after the addition: $\Delta\gamma = \Delta\alpha + \Delta\beta$. Let $c_1 = a_1 + b_1$ be any output from the addition (Fig. 2). The additive differential probability of ARX is defined as the number of pairs (c_1, d_1) for which $\Delta e = \Delta\eta$, divided by all pairs (c_1, d_1):

$$\text{adp}^{\text{ARX}}(\Delta\gamma, \Delta\lambda \overset{r}{\rightarrow} \Delta\eta) \overset{\text{def}}{=} \frac{|\{(c_1, d_1) : \Delta e = \Delta\eta\}|}{|\{(c_1, d_1)\}|} \ . \tag{19}$$

An estimation of adp^{ARX} can be obtained as the product of the probabilities of rotation and XOR. We designate the probability computed in this way by P_{rotxor}:

$$P_{\text{rotxor}} = \sum_{j=0}^{4} (\text{adp}^{\lll}(\Delta\gamma \overset{r}{\rightarrow} \Delta\rho_j) \cdot \text{adp}^{\oplus}(\Delta\rho_j, \Delta\lambda \rightarrow \Delta\eta)) \ , \tag{20}$$

where $\Delta\rho_j, 0 \le j < 4$ are the four possible output differences after the rotation (2). Equation (20) would be an accurate evaluation of adp^{ARX} if the inputs to the rotation and the inputs to the XOR operation were independent. In reality they are not, as illustrated by the following example.

Example 1. Let $n = 4$, $r = 1$, $\Delta\gamma = 1000_2$, $\Delta\lambda = 0000_2$, $\Delta\eta = 0001_2$. Two output differences after the rotation are possible: $\Delta\rho_0 = 0001_2$ and $\Delta\rho_2 = 1111_2$, each with probability 2^{-1}. They both propagate through the XOR operation with probability $2^{-1.54}$. The total probability P_{rotxor} is

$$P_{\text{rotxor}} = \text{adp}^{\lll}(1000_2 \xrightarrow{1} 0001_2) \cdot \text{adp}^{\oplus}(0001_2, 0000_2 \to 0001_2)$$
$$+ \text{adp}^{\lll}(1000_2 \xrightarrow{1} 1111_2) \cdot \text{adp}^{\oplus}(1111_2, 0000_2 \to 0001_2)$$
$$= 2^{-1} \cdot 2^{-1.54} + 2^{-1} \cdot 2^{-1.54} = 2^{-1.54} \ . \tag{21}$$

The actual probability is, however, higher than P_{rotxor} and is $P_{\text{exper}} = 2^{-1}$. The reason for the discrepancy is the fact that there exist pairs of inputs to XOR that satisfy the differences $\Delta\rho_0$ or $\Delta\rho_2$, but when they are rotated back they do not satisfy the difference $\Delta\gamma$. One such input pair is $(q_1, q_2) = (2, 1)$. This pair satisfies the difference $\Delta\rho_2$: $q_2 - q_1 = (1 - 2) \mod 16 = 15 = 1111_2$. Yet, it does not satisfy the difference $\Delta\gamma$: $(q_2 \ggg 1) - (q_1 \ggg 1) = (8 - 1) \mod 16 = 7 = 0111_2 \ne 1000_2$. There are 8 such pairs in total: $(0, 15)$, $(2, 1)$, $(4, 3)$, $(6, 5)$, $(8, 7)$, $(10, 9)$, $(12, 11)$, $(14, 13)$. Given the output difference $\Delta\rho_2$, these pairs are impossible. Thus the total number of possible inputs to the XOR is reduced from 256 to 128. The reason is, that for every impossible pair (q_1, q_2), there are 16 possibilities for the second input pair $(d_1 + \Delta\lambda, d_1)$. Of those 128 pairs, 64 satisfy the output difference $\Delta\eta$. Thus the actual probability is adp$^{\oplus}(1111_2, 0000_2 \to 0001_2) = 64/128 = 2^{-1}$ and not $88/256 = 2^{-1.54}$. We have a similar situation for the difference $\Delta\rho_0$. In that case the impossible pairs are $(1, 2)$, $(3, 4)$, $(5, 6)$, $(7, 8)$, $(9, 10)$, $(11, 12)$, $(13, 14)$, $(15, 0)$ and the adp$^{\oplus}$ probability is again 2^{-1}. Thus the final probability adp^{ARX} is 2^{-1}.

5   Computation of adp^{ARX}

In Example 1, we showed that the inputs to the rotation and to the XOR operation are not independent. This causes the additive differential probability of ARX, estimated by the multiplication of the probabilities of the rotation and the XOR, to differ from the actual probability. This problem can be solved if the intermediate differences $\Delta\rho_j, 0 \le j < 4$ are not computed explicitly. Consider the ARX operation (17). Let $a_1 + b_1 = c_1$, $q_1 = (c_1 \lll r)$ and $e_1 = \text{ARX}(a_1, b_1, d_1, r)$, $e_2 = \text{ARX}(a_2, b_2, d_2, r)$, as shown in Fig. 2. Note that $c_1[i] = q_1[i + r]$. Therefore $q_1[i + r] \oplus d_1[i + r] = e_1[i + r]$ is equivalent to

$$c_1[i] \oplus d_1[i + r] = e_1[i + r] \ . \tag{22}$$

Using this representation, we can compute the bits of the output e_1 without using the intermediate variable q_1. Consequently, we can compute the output difference $\Delta e = e_2 - e_1$ without using the intermediate differences $\Delta \rho_i$:

$$c_2[i] = c_1[i] \oplus \Delta\gamma[i] \oplus s_1[i] \ , \tag{23}$$
$$d_2[i+r] = d_1[i+r] \oplus \Delta\lambda[i+r] \oplus s_2[i+r] \ , \tag{24}$$
$$\Delta e[i+r] = e_1[i+r] \oplus e_2[i+r] \oplus s_3[i+r] \ , \tag{25}$$

where

$$s_1[i] = (c_1[i-1] + \Delta\gamma[i-1] + s_1[i-1]) \gg 1 \ , \tag{26}$$
$$s_2[i+r] = (d_1[i+r-1] + \Delta\lambda[i+r-1] + s_2[i+r-1]) \gg 1 \ , \tag{27}$$
$$s_3[i+r] = (e_2[i+r-1] - e_1[i+r-1] + s_3[i+r-1]) \gg 1 \ . \tag{28}$$

The S-function for $\mathrm{adp}^{\mathrm{ARX}}$ is defined as

$$(\Delta e[i+r], S[i+1]) = f(c_1[i], d_1[i+r], \Delta\gamma[i], \Delta\lambda[i+r], S[i]),$$
$$0 \le i < n \ . \tag{29}$$

The definitions of the S-functions for $\mathrm{adp}^{\mathrm{ARX}}$ (29) and $\mathrm{adp}^{\oplus}$ (15) are very similar. Yet the computation of the two differ in several aspects. We describe these differences below.

5.1   The Initial State

As described in Sect. 3 for $\mathrm{adp}^{\oplus}$, the state is composed of two carries and one borrow arising from the three modular operations involved in computing the output (10). At position $i = 0$, these values are all zero. Therefore, the initial state is $S[0] = (s_1[0], s_2[0], s_3[0]) = (0,0,0)$. In the case of $\mathrm{adp}^{\mathrm{ARX}}$ the situation is slightly different. The reason is that when we perform the ARX operation bitwise, at position 0, we compute the 0-th bit of c_2 and the r-th bits of d_2 and Δe (23)-(25). Similarly to $\mathrm{adp}^{\oplus}$, the carry $s_1[0]$ is zero. However the carry $s_2[r]$ and the borrow $s_3[r]$ are not necessarily zero:

$$s_1[0] = 0 \ , \tag{30}$$
$$s_2[r] = (d_1[r-1] + \Delta\lambda[r-1] + s_2[r-1]) \gg 1 \ , \tag{31}$$
$$s_3[r] = (e_2[r-1] - e_1[r-1] + s_3[r-1]) \gg 1 \ . \tag{32}$$

Thus the initial state of the $\mathrm{adp}^{\mathrm{ARX}}$ S-function is $S[0] = (s_1[0], s_2[r], s_3[r])$. Because $s_2[r] \in \{0,1\}$ and $s_3[r] \in \{-1,0\}$, there are four possibilities for $S[0]$. Each of them corresponds to one of the 3-tuples $(0,0,-1), (0,1,-1), (0,0,0), (0,1,0)$. We map all 8 possible values of any state $S[i] = (s_1[i], s_2[i+r], s_3[i+r])$ to the set of integers $\{0,1,...,7\}$ as shown in Table 2. Following this convention, $S[0] \in \{0,2,4,6\}$.

5.2   The Final State

From (30)-(32), it follows that in order to compute $S[0]$ we have to know $s_2[r-1]$ and $s_3[r-1]$. In other words, in order to compute the initial state of the $\mathrm{adp}^{\mathrm{ARX}}$

Table 2. Mapping between the 8 states of the adp^{ARX} S-function and the set of integers $\{0, \ldots, 7\}$

$\mathbf{S[i]}$	0	1	2	3	4	5	6	7
$(\mathbf{s_1[i]}, \mathbf{s_2[i+r]}, \mathbf{s_3[i+r]})$	(0,0,-1)	(1,0,-1)	(0,1,-1)	(1,1,-1)	(0,0,0)	(1,0,0)	(0,1,0)	(1,1,0)

S-function we need information from the final state $S[n-1] = (s_1[n-1], s_2[r-1], s_3[r-1])$. However, at the start of the computation ($i = 0$) we do not know the output of the S-function at position $i = n-1$ yet. We solve this problem by iterating over all four values of $(s_2[r-1], s_3[r-1])$ at $i = 0$. For each of them, we compute $S[0]$ and we proceed with the computation of the S-function. From the set of final output states $S[n-1]$, we accept as valid only those that match the values of $(s_2[r-1], s_3[r-1])$ at position $i = 0$. Each value of the tuple $(s_2[r-1], s_3[r-1])$ will match exactly two of all eight final states $S[n-1]$ corresponding to the two possibilities for $c_1[n-1] \in \{0, 1\}$. For example the initial state $(0, 0, -1)$ will be matched by final states $(0, 0, -1)$ and $(1, 0, -1)$. In general, following the mapping in Table 2, an initial state $S[0] = j \in \{0, 2, 4, 6\}$ will match final states $S[n-1] = j$ and $S[n-1] = j+1$.

5.3 A Special Intermediate State

There is one final issue that should be taken care of, before we are able to compute adp^{ARX}. Consider step $i = n - r - 1$ of the computation of the S-function of adp^{ARX}. At this step, we are operating on bits at position $n-1$ in order to compute $s_2[0]$ and $s_3[0]$. Since these are the most-significant input bits, the carries and borrows that they generate should be discarded. Consequently, $s_2[0]$ and $s_3[0]$ should be set to zero at this step:

$$s_1[n-r] = (c_1[n-r-1] + \Delta\gamma[n-r-1] + s_1[n-r-1]) \gg 1 , \tag{33}$$
$$s_2[0] = 0 , \tag{34}$$
$$s_3[0] = 0 . \tag{35}$$

Therefore state $S[n-r] = (s_1[n-r], s_2[0], s_3[0])$ is a special intermediate state for which the only permissible values are $(0, 0, 0)$ and $(1, 0, 0)$ i.e. $S[n-r] \in \{4, 5\}$. Because of this special state, it is necessary to construct an 8×8 projection matrix R in addition to the matrices A_q, $0 \le q < 8$ used in the computation of adp$^{\oplus}$ (16). By multiplying the matrix $A_{w[n-r-1]}$ at position $n - r - 1$ to the left by R, the transition from the set of output states corresponding to the value of the 3-tuple $(\Delta\gamma[n-r], \Delta\lambda[0], \Delta\eta[0])$ to the set of reachable output states is performed (cf. Sect. 3). This operation effectively transforms every state $S[n-r] = (s_1[n-r], s_2[0], s_3[0])$ to the permissible value for the special state $S[n-r] = (s_1[n-r], 0, 0)$.

5.4   Computing adp^{ARX}

The probability adp^{ARX} can be computed as follows:

$$\text{adp}^{\text{ARX}}(\Delta\gamma, \Delta\lambda \xrightarrow{r} \Delta\eta) =$$
$$2^{-2n} \sum_{j \in \{0,2,4,6\}} (L_j A_{w[n-1]} \cdots A_{w[n-r]} R A_{w[n-r-1]} \cdots A_{w[1]} A_{w[0]} C_j) \ . \qquad (36)$$

In (36), $j \in \{0, 2, 4, 6\}$ iterates over the four possible initial states. The binary column vector C_j of dimension 8×1 indicates the initial state. It has 1 at position j and 0 elsewhere. The vector L_j is a 1×8 binary row vector that has 1 at positions j and $j+1$ and has 0 elsewhere. By multiplying the result of the matrix multiplication by L_j, we are effectively adding only the two final states that correspond to the initial state j (cf. Sect. 5.2). The indices $w[0], \ldots, w[n-1]$ are in the set $\{0, 1, \ldots, 7\}$. Index $w[i]$ is obtained by concatenating the corresponding bits of the differences: $w[i] = \Delta\gamma[i] \parallel \Delta\lambda[i+r] \parallel \Delta\eta[i+r]$. For every bit position $0 \leq i < n$, index $w[i]$ selects one of the eight 8×8 adjacency matrices A_q, $0 \leq q < 8$. For position $i = n - r - 1$, matrix $A_{w[n-r-1]}$ is additionally multiplied to the left by the projection matrix R. The matrices A_q are the same as the the ones used in the computation of adp$^{\oplus}$. Matrices A_q and R are given in Appendix A.

The computation of adp^{ARX} (36) is slightly different from the computation of adp$^{\oplus}$ (16). The main difference is that there are four evaluations of the S-function. From each of them, two of the eight final states are selected. The second difference is the presence of the additional projection matrix R.

In Example 2, we demonstrate the computation of adp^{ARX} for the additive differences given in Example 1.

Example 2. For $n = 4$, $r = 1$, $\Delta\gamma = 1000_2$, $\Delta\lambda = 0000_2$, $\Delta\eta = 0001_2$, we want to compute adp$^{\text{ARX}}(\Delta\gamma, \Delta\lambda \xrightarrow{r} \Delta\eta)$. First we compute the indices $w[i] = \Delta\gamma[i] \parallel \Delta\lambda[i+1] \parallel \Delta\eta[i-1]$, $0 \leq i < 4$:

$$w[0] = \Delta\gamma[0] \parallel \Delta\lambda[1] \parallel \Delta\eta[1] = 000 \ ,$$
$$w[1] = \Delta\gamma[1] \parallel \Delta\lambda[2] \parallel \Delta\eta[2] = 000 \ ,$$
$$w[2] = \Delta\gamma[2] \parallel \Delta\lambda[3] \parallel \Delta\eta[3] = 000 \ ,$$
$$w[3] = \Delta\gamma[3] \parallel \Delta\lambda[0] \parallel \Delta\eta[0] = 101 \ .$$

Indices $w[0], w[1], w[2]$ select matrix A_{000}; index $w[3]$ selects matrix A_{101}. The probability adp^{ARX} is computed as

$$\text{adp}^{\text{ARX}}(1000_2, 0000_2 \xrightarrow{1} 0001_2)$$
$$= 2^{-8} \sum_{j \in \{0,2,4,6\}} L_j A_{101} R A_{000} A_{000} A_{000} C_j = 2^{-1} \ ,$$

where

$$C_0 = \begin{pmatrix} 1\,0\,0\,0\,0\,0\,0\,0 \end{pmatrix}^T , \quad L_0 = \begin{pmatrix} 1\,1\,0\,0\,0\,0\,0\,0 \end{pmatrix} ,$$
$$C_2 = \begin{pmatrix} 0\,0\,1\,0\,0\,0\,0\,0 \end{pmatrix}^T , \quad L_2 = \begin{pmatrix} 0\,0\,1\,1\,0\,0\,0\,0 \end{pmatrix} ,$$
$$C_4 = \begin{pmatrix} 0\,0\,0\,0\,1\,0\,0\,0 \end{pmatrix}^T , \quad L_4 = \begin{pmatrix} 0\,0\,0\,0\,1\,1\,0\,0 \end{pmatrix} ,$$
$$C_6 = \begin{pmatrix} 0\,0\,0\,0\,0\,0\,1\,0 \end{pmatrix}^T , \quad L_6 = \begin{pmatrix} 0\,0\,0\,0\,0\,0\,1\,1 \end{pmatrix} .$$

6   Proof of Correctness

With the following theorem we state that the computation of the probability adp^{ARX} (36) is correct. To prove this, we use arguments similar to [11, §3.4, Theorem 1]. The full proof is given in Appendix B. In this section, we provide the intuition behind it.

Theorem 1.

$$2^{-2n} \sum_{j \in \{0,2,4,6\}} (L_j A_{w[n-1]} \cdots A_{w[n-r]} R A_{w[n-r-1]} \cdots A_{w[1]} A_{w[0]} C_j) =$$

$$\frac{|\{(c_1, d_1) : \Delta e = \Delta \eta\}|}{|\{(c_1, d_1)\}|} . \tag{37}$$

Proof. The full proof is given in Appendix B.

Theorem 1 states that the probability computed using the proposed method (36) is equal to the probability adp^{ARX} as defined by (19). The most natural way to see this is by using a graph representation of the S-function [11]. In [11, §3.4] it is shown that an S-function can be represented as a directed acyclic graph composed of bipartite subgraphs. In this representation, all pairs of inputs that satisfy the input differences are equal to the number of paths through the graph. Each path connects a valid initial state to a valid final state. A subset of these paths corresponds to the set of input pairs that satisfy both the input and the output differences. Therefore, the computation of the S-function is equivalent to counting the number of paths in the subset and dividing the result by the number of all paths. Since, in the process, no path is counted more than once, the result is exactly equal to adp^{ARX} as defined by (19).

7   Experiments

In this section, we confirm the correctness of the computation of adp^{ARX} (36) experimentally. We performed two sets of experiments: one for 4-bit words and one for 32-bit words. In both sets, we compare three computations of the additive differential probability of ARX:

- P_{exper}: the probability computed experimentally, using (19), over a certain number of inputs that satisfy the input differences

Table 3. Comparing three ways of computing the additive differential probability of ARX for 4-bit words. Shown are the 24 cases for which the deviation of P_{rotxor} (20) from the experimentally obtained value P_{exper} (19) is highest: $|P_{\text{exper}} - P_{\text{rotxor}}| > 0.1$. The probability $\text{adp}^{\text{ARX}}(\Delta\gamma, \Delta\lambda \xrightarrow{r} \Delta\eta)$ (36) matches exactly the probability obtained experimentally P_{exper}. The latter is computed over all 2^8 possible inputs. The values of the differences are in binary format.

#	$\Delta\gamma$	$\Delta\lambda$	$\Delta\eta$	r	P_{exper}	P_{ARX}	P_{rotxor}
1	1000	0000	0001	1	0.500	0.500	0.344
2	1000	0000	0010	2	0.500	0.500	0.375
3	1000	0000	0110	2	0	0	0.125
4	1000	0000	1010	2	0	0	0.125
5	1000	0000	1110	2	0.500	0.500	0.375
6	1000	0000	1111	1	0.500	0.500	0.344
7	1000	0001	0000	1	0.500	0.500	0.344
8	1000	0010	0000	2	0.500	0.500	0.375
9	1000	0010	1000	2	0	0	0.125
10	1000	0110	0000	2	0	0	0.125
11	1000	0110	1000	2	0.500	0.500	0.375
12	1000	0111	1000	1	0.500	0.500	0.344
13	1000	1000	0010	2	0	0	0.125
14	1000	1000	0110	2	0.500	0.500	0.375
15	1000	1000	0111	1	0.500	0.500	0.344
16	1000	1000	1001	1	0.500	0.500	0.344
17	1000	1000	1010	2	0.500	0.500	0.375
18	1000	1000	1110	2	0	0	0.125
19	1000	1001	1000	1	0.500	0.500	0.344
20	1000	1010	0000	2	0	0	0.125
21	1000	1010	1000	2	0.500	0.500	0.375
22	1000	1110	0000	2	0.500	0.500	0.375
23	1000	1110	1000	2	0	0	0.125
24	1000	1111	0000	1	0.500	0.500	0.344

- adp^{ARX}: the probability computed using the proposed method (36)
- P_{rotxor}: the probability computed as a product of the probabilities $\text{adp}^{\lll}$ and $\text{adp}^{\oplus}$ (20)

In the set of experiments on 4-bit words, we exhaustively searched over all possible combinations of input and output differences $\Delta\gamma, \Delta\lambda, \Delta\eta$ and over all non-zero rotation constants $r \in \{1, 2, 3\}$. We performed $12,288$ experiments in total. For each of them we computed P_{exper}, adp^{ARX} and P_{rotxor}. The probability P_{exper} was computed over all 2^8 possible input words. In each experiment, the probability adp^{ARX} was equal to P_{exper}, while P_{rotxor} often deviated. The 24 cases in which the absolute deviation is higher than 0.1 are shown in Table 3.

We experimented over random 32-bit input and output differences of relatively low weight (less than 16). The probability P_{exper} was computed over 2^{22} random inputs. We performed 2^{10} experiments in total. For all of them, the estimation

Table 4. Comparing three ways of computing the additive differential probability ARX for 32-bit words. Shown are 11 selected cases for which the deviation of P_{rotxor} (20) from the experimentally obtained value P_{exper} (19) is high. The probability $\text{adp}^{\text{ARX}}(\Delta\gamma, \Delta\lambda \xrightarrow{r} \Delta\eta)$ (36) closely follows the experimentally obtained value P_{exper}. The latter is computed over 2^{22} random inputs. Base-2 logarithms of the probabilities are given.

#	$\Delta\gamma$	$\Delta\lambda$	$\Delta\eta$	r	P_{exper}	P_{ARX}	P_{rotxor}
0	0x80000100	0x00000000	0x0007fc00	11	-2.58331	-2.58496	-4.17006
1	0x40000008	0x00000000	0x000001d0	6	-4.58591	-4.58496	-5.59061
2	0x80000008	0x04000000	0xfc000f00	9	-4.16817	-4.16711	-5.70768
3	0x40010001	0x04000000	0xd3ffc000	30	-5.90603	-5.91254	-6.60771
4	0xa2005800	0x00400000	0xf4000b00	29	-7.53935	-7.54954	-8.57279
5	0x45003700	0x00000000	0xc8ffbb00	16	-8.77902	-8.76145	-9.37302
6	0x4007800d	0x03800300	0x01e803f0	21	-11.14047	-11.17440	-11.86110
7	0xbf006400	0x00900050	0xf37ff9f0	28	-11.85917	-11.82987	-12.81410
8	0x8d00ec00	0x00a000f0	0xfbf7f870	27	-12.04435	-12.05139	-13.47130
9	0x7c005e00	0x00700080	0xffb3fe78	9	-9.96830	-9.98809	-11.36139
10	0xda008200	0x001000d0	0xe01d9f38	20	-15.05749	-15.10578	-15.77518
11	0xe4006600	0x00f00040	0xf0cff9e0	28	-15.04580	-15.04912	-15.32160

of the probability adp^{ARX} was closer to the experimentally obtained value P_{exper} than to P_{rotxor}. A selection of 11 cases for which the absolute deviation from P_{rotxor} was observed to be relatively high is shown in Table 4.

8 Conclusions

In this paper, we analyzed the probability adp^{ARX} with which additive differences propagate through the sequence of operations: modular addition, bit rotation and XOR. We proposed a method for the computation of adp^{ARX}, based on the recently proposed concept of S-functions. The time complexity of our algorithm is linear in the word size n. To the best of our knowledge, our algorithm is the first to calculate adp^{ARX} efficiently for large n.

In Sect. 7, we observed that the estimated probability obtained by analyzing the components of ARX separately, can differ significantly from the actual probability. In our method, we analyze the three operations as a single operation (ARX). In this way, we obtain the exact probability adp^{ARX}. Our algorithm can be used to evaluate the probability of differential characteristics for cryptographic algorithms more accurately.

An interesting topic for future research, is therefore to use our technique in the search of differential characteristics. Possible targets include several hash functions from NIST's ongoing SHA-3 competition, as well as stream ciphers (e.g. Salsa20), or block ciphers such as XTEA.

Acknowledgments. The authors would like to thank Vincent Rijmen for reviewing the draft paper and for providing important suggestions and comments. We also thank our colleagues at COSIC for the useful discussions, as well as the anonymous reviewers for their detailed comments.

References

1. Aumasson, J.-P., Henzen, L., Meier, W., Phan, R.C.-W.: SHA-3 proposal BLAKE. Submission to the NIST SHA-3 Competition, Round 2 (2008)
2. Bernstein, D.J.: The Salsa20 Family of Stream Ciphers. In: Robshaw, M.J.B., Billet, O. (eds.) [14], pp. 84–97
3. Bernstein, D.J.: CubeHash specification (2.B.1). Submission to the NIST SHA-3 Competition, Round 2 (2009)
4. Bresson, E., Canteaut, A., Chevallier-Mames, B., Clavier, C., Fuhr, T., Gouget, A., Icart, T., Misarsky, J.-F., Naya-Plasencia, M., Paillier, P., Pornin, T., Reinhard, J.-R., Thuillet, C., Videau, M.: Shabal, a Submission to NIST's Cryptographic Hash Algorithm Competition. Submission to the NIST SHA-3 Competition, Round 2 (2008)
5. Daum, M.: Cryptanalysis of Hash Functions of the MD4-Family. PhD thesis, Ruhr-Universität Bochum (2005)
6. Ferguson, N., Lucks, S., Schneier, B., Whiting, D., Bellare, M., Kohno, T., Callas, J., Walker, J.: The Skein Hash Function Family. Submission to the NIST SHA-3 Competition, Round 2 (2009)
7. Gligoroski, D., Klima, V., Knapskog, S.J., El-Hadedy, M., Amundsen, J., Mjølsnes, S.F.: Cryptographic Hash Function BLUE MIDNIGHT WISH. Submission to the NIST SHA-3 Competition, Round 2 (2009)
8. Leurent, G., Bouillaguet, C., Fouque, P.-A.: SIMD Is a Message Digest. Submission to the NIST SHA-3 Competition, Round 2 (2009)
9. Lipmaa, H., Moriai, S.: Efficient Algorithms for Computing Differential Properties of Addition. In: Matsui, M. (ed.) FSE 2001. LNCS, vol. 2355, pp. 336–350. Springer, Heidelberg (2002)
10. Lipmaa, H., Wallén, J., Dumas, P.: On the Additive Differential Probability of Exclusive-Or. In: Roy, B.K., Meier, W. (eds.) FSE 2004. LNCS, vol. 3017, pp. 317–331. Springer, Heidelberg (2004)
11. Mouha, N., Velichkov, V., De Cannière, C., Preneel, B.: The Differential Analysis of S-Functions. In: Biryukov, A., Gong, G., Stinson, D.R. (eds.) SAC 2010. LNCS, vol. 6544, pp. 36–56. Springer, Heidelberg (2011)
12. National Institute of Standards and Technology. Announcing Request for Candidate Algorithm Nominations for a New Cryptographic Hash Algorithm (SHA-3) Family. Federal Register 27(212), 62212–62220 (2007), http://csrc.nist.gov/groups/ST/hash/documents/FR_Notice_Nov07.pdf (October 17, 2008)
13. Needham, R.M., Wheeler, D.J.: Tea extensions. Computer Laboratory, Cambridge University, England (1997), http://www.movable-type.co.uk/scripts/xtea.pdf
14. Robshaw, M.J.B., Billet, O. (eds.): New Stream Cipher Designs - The eSTREAM Finalists. LNCS, vol. 4986. Springer, Heidelberg (2008)
15. Staffelbach, O., Meier, W.: Cryptographic Significance of the Carry for Ciphers Based on Integer Addition. In: Menezes, A., Vanstone, S.A. (eds.) CRYPTO 1990. LNCS, vol. 537, pp. 601–613. Springer, Heidelberg (1991)
16. Wu, H.: The Stream Cipher HC-128. In: Robshaw, M.J.B., Billet, O. (eds.) [14], pp. 39–47

A   Appendix

$$A_{000} = \begin{bmatrix} 0 & 1 & 1 & 0 & 0 & 0 & 0 & 0 \\ 0 & 1 & 0 & 0 & 0 & 0 & 0 & 0 \\ 0 & 0 & 1 & 0 & 0 & 0 & 0 & 0 \\ 0 & 0 & 0 & 0 & 0 & 0 & 0 & 0 \\ 0 & 1 & 1 & 0 & 4 & 0 & 0 & 1 \\ 0 & 1 & 0 & 0 & 0 & 0 & 0 & 1 \\ 0 & 0 & 1 & 0 & 0 & 0 & 0 & 1 \\ 0 & 0 & 0 & 0 & 0 & 0 & 0 & 1 \end{bmatrix}, A_{001} = \begin{bmatrix} 4 & 0 & 0 & 1 & 0 & 1 & 1 & 0 \\ 0 & 0 & 0 & 1 & 0 & 1 & 0 & 0 \\ 0 & 0 & 0 & 1 & 0 & 0 & 1 & 0 \\ 0 & 0 & 0 & 1 & 0 & 0 & 0 & 0 \\ 0 & 0 & 0 & 0 & 0 & 1 & 1 & 0 \\ 0 & 0 & 0 & 0 & 0 & 1 & 0 & 0 \\ 0 & 0 & 0 & 0 & 0 & 0 & 1 & 0 \\ 0 & 0 & 0 & 0 & 0 & 0 & 0 & 0 \end{bmatrix}, A_{010} = \begin{bmatrix} 1 & 0 & 0 & 0 & 0 & 0 & 0 & 0 \\ 0 & 0 & 0 & 0 & 0 & 0 & 0 & 0 \\ 1 & 0 & 0 & 1 & 0 & 0 & 0 & 0 \\ 0 & 0 & 0 & 1 & 0 & 0 & 0 & 0 \\ 1 & 0 & 0 & 0 & 0 & 1 & 0 & 0 \\ 0 & 0 & 0 & 0 & 0 & 1 & 0 & 0 \\ 1 & 0 & 0 & 1 & 0 & 1 & 4 & 0 \\ 0 & 0 & 0 & 1 & 0 & 1 & 0 & 0 \end{bmatrix},$$

$$A_{011} = \begin{bmatrix} 0 & 1 & 0 & 0 & 1 & 0 & 0 & 0 \\ 0 & 1 & 0 & 0 & 0 & 0 & 0 & 0 \\ 0 & 1 & 4 & 0 & 1 & 0 & 0 & 1 \\ 0 & 1 & 0 & 0 & 0 & 0 & 0 & 1 \\ 0 & 0 & 0 & 0 & 1 & 0 & 0 & 0 \\ 0 & 0 & 0 & 0 & 0 & 0 & 0 & 0 \\ 0 & 0 & 0 & 0 & 1 & 0 & 0 & 1 \\ 0 & 0 & 0 & 0 & 0 & 0 & 0 & 1 \end{bmatrix}, A_{100} = \begin{bmatrix} 1 & 0 & 0 & 0 & 0 & 0 & 0 & 0 \\ 1 & 0 & 0 & 1 & 0 & 0 & 0 & 0 \\ 0 & 0 & 0 & 0 & 0 & 0 & 0 & 0 \\ 0 & 0 & 0 & 1 & 0 & 0 & 0 & 0 \\ 1 & 0 & 0 & 0 & 0 & 0 & 1 & 0 \\ 1 & 0 & 0 & 1 & 0 & 4 & 1 & 0 \\ 0 & 0 & 0 & 0 & 0 & 0 & 1 & 0 \\ 0 & 0 & 0 & 1 & 0 & 0 & 1 & 0 \end{bmatrix}, A_{101} = \begin{bmatrix} 0 & 0 & 1 & 0 & 1 & 0 & 0 & 0 \\ 0 & 4 & 1 & 0 & 1 & 0 & 0 & 1 \\ 0 & 0 & 1 & 0 & 0 & 0 & 0 & 0 \\ 0 & 0 & 1 & 0 & 0 & 0 & 0 & 1 \\ 0 & 0 & 0 & 0 & 1 & 0 & 0 & 0 \\ 0 & 0 & 0 & 0 & 1 & 0 & 0 & 1 \\ 0 & 0 & 0 & 0 & 0 & 0 & 0 & 0 \\ 0 & 0 & 0 & 0 & 0 & 0 & 0 & 1 \end{bmatrix},$$

$$A_{110} = \begin{bmatrix} 0 & 0 & 0 & 0 & 0 & 0 & 0 & 0 \\ 0 & 1 & 0 & 0 & 0 & 0 & 0 & 0 \\ 0 & 0 & 1 & 0 & 0 & 0 & 0 & 0 \\ 0 & 1 & 1 & 0 & 0 & 0 & 0 & 0 \\ 0 & 0 & 0 & 0 & 1 & 0 & 0 & 0 \\ 0 & 1 & 0 & 0 & 1 & 0 & 0 & 0 \\ 0 & 0 & 1 & 0 & 1 & 0 & 0 & 0 \\ 0 & 1 & 1 & 0 & 1 & 0 & 0 & 4 \end{bmatrix}, A_{111} = \begin{bmatrix} 1 & 0 & 0 & 0 & 0 & 0 & 0 & 0 \\ 1 & 0 & 0 & 0 & 0 & 1 & 0 & 0 \\ 1 & 0 & 0 & 0 & 0 & 0 & 1 & 0 \\ 1 & 0 & 0 & 4 & 0 & 1 & 1 & 0 \\ 0 & 0 & 0 & 0 & 0 & 0 & 0 & 0 \\ 0 & 0 & 0 & 0 & 0 & 1 & 0 & 0 \\ 0 & 0 & 0 & 0 & 0 & 0 & 1 & 0 \\ 0 & 0 & 0 & 0 & 0 & 1 & 1 & 0 \end{bmatrix}, R = \begin{bmatrix} 0 & 0 & 0 & 0 & 0 & 0 & 0 & 0 \\ 0 & 0 & 0 & 0 & 0 & 0 & 0 & 0 \\ 0 & 0 & 0 & 0 & 0 & 0 & 0 & 0 \\ 0 & 0 & 0 & 0 & 0 & 0 & 0 & 0 \\ 1 & 0 & 1 & 0 & 1 & 0 & 1 & 0 \\ 0 & 1 & 0 & 1 & 0 & 1 & 0 & 1 \\ 0 & 0 & 0 & 0 & 0 & 0 & 0 & 0 \\ 0 & 0 & 0 & 0 & 0 & 0 & 0 & 0 \end{bmatrix}.$$

B   Proof of Correctness of the Computation of adp^{ARX}

In this section, we provide the full proof of Theorem 1. We use the graph representation of an S-function [11]. For input words of size n, an S-function can be represented as a directed acyclic graph, composed of n bipartite subgraphs. Each bipartite subgraph corresponds to one of the eight adjacency matrices A_q, $q \in \{0, 1, \ldots, 7\}$. The vertices of the i-th subgraph are composed of the two disjoint sets of eight input states $S[i] \in \{0, 1, \ldots, 7\}$ and eight output states $S[i+1] \in \{0, 1, \ldots, 7\}$. Furthermore, the output states of the i-th subgraph are input states for the $(i+1)$-th subgraph. An edge between a vertex in $S[i]$ and a vertex in $S[i+1]$ corresponds to a value of the tuple $(c_1[i], d_1[i+r])$ that results in the fixed output difference $\Delta e[i+r] = \Delta \eta[i+r]$. With this representation in mind, we state the following two lemmas before we proceed to the main theorem.

Lemma 1. *Let input differences $\Delta \gamma[i]$, $\Delta \lambda[i+r]$ be given. Then, for every input value $(c_1[i], d_1[i+r])$ and input state $S[i]$, the output value $\Delta e[i+r]$ and the output state $S[i+1]$ are uniquely determined.*

Proof. The proof follows directly from (23)-(25).   □

Lemma 2. *The i-th subgraph in the graph representation of the* adp^{ARX} *S-function (29) contains an edge if and only if* $\Delta e[i+r] = \Delta \eta[i+r]$

Proof. The statement holds by construction of the subgraphs.   □

Theorem 1

$$2^{-2n} \sum_{j \in \{0,2,4,6\}} (L_j A_{w[n-1]} \cdots A_{w[n-r]} R A_{w[n-r-1]} \cdots A_{w[1]} A_{w[0]} C_j) =$$

$$\frac{|\{(c_1, d_1) : \Delta e = \Delta \eta\}|}{|\{(c_1, d_1)\}|} . \tag{38}$$

Proof. Proving the statement of the theorem is equivalent to proving that the result computed by formula (36) is equal to the definition of adp^{ARX} (19). Consider the S-function for adp^{ARX} (29) and the i-th subgraph of its graph representation. Fix the inputs $\Delta \gamma[i], \Delta \lambda[i+r]$. From Lemma 1, it follows that every edge in the subgraph corresponds to a distinct pair of inputs $(c_1[i], d_1[i+r]), (c_2[i], d_2[i+r])$ that satisfies the input differences $(\Delta \gamma[i], \Delta \lambda[i+r])$. From Lemma 2, it follows that the subgraph contains only those among all edges, for which the pair of inputs satisfies also the output difference $\Delta \eta[i+r]$. Consider next the graph composed of all n subgraphs. A path in this graph is composed of n edges: one edge from each subgraph. For bit position i, one edge corresponds to distinct pairs $(c_1[i], d_1[i+r]), (c_2[i], d_2[i+r])$ that satisfy differences $\Delta \gamma[i], \Delta \lambda[i+r], \Delta \eta[i+r]$. Therefore, a path composed of n edges will correspond to distinct pairs $(c_1, d_1), (c_2, d_2)$ that satisfy the n-bit differences $\Delta \gamma, \Delta \lambda, \Delta \eta$. It follows that the number of paths in the S-function graph is equal to the number of pairs of inputs that satisfy both the input and the output differences. The number of paths that connect input state $S[0] = u \in \{0, \ldots, 7\}$ to output state $S[n-1] = v \in \{0, \ldots, 7\}$ is equal to the value of the element in column u and row v of the matrix A, denoted by $A_{u,v}$ with indexing starting from zero. The matrix A is obtained by multiplying the n adjacency matrices corresponding to each of the n subgraphs

$$A = A_{w[n-1]} \cdots A_{w[n-r]} R A_{w[n-r-1]} \cdots A_{w[1]} A_{w[0]} , \tag{39}$$

where R is the projection matrix derived in Sect. 5. In Sect. 5, it was shown that due to the bit rotation in the ARX operation, the only valid initial states for the S-function are $S[0] = u \in \{0, 2, 4, 6\}$. Their corresponding valid final states are $S[n-1] = u$ and $S[n-1] = u+1$. Therefore the number of paths connecting valid input and output states is equal to the sum of elements $A_{u,v}$ $u \in \{0, 2, 4, 6\}, v \in \{u, u+1\}$ of A:

$$\sum_{u \in \{0,2,4,6\}} \sum_{v \in \{u,u+1\}} A_{u,v} = \sum_{j \in \{0,2,4,6\}} L_j A C_j , \tag{40}$$

where C_j and L_j are the same as in (36). It remains to prove that (40) is equal to $|\{(c_1, d_1) : \Delta e = \Delta \eta\}|$. For this it is enough to show that none of the paths corresponding to $A_{u,v}$ overlap. This is indeed the case since the four initial states u do not overlap (no two values of u are equal) and each of them ends in a set of final states so that no two sets $\{u, u + 1\}$ overlap. From this, and because $|\{(c_1, d_1)\}| = 2^{2n}$, it follows that

$$2^{-2n} \sum_{j \in \{0,2,4,6\}} L_j A C_j = \frac{|\{(c_1, d_1) : \Delta e = \Delta \eta\}|}{|\{(c_1, d_1)\}|} = \mathrm{adp}^{\mathrm{ARX}}(\Delta \gamma, \Delta \lambda \xrightarrow{r} \Delta \eta).$$

□

Linear Approximations of Addition Modulo 2^n-1[*]

Chunfang Zhou[1,2], Xiutao Feng[1], and Chuankun Wu[1]

[1] State Key Laboratory of Information Security, Institute of Software,
Chinese Academy of Sciences, Beijing, 100190, China
[2] Graduate University of the Chinese Academy of Science, Beijing, 100049, China
{cfzhou,fengxt,ckwu}@is.iscas.ac.cn

Abstract. Addition modulo $2^{31} - 1$ is a basic arithmetic operation in the stream cipher ZUC. For evaluating ZUC's resistance against linear cryptanalysis, it is necessary to study properties of linear approximations of the addition modulo $2^{31} - 1$. In this paper we discuss linear approximations of the addition of k inputs modulo $2^n - 1$ for $n \geq 2$. As a result, an explicit expression of the correlations of linear approximations of the addition modulo $2^n - 1$ is given when $k = 2$, and an iterative expression when $k > 2$. For a class of special linear approximations with all masks being equal to 1, we further discuss the limit of their correlations when n goes to infinity. It is shown that when k is even, the limit is equal to zero, and when k is odd, the limit is bounded by a constant depending on k.

Keywords: Linear approximation, modular additions, linear cryptanalysis.

1 Introduction

Linear cryptanalysis [1] is one of the most powerful and general cryptanalytic methods. Its main task is to find linear relations between the inputs and outputs of target functions. In block ciphers, we usually find some linear relations among keys, plaintexts and ciphertexts that hold with certain probability. If some plaintext/ciphertext pairs are known, some bits of the key can be recovered with high probability [1, 2]. In stream ciphers, linear cryptanalysis is usually combined with distinguishing cryptanalysis together, and its goal is to establish a linear distinguisher to distinguish the keystream generated by the target algorithm from a random sequence [3, 4].

For both block ciphers and stream ciphers, it is important to find an efficient method to evaluate their resistance against linear cryptanalysis. Most cryptographic algorithms are usually designed by composing distinct and well chosen components and operations. Hence we should calculate linear approximations

---

[*] This work was supported by the Natural Science Foundation of China (Grant No. 60833008 and 60902024) and the National 973 Program (Grant No. 2007CB807902).

A. Joux (Ed.): FSE 2011, LNCS 6733, pp. 359–377, 2011.

of those components or operations. The addition modulo 2^n, especially when n is equal to the length of a computer word, e.g., 8, 16 or 32, is one of the most common operations, and is widely used in the design of cryptographic algorithms [5–8]. Many results on the addition modulo 2^n have been obtained, see [9–15].

The addition modulo 2^n-1 is another important arithmetic operation [16, 17]. Some properties of the addition modulo $2^n - 1$ have been explored in [18, 19]. However few results on linear approximations on the addition modulo $2^n - 1$ can be found from public literature. Recently a new stream cipher named ZUC [20], together with 128-EEA3 and 128-EIA3, has been proposed as the third suite of LTE encryption and integrity candidates, see [21] for details. In ZUC, the addition modulo $2^{31}-1$ is a basic operation since the linear feedback shift register (LFSR) of ZUC is defined over the prime field $\mathbb{F}_{2^{31}-1}$. For evaluating ZUC's resistance against linear cryptanalysis, it is necessary to study the properties of linear approximations of the addition modulo $2^{31} - 1$. In this paper, by means of known results on the addition modulo 2^n, we directly derive an expression for the correlations of arbitrary linear approximations of the addition modulo $2^n - 1$ with two inputs. For the case where more than two inputs are involved, we give an iterative expression. Moreover, for a class of special linear approximations with all masks being equal to 1, we discuss the limit of their correlations when n goes to infinity. Let k be the number of inputs of the addition modulo $2^n - 1$. It is shown that when k is even, the limit is equal to zero, and when k is odd, the limit is a constant depending on k.

The rest of the paper is organized as follows: in section 2, we give the definitions of linear approximations and their correlations and recall some properties of the addition modulo 2^n briefly. In section 3 some basic properties of linear approximation of the addition modulo $2^n - 1$ are given, and more properties for the case $k = 2$ are given in section 4. In section 5 we further discuss the limit of linear approximations with all masks being equal to 1. Finally we conclude in section 6.

2 Preliminaries

2.1 Linear Approximation and Its Correlation

Let n be a positive integer. Denote Z_{2^n} the set of integers x such that $0 \le x \le 2^n - 1$. Given an integer $x \in Z_{2^n}$, let

$$x = x^{(n-1)}x^{(n-2)}\cdots x^{(0)} = \sum_{i=0}^{n-1} x^{(i)}2^i$$

be the binary representation of x, where $x^{(i)} \in \{0,1\}$. We call $x^{(i)}$ the i-th bit of x, $0 \le i \le n - 1$. In the rest of the paper, without further specification, we always denote by $x^{(i)}$ the i-th bit of the integer x in its binary representation. For arbitrary two integers $w, x \in Z_{2^n}$, the inner product of w and x is defined as

$$w \cdot x = \bigoplus_{i=0}^{n-1} w^{(i)} x^{(i)}.$$

Let J be a nonempty subset of Z_{2^n}, k be a positive integer and f be a function from J^k to J. Given $k+1$ constants $u, w_1, \cdots, w_k \in Z_{2^n}$, the linear approximation of the function f associated with $u, w_1, \cdots, w_k$ is an approximate relation of the form

$$u \cdot f(x_1, \cdots, x_k) = \bigoplus_{i=1}^{k} w_i \cdot x_i, \tag{1}$$

and the $(k+1)$-tuple $(u, w_1, \cdots, w_k)$ is called a *linear mask* of f. The efficiency of the linear approximation (1) is measured by its correlation which is defined as

$$\mathbf{cor}_f(u; w_1, \cdots, w_k) = 2 \Pr(u \cdot f(x_1, \cdots, x_k) = \bigoplus_{i=1}^{k} w_i \cdot x_i) - 1$$

$$= \frac{1}{|J|^k} \sum_{(x_1, \cdots, x_k) \in J^k} (-1)^{u \cdot f(x_1, \cdots, x_k) \oplus \bigoplus_{i=1}^{k} w_i \cdot x_i}, \tag{2}$$

where the probability is taken over uniformly distributed $x_1, \cdots, x_k$ over J, and $|J|$ denotes the cardinality of the set J.

2.2 Linear Approximations of the Addition Modulo 2^n

In this section we recall some properties of linear approximations of the addition modulo 2^n briefly, for more details please refer to [9, 10].

Denote by $\boxplus$ the addition modulo 2^n, that is, for any $x_1, x_2 \in Z_{2^n}$, we have $x_1 \boxplus x_2 = (x_1 + x_2) \mod 2^n$. Let (u, w_1, w_2) be a linear mask of the addition $\boxplus$, and denote by $\mathbf{cor}_{\boxplus}(u; w_1, w_2)$ the correlation of the linear approximation $u \cdot (x_1 \boxplus x_2) = w_1 \cdot x_1 \oplus w_2 \cdot x_2$. From the linear mask (u, w_1, w_2) we derive a sequence $\underline{z} = z_{n-1} \cdots z_0$ as follows

$$z_i = u^{(i)} 2^2 + w_2^{(i)} 2 + w_1^{(i)}, \quad i = 0, 1, \cdots, n-1. \tag{3}$$

It's easy to see that $0 \le z_i \le 7$ for all $0 \le i \le n-1$. Define

$$M_n(u, w_1, w_2) = \prod_{i=0}^{n-1} A_{z_i}, \tag{4}$$

where A_j $(j = 0, 1, \cdots, 7)$ are constant matrices of size 2×2 and defined as follows

$$A_0 = \frac{1}{4} \begin{pmatrix} 3 & 1 \\ 1 & 3 \end{pmatrix}, A_1 = A_2 = -A_4 = \frac{1}{4} \begin{pmatrix} 1 & 1 \\ -1 & -1 \end{pmatrix},$$

$$-A_3 = A_5 = A_6 = \frac{1}{4} \begin{pmatrix} 1 & -1 \\ -1 & 1 \end{pmatrix}, A_7 = \frac{1}{4} \begin{pmatrix} 3 & -1 \\ 1 & -3 \end{pmatrix}.$$

Then we have

Theorem 1 ([9]). *For any given linear mask* (u, w_1, w_2), *let* $M_n(u, w_1, w_2)$ *be defined as above. Set* $M_n(u, w_1, w_2) = (M_{i,j})_{0 \leq i,j \leq 1}$. *Then we have*

$$M_{i,j} = \Pr(u \cdot (x_1 \boxplus x_2) = w_1 \cdot x_1 \oplus w_2 \cdot x_2 \wedge c_n = i \wedge c_0 = j)$$
$$- \Pr(u \cdot (x_1 \boxplus x_2) \neq w_1 \cdot x_1 \oplus w_2 \cdot x_2 \wedge c_n = i \wedge c_0 = j),$$

where c_0 *is an initial carry bit, and* c_n *is the n-th carry bit of the addition of* x_1 *and* x_2 *with the initial carry bit* c_0. *By convention* $c_0 = 0$, *and we have*

$$\mathbf{cor}_{\boxplus}(u; w_1, w_2) = M_{0,0} + M_{1,0}. \tag{5}$$

Note that for any integers x_1 and x_2, if $c_0 = 1$, then the addition of x_1 and x_2 modulo 2^n with the initial carry c_0 is equivalent to $(x_1 + x_2 + 1) \mod 2^n$. Therefore we get the following corollary.

Corollary 1. *Let* $x_1 \overline{\boxplus} x_2 = x_1 \boxplus x_2 \boxplus 1$ *and* (u, w_1, w_2) *be a linear mask of* $\boxplus$. *Denote by* $\mathbf{cor}_{\overline{\boxplus}}(u; w_1, w_2)$ *the correlation of the linear approximation* $u \cdot (x_1 \overline{\boxplus} x_2) = w_1 \cdot x_1 \oplus w_2 \cdot x_2$. *Then we have*

$$\mathbf{cor}_{\overline{\boxplus}}(u; w_1, w_2) = M_{0,1} + M_{1,1}. \tag{6}$$

3 Some Properties on Linear Approximations of the Addition Modulo $2^n - 1$

In this section we will discuss some properties of linear approximations of the addition modulo $2^n - 1$ with k inputs, where we always assume that $n \geq 2$ and $k \geq 2$. For consistency with the definition of the addition of the prime field $\mathbb{F}_{2^n-1}$ in ZUC [20], here we make the convention that the set of representatives of the residue class modulo $2^n - 1$ are $\{1, 2, \cdots, 2^n - 1\}$ instead of $\{0, 1, \cdots, 2^n - 2\}$. It should be pointed out that all results in this paper can induce the corresponding ones in $\{0, 1, \cdots, 2^n - 2\}$ directly.

Let $J = \{1, 2, \cdots, 2^n - 1\}$, and denote by $\hat{\boxplus}$ the addition modulo $2^n - 1$ as defined in ZUC, more precisely, for any $x_1, x_2 \in J$, we have

$$x_1 \hat{\boxplus} x_2 = \begin{cases} x_1 + x_2 & \text{if } x_1 + x_2 < 2^n, \\ (x_1 + x_2 + 1) \mod 2^n & \text{if } x_1 + x_2 \geq 2^n. \end{cases} \tag{7}$$

For example, set $n = 3$, then $J = \{1, 2, \cdots, 7\}$, and $2\hat{\boxplus}6 = 1$, $3\hat{\boxplus}4 = 7$.

In the following we consider the addition modulo $2^n - 1$ over J with k inputs. For any given linear mask $(u, w_1, \cdots, w_k)$, we denote by $\mathbf{cor}_{\hat{\boxplus}}(u; w_1, \cdots, w_k)$ the correlation of the linear approximation

$$u \cdot (x_1 \hat{\boxplus} \cdots \hat{\boxplus} x_k) = \bigoplus_{i=1}^{k} w_i \cdot x_i.$$

For simplicity we write $\mathbf{cor}_{\hat{\boxplus}}(u; w_1, \cdots, w_k)$ as $\mathbf{cor}(u; w_1, \cdots, w_k)$.
The following two theorems can easily be derived.

Theorem 2. *For any given linear mask* $(u, w_1, \cdots, w_k)$ *and any permutation* $(i_1, \cdots, i_k)$ *of* $(1, \cdots, k)$, *we have*

$$\mathbf{cor}(u; w_1, \cdots, w_k) = \mathbf{cor}(u; w_{i_1}, \cdots, w_{i_k}). \tag{8}$$

Proof. Define

$$J(u; w_1, \cdots, w_k) = \{ (x_1, \cdots, x_k) \in J^k \mid u \cdot (x_1 \hat{\boxplus} \cdots \hat{\boxplus} x_k) = \bigoplus_{i=1}^{k} w_i \cdot x_i \}.$$

By the definition of the correlation (see Eqn. (2)), we only need to prove that

$$|J(u; w_1, \cdots, w_k)| = |J(u; w_{i_1}, \cdots, w_{i_k})|. \tag{9}$$

For any $(x_1, \cdots, x_k) \in J(u; w_1, \cdots, w_k)$, we have

$$u \cdot (\hat{\boxplus}_{j=1}^{k} x_{i_j}) = u \cdot (\hat{\boxplus}_{i=1}^{k} x_i) = \bigoplus_{i=1}^{k} w_i \cdot x_i = \bigoplus_{j=1}^{k} w_{i_j} \cdot x_{i_j},$$

which shows $(x_{i_1}, \cdots, x_{i_k}) \in J(u; w_{i_1}, \cdots, w_{i_k})$, and vice versa. So Eqn. (9) holds. ∎

Theorem 3. *For any given linear mask* $(u, w_1, \cdots, w_k)$ *and integer* ℓ *such that* $1 \le \ell \le n - 1$, *we have*

$$\mathbf{cor}(u; w_1, \cdots, w_k) = \mathbf{cor}(u \lll \ell; w_1 \lll \ell, \cdots, w_k \lll \ell), \tag{10}$$

where $x \lll \ell$ *denotes the cyclic shift of* x ℓ *bits to the left* .

Proof. Similarly to the proof of Theorem 2, we only need to prove that

$$|J(u; w_1, \cdots, w_k)| = |J(u \lll \ell; w_1 \lll \ell, \cdots, w_k \lll \ell)|. \tag{11}$$

It is easy to see that $x \lll \ell \equiv 2^\ell x \mod (2^n - 1)$ holds for any $x \in J$, which means that for any $x_1, \cdots, x_k \in J$, we have

$$(\sum_{i=1}^{k} x_i) \lll \ell \equiv 2^\ell \sum_{i=1}^{k} x_i \equiv \sum_{i=1}^{k} 2^\ell x_i \equiv \sum_{i=1}^{k} (x_i \lll \ell) \mod (2^n - 1),$$

namely, $(\hat{\boxplus}_{i=1}^{k} x_i) \lll \ell = \hat{\boxplus}_{i=1}^{k} (x_i \lll \ell)$.

So for any $(x_1, \cdots, x_k) \in J(u; w_1, \cdots, w_k)$, we have

$$(u \lll \ell) \cdot (\hat{\boxplus}_{i=1}^{k} (x_i \lll \ell)) = (u \lll \ell) \cdot ((\hat{\boxplus}_{i=1}^{k} x_i) \lll \ell) = (u \cdot (\hat{\boxplus}_{i=1}^{k} x_i)) \lll \ell$$

$$= (\bigoplus_{i=1}^{k} w_i \cdot x_i) \lll \ell = \bigoplus_{i=1}^{k} (w_i \lll \ell) \cdot (x_i \lll \ell).$$

It follows that $(x_1 \lll \ell, \cdots, x_k \lll \ell) \in J(u \lll \ell; w_1 \lll \ell, \cdots, w_k \lll \ell)$, that is, $|J(u; w_1, \cdots, w_k)| \le |J(u \lll \ell; w_1 \lll \ell, \cdots, w_k \lll \ell)|$. Note that $(x \lll \ell) \lll (n - \ell) = x$ for any $x \in J$. By shifting each mask cyclicly $n - \ell$ bits to the left, we have

$$|J(u \lll \ell; w_1 \lll \ell, \cdots, w_k \lll \ell)| \le |J(u; w_1, \cdots, w_k)|.$$

So Eqn. (11) follows. ∎

3.1 Addition of Two Inputs in $\mathbb{F}_{2^n-1}$

In this section we will derive an explicit expression of $\mathbf{cor}(u; w_1, w_2)$ for any linear mask (u, w_1, w_2) from Theorem 1. For any given linear mask (u, w_1, w_2), we keep the notations $\underline{z}$, $M_n(u, w_1, w_2)$ and $M_{i,j}$ $(0 \le i, j \le 1)$ defined in section 2.

One can notice that when $x_1 + x_2 < 2^n$, we have $x_1 \hat{\boxplus} x_2 = x_1 \boxplus x_2$, and when $x_1 + x_2 \ge 2^n$, we have $x_1 \hat{\boxplus} x_2 = x_1 \boxplus x_2 \boxplus 1$. Thus by Theorem 1 and Corollary 1, it seems that $\mathbf{cor}(u; w_1, w_2)$ is almost equal to $M_{0,0} + M_{1,1}$ if the difference between Z_{2^n} and J is ignored. Below we give an explicit expression for $\mathbf{cor}(u; w_1, w_2)$.

Theorem 4. *Let (u, w_1, w_2) be a linear mask of the addition $\hat{\boxplus}$ modulo $2^n - 1$, and $M_n(u, w_1, w_2) = (M_{i,j})_{0 \le i,j \le 1}$ be defined as above. Then we have*

$$\mathbf{cor}(u; w_1, w_2) = \frac{2^{2n}(M_{0,0} + M_{1,1}) + 2^n \cdot c + 1}{(2^n - 1)^2}, \qquad (12)$$

where

$$c = \begin{cases} -3, & \text{if } u = w_1 = w_2 \text{ and } w_H(w_2) \text{ is even,} \\ 1, & \text{if } u \ne w_1 = w_2 \text{ and } w_H(w_2) \text{ is odd,} \\ 0, & \text{if } u, w_1 \text{ and } w_2 \text{ are pairwise different,} \\ -1, & \text{otherwise,} \end{cases}$$

and $w_H(w_2)$ denotes the hamming weight of w_2 in its binary representation.

Proof. For any given $x_1, x_2 \in J$, we consider $x_1 \hat{\boxplus} x_2$ from the following two aspects.

First, when $0 < x_1 + x_2 < 2^n$, it is known that $x_1 \hat{\boxplus} x_2 = x_1 \boxplus x_2$. By Theorem 1, we have

$$\begin{aligned} M_{0,0} = & \Pr(u \cdot (x_1 \boxplus x_2) = w_1 \cdot x_1 \oplus w_2 \cdot x_2 \wedge 0 \le x_1 + x_2 < 2^n) \\ & - \Pr(u \cdot (x_1 \boxplus x_2) \ne w_1 \cdot x_1 \oplus w_2 \cdot x_2 \wedge 0 \le x_1 + x_2 < 2^n). \end{aligned}$$

Since

$$\begin{aligned} & \Pr(u \cdot (x_1 \boxplus x_2) = w_1 \cdot x_1 \oplus w_2 \cdot x_2 \wedge 0 \le x_1 + x_2 < 2^n) \\ & + \Pr(u \cdot (x_1 \boxplus x_2) \ne w_1 \cdot x_1 \oplus w_2 \cdot x_2 \wedge 0 \le x_1 + x_2 < 2^n) \\ & = \Pr(x_1 + x_2 < 2^n) = \frac{2^n + 1}{2^{n+1}}, \end{aligned}$$

thus we have

$$\Pr(u \cdot (x_1 \boxplus x_2) = w_1 \cdot x_1 \oplus w_2 \cdot x_2 \wedge 0 \le x_1 + x_2 < 2^n) = \frac{1}{2} M_{0,0} + \frac{2^n + 1}{2^{n+2}}.$$

It follows that there are $2^{n-2}(2^n + 1) + 2^{2n-1} M_{0,0}$ pairs (x_1, x_2) satisfying $u \cdot (x_1 \boxplus x_2) = w_1 \cdot x_1 \oplus w_2 \cdot x_2$ and $0 \le x_1 + x_2 < 2^n$. We consider those pairs of the form $(0, x_2)$. When $x_1 = 0$, we get $(u \oplus w_2) \cdot x_2 = 0$ due to $u \cdot x_2 = w_2 \cdot x_2$. It follows that there are 2^{n-1} solutions x_2 if $u \ne w_2$ and 2^n solutions if $u = w_2$.

Hence there are 2^{n-1} pairs of the form $(0, x_2)$ among all the above pairs not in $J \times J$ if $u \neq w_2$ and 2^n pairs not in $J \times J$ if $u = w_2$. By the symmetric position of x_1 and x_2, we have the same conclusion for $x_2 = 0$. In addition, the pair $(0,0)$ always satisfies $u \cdot (x_1 \boxplus x_2) = w_1 \cdot x_1 \oplus w_2 \cdot x_2$ but is not in $J \times J$.

Second, when $x_1 + x_2 \geq 2^n$, we have $x_1 \hat{\boxplus} x_2 = x_1 \boxplus x_2 \boxplus 1$. Similar to the above case, there are totally $2^{n-2}(2^n + 1) + 2^{2n-1}M_{1,1}$ pairs (x_1, x_2) satisfying both $x_1 + x_2 + 1 \geq 2^n$ and $u \cdot (x_1 \boxplus x_2 \boxplus 1) = w_1 \cdot x_1 \oplus w_2 \cdot x_2$. Now we consider how to remove some pairs (x_1, x_2) satisfying $x_1 + x_2 + 1 = 2^n$ from the above pairs. Note that $x_1 \boxplus x_2 \boxplus 1 = 0$, thus we only need to count pairs (x_1, x_2) such that $x_1 + x_2 = 2^n - 1$ and $w_1 \cdot x_1 = w_2 \cdot x_2$. Since $x_1 + x_2 = 2^n - 1 = x_1 \oplus x_2$, it follows that

$$(w_1 \oplus w_2) \cdot x_1 = w_2 \cdot (2^n - 1). \tag{13}$$

If $w_1 \neq w_2$, Eqn. (13) has 2^{n-1} solutions; if $w_1 = w_2$, when $w_H(w_2)$ is an odd number, Eqn. (13) has no solutions, and when $w_H(w_2)$ is an even number, Eqn. (13) has 2^n solutions.

Denote by d the number of pairs $(x_1, x_2) \in Z_{2^n} \times Z_{2^n}$ which satisfy the linear approximation defined by mask (u, w_1, w_2) and $x_1 = 0$ or $x_2 = 0$ or $x_1 + x_2 = 2^n - 1$. Combine the above two cases, we have

$$d = \begin{cases} 3 \cdot 2^n - 1, & \text{if } u = w_1 = w_2 \text{ and } w_H(w_2) \text{ is even,} \\ 2^n - 1, & \text{if } u \neq w_1 = w_2 \text{ and } w_H(w_2) \text{ is odd,} \\ 3 \cdot 2^{n-1} - 1, & \text{if } u, w_1 \text{ and } w_2 \text{ are pairwise different,} \\ 2 \cdot 2^n - 1, & \text{otherwise.} \end{cases}$$

By the definition of correlation, we have

$$\mathbf{cor}(u; w_1, w_2)$$
$$= 2 \cdot \frac{(2^{n-2}(2^n + 1) + 2^{2n-1}M_{0,0}) + (2^{n-2}(2^n + 1) + 2^{2n-1}M_{1,1}) - d}{(2^n - 1)^2} - 1$$
$$= \frac{2^{2n}(M_{0,0} + M_{1,1}) + 3 \cdot 2^n - 1 - 2d}{(2^n - 1)^2}.$$

Then we can get the desired conclusion. ∎

3.2 Addition of More than Two Inputs in $\mathbb{F}_{2^n-1}$

In this section we will derive an iterative expression of $\mathbf{cor}(u; w_1, \cdots, w_k)$ for any linear mask $(u, w_1, \cdots, w_k)$. The addition of k inputs $x_1, \cdots, x_k$ can be seen as the addition of $x_1 \hat{\boxplus} \cdots \hat{\boxplus} x_{k-1}$ and x_k.

Theorem 5. *For any given linear mask $(u, w_1, \cdots, w_k)$ and integer $k > 2$, we have*

$$\mathbf{cor}(u; w_1, \cdots, w_k) = \frac{2^n - 1}{2^n} \sum_{w=0}^{2^n - 1} \mathbf{cor}(w; w_1, \cdots, w_{k-1})\mathbf{cor}(u; w, w_k). \tag{14}$$

Proof. By Eqn. (2), we have

$$\mathbf{cor}(u; w_1, \cdots, w_k) = \frac{1}{(2^n - 1)^k} \sum_{(x_1, \cdots, x_k) \in J^k} (-1)^{u \cdot (\boxplus_{i=1}^k x_i) \oplus \bigoplus_{i=1}^k w_i \cdot x_i}.$$

Denote $y = x_1 \hat{\boxplus} \cdots \hat{\boxplus} x_{k-1}$. Then we have

$$\sum_{w=0}^{2^n - 1} \mathbf{cor}(w; w_1, \cdots, w_{k-1}) \mathbf{cor}(u; w, w_k)$$

$$= \frac{1}{(2^n - 1)^{k+1}} \sum_{w=0}^{2^n - 1} \sum_{\substack{(x_1, \cdots, x_{k-1}) \in J^{k-1} \\ (z, x_k) \in J^2}} (-1)^{w \cdot y \oplus \bigoplus_{i=1}^{k-1} w_i \cdot x_i} (-1)^{u \cdot (z \hat{\boxplus} x_k) \oplus w \cdot z \oplus w_k \cdot x_k}$$

$$= \frac{1}{(2^n - 1)^{k+1}} \sum_{(x_1, \cdots, x_k, z) \in J^{k+1}} (-1)^{u \cdot (z \hat{\boxplus} x_k) \oplus \bigoplus_{i=1}^k w_i \cdot x_i} \sum_{w=0}^{2^n - 1} (-1)^{w \cdot z \oplus w \cdot y}$$

Note that

$$\sum_{w=0}^{2^n - 1} (-1)^{w \cdot z \oplus w \cdot y} = \begin{cases} 2^n, & \text{if } z = y, \\ 0, & \text{if } z \neq y. \end{cases}$$

Then we have

$$\sum_{w=0}^{2^n - 1} \mathbf{cor}(w; w_1, \cdots, w_{k-1}) \mathbf{cor}(u; w, w_k)$$

$$= \frac{2^n}{(2^n - 1)^{k+1}} \sum_{(x_1, \cdots, x_k) \in J^k} (-1)^{u \cdot (y \hat{\boxplus} x_k) \oplus \bigoplus_{i=1}^k w_i \cdot x_i}$$

$$= \frac{2^n}{2^n - 1} \mathbf{cor}(u; w_1, \cdots, w_k). \qquad \blacksquare$$

4   More Properties of Linear Approximations of the Addition Modulo $2^n - 1$ with Two Inputs

In this section we will provide more properties of linear approximations of the addition modulo $2^n - 1$ with two inputs, that is, $k = 2$. First we introduce some notations and concepts.

Let $\mathbb{Q}$ be the rational field. Define

$$\mathrm{I} = \left\{ \begin{pmatrix} a & b \\ b & a \end{pmatrix} \,|\, a, b \in \mathbb{Q} \right\},$$

$$\mathrm{II} = \left\{ \begin{pmatrix} a & -b \\ b & -a \end{pmatrix} \,|\, a, b \in \mathbb{Q} \right\},$$

and call a matrix in the set I (or II) to be type-I (or type-II). It is easy to see that $A_0, A_3, A_5, A_6 \in$ I and $A_1, A_2, A_4, A_7 \in$ II (which are defined in section 2). The following two properties can easily be verified.

Lemma 1. *The product of arbitrary two type-I (or type-II) matrices is a type-I matrix.*

Lemma 2. *The product of a type-I matrix and a type-II matrix is a type-II matrix.*

By the definition of $M_n(u, w_1, w_2)$ and Lemmas 1 and 2, we have

Lemma 3. *For any given linear mask (u, w_1, w_2), $M_n(u, w_1, w_2)$ is either type-I or type-II.*

For any given square matrix M, denote by $\mathbf{Tr}(M)$ the trace of the matrix M, that is, the sum of elements on the main diagonal of M. Since the trace of an arbitrary type-II matrix is zero, thus the following conclusions hold.

Corollary 2. *For any given linear mask (u, w_1, w_2), let $\underline{z} = z_{n-1} \cdots z_0$ be the sequence derived from (u, w_1, w_2) by the formula (3). If the number of elements z_i such that $z_i \in \{1, 2, 4, 7\}$ is odd, $i = 0, 1, \cdots, n-1$, then $\mathbf{Tr}(M_n(u, w_1, w_2)) = 0$.*

Corollary 3. *Let $u \in Z_{2^n}$ and $w_H(u)$ be odd. Then $\mathbf{Tr}(M_n(u, u, u)) = 0$. Thus we have*

$$\mathbf{cor}(u; u, u) = -\frac{1}{2^n - 1}$$

and

$$\lim_{n \to \infty} \mathbf{cor}(u; u, u) = 0.$$

Corollary 4. *Let $u \in Z_{2^n}$ and $w_H(u)$ be even. Then $M_n(u, u, u)$ is type-I, that is, $M_{0,0} = M_{1,1}$. Thus we have*

$$\mathbf{cor}(u; u, u) = \frac{2^{2n} \cdot 2M_{0,0} - 3 \cdot 2^n + 1}{(2^n - 1)^2}.$$

If all 1's of u in the binary representation are adjacent, then we have

$$\mathbf{cor}(u; u, u) = \frac{2^{2n} \cdot (2^{\frac{w_H(u)}{2} - n} + 2^{-\frac{w_H(u)}{2}}) - 3 \cdot 2^n + 1}{(2^n - 1)^2}$$

and

$$\lim_{n \to \infty} \mathbf{cor}(u; u, u) = 2^{-\frac{w_H(u)}{2}}.$$

Proof. By Theorem 3, we only need to consider the masks whose binary expression be of the form $(\underbrace{0, \cdots, 0}_{n - w_H(u)} \underbrace{1, \cdots, 1}_{w_H(u)})$. Then $M_n(u, u, u) = A_0^{n - w_H(u)} A_7^{w_H(u)}$.

Denote by $\mathbf{I}_2$ the 2×2 identity matrix. It is easy to see that $A_7^2 = \frac{1}{2}\mathbf{I}_2$. Since

$w_H(u)$ is even, we have $A_7^{w_H(u)} = 2^{-\frac{w_H(u)}{2}} I_2$. So $M_n(u, u, u) = 2^{-\frac{w_H(u)}{2}} A_0^{n-w_H(u)}$.

Since A_0 is a symmetric matrix of the form $\begin{pmatrix} a & b \\ b & a \end{pmatrix}$, it is easily proved by induction that

$$\begin{pmatrix} a & b \\ b & a \end{pmatrix}^t = \frac{1}{2} \begin{pmatrix} (a+b)^t + (a-b)^t & (a+b)^t - (a-b)^t \\ (a+b)^t - (a-b)^t & (a+b)^t + (a-b)^t \end{pmatrix}$$

for $t \geq 1$. Then we have $\mathbf{Tr}(A_0^{n-w_H(u)}) = 1 + 2^{w_H(u)-n}$. So $\mathbf{Tr}(M_n(u, u, u)) = 2^{-\frac{w_H(u)}{2}} \mathbf{Tr}(A_0^{n-w_H(u)}) = 2^{-\frac{w_H(u)}{2}} + 2^{\frac{w_H(u)}{2}-n}$, and the conclusion follows. ∎

Below we give some facts on A_i, $0 \leq i \leq 7$, which will be used later.

Lemma 4. 1. $A_0 A_i = \frac{1}{2} A_i$, for $\forall\, i \in \{1, 2, 3, 4, 5, 6\}$;
  2. $A_i A_0 = A_i$ if $i \in \{1, 2, 4\}$ and $A_i A_0 = \frac{1}{2} A_i$ if $i \in \{3, 5, 6\}$;
  3. $A_i A_j = 0$, $i \in \{1, 2, 4\}$ and $j \in \{1, 2, 3, 4, 5, 6\}$;
  4. $A_1 A_7 = A_2 A_7 = -A_4 A_7 = A_6$.

Now we consider a class of special linear masks $(u, 1, w)$. Let $\underline{z} = z_{n-1} \cdots z_0$ be the sequence derived from $(u, 1, w)$. It is easy to see that $z_0 \in \{1, 3, 5, 7\}$ and $z_i \in \{0, 2, 4, 6\}$, $1 \leq i \leq n - 1$. In the rest of the paper we simply write M instead of $M_n(u, 1, w)$.

Lemma 5. For any integers $u, w \in Z_{2^n}$, if $\mathbf{Tr}(M) \neq 0$, then the sequence $\underline{z}$ is of the form either $\{0, 6\}^{n-1}\{3, 5\}$ or $\{0, 6\}^*\{2, 4\}0^*7$.

Proof. Let r be the number of z_i such that $z_i \in \{2, 4\}$, $i = 1, 2, \cdots, n - 1$. We first prove that $r \leq 1$. Assume that $r > 1$. Then there exist two indexes i and j such that $z_i, z_j \in \{2, 4\}$, $1 \leq i < j \leq n - 1$. By Items 2 and 3 of Lemma 4, we have $A_{z_i} \cdots A_{z_j} = 0$. It follows that $M = 0$, which contradicts $\mathbf{Tr}(M) \neq 0$.

 When $r = 0$, if $z_0 \in \{1, 7\}$, by Corollary 2, it is known that the matrix M is of type-II, which contradicts $\mathbf{Tr}(M) \neq 0$ as well. Thus $z_0 \in \{3, 5\}$. So $\underline{z}$ is of the form $\{0, 6\}^{n-1}\{3, 5\}$.

 When $r = 1$, let $z_j \in \{2, 4\}$, where $1 \leq j \leq n - 1$. First we claim that $z_i = 0$ for all $1 \leq i < j$. If there exists some index i such that $z_i \neq 0$, by Items 2 and 3 of Lemma 4, we have $A_{z_i} \cdots A_{z_j} = 0$, furthermore $M = 0$, which is a contradiction. Second, if $z_0 \in \{1, 3, 5\}$, by Items 2 and 3 of Lemma 4, we have $A_{z_0} \cdots A_{z_i} = 0$. So $\underline{z}$ is of the form $\{0, 6\}^*\{2, 4\}0^*7$. ∎

Theorem 6. For any integers $u, w \in Z_{2^n}$, $\mathbf{Tr}(M) \neq 0$ if and only if $u = w \oplus 2^i$, where $0 \leq i \leq LNB(w \oplus 1)$, $LNB(x)$ denotes the least index where 1 appears in the binary representation of x if $x \neq 0$, and $LNB(0) = n - 1$.

Proof. The necessity follows directly from Lemma 5. Below we prove the sufficiency. First we prove that $\mathbf{Tr}(A_6^t) = 2^{-t}$ for any $t \geq 1$. In fact, it is easy to calculate two characteristic roots 0 and 2^{-1} of A_6. Thus we have $\mathbf{Tr}(A_6^t) = 0^t + (2^{-1})^t = 2^{-t}$.

If $i = 0$, i.e., $u = w \oplus 1$, then $\underline{z}$ is of the form $\{0,6\}^{n-1}\{3,5\}$. Let t be the number of z_i such that $z_i = 6$, $i = 1, 2, \cdots, n-1$. Then 0 occurs in $z_{n-1} \cdots z_1$ for $n-1-t$ times. Thus by Lemma 4, we have

$$\mathbf{Tr}(M) = \mathbf{Tr}(A_{z_{n-1}} \cdots \cdot A_{z_0})$$
$$= \mathbf{Tr}(2^{-(n-1-t)} A_6^t A_{z_0})$$
$$= (-1)^w 2^{-(n-1-t)} \mathbf{Tr}(A_6^{t+1})$$
$$= (-1)^w 2^{-(n-1-t)} 2^{-(t+1)}$$
$$= (-1)^w 2^{-n}.$$

If $i > 0$, then $\underline{z}$ is of the form $\{0,6\}^*\{2,4\}0^*7$ and $z_i \in \{2,4\}$. Let t be the number of repetitions of 6 appearing in $z_{n-1} \cdots z_{i+1}$. Then by Lemma 4, we have

$$\mathbf{Tr}(M) = \mathbf{Tr}(A_{z_{n-1}} \cdots \cdot A_{z_0})$$
$$= \mathbf{Tr}(2^{-(n-1-i-t)} A_6^t A_{z_i} A_7)$$
$$= (-1)^s 2^{-(n-1-i-t)} \mathbf{Tr}(A_6^{t+1})$$
$$= (-1)^s 2^{-(n-1-i-t)} 2^{-(t+1)}$$
$$= (-1)^s 2^{-(n-i)},$$

where $s = w^{(i)} \oplus 1$.   ∎

Theorem 6 gives a sufficient and necessary condition for judjing whether or not M is of type-II for any linear mask $(u, 1, w)$. From its proof we can get the following result.

Corollary 5. *For any integers $u, w \in Z_{2^n}$ such that $u = w \oplus 2^i$, where $0 \leq i \leq LNB(w \oplus 1)$, we have $\mathbf{Tr}(M) = (-1)^s 2^{-(n-i)}$, where*

$$s = \begin{cases} 0 & \text{if } i = 0 \text{ and } w^{(0)} = 0 \text{ or } i > 0 \text{ and } w^{(i)} = 1, \\ 1 & \text{otherwise.} \end{cases}$$

By Theorem 4 and Corollary 5, we can derive the following corollary.

Corollary 6. *The correlation of the linear approximation of addition in $\mathbb{F}_{2^n-1}$ with a mask of the form $(w, 1, 1)$ is given by*

$$\mathbf{cor}(w; 1, 1) = \begin{cases} \frac{1}{(2^n-1)^2} & \text{if } w = 0, \\ -\frac{1}{2^n-1} & \text{if } w = 1, \\ \frac{-2^{n+i}+2^n+1}{(2^n-1)^2} & \text{if } w = 2^i + 1, 1 \leq i \leq n-1, \\ \frac{2^n+1}{(2^n-1)^2} & \text{otherwise.} \end{cases}$$

When the mask is of the form $(1, w, 1)$, the correlation is given by

$$\mathbf{cor}(1; w, 1) = \begin{cases} \frac{1}{(2^n-1)^2} & \text{if } w = 0, \\ \frac{2^{n+i}-2^n+1}{(2^n-1)^2} & \text{if } w = 2^i + 1, 1 \leq i \leq n-1, \\ -\frac{1}{2^n-1} & \text{otherwise.} \end{cases}$$

Finally we give an upper bound of $|\mathbf{cor}(u; 1, w)|$. For any given integer $x \in Z_{2^n}$, define

$$J_x = \{x \oplus 2^i | 1 \leq i \leq LNB(x \oplus 1)\}.$$

Theorem 7. *For any integers $u, w \in Z_{2^n}$, if $w \notin J_u$, then*

$$|\mathbf{cor}(u; 1, w)| < \frac{3}{2^n - 1}. \tag{15}$$

Proof. If $w \neq u \oplus 1$, by Theorem 6, we have $\mathbf{Tr}(M) = 0$, that is, $M_{0,0} + M_{1,1} = 0$. If $u = w = 1$, Eqn. (15) follows directly from Corollary 6. If $u \neq w$, by Theorem 4 we have

$$|\mathbf{cor}(u; 1, w)| \leq \frac{2^n + 1}{(2^n - 1)^2} < \frac{3}{2^n - 1}.$$

If $w = u \oplus 1$, by Corollary 5 and Theorem 4, we have

$$|\mathbf{cor}(u; 1, w)| \leq \frac{2^{2n} \cdot 2^{-n} + 2^n + 1}{(2^n - 1)^2} = \frac{2 \cdot 2^n + 1}{(2^n - 1)^2} < \frac{3}{2^n - 1}. \qquad \blacksquare$$

5   The Limit of $\mathbf{cor}(1; 1^k)$ for the Addition in $\mathbb{F}_{2^n-1}$ when $n \to \infty$

In this section we will discuss the limit of correlations $\mathbf{cor}(u; \underbrace{u, \cdots, u}_{k})$ when n goes to infinity, where $w_H(u) = 1$. By Theorem 3, it is known that $\mathbf{cor}(u; \underbrace{u, \cdots, u}_{k}) = \mathbf{cor}(1; \underbrace{1, \cdots, 1}_{k})$. So below we only consider $\mathbf{cor}(1; \underbrace{1, \cdots, 1}_{k})$. For simplicity, we denote it by $\mathbf{cor}(1; 1^k)$.

Lemma 6. *For any integers $n \geq 2$ and $k \geq 2$, we have*

$$\sum_{u \in Z_{2^n}} |\mathbf{cor}(u; 1^k)| < (n+3)^{k-1}.$$

Proof. Note that $|J_x| \leq n$ for all $x \in Z_{2^n}$. When $k = 2$, by Theorem 7, we have

$$\sum_{u \in Z_{2^n}} |\mathbf{cor}(u; 1, 1)| = \sum_{u \in J_1} |\mathbf{cor}(u; 1, 1)| + \sum_{u \notin J_1} |\mathbf{cor}(u; 1, 1)|$$

$$\leq \sum_{u \in J_1} 1 + \frac{3}{2^n - 1} \sum_{u \notin J_1} 1 < n + 3.$$

Suppose that when $k = k_0$, we have $\sum_{u \in Z_{2^n}} |\text{cor}(u; 1^{k_0})| < (n+3)^{k_0-1}$. Then

$$\sum_{u \in Z_{2^n}} |\text{cor}(u; 1^{k_0+1})|$$

$$= \frac{2^n - 1}{2^n} \sum_{u \in Z_{2^n}} | \sum_{w \in Z_{2^n}} \text{cor}(w; 1^{k_0}) \text{cor}(u; w, 1)|$$

$$< \sum_{u \in Z_{2^n}} \sum_{w \in Z_{2^n}} |\text{cor}(w; 1^{k_0}) \text{cor}(u; w, 1)|$$

$$= \sum_{u \in Z_{2^n}} (\sum_{w \in J_u} |\text{cor}(w; 1^{k_0}) \text{cor}(u; w, 1)| + \sum_{w \notin J_u} |\text{cor}(w; 1^{k_0}) \text{cor}(u; w, 1)|)$$

$$< \sum_{u \in Z_{2^n}} \sum_{w \in J_u} |\text{cor}(w; 1^{k_0})| + \frac{3}{2^n - 1} \sum_{u \in Z_{2^n}} \sum_{w \notin J_u} |\text{cor}(w; 1^{k_0})|$$

$$< n \cdot (n+3)^{k_0-1} + \frac{3}{2^n - 1} \cdot (2^n - 1) \cdot (n+3)^{k_0-1}$$

$$= (n+3)^{k_0}.$$

By induction the conclusion of the theorem holds. ∎

Lemma 7. *For any integer $t \geq 1$ and $i \geq 2$, we have*

$$\lim_{n \to \infty} \sum_{u_1 \in J_1} \sum_{u_2 \in J_{u_1}} \cdots \sum_{u_{t-1} \in J_{u_{t-2}}} \sum_{u_t \notin J_{u_{t-1}}} \text{cor}(u_t; 1^i) \prod_{j=1}^{t} \text{cor}(u_{j-1}; u_j, 1) = 0,$$

where $u_0 = 1$.

Proof. By Lemma 6 and Theorem 7, we have

$$\left| \sum_{u_1 \in J_1} \sum_{u_2 \in J_{u_1}} \cdots \sum_{u_{t-1} \in J_{u_{t-2}}} \sum_{u_t \notin J_{u_{t-1}}} \text{cor}(u_t; 1^i) \prod_{j=1}^{t} \text{cor}(u_{j-1}; u_j, 1) \right|$$

$$< \frac{3}{2^n - 1} \sum_{u_1 \in J_1} \sum_{u_2 \in J_{u_1}} \cdots \sum_{u_{t-1} \in J_{u_{t-2}}} \sum_{u_t \notin J_{u_{t-1}}} |\text{cor}(u_t; 1^i)| \prod_{j=1}^{t-1} \text{cor}(u_{j-1}; u_j, 1)|$$

$$\leq \frac{3}{2^n - 1} \sum_{u_1 \in J_1} \sum_{u_2 \in J_{u_1}} \cdots \sum_{u_{t-1} \in J_{u_{t-2}}} \sum_{u_t \notin J_{u_{t-1}}} |\text{cor}(u_t; 1^i)|$$

$$< \frac{3}{2^n - 1}(n+3)^{i-1} \sum_{u_1 \in J_1} \sum_{u_2 \in J_{u_1}} \cdots \sum_{u_{t-1} \in J_{u_{t-2}}} 1$$

$$< \frac{3}{2^n - 1}(n+3)^{i-1} n^{t-1}.$$

Since $\frac{3}{2^n-1}(n+3)^{i-1}n^{t-1}$ approaches 0 when n approaches infinity, thus the conclusion holds. ∎

Lemma 8. *For any integer $k \geq 3$, if $\lim\limits_{n \to \infty} \mathbf{cor}(1; 1^k)$ exists, then*

$$\lim_{n \to \infty} \mathbf{cor}(1; 1^k) = \lim_{n \to \infty} \sum_{u_1 \in J_1} \sum_{u_2 \in J_{u_1}} \cdots \sum_{u_{k-2} \in J_{u_{k-3}}} \prod_{j=1}^{k-1} \mathbf{cor}(u_{j-1}; u_j, 1),$$

where $u_0 = u_{k-1} = 1$.

Proof.

$$\lim_{n \to \infty} \mathbf{cor}(1; 1^k)$$

$$= \lim_{n \to \infty} \sum_{u_1 \in Z_{2^n}} \mathbf{cor}(u_1; 1^{k-1})\mathbf{cor}(1; u_1, 1)$$

$$= \lim_{n \to \infty} \left(\sum_{u_1 \in J_1} + \sum_{u_1 \notin J_1} \right) \mathbf{cor}(u_1; 1^{k-1})\mathbf{cor}(1; u_1, 1)$$

$$= \lim_{n \to \infty} \sum_{u_1 \in J_1} \mathbf{cor}(u_1; 1^{k-1})\mathbf{cor}(1; u_1, 1) \quad \text{(by Lemma 7)}$$

$$= \lim_{n \to \infty} \sum_{u_1 \in J_1} \sum_{u_2 \in Z_{2^n}} \mathbf{cor}(u_2; 1^{k-2})\mathbf{cor}(u_1; u_2, 1)\mathbf{cor}(1; u_1, 1)$$

$$= \lim_{n \to \infty} \sum_{u_1 \in J_1} \left(\sum_{u_2 \in J_{u_1}} + \sum_{u_2 \notin J_{u_1}} \right) \mathbf{cor}(u_2; 1^{k-2})\mathbf{cor}(u_1; u_2, 1)\mathbf{cor}(1; u_1, 1)$$

$$= \lim_{n \to \infty} \sum_{u_1 \in J_1} \sum_{u_2 \in J_{u_1}} \mathbf{cor}(u_2; 1^{k-2})\mathbf{cor}(u_1; u_2, 1)\mathbf{cor}(1; u_1, 1) \quad \text{(by Lemma 7)}$$

$$= \lim_{n \to \infty} \sum_{u_1 \in J_1} \sum_{u_2 \in J_{u_1}} \cdots \sum_{u_{k-2} \in J_{u_{k-3}}} \prod_{j=1}^{k-1} \mathbf{cor}(u_{j-1}; u_j, 1). \quad \blacksquare$$

Theorem 8. *For any integer $k \geq 3$, if $\lim\limits_{n \to \infty} \mathbf{cor}(1; 1^k)$ exists, then*

$$\lim_{n \to \infty} \mathbf{cor}(1; 1^k) = \lim_{n \to \infty} \sum_{u_1 \in J_1} \sum_{u_2 \in J_{u_1}} \cdots \sum_{u_{k-1} \in J_{u_{k-2}}} \prod_{j=1}^{k-1} \mathbf{Tr}(M_n(u_{j-1}, u_j, 1)),$$

where $u_0 = u_{k-1} = 1$.

Proof. Recall that $A_1 = A_2$ and $A_5 = A_6$, then it is easily proved that for arbitrary two integers $u, w \in Z_{2^n}$, the matrices sequence derived from $(u, 1, w)$ is the same with the matrices sequence derived from $(u, w, 1)$. So we have $M_n(u, 1, w) = M_n(u, w, 1)$. By Theorem 4, Theorem 6 and Corollary 5, we have

$$\mathbf{cor}(u; w, 1) = \mathbf{Tr}(M_n(u, w, 1)) + \frac{\delta(u, w, 1)}{2^n - 1},$$

where

$$|\delta(u, w, 1)| = \left| \frac{(2^{n+1} - 1)\mathbf{Tr}(M_n(u, w, 1)) + 2^n \cdot c + 1}{2^n - 1} \right|$$

$$\leq \frac{(2^{n+1} - 1)|\mathbf{Tr}(M_n(u, 1, w))| + 2^n \cdot |c| + 1}{2^n - 1}$$

$$\leq \frac{(2^{n+1} - 1) + 2^n \cdot 3 + 1}{2^n - 1}$$

$$< 7.$$

Then

$$\sum_{u_1 \in J_1} \sum_{u_2 \in J_{u_1}} \cdots \sum_{u_{k-2} \in J_{u_{k-3}}} \prod_{j=1}^{k-1} \mathbf{cor}(u_{j-1}; u_j, 1)$$

$$= \sum_{u_1 \in J_1} \sum_{u_2 \in J_{u_1}} \cdots \sum_{u_{k-2} \in J_{u_{k-3}}} \left(\mathbf{Tr}(M_n(u_0, u_1, 1)) + \frac{\delta(u_0, u_1, 1)}{p} \right) \prod_{j=2}^{k-1} \mathbf{cor}(u_{j-1}; u_j, 1)$$

$$= A + B,$$

where

$$A = \sum_{u_1 \in J_1} \sum_{u_2 \in J_{u_1}} \cdots \sum_{u_{k-2} \in J_{u_{k-3}}} \mathbf{Tr}(M_n(u_0, u_1, 1)) \prod_{j=2}^{k-1} \mathbf{cor}(u_{j-1}; u_j, 1)$$

and

$$B = \sum_{u_1 \in J_1} \sum_{u_2 \in J_{u_1}} \cdots \sum_{u_{k-2} \in J_{u_{k-3}}} \frac{\delta(u_0, u_1, 1)}{2^n - 1} \prod_{j=2}^{k-1} \mathbf{cor}(u_{j-1}; u_j, 1).$$

Since

$$|B| \leq \frac{7}{2^n - 1} \sum_{u_1 \in J_1} \sum_{u_2 \in J_{u_1}} \cdots \sum_{u_{k-2} \in J_{u_{k-3}}} |\prod_{j=2}^{k-2} \mathbf{cor}(u_{j-1}; u_j, 1)|$$

$$\leq \frac{7}{2^n - 1} \sum_{u_1 \in J_1} \sum_{u_2 \in J_{u_1}} \cdots \sum_{u_{k-2} \in J_{u_{k-3}}} 1$$

$$\leq \frac{7}{2^n - 1} n^k \xrightarrow{n \to \infty} 0,$$

thus we have

$$\lim_{n \to \infty} \mathbf{cor}(1; 1^k) = \lim_{n \to \infty} A.$$

Repeat the above procedure, and we always strip $\frac{\delta(u_{j-1}, u_j, 1)}{2^n - 1}$ from $\mathbf{cor}(u_{j-1}; u_j, 1)$, $j = 2, 3, \cdots, k - 1$. Then finally we can get the desired conclusion. ■

Corollary 7. $\lim\limits_{n\to\infty}\mathbf{cor}(1;1^2) = 0$ *and* $\lim\limits_{n\to\infty}\mathbf{cor}(1;1^3) = -\frac{1}{3}$.

Proof. Since $M_n(1,1,1) = A_0^{n-1}A_7$ is of type-II, thus $\mathbf{Tr}(M_n(1,1,1)) = 0$, furthermore we have $\lim\limits_{n\to\infty}\mathbf{cor}(1;1,1) = 0$. By Theorem 8 and Corollary 6, we have

$$\lim_{n\to\infty}\mathbf{cor}(1;1^3)$$
$$= \lim_{n\to\infty}\sum_{u\in J_1}\mathbf{Tr}(M_n(u,1,1))\mathbf{Tr}(M_n(1,u,1))$$
$$= \lim_{n\to\infty}\sum_{i=1}^{n-1}\mathbf{Tr}(M_n(2^i+1,1,1))\mathbf{Tr}(M_n(1,2^i+1,1))$$
$$= \lim_{n\to\infty}\sum_{i=1}^{n-1}(-2^{-(n-i)})\cdot 2^{-(n-i)}$$
$$= -\lim_{n\to\infty}\sum_{i=1}^{n-1}4^{-(n-i)} = -\frac{1}{3}. \qquad\blacksquare$$

In order to deal with the general case $\lim\limits_{n\to\infty}\mathbf{cor}(1;1^k)$, for a given integer $k \geq 3$, we define

$$U_k = \{u_0u_1u_2\cdots u_{k-2}u_{k-1}|u_j \in J_{u_{j-1}}, 1 \leq j \leq k-1, u_{k-1} = u_0 = 1\}. \quad (16)$$

Then Theorem 8 can also be written as:

Theorem 9. *For given integer $k \geq 3$, if $\lim\limits_{n\to\infty}\mathbf{cor}(1;1^k)$ exists, then*

$$\lim_{n\to\infty}\mathbf{cor}(1;1^k) = \lim_{n\to\infty}\sum_{u_0u_1\cdots u_{k-1}\in U_k}\prod_{j=1}^{k-1}\mathbf{Tr}(M_n(u_{j-1},u_j,1)).$$

For any string $u_0u_1u_2\cdots u_{k-2}u_{k-1} \in U_k$, by the definition of $J_{u_{j-1}}$, we have $u_j > 0$ for $0 \leq j \leq k-1$, and there is only one bit in u_j different from u_{j-1}, that is, $w_H(u_{j-1}) - w_H(u_j) = \pm 1$. Note that $w_H(u_0) = 1$ is odd, thus $w_H(u_2), w_H(u_4), \cdots$ are all odd and $w_H(u_1), w_H(u_3), \cdots$ are all even.

When k is even, it is known that $w_H(u_{k-1})$ is even, which contradicts $u_{k-1} = 1$. It follows that $U_k = \emptyset$. Hence we have the following conclusion.

Theorem 10. *For any even positive integer k, we have $\lim\limits_{n\to\infty}\mathbf{cor}(1;1^k) = 0$.*

When k is odd, set $u_{2j} = 1$ and $u_{2j+1} = 2^{n-1}+1$ for $0 \leq j \leq \frac{k-1}{2}$. Then $u_0\cdots u_{k-2}u_{k-1} \in U_k$. It shows that $U_k \neq \emptyset$. For all odd integer k, we define

$$I_k = \{i_1i_2\cdots i_{k-1}|2^{i_j} = u_j \oplus u_{j-1}, u_0\cdots u_{k-2}u_{k-1} \in U_k\},$$
$$I_{k,d} = \{i_1i_2\cdots i_{k-1}|d = \sum_{j=1}^{k-1}i_j, i_1i_2\cdots i_{k-1} \in I_k\},$$

and denote $N_{k,d} = |I_{k,d}|$.

Theorem 11. *For any odd integer $k \geq 3$, we have*

$$\sum_{u_0 u_1 \cdots u_{k-1} \in U_k} \prod_{j=1}^{k-1} \mathbf{Tr}(M_n(u_{j-1}, u_j, 1)) = (-1)^{\frac{k-1}{2}} \cdot 2^{-(k-1)n} \sum_{d=k-1}^{(k-1)(n-1)} N_{k,d} \cdot 2^d.$$

Proof. For any $u_0 \cdots u_{k-1} \in U_k$, by Corollary 5, when $w_H(u_j) - w_H(u_{j-1}) = 1$, the sign of $Tr(M_n(u_{j-1}, u_j, 1))$ is positive, and when $w_H(u_j) - w_H(u_{j-1}) = -1$, the sign of $Tr(M_n(u_{j-1}, u_j, 1))$ is negative. So the sign of $\prod_{j=1}^{k-1} Tr(M_n(u_{j-1}, u_j, 1))$ is the same as that of $\prod_{j=1}^{k-1}(w_H(u_j) - w_H(u_{j-1}))$. Note that $\sum_{j=1}^{k-1}(w_H(u_j) - w_H(u_{j-1})) = 0$. It follows that the number of j such that $w_H(u_j) - w_H(u_{j-1}) = 1$ is equal to that of j such that $w_H(u_j) - w_H(u_{j-1}) = -1$. Thus the sign of $\prod_{j=1}^{k-1} Tr(M_n(u_{j-1}, u_j, 1))$ equals $(-1)^{\frac{k-1}{2}}$. Then we have

$$\sum_{u_0 \cdots u_{k-1} \in U_k} \prod_{j=1}^{k-1} Tr(M_n(u_{j-1}, u_j, 1))$$

$$= (-1)^{\frac{k-1}{2}} \sum_{i_1 i_2 \cdots i_{k-1} \in I_k} \prod_{j=1}^{k-1} 2^{-(n-i_j)}$$

$$= (-1)^{\frac{k-1}{2}} \cdot 2^{-(k-1)n} \sum_{d=k-1}^{(k-1)(n-1)} N_{k,d} \cdot 2^d. \qquad \blacksquare$$

Theorem 12. *For any odd integer $k \geq 3$, if $\lim_{n \to \infty} \mathbf{cor}(1; 1^k)$ exists, then*

1. $\lim_{n \to \infty} \mathbf{cor}(1; 1^k) \geq \frac{1}{3} 2^{-(k-3)}$, *if $k \equiv 1 \mod 4$;*
2. $\lim_{n \to \infty} \mathbf{cor}(1; 1^k) \leq -\frac{1}{3} 2^{-(k-3)}$, *if $k \equiv 3 \mod 4$.*

Proof. For any given $u_0 \cdots u_{k-1} \in U_k$, denote $2^{i_j} = u_j \oplus u_{j-1}$, $1 \leq j \leq k-1$. Then $i_1 i_2 \cdots i_{k-1} \in I_k$. Note that $2^{i_1} \oplus 2^{i_2} \oplus \cdots \oplus 2^{i_{k-1}} = \bigoplus_{j=1}^{k-1}(u_j \oplus u_{j-1}) = 0$, which means that $i_1, i_2, \cdots, i_{k-1}$ can be divided into two identical sets. So $d = \sum_{j=1}^{k-1} i_j$ is always even. Note that $1 \leq i_j \leq n-1$, thus $k-1 \leq d \leq (k-1)(d-1)$. In addition, by the definition of I_k and $I_{k,d}$, for any even integer $k-1 \leq d \leq (n-1)(k-1)$, it is easy to verify that there exist $i_1, i_2, \cdots, i_{k-1}$ such that $i_1 i_2 \cdots i_{k-1} \in I_{k,d}$, that is, $N_{k,d} \geq 1$. For example, when $d = k-1$, set $i_j = 1$ for $1 \leq j \leq k-1$, then $i_1 \cdots i_{k-1} \in I_{k,k-1}$; when $d = (k-1)(n-1)$, set $i_j = n-1$ for $1 \leq j \leq k-1$, then $i_1 \cdots i_{k-1} \in I_{k,(k-1)(n-1)}$. By Theorem 11, we have

$$|\lim_{n \to \infty} \mathbf{cor}(1; 1^k)|$$

$$= \lim_{n \to \infty} 2^{-(k-1)n} \sum_{d=(k-1)/2}^{(k-1)(n-1)/2} N_{k,2d} 2^{2d}$$

$$\geq \lim_{n\to\infty} 2^{-(k-1)n} \sum_{d=(k-1)/2}^{(k-1)(n-1)/2} 2^{2d}$$

$$= \lim_{n\to\infty} 2^{-(k-1)n} \frac{2^{(k-1)(n-1)+2} - 2^{k-1}}{2^2 - 1}$$

$$= \frac{1}{3} 2^{-(k-3)}. \qquad \blacksquare$$

6   Conclusion

In this paper we discussed some properties of linear approximations of the addition modulo $2^n - 1$. We presented an explicit expression for the case when two inputs are involved, and an iterative expression for the case when more than two inputs are involved. For a class of special linear approximations with all masks being equal to 1, we further discussed the limit of their correlations when n approaches infinity. More precisely, let k be the number of inputs of the addition modulo $2^n - 1$, we show that when k is even, the limit is equal to zero, and when k is odd, the limit is bounded by a constant depending on k.

Finally when both n and k approach infinity, we have a conjecture on $\mathbf{cor}(1; 1^k)$.

Conjecture 1. $\lim_{k\to\infty}\lim_{n\to\infty} \mathbf{cor}(1; 1^k) = 0.$

Acknowledgement

The authors are grateful to Marion Videau and the anonymous reviewers for their constructive comments. We also would like to thank professor Dengguo Feng and professor Dongdai Lin for their instructions.

References

1. Matsui, M.: Linear Cryptanalysis Method for DES Cipher. In: Helleseth, T. (ed.) EUROCRYPT 1993. LNCS, vol. 765, pp. 386–397. Springer, Heidelberg (1994)
2. Nyberg, K.: Linear Approximation of Block Ciphers. In: De Santis, A. (ed.) EUROCRYPT 1994. LNCS, vol. 950, pp. 439–444. Springer, Heidelberg (1995)
3. Coppersmith, D., Halevi, S., Jutla, C.: Cryptanalysis of Stream Ciphers with Linear Masking. In: Yung, M. (ed.) CRYPTO 2002. LNCS, vol. 2442, pp. 515–532. Springer, Heidelberg (2002)
4. Watanabe, D., Biryukov, A., Cannière, C.D.: A Distiguishing Attack of SNOW 2.0 with Linear Masking Method. In: Matsui, M., Zuccherato, R.J. (eds.) SAC 2003. LNCS, vol. 3006, pp. 222–233. Springer, Heidelberg (2004)
5. Lai, X.: On the Design and Security of Block Ciphers. ETH Series in Information Processing. Hartung-Gorre Verlag, Konstanz (1992)
6. GOST 28147-89. Cryptographic Protection for Data Processing Systems, Government Committee of the USSR for Standards (1989)

7. Rivest, R.: The MD5 Message-Digest Algorithm. RFC 1321, MIT and RSA Data Security, Inc. (April 1992)
8. Ekdahl, P., Johansson, T.: A New Version of the Stream Cipher SNOW. In: Nyberg, K., Heys, H.M. (eds.) SAC 2002. LNCS, vol. 2595, pp. 47–61. Springer, Heidelberg (2003)
9. Nyberg, K., Wallén, J.: Improved Linear Distinguishers for SNOW 2.0. In: Robshaw, M.J.B. (ed.) FSE 2006. LNCS, vol. 4047, pp. 144–162. Springer, Heidelberg (2006)
10. Wallén, J.: Linear Approximations of Addition Modulo 2^n. In: Johansson, T. (ed.) FSE 2003. LNCS, vol. 2887, pp. 261–273. Springer, Heidelberg (2003)
11. Berson, T.A.: Differential Cryptanalysis Mod 2^{32} with Applications to MD5. In: Rueppel, R.A. (ed.) EUROCRYPT 1992. LNCS, vol. 658, pp. 71–80. Springer, Heidelberg (1993)
12. Lipmaa, H., Moriai, S.: Efficient Algorithms for Computing Differential Properties of Addition. In: Matsui, M. (ed.) FSE 2001. LNCS, vol. 2355, pp. 336–350. Springer, Heidelberg (2002)
13. Courtois, N.T., Debraize, B.: Algebraic Description and Simultaneous Linear Approximations of Addition in Snow 2.0. In: Chen, L., Ryan, M.D., Wang, G. (eds.) ICICS 2008. LNCS, vol. 5308, pp. 328–344. Springer, Heidelberg (2008)
14. Maximov, A., Johansson, T.: Fast Computation of Large Distributions and Its Cryptographic Applications. In: Roy, B. (ed.) ASIACRYPT 2005. LNCS, vol. 3788, pp. 313–332. Springer, Heidelberg (2005)
15. Nyberg, K.: Correlation Theorems in Cryptanalysis. Discrete Applied Mathematics 111(1-2), 177–188 (2001)
16. Tu, Z., Deng, Y.: A Conjecture on Binary String and Its Applications on Constructing Boolean Functions of Optimal Algebraic Immunity. Cryptology ePrint Archive, Report 2009/272 (2009), http://eprint.iacr.org/2009/272
17. Tu, Z., Deng, Y.: A Class of 1-Resilient Function with High Nonlinearity and Algebraic Immunity. Cryptology ePrint Archive, Report 2010/179 (2010), http://eprint.iacr.org/2010/179
18. Zimmermann, R.: Efficient VLSI Implementation of Modulo $2^n \pm 1$ Addition and Multiplication. In: Proceedings of 14th IEEE Symposium on Computer Arithmetic, pp. 158–167 (1999)
19. Flori, J.P., Randriam, H., Cohen, G., Mesnager, S.: On a Conjecture about Binary Strings Distribution. In: Carlet, C., Pott, A. (eds.) SETA 2010. LNCS, vol. 6338, pp. 346–358. Springer, Heidelberg (2010)
20. Specification of the 3GPP Confidentiality and Integrity Algorithms 128-EEA3 & 128-EIA3, Document 2: ZUC Specification, http://www.gsmworld.com/our-work/programmes-and-initiatives/fraud-and-security/gsm_security_algorithms.htm
21. GSM Algorithms, http://www.gsmworld.com/our-work/programmes-and-initiatives/fraud-and-security/gsm_security_algorithms.htm

Meet-in-the-Middle Preimage Attacks on AES Hashing Modes and an Application to Whirlpool

Yu Sasaki

NTT Information Sharing Platform Laboratories, NTT Corporation
3-9-11 Midoricho, Musashino-shi, Tokyo 180-8585 Japan
sasaki.yu@lab.ntt.co.jp

Abstract. We study the security of AES in the open-key setting by showing an analysis on hash function modes instantiating AES including Davies-Meyer, Matyas-Meyer-Oseas, and Miyaguchi-Preneel modes. In particular, we propose preimage attacks on these constructions, while most of previous work focused their attention on collision attacks or distinguishers using non-ideal differential properties. This research is based on the motivation that we should evaluate classical and important security notions for hash functions and avoid complicated attack models that seem to have little relevance in practice. We apply a recently developed meet-in-the-middle preimage approach. As a result, we obtain a preimage attack on 7 rounds of Davies-Meyer AES and a second preimage attack on 7 rounds of Matyas-Meyer-Oseas and Miyaguchi-Preneel AES. Considering that the previous best collision attack only can work up to 6 rounds, the number of attacked rounds reaches the best in terms of the classical security notions. In our attacks, the key is regarded as a known constant, and the attacks thus can work for any key length in common.

Keywords: AES, hash function, Davies-Meyer, Matyas-Meyer-Oseas, Miyaguchi-Preneel, PGV, preimage, meet-in-the-middle, Whirlpool.

1 Introduction

Block ciphers are taking important roles in various aspects of our life. Currently, one of the most widely used block-ciphers all over the world is AES [12,34].

Since 2009, great progress in the cryptanalysis on AES has been made. Related-key attacks against full-round AES-256 [6,7], full-round AES-192 [6], 7-round AES-128 [9], and 10-round AES-256 with a practical complexity [5] have been proposed. Regarding AES-128, besides the above related-key boomerang attack [9] several non-marginal single-key attacks have been proposed; an impossible differential attack [25] and a single-key attack [15] based on a collision attack [16]. In any attack, the maximum number of attacked rounds is 7.

On the other hand, block ciphers are sometimes used as hash functions through mode-of-operations. For example, if one needs both a block-cipher and a hash function in a resource-restricted environment such as RFID Tag, only a block-cipher is implemented and a hash function is built using it. Besides, many

A. Joux (Ed.): FSE 2011, LNCS 6733, pp. 378–396, 2011.

Tag-based applications, such as authentication or anonymity/privacy, do not need the collision resistance [10]. Hence, building a 128-bit hash function with AES is a possible candidate. In fact, [10] proposed 80-bit and 64-bit hash functions using block-cipher PRESENT. Another concern is that many hash functions, even in the SHA-3 competition [35], are designed based on block-ciphers. Hence, block-ciphers' security in hashing modes is an interesting research object.

The known-key attack proposed by Knudsen and Rijmen [21] is the framework for this context. In this model, a secret key is randomly chosen and given to attackers. Then, attackers aim to efficiently detect a certain property of a random instance of the block cipher, where the same property cannot be observed for a random permutation with the same complexity. The attack can be extended to the chosen-key model. e.g. [7]. In the known-key model, the key size is irrelevant to the attack. In other words, all key sizes are simultaneously attacked. While, in the chosen-key model, the attack depends on the key-schedule algorithm. Hence, different strategies is necessary for different key sizes.

The first known-key attack was presented by Knudsen and Rijmen [21], which found a non-ideal property of 7-round AES. Then, Mendel et al. presented the known-key attack on 7-round AES [26] based on the rebound attack proposed by Mendel et al. [27]. Finally, Gilbert and Peyrin [17] and Lamberger et al. [22] independently applied Super-Sbox analysis to the rebound attack. Gilbert and Peyrin [17] showed that 8-round AES was not ideal in the known-key setting. Regarding the chosen-key attack, Biryukov et al. [7] presented a chosen-key distinguisher on full-round AES-256, which is converted to a q-pseudo-collision attack on AES-256 based compression functions. Biryukov and Nikolić also discovered a chosen-key distinguisher on 8-round AES-128 [8].

Although the above results led to significant progress for theoretical cryptanalysis in the secret-, known-, and chosen-key settings, one major drawback is the use of complicated attack models, which are sometimes too theoretic such as related-subkey attacks on block ciphers and distinguishers on block-cipher based compression functions. From this background, several researchers recently have attempted to analyze AES in a simple attack model. For example, Dunkelman et al. [15] and Wei et al. [36] avoided the related-key model and proposed attacks on 8 round AES-256 or AES-192 in the single-key model.

In this paper, we follow the similar principle as Dunkelman et al. [15] and Wei et al. [36]. That is to say, we analyze the security of hashing modes instantiating AES in terms of the classical security notions of hash functions, which are actually important for their applications. In particular, we study the preimage resistance of hash functions rather than compression functions.

For hash functions, three security notions are classically considered to be important; collision resistance, second-preimage resistance, and preimage resistance. Among these three, the collision resistance of reduced-round AES can be attacked by applying the techniques used in the rebound attack [27]. In fact, Lamberger et al. [23, Section 5.3] describe a collision attack on an AES-based hash function Whirlpool [30] reduced to 5.5 rounds, which is trivially converted to a collision attack on the Matyas-Meyer-Oseas mode [28, Algorithm 9.41]

Table 1. Comparison of attacks. 0.5 round of collision and near-collision attacks on Whirlpool by [23] is omitted.

Attack	Rounds	Key-size	Mode	Comp. Func. (Time, Mem.)	Hash (Time, Mem.)	Ref.
Attacks on AES Hasing modes						
Collision	6	128/192/256	MMO,MP	$(2^{56}, 2^{32})$	$(2^{56}, 2^{32})$	[23]
2nd preimage	6	128/192/256	MMO,MP	$(2^{112}, 2^{16})$	$(2^{112}, 2^{16})$	Ours
2nd preimage	7	128/192/256	MMO,MP	$(2^{120}, 2^{8})$	$(2^{120}, 2^{8})$	Ours
Preimage	6	128/192/256	DM	$(2^{112}, 2^{16})$	$(2^{121}, 2^{16})$	Ours
Preimage	7	128/192/256	DM	$(2^{120}, 2^{8})$	$(2^{125}, 2^{8})$	Ours
Near collision	7	128/192/256	MMO,MP	$(2^{32}, 2^{32})$	$(2^{32}, 2^{32})$	[23]
Distinguisher	8	128/192/256	MMO,MP	$(2^{48}, 2^{32})$	-	[17]
q-multicollision	14	256	DM	$(q \cdot 2^{67}, \text{negl.})$	-	[7]
Attacks on Whirlpool						
Collision	5	-	-	$(2^{120}, 2^{64})$	$(2^{120}, 2^{64})$	[23]
2ne Preimage	5	-	-	$(2^{504}, 2^{8})$	$(2^{504}, 2^{8})$	Ours
Near collision	7	-	-	$(2^{112}, 2^{64})$	$(2^{112}, 2^{64})$	[23]
Collision	7	-	-	$(2^{184}, 2^{8})$	-	[23]
Near collision	9	-	-	$(2^{176}, 2^{8})$	-	[23]
Distinguisher	10	-	-	$(2^{176}, 2^{8})$	-	[23]

instantiating 6-round AES. As far as we know, there is no result that attacks second-preimage resistance or preimage resistance of such an AES usage. Note that the attack by [23] can generate near-collisions on some PGV compression functions with 7-round AES, which might be a valid security notion.

Our Contributions. In this paper, we propose preimage attacks on AES hashing modes including Davies-Meyer (DM) [28, Algorithm 9.42], Matyas-Meyer-Oseas (MMO), and Miyaguchi-Preneel (MP) [28, Algorithm 9.43] modes. As a result, we obtain a preimage attack on 7 rounds of DM-AES with a complexity of 2^{125} 7-round AES computations and the memory of 2^{8} AES state. We also obtain a second preimage attack on 7 rounds of MMO-AES and MP-AES with a complexity of 2^{120} 7-round AES computations and the memory of 2^{8} AES state. Our attack can also generate second preimages of 5-round Whirlpool with a complexity of 2^{504}. The attack results are summarized in Table 1.

We apply a meet-in-the-middle preimage approach developed by Aoki and Sasaki [3]. This approach has successfully been applied to many hash functions e.g. MD5 [32] and Tiger [18]. All of previously analyzed hash functions adopt the DM mode with a relatively weak message schedule, and the weak message schedule is in fact exploited by the attack. However, for AES, the situation is very different because AES has a heavy round function and key schedule. Moreover, it is unclear how to perform preimage attacks against MMO and MP modes.

In our attacks, we fix the value of key-input to a randomly chosen value and search for a plaintext-input that achieves the given hash target. This allows us to attack All PGV modes [29] in the same procedure. We then show that the splice-and-cut technique proposed by [3] and the omission of a MixColumns operation in the last round can be combined well and lead to a significant improvement for the preimage attack. Intuitively, this is because the round function without MixColumns is computed as a middle round. This breaks the AES design principle, where AES two rounds achieve the full diffusion, and leads to an attack improvement. Finally, we optimize several techniques of the meet-in-the-middle preimage attack for AES. Specifically, the initial-structure and matching through MixColumns contribute to increase the number of attacked rounds.

Paper Outline. In Sect. 2, we describe AES. In Sect. 3, we introduce previous work. In Sect. 4, we explain a basic idea of our attack. In Sect. 5, we present a preimage attack on 7-round AES. In Sect. 6, we give observations on our attack and apply it to 5-round Whirlpool. Finally, we conclude this paper in Sect. 7.

2   Specifications

Advanced Encryption Standard (AES) [34,12] is a 128-bit block cipher supporting three different key sizes; 128, 192, 256 bits. AES computes 10, 12, and 14 rounds for AES-128, -192, and -256, respectively.

By using the key schedule function, round keys are generated from the original secret key. We omit its description because our attacks regard round keys as given constant numbers and thus they are irrelevant to our attacks.

When the data is processed, the internal state is represented by a $4 * 4$ byte array. At the first, the original secret key is XORed to the plaintext, and then, a round function consisting of the following four operations is iteratively applied.

- SubBytes(SB): substitute each byte according to an S-box table.
- ShiftRows(SR): apply the j-byte left rotation to each byte at row j.
- MixColumns(MC): multiply each column by an MDS matrix. MDS guarantees that the sum of active bytes in the input and output of the MixColumns operation is at least 5 unless all bytes are non-active. The matrices for the encryption and decryption are shown below. Note that $X[j]$ is the input value and $Y[j]$ is the updated value. Numbers with $_x$ are hexadecimal numbers.

$$\begin{pmatrix} Y[0] \\ Y[1] \\ Y[2] \\ Y[3] \end{pmatrix} = \begin{pmatrix} 2 & 3 & 1 & 1 \\ 1 & 2 & 3 & 1 \\ 1 & 1 & 2 & 3 \\ 3 & 1 & 1 & 2 \end{pmatrix} \begin{pmatrix} X[0] \\ X[1] \\ X[2] \\ X[3] \end{pmatrix}, \quad \begin{pmatrix} Y[0] \\ Y[1] \\ Y[2] \\ Y[3] \end{pmatrix} = \begin{pmatrix} _xe & _xb & _xd & _x9 \\ _x9 & _xe & _xb & _xd \\ _xd & _x9 & _xe & _xb \\ _xb & _xd & _x9 & _xe \end{pmatrix} \begin{pmatrix} X[0] \\ X[1] \\ X[2] \\ X[3] \end{pmatrix} \quad (1)$$

- AddRoundKey(AK): apply bit-wise exclusive-or with a round key.

Note that the MixColumns operation is not computed at the last round.

Byte positions in a state S are denoted by integer numbers $B, B \in \{0, 1, 2, \ldots, 15\}$, as shown in Fig. 1, where the byte $4j + i$ corresponds to the byte in the i-th row and j-th column of S, and is denoted by $S[4 \cdot j + i]$. We often denote several bytes of state S by $S[a, b, \ldots]$, e.g. 4 bytes in the right most column are denoted by $S[12, 13, 14, 15]$.

0	4	8	12
1	5	9	13
2	6	10	14
3	7	11	15

Fig. 1. Byte positions

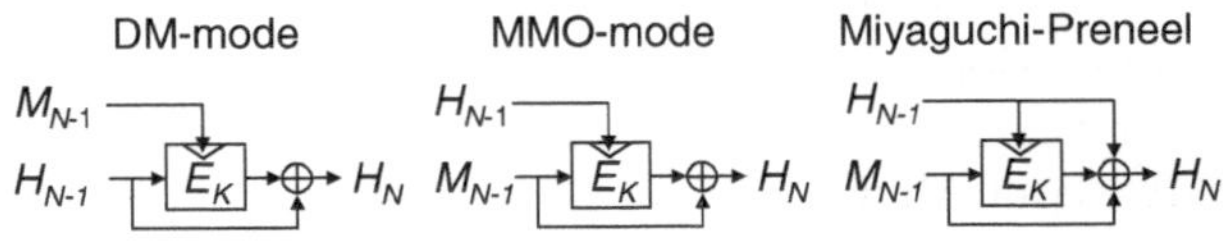

Fig. 2. Illustrations for DM, MMO, and MP modes

Hash Functions Based on Block Ciphers. To build a hash function, we need a domain extension for iteratively applying the compression function. The Merkle-Damgård domain extension is probably mostly used one in practice. It applies the padding to the input message M so that the last block includes the original message length, and splits the padded message to $M_0\|M_1\|\cdots\|M_{L-1}$, where the size of each M_N is the block length. An initial value H_0 is defined, and $H_N = \mathrm{CF}(H_{N-1}, M_{N-1})$ is iteratively applied for $N = 1, 2, \ldots, L$. Finally, H_L is output as a hash value of M. This paper assumes that the Merkle-Damgård domain extension is used as a domain extender.

The PGV modes [29] are mode-of-operations to build a compression function from a block cipher. In fact, many hash functions, e.g. MD5, SHA-2, and several SHA-3 candidates, use the PGV modes. Among PGV modes, the DM, MMO, and MP modes are used in practice. Let us denote a block cipher E with a key K by E_K. The construction of each mode is as follows, which is shown in Fig. 2.

DM mode: $\mathrm{CF}(H_{N-1}, M_{N-1}) = E_{M_{N-1}}(H_{N-1}) \oplus H_{N-1},$
MMO mode: $\mathrm{CF}(H_{N-1}, M_{N-1}) = E_{H_{N-1}}(M_{N-1}) \oplus M_{N-1},$
MP mode: $\mathrm{CF}(H_{N-1}, M_{N-1}) = E_{H_{N-1}}(M_{N-1}) \oplus M_{N-1} \oplus H_{N-1}.$

3 Previous Work

3.1 Meet-in-the-Middle Preimage Attacks

To mount the preimage attack, we apply a meet-in-the-middle (MitM) preimage approach developed by Aoki and Sasaki [3], which is based on the pioneering work by Leurent [24][1]. In this approach, the compression function is divided into two sub-functions so that a portion of bits of the input message only affect one sub-function and another portion of bits of the input message only affect the other sub-function as shown in Fig. 3. This allows attackers to mount the meet-in-the-middle attack. Sub-functions are called *chunks* (stands for chunks of steps or rounds) and bits affecting only one chunk are called *neutral*. In this paper, we call the chunk that computes the round function in the forward direction *forward chunk* and in the backward direction *backward chunk*.

In addition to the basic concept, several techniques have been proposed to extend the attack framework. The *splice-and-cut* technique [3] regards that the first and last steps are consecutive, and thus any step can be the start point or

---

[1] A basic idea of the MitM preimage finding technique can also be seen in [4] and [19].

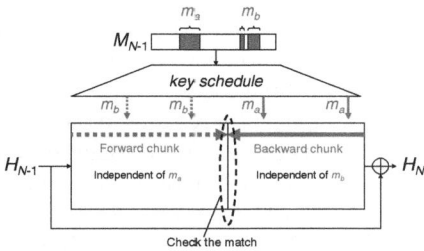

Fig. 3. Basic MitM attack on DM mode

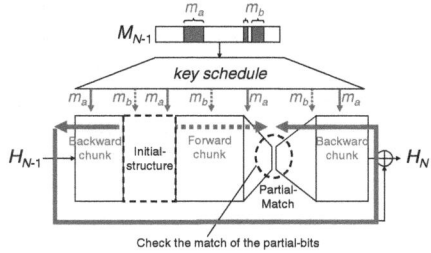

Fig. 4. Advanced techniques for MitM

matching point of the MitM attack. However, as a side-effect, generated items become pseudo-preimages rather than preimages. The *local-collision* technique [31], *initial-structure* technique [32] and *probabilistic initial-structure* technique [18] ignore the order of message words at the start point of the MitM attack. For example in Fig. 4, the order of neutral words m_a and m_b is reversed between the start points of the forward and backward chunks. These techniques enables MitM attacks even in such a situation. Finally, the *partial-matching/-fixing* techniques [3] and *indirect partial-matching* technique [1] match two chunks partially and efficiently. A framework with these techniques is illustrated in Fig. 4.

In n-bit narrow-pipe iterated hash functions, pseudo-preimage attacks with a complexity of 2^m, where $m < n - 2$, can be converted to preimage attacks with a complexity of $2^{\frac{m+n}{2}+1}$ in generic [28, Fact9.99]. Several researchers showed that pseudo-preimage attacks satisfying certain special properties can be converted to preimage attacks more efficiently than the generic approach [11,18,24]. Because our attacks cannot satisfy such properties, we omit their details.

The MitM preimage approach has been applied to many hash functions such as HAVAL [31], MD5 [32], reduced SHA-0/-1 [2], reduced SHA-2 [1], and Tiger [18]. All of previously attacked hash functions adopt the DM mode and their weak key-schedules are exploited by the attack. This strategy cannot work for AES because, in the AES key-schedule, the impact of any change on the secret key or a subkey always propagate to all other subkeys. This indicates that neutral words such as described in Fig. 3 or Fig. 4 do not exist for AES. Moreover, if the message is input as a plaintext in the MMO and MP modes, no input value is available to separate the compression function into two parts.

3.2   Previous Analysis on AES

The security of AES in hash function modes was first evaluated by Knudsen and Rijmen [21]. They showed a non-ideal property of 7-round AES. Lamberger *et al.* showed a collision attack on 5.5-round Whirlpool based on the rebound attack [27], which is trivially converted to a collision attack on 6-round AES. As far as we know, no result is known on the second-preimage or preimage resistance.

Note that current collision attacks can be applied only if the mode-of-operation is MMO or MP, where attackers fix H_{N-1} and search for (M_{N-1}, M'_{N-1}). If the DM mode is used instead, attackers fix M_{N-1} and search for (H_{N-1}, H'_{N-1}). Hence, only pseudo-collisions on the compression function can be generated.

As analysis methods against the AES block cipher, there exist attacks named collision attack [16] and Meet-in-the-Middle attack [13] (and their extension [15]). These attacks are not related to attacks on hash functions. These attacks based on an observation that a function from a certain input byte to a certain output byte after 4 rounds can be described by 25-byte parameters. Hence, this collision attack does not find paired messages producing an identical state, or this Meet-in-the-Middle attack does not separate the cipher into two independent sub-functions. The goal of these attacks is recovering the secret key of the AES block cipher. Their applicability to hashing modes is not understood well.

4   Basic Idea of Our Attack and Techniques for Extension

We first explain a basic idea of our attack by using 4-round AES as an example (Sect. 4.1). We fix the block-cipher's key to a constant. This approach is different from previous work in Sect. 3.1 which utilize the independence among subkeys. In this attack, for simplicity, we only apply the splice-and-cut technique. We then explain several techniques to extend the number of attacked rounds (Sect. 4.2).

4.1   Basic Attack for 4-Round AES

Goal of the Attack. We fix the key-input when we perform the MitM attack, and the goal is to find the plaintext-input that provides the given target. This approach is irrelevant to the mode-of-operation used. That is, in the DM-mode, we fix a message M_{N-1} to some constant and try to find a chaining variable H_{N-1} that produces the given target H_N. Similarly, in the MMO- and MP-modes, we fix a chaining variable H_{N-1} and search for a message M_{N-1}. Generated pseudo-preimages are later converted to preimages with a technique in Sect. 3.1.

Chunk Separation. We separate 4-round operations into two chunks as shown in Fig. 5. The start point of each chunk is state #9. We choose #9[0] as a neutral byte for the forward chunk and #9[12] for the backward chunk. We fix the other bytes, i.e. #9[1, 2, ..., 11, 13, 14, 15], to randomly chosen values. The backward chunk covers the computation from state #9 to #5 and the forward chunk covers from state #9 to #16 and #0 to #5. Results from two chunks will match at #5.

Forward Computation. The forward computation starts from #9. Because #9[12] is a neutral byte for the backward chunk, we regard #9[12] to be unknown during the forward computation. Hence, one byte is unknown at #11 and the unknown byte is expanded to 4 bytes at #12. Similarly, by simply tracing the computation, we obtain 8-byte information at #5 (#5[0,1,2,3,8,9,10,11]).

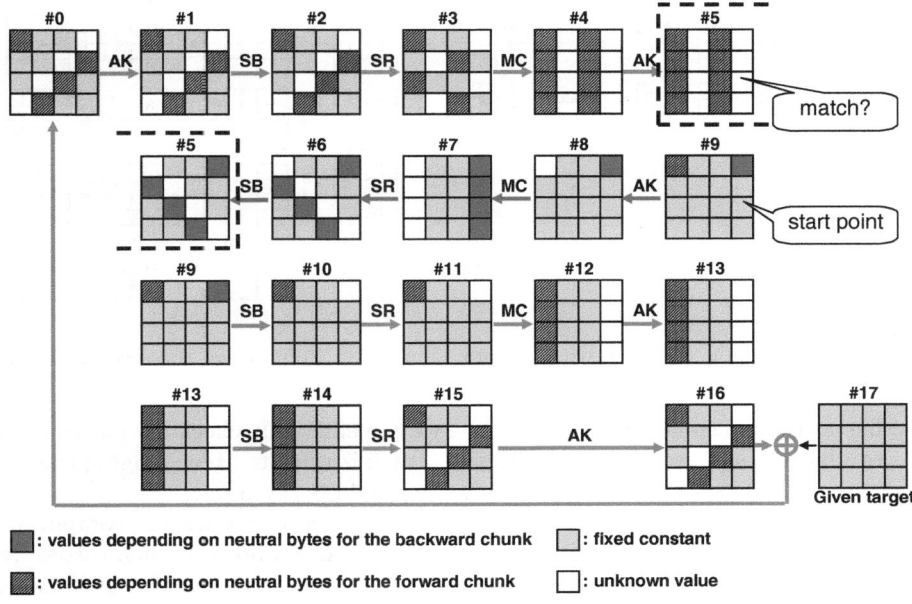

Fig. 5. Basic attack idea: chunk separation for 4-round AES

An important observation is that the omission of the MixColumns operation in the last round extends the number of rounds that can be computed independently. The diffusion of AES is designed so that the full diffusion can be achieved after the 2-round operation. In fact, if MixColumns exists between #15 and #16, all bytes become unknown after this operation and it limits the attack efficiency strongly. However, the omission of MixColumns yields 12 known bytes at #16, and to make things worse, the positions of unknown bytes will overlap by the next ShiftRows operation, and thus attackers can compute MixColumns for another round. As a conclusion, we can summarize this property as follows;

AES 2-rounds achieve the full diffusion, however, if MixColumns *is omitted in the second round, 4 rounds are needed to achieve the full diffusion.*

This property is illustrated in Fig. 6. This situation does not seem to occur for the AES block cipher. However, in hash function modes, the splice-and-cut can exploit it by starting the forward chunk from the second last round.

Overall Attack Procedure. Because the backward computation is similar and straight-forward, we omit its explanation. Overall, if we fix 14 bytes at #9 as shown in Fig. 5, we can compute the forward chunk for 2^8 values of #9[0] and obtain 8-byte information at #5. The obtained results are stored in a table and sorted. Similarly, we can compute the backward chunk for 2^8 values of #9[12] and obtain 12-byte information at #5. Because 2 bytes (#5[1,11]) are overlapped

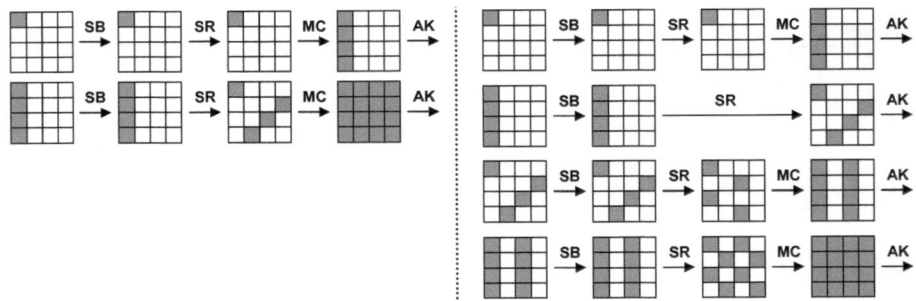

Fig. 6. Slow diffusion with the omission of MixColumns in the second round

between the results from both chunks, we can efficiently check the match of those results. If the match is not found, the attack is repeated by changing the values of fixed 14 bytes at #9 or the values for the AES key-input.

Each chunk can be built using 2^8 possible neutral values with 2^8 computations. The results are stored in a table of the size 2^8 AES state and then sorted. After that, 2^{16} pairs are tested in the 2-byte match and 1 pair will succeed. Hence, if we repeat the attack 2^{112} times, we will find a pair that also matches other 14 bytes. The final complexity for generating pseudo-preimages is $2^8 \cdot 2^{112} = 2^{120}$. This is converted to a preimage attack on the hash function with a complexity of $2^{\frac{120+128}{2}+1} = 2^{125}$ using a generic conversion in Sect. 3.1.

Note that the attack efficiency is not optimized because the purpose of this attack is to demonstrate the basic idea of our attack. Also note that, the impact of the change of #9[12] does not propagate to all bytes at the matching stage. This is because the backward chunk is too short. If the number of attacked rounds is extended as explained in Sect. 5, the impact will propagate to all bytes.

4.2 Techniques for Attacking More Rounds

We show that a technique similar to the initial-structure [32] can extend the number of rounds in each chunk by one round (in total 2 rounds), and by considering the MixColumns operation deeply, we can include one more round during the matching stage. These techniques are directly applied to our 7-round attack that will be explained in Sect. 5. Specifically, the differential path described in Fig. 7 and Fig. 8 are the copy of a part of differential path in Fig. 9.

Initial-Structure. The idea is choosing several bytes as neutral bytes, and determining these values so that several output bytes of the MixColumns or InverseMixColumns operations can be constant values. This minimizes the number of unknown bytes after the first MixColumns operation in each chunk, and thus, the number of attacked rounds is extended by one round in each chunk. The construction of the initial-structure is shown in Fig. 7.

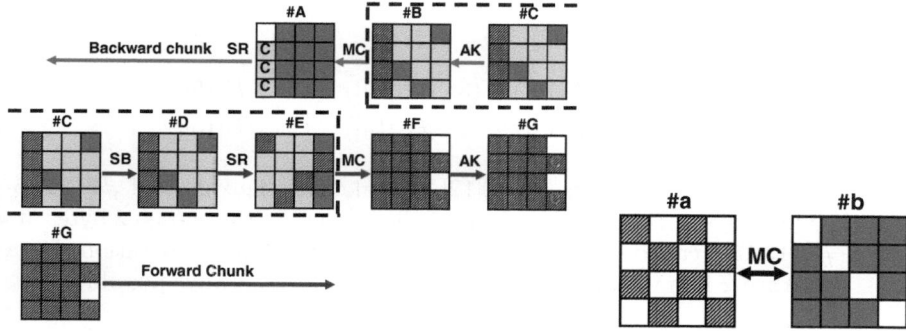

Fig. 7. Initial-structure technique for AES

Fig. 8. Known-byte patterns for matching through MixColumns

The neutral bytes for the forward chunk are 4 bytes #B[0,1,2,3]. We choose 2^8 values of these bytes and use them to compute the forward chunk independently of the backward chunk. These values are determined as follows;

1. Randomly choose constant values for 3 bytes at #A (#A[1,2,3]).
2. For all possible 2^8 values of #B[0], we calculate the values of #B[1,2,3] so that the chosen 3 bytes at #A (#A[1,2,3]) can be achieved through the InverseMixColumns operation. Note that, according to Eq. (1), #A[1,2,3] is computed by Eq.(2). Because there are 3 free variables to control 3 bytes, this is possible by solving a system of equations.

$$\begin{pmatrix} \#A[1] \\ \#A[2] \\ \#A[3] \end{pmatrix} = \begin{pmatrix} x9 & xe & xb & xd \\ xd & x9 & xe & xb \\ xb & xd & x9 & xe \end{pmatrix} \begin{pmatrix} \#B[0] \\ \#B[1] \\ \#B[2] \\ \#B[3] \end{pmatrix} \qquad (2)$$

As a result, for any 2^8 neutral values of #B[0,1,2,3], #A[1,2,3] become constant, and thus, the backward computation from #A can start with 15 known bytes and only 1 unknown byte.

The neutral bytes for the backward chunk are 3 bytes #E[12,14,15]. We choose 2^8 values of these bytes as follows;

1. Randomly choose constant values for 2 bytes, which will be the impact from #E[12,14,15] to the chosen 2 bytes at #F (#F[13,15]). In details, by considering the MixColumns operation in Eq. (1), #F[13,15] are written as follows:

$$\#F[13] = (1 \cdot \#E[12]) \oplus (2 \cdot \#E[13]) \oplus (3 \cdot \#E[14]) \oplus (1 \cdot \#E[15]), \qquad (3)$$

$$\#F[15] = (3 \cdot \#E[12]) \oplus (1 \cdot \#E[13]) \oplus (1 \cdot \#E[14]) \oplus (2 \cdot \#E[15]). \qquad (4)$$

The impacts on #F[13] and #F[15] from #E[12,14,15] mean the following values respectively.

$$(1 \cdot \#E[12]) \oplus (3 \cdot \#E[14]) \oplus (1 \cdot \#E[15]), \tag{5}$$
$$(3 \cdot \#E[12]) \oplus (1 \cdot \#E[14]) \oplus (2 \cdot \#E[15]). \tag{6}$$

2. For all possible 2^8 values of #E[12], we calculate the values of #E[14,15] by solving a system of equations so that the impact on the chosen 2 bytes at #F (#F[13,15]) can be achieved through the MixColumns operation. Because there are 2 free variables to control 2 bytes, this is always possible.

As a result, for any 2^8 neutral values of #E[12,14,15], the impact from these values to #F[13,15] becomes the determined constant. Note that #F[13,15] are also influenced by #E[13], and thus, final values of #F[13,15] are exclusive-or of the determined constant and values depending on #E[13]. Finally, the forward computation from #F can start with 14 known bytes and only 2 unknown bytes.

Match through MixColumns. Assume that many values of the partially known states of the form #a and #b in Fig. 8 are stored in tables. The goal of this match is efficiently finding paired values (#a, #b) that match through the MixColumns operation. Because MixColumns is applied column by column, the match is also tested column by column. We explain the match for the first column as an example. The other columns can be tested in the same procedure.

Let us consider the InverseMixColumns operation from #b to #a. From Eq. (1), #a[0] and #a[2] are expressed as follows;

$$\#a[0] = (\ _xe \cdot \#b[0]\ ) \oplus (\ _xb \cdot \#b[1]\ ) \oplus (\ _xd \cdot \#b[2]\ ) \oplus (\ _x9 \cdot \#b[3]\ ) \tag{7}$$
$$\#a[2] = (\ _xd \cdot \#b[0]\ ) \oplus (\ _x9 \cdot \#b[1]\ ) \oplus (\ _xe \cdot \#b[2]\ ) \oplus (\ _xb \cdot \#b[3]\ ) \tag{8}$$

Considering that #b[1,2,3] are known values, the equations can be transformed by using some constant numbers C_0 and C_1 as follows;

$$\#a[0] \oplus C_0 = \ _xe \cdot \#b[0], \qquad \#a[2] \oplus C_1 = \ _xd \cdot \#b[0]. \tag{9}$$

Whether or not these equations are satisfied can be checked efficiently by using the idea based on the indirect partial-matching [1]. Namely, $_xd \cdot (\#a[0] \oplus C_0) = _xe \cdot (\#a[2] \oplus C_1)$ is obtained from Eq. (9), and then obtain the following equation:

$$_xd \cdot \#a[0] \ \oplus \ _xe \cdot \#a[2] = \ _xd \cdot C_0 \ \oplus \ _xe \cdot C_1. \tag{10}$$

Let us denote $_xd \cdot \#a[0] \ \oplus \ _xe \cdot \#a[2]$ and $_xd \cdot C_0 \ \oplus \ _xe \cdot C_1$ by C_{for} and C_{back}, respectively. By computing C_{for} and C_{back} in the computation for each chunk, we can perform the match by just comparing these values.

Note that AES has 4 columns in a state. Because the number of candidates for the match can be reduced by a factor of 2^{-8} per column, for 4 columns, the number of candidates is reduced by a factor of 2^{-32}.

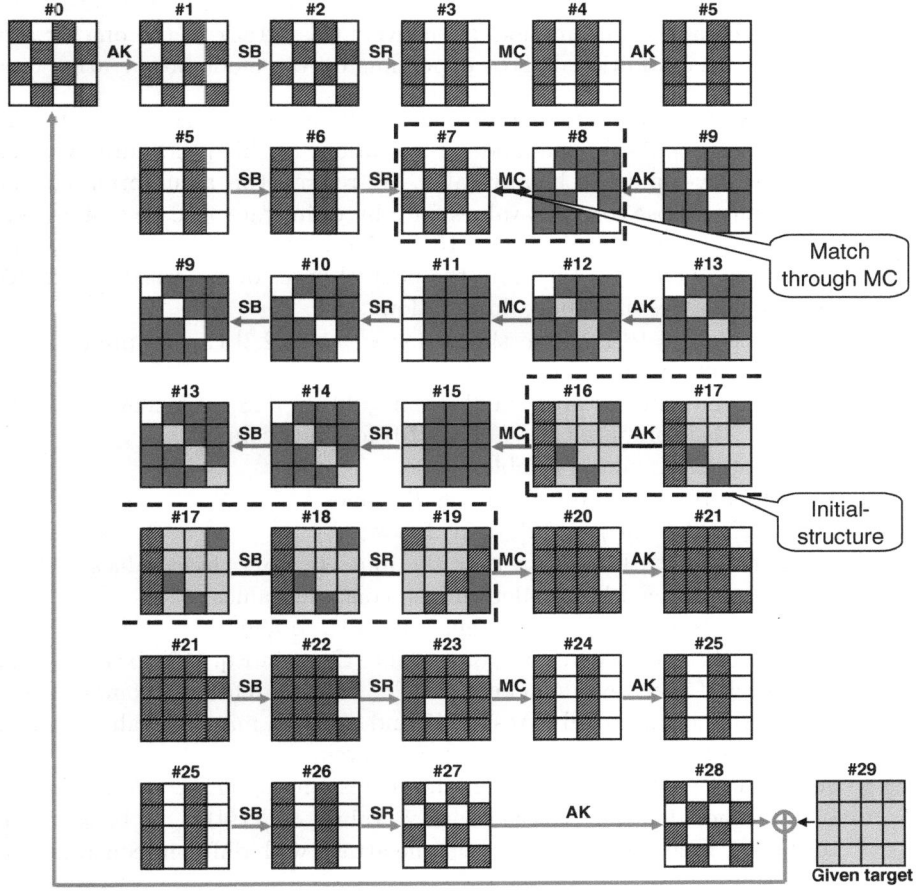

Fig. 9. Chunk separation for 7-round AES

5   Preimage Attack against 7-Round AES

By considering the techniques explained in Sect. 4.2, we can attack up to 7 rounds of AES. The chunk separation for this attack is depicted in Fig. 9.

In this attack, states #16 to #19 are chosen as the initial-structure and we apply the match between states #7 and #8. The neutral bytes for the forward computation are 4 bytes at #16, namely, #16[0,1,2,3]. The neutral bytes for the backward computation are 3 bytes at #19, namely, #19[12,14,15].

To make the initial-structure work, we choose neutral bytes for the forward chunk so that 3 bytes #15[1,2,3] can be pre-determined constant values. Similarly, neutral bytes for the backward chunk are computed so that impacts on 2 bytes #20[13,15] can be pre-determined constant values. The matching procedure is exactly the same as the one explained in Section 4.2. The detailed attack procedure is explained below. Note that the procedure below is a preimage attack on the compression function. To convert this attack to the one for a hash function, we need additional effort depending on the mode-of-operation used.

1. Choose a value for the key-input, and compute all sub-keys. How to choose the value depends on the mode-of-operation. In this procedure, we assume that the key-input can be any value. We remove this assumption later.
2. Randomly choose constant values for 9 bytes in state #16 (#16[4,5,7,8,9,10, 13,14,15]), for 3 bytes in state #15 (#15[1,2,3]), and for impacts on 2 bytes in state #20 (#20[13,15]) from the neutral bytes of the backward chunk.
3. For all 2^8 values of #16[0], do as follows.
   (a) Calculate #16[1,2,3] so that 3 bytes #15[1,2,3] can be pre-determined constant values.
   (b) Compute the forward chunk from #20 to #28, and then, from #0 to #7. Also compute C_{for} in Eq. (10) for each column.
   (c) Store the results in a table T_{for}.
4. Sort the table T_{for} after obtaining 2^8 values.
5. For all 2^8 values of #19[12], do as follows.
   (a) Calculate #19[14,15] so that the impacts from these values to 2 bytes #20[13,15] can be pre-determined constant values.
   (b) Compute the backward chunk from #15 to #8.
   (c) From 12 known bytes of #8, compute C_{back} in Eq. (10) for each column.
   (d) Check if there exists an entry in T_{for} that matches the computed C_{back}.
   (e) If exits, compute all bytes of #7 and #8 with matched values and check if all 128-bit values of #7 and #8 match.
   (f) If all 128-bits match, output the corresponding (H_{N-1}, M_{N-1}).
6. If the attack does not succeed with 2^8 values of #19[12], go back to Step 2 (or Step 1 if necessary) and repeat the attack with different constant values.

Complexity Evaluation. In this attack, the sum of the complexity for computing 2^8 results of #7 and #8 is roughly 2^8 7-round AES computations. At Step 3c, we need a memory to store $2^8 \cdot$ 4-byte information for C_{for}. At Step 5d we search for the table of the size 2^8. Hence, for all 2^8 values of #19[12], the cost is about $2^8 \cdot \log 2^8$ memory access, which is enough small compared to 2^8 7-round AES for computing #7 and #8. The success probability of the match is 2^{-8} for each column, and thus 2^{-32} for 4 columns. Hence, after generating 2^8 results for #7 and 2^8 results for #8, $2^8 \cdot 2^8 \cdot 2^{-32} = 2^{-16}$ candidate will remain, where a remaining candidate satisfies 4-byte linear relations in a state. Therefore, if we iterate the above procedure 2^{112} times, we obtain $2^{112} \cdot 2^{-16} = 2^{96}$ candidates satisfying the match and one of them will satisfy the other 12-byte linear relations in a state, in other words, a preimage on the compression function is found. Note that, at Step 6, the algorithm can go back to Step 2 up to 2^{112} times for a fixed key-input. To repeat the attack more, we need to change the key-input at Step 1. The final complexity of the attack is $2^8 \cdot 2^{112} = 2^{120}$ AES 7-round computations and we need a memory for storing $2^8 \cdot$ 4-byte information.

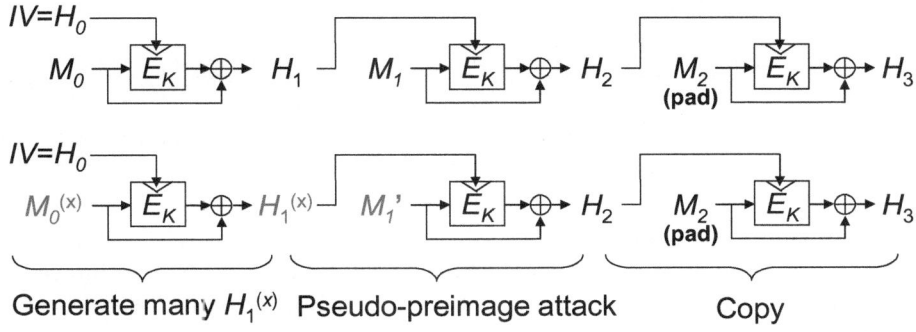

Fig. 10. (Top) Given first-preimage (Bottom) Second-preimage construction

Conversion to Hash Function Scenario. How to convert this attack to the hash function scenario depends on the mode-of-operation used. For the DM mode, this attack finds (H_{N-1}, M_{N-1}), where the value of M_{N-1} is chosen at Step 1 and the value of H_{N-1} is determined randomly during the attack. In this scenario, we can choose the message so that the padding string can be satisfied. Instead, H_{N-1} cannot be fixed to IV. Hence, we choose M_{N-1} at Step 1 so that the padding string for 2-block messages is satisfied, and convert pseudo-preimages into preimages with the conversion in Sect. 3.1. Finally, the attack generates preimages of 2-block long with a complexity of $2^{1+(120+128)/2} = 2^{125}$.

For the MMO or MP modes, the value of H_{N-1} is chosen at Step 1 and the value of M_{N-1} is determined randomly during the attack. Therefore, we can always start from the IV, but cannot satisfy the padding string. Because of the padding problem, this attack cannot generate preimages. Hence, we aim to generate second preimages. Assume that the given first preimage is 3-block long. The attack is depicted in Fig. 10. Our attack copies the last message block in which the padding string is included. For the first block, we choose several different message values $M_0^{(x)}$ to find several different chaining variables $H_1^{(x)}$. Finally, for the second block, we choose one of generated $H_1^{(x)}$ and search for a message M_1' that satisfies $\mathrm{CF}(M_1', H_1^{(x)}) = H_2$ using the pseudo-preimage attack. Hence, second preimages are generated with a complexity of 2^{120}. Note that, a fixed H_2 cannot be mapped from a fixed $H_1^{(x)}$ for any M_1' with a probability of $1 - e^{-1}$. In such a case, we choose another $H_1^{(x)}$ and repeat the attack. After several trials of $H_1^{(x)}$, we will find a valid M_1' with a probability almost 1.

6   Discussion

Other PGV Modes. In PGV, 12 schemes in Table 2 are secure. Our attacks can be applied to all 12 schemes. In our attack, the key is chosen and fixed. Therefore, as long as the key is equivalent to H_i, the same attack as the

Table 2. Twelve secure PGV constructions. X_i represents $H_i \oplus M_i$.

No.	Computation	No.	Computation	No.	Computation	No.	Computation
1	$E_{H_i}(M_i) \oplus M_i$	2	$E_{H_i}(X_i) \oplus X_i$	3	$E_{H_i}(M_i) \oplus X_i$	4	$E_{H_i}(X_i) \oplus M_i$
5	$E_{M_i}(H_i) \oplus H_i$	6	$E_{M_i}(X_i) \oplus X_i$	7	$E_{M_i}(H_i) \oplus X_i$	8	$E_{M_i}(X_i) \oplus H_i$
9	$E_{X_i}(M_i) \oplus M_i$	10	$E_{X_i}(H_i) \oplus H_i$	11	$E_{X_i}(M_i) \oplus H_i$	12	$E_{X_i}(H_i) \oplus M_i$

MMO-mode can be performed. Hence, on No.1 to No.4, a second-preimage attack with 2^{120} computations is possible. Similarly, on No.5 to No.8, a preimage attack with 2^{125} computations is possible. On No.9 to No.12, the key X_i is chosen but either H_i or M_i cannot be chosen. Hence, only a second-preimage attack with 2^{125} computations is possible. Considering the long message attack [20], our attack has an advantage only if the length of the given message is 3- to 7-blocks.

As a further generalization, applications to the generalized PGV construction proposed by [33] seems interesting. We leave this work as an open problem.

Complexity on AES 6-Rounds. To demonstrate the change of the complexity with a different number of rounds, we attacked 6-rounds. The idea is omitting the match through the MC and apply the direct match instead. This enables attackers to reduce the number of known bytes in the backward chunk, and thus to keep 2^{16} neutral values for each chunk. The final results are listed in Table 1.

Known-Key Attack on 7-Round AES. Our attack can be regarded as a new approach of the known-key attack on AES, which finds fixed points on 7-round AES. The success probability of the attack is $1 - e^{-1}$.

The attacker is given a randomly chosen key k. Then, she carries out the pseudo-preimage attack on 7-round AES in Sect. 5 with setting the given target hash value to 0. With a complexity of 2^{120}, a plaintext p s.t. $p = E_k(p)$ will be found, while finding such p will cost 2^{128} for a 128-bit random permutation.

Note that Gilbert and Peyrin proposed a known-key distinguisher on 8-round AES [17]. They find some non-ideal differential property, while our attack finds a fixed point which has been discussed for a long time. The application of our known-key distinguisher to the hash function scenario is meaningful, which finds a preimage of 0 in several PGV constructions. However, [17] attacks 8-round with a feasible complexity, while ours attacks 7-round with an infeasible complexity.

Difficulties in Chosen-Key Setting. In our attack, the key-input is fixed to a constant. It might be possible to extend the attack by actively choosing the key-input. However, this is not trivial. Firstly, the splice-and-cut technique cannot be used for the key-schedule function, namely, most part of the first round key cannot be obtained from the last round key without computing the inversion. Hence, the MitM attack would be difficult. Secondly, the previous related-key attacks on AES focused their attention on differential properties. It is unclear how to use the weak property of the key-schedule function to build preimages.

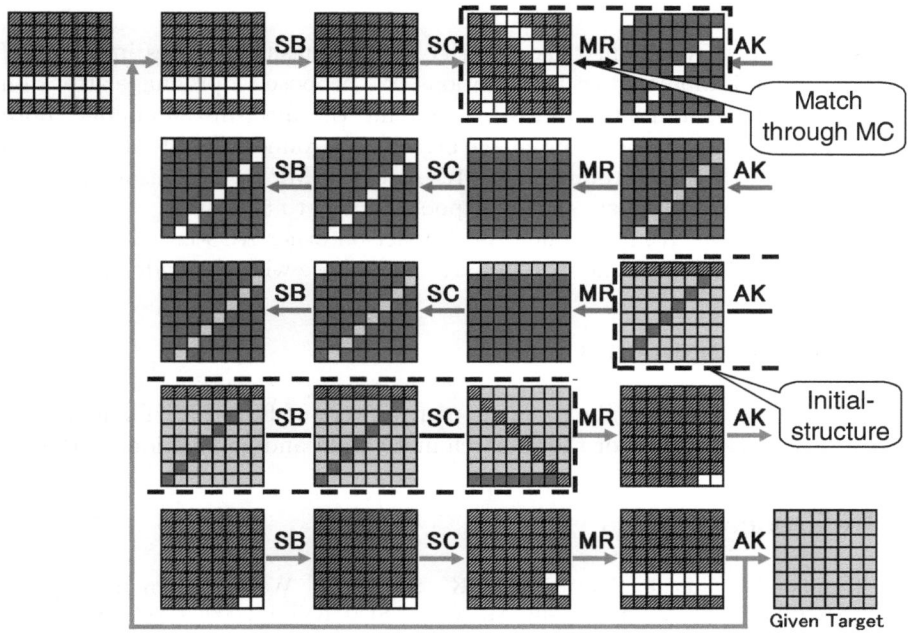

Fig. 11. Chunk separation for second preimage attack on 5-round Whirlpool. Whirlpool computes ShiftColumns(SC) and MixRows(MR) instead of SR and MC.

Application for Whirlpool. Our attack exploits the omission of MixColumns in the last round. Here, we discuss a variant of AES, which computes MixColumns in the last round[2]. To check the number of attacked rounds, we again use Fig. 9. Due to MixColumns in the last round, we need to remove the 7th round from our target. Note that known byte positions at states #25 and #5 are identical. This indicates that we also need to remove the first round, and state #25 should be connected to state #6. As a result, we can attack only 5 rounds.

Whirlpool [30] is a hash function which is deeply based on AES with a larger state size. Because Whirlpool computes MixColumns during the last round, the above analysis can be directly applied as a preimage attack on reduced Whirlpool. Whirlpool uses $8*8$-byte (512-bit) state and consists of 10 naturally-expanded AES rounds. It produces the compression function output with the MP mode. In Fig. 11, we show the chunk separation for attacking 5-round Whirlpool.

The attack strategy is the same as 7-round attack on AES. Hence we omit the details. Pseudo-preimages can be generated faster than the brute force attack by a factor of 2^8, which is 2^{504}. Because Whirlpool uses the MP mode, this can be a second-preimage attack with a complexity of 2^{504} and a memory of 2^8.

---

[2] One may note another study on the effects of the omission of MixColumns [14].

7   Concluding Remarks

In this paper, we studied the security of AES hashing modes in terms of the classical and important security notions. We proposed a preimage attack on the PGV modes instantiating AES by applying the meet-in-the-middle approach. As a result, we obtained a preimage attack on 7 rounds of DM-AES and second-preimage attack on 7 rounds of MMO-AES and MP-AES. This attack can also generate second preimages of Whirlpool reduced to 5 rounds.

Note that our results do not give impact on other AES-based hash functions e.g. several SHA-3 candidates, in particular, those with the wide-pipe structure.

Acknowledgements

I would like to thank the anonymous reviewers of FSE 2011 for many fruitful comments, especially for the research motivation and applications of the attack.

References

1. Aoki, K., Guo, J., Matusiewicz, K., Sasaki, Y., Wang, L.: Preimages for Step-Reduced SHA-2. In: Matsui, M. (ed.) ASIACRYPT 2009. LNCS, vol. 5912, pp. 578–597. Springer, Heidelberg (2009)
2. Aoki, K., Sasaki, Y.: Meet-in-the-middle preimage attacks against reduced SHA-0 and SHA-1. In: Halevi, S. (ed.) CRYPTO 2009. LNCS, vol. 5677, pp. 70–89. Springer, Heidelberg (2009)
3. Aoki, K., Sasaki, Y.: Preimage attacks on one-block MD4, 63-step MD5 and more. In: Avanzi, R.M., Keliher, L., Sica, F. (eds.) SAC 2008. LNCS, vol. 5381, pp. 103–119. Springer, Heidelberg (2009)
4. Aumasson, J.-P., Meier, W., Mendel, F.: Preimage Attacks on 3-Pass HAVAL and Step-Reduced MD5. In: Avanzi, R.M., Keliher, L., Sica, F. (eds.) SAC 2008. LNCS, vol. 5381, pp. 120–135. Springer, Heidelberg (2009)
5. Biryukov, A., Dunkelman, O., Keller, N., Khovratovich, D., Shamir, A.: Key recovery attacks of practical complexity on AES-256 variants with up to 10 rounds. In: Gilbert, H. (ed.) EUROCRYPT 2010. LNCS, vol. 6110, pp. 299–319. Springer, Heidelberg (2010)
6. Biryukov, A., Khovratovich, D.: Related-key cryptanalysis of the full AES-192 and AES-256. In: Matsui, M. (ed.) ASIACRYPT 2009. LNCS, vol. 5912, pp. 1–18. Springer, Heidelberg (2009)
7. Biryukov, A., Khovratovich, D., Nikolić, I.: Distinguisher and related-key attack on the full AES-256. In: Halevi, S. (ed.) CRYPTO 2009. LNCS, vol. 5677, pp. 231–249. Springer, Heidelberg (2009)
8. Biryukov, A., Nikolić, I.: A new security analysis of AES-128. In: Rump Session of CRYPTO 2009 (2009), http://rump2009.cr.yp.to/
9. Biryukov, A., Nikolić, I.: Automatic search for related-key differential characteristics in byte-oriented block ciphers: Application to AES, camellia, khazad and others. In: Gilbert, H. (ed.) EUROCRYPT 2010. LNCS, vol. 6110, pp. 322–344. Springer, Heidelberg (2010)

10. Bogdanov, A., Leander, G., Paar, C., Poschmann, A., Robshaw, M.J.B., Seurin, Y.: Hash functions and RFID tags: Mind the gap. In: Oswald, E., Rohatgi, P. (eds.) CHES 2008. LNCS, vol. 5154, pp. 283–299. Springer, Heidelberg (2008)

11. De Cannière, C., Rechberger, C.: Preimages for reduced SHA-0 and SHA-1. In: Wagner, D. (ed.) CRYPTO 2008. LNCS, vol. 5157, pp. 179–202. Springer, Heidelberg (2008)

12. Daemen, J., Rijmen, V.: The design of Rijndeal: AES – the Advanced Encryption Standard (AES). Springer, Heidelberg (2002)

13. Demirci, H., Selçuk, A.A.: A meet-in-the-middle attack on 8-round AES. In: Nyberg, K. (ed.) FSE 2008. LNCS, vol. 5086, pp. 116–126. Springer, Heidelberg (2008)

14. Dunkelman, O., Keller, N.: The effects of the omission of last round's MixColumns on AES. Cryptology ePrint Archive, Report 2010/041 (2010), http://eprint.iacr.org/2010/041

15. Dunkelman, O., Keller, N., Shamir, A.: Improved single-key attacks on 8-round AES-192 and AES-256. In: Abe, M. (ed.) ASIACRYPT 2010. LNCS, vol. 6477, pp. 158–176. Springer, Heidelberg (2010)

16. Gilbert, H., Minier, M.: A collision attack on 7 rounds Rijndael. In: Third AES Candidate Conference (AES3), pp. 230–241. Springer, Heidelberg (2000)

17. Gilbert, H., Peyrin, T.: Super-sbox cryptanalysis: Improved attacks for AES-like permutations. In: Hong, S., Iwata, T. (eds.) FSE 2010. LNCS, vol. 6147, pp. 365–383. Springer, Heidelberg (2010)

18. Guo, J., Ling, S., Rechberger, C., Wang, H.: Advanced meet-in-the-middle preimage attacks: First results on full tiger, and improved results on MD4 and SHA-2. In: Abe, M. (ed.) ASIACRYPT 2010. LNCS, vol. 6477, pp. 56–75. Springer, Heidelberg (2010)

19. Indesteege, S., Preneel, B.: Preimages for reduced-round tiger. In: Lucks, S., Sadeghi, A.-R., Wolf, C. (eds.) WEWoRC 2007. LNCS, vol. 4945, pp. 90–99. Springer, Heidelberg (2008)

20. Kelsey, J., Schneier, B.: Second preimages on n-bit hash functions for much less than 2^n work. In: Cramer, R. (ed.) EUROCRYPT 2005. LNCS, vol. 3494, pp. 474–490. Springer, Heidelberg (2005)

21. Knudsen, L.R., Rijmen, V.: Known-key distinguishers for some block ciphers. In: Kurosawa, K. (ed.) ASIACRYPT 2007. LNCS, vol. 4833, pp. 315–324. Springer, Heidelberg (2007)

22. Lamberger, M., Mendel, F., Rechberger, C., Rijmen, V., Schläffer, M.: Rebound distinguishers: Results on the full whirlpool compression function. In: Matsui, M. (ed.) ASIACRYPT 2009. LNCS, vol. 5912, pp. 126–143. Springer, Heidelberg (2009)

23. Lamberger, M., Mendel, F., Rechberger, C., Rijmen, V., Schläffer, M.: The rebound attack and subspace distinguishers: Application to Whirlpool. Cryptology ePrint Archive, Report 2010/198 (2010), http://eprint.iacr.org/2010/198

24. Leurent, G.: MD4 is not one-way. In: Nyberg, K. (ed.) FSE 2008. LNCS, vol. 5086, pp. 412–428. Springer, Heidelberg (2008)

25. Lu, J., Dunkelman, O., Keller, N., Kim, J.-S.: New impossible differential attacks on AES. In: Chowdhury, D.R., Rijmen, V., Das, A. (eds.) INDOCRYPT 2008. LNCS, vol. 5365, pp. 279–293. Springer, Heidelberg (2008)

26. Mendel, F., Peyrin, T., Rechberger, C., Schläffer, M.: Improved cryptanalysis of the reduced grøstl compression function, ECHO permutation and AES block cipher. In: Jacobson Jr., M.J., Rijmen, V., Safavi-Naini, R. (eds.) SAC 2009. LNCS, vol. 5867, pp. 16–35. Springer, Heidelberg (2009)

27. Mendel, F., Rechberger, C., Schläffer, M., Thomsen, S.S.: The rebound attack: Cryptanalysis of reduced whirlpool and grøstl. In: Dunkelman, O. (ed.) FSE 2009. LNCS, vol. 5665, pp. 260–276. Springer, Heidelberg (2009)

28. Menezes, A.J., van Oorschot, P.C., Vanstone, S.A.: Handbook of Applied Cryptography. CRC Press, Boca Raton (1997)

29. Preneel, B., Govaerts, R., Vandewalle, J.: Hash functions based on block ciphers: A synthetic approach. In: Stinson, D.R. (ed.) CRYPTO 1993. LNCS, vol. 773, pp. 368–378. Springer, Heidelberg (1994)

30. Rijmen, V., Barreto, P.S.L.M.: The WHIRLPOOL hashing function. Submitted to NISSIE (September 2000)

31. Sasaki, Y., Aoki, K.: Preimage attacks on 3, 4, and 5-pass HAVAL. In: Pieprzyk, J. (ed.) ASIACRYPT 2008. LNCS, vol. 5350, pp. 253–271. Springer, Heidelberg (2008)

32. Sasaki, Y., Aoki, K.: Finding preimages in full MD5 faster than exhaustive search. In: Joux, A. (ed.) EUROCRYPT 2009. LNCS, vol. 5479, pp. 134–152. Springer, Heidelberg (2009)

33. Stam, M.: Blockcipher-based hashing revisited. In: Dunkelman, O. (ed.) FSE 2009. LNCS, vol. 5665, pp. 67–83. Springer, Heidelberg (2009)

34. U.S. Department of Commerce, National Institute of Standards and Technology. Specification for the ADVANCED ENCRYPTION STANDARD (AES) (Federal Information Processing Standards Publication 197) (2001), http://csrc.nist.gov/encryption/aes/index.html#fips

35. U.S. Department of Commerce, National Institute of Standards and Technology. Federal Register /Vol. 72, No. 212/Friday, November 2, 2007/Notices (2007), http://csrc.nist.gov/groups/ST/hash/documents/FR_Notice_Nov07.pdf

36. Wei, Y., Lu, J., Hu, Y.: Meet-in-the-middle attack on 8 rounds of AES block cipher under 192 key bits. Cryptology ePrint Archive, Report 2010/537 (2010), http://eprint.iacr.org/2010/537 (appeared in the accepted papers list of ISPEC 2011)

Known-Key Distinguishers on 11-Round Feistel and Collision Attacks on Its Hashing Modes

Yu Sasaki and Kan Yasuda

NTT Information Sharing Platform Laboratories, NTT Corporation, Japan
{sasaki.yu,yasuda.kan}@lab.ntt.co.jp

Abstract. We present new attacks on the Feistel network, where each round function consists of a subkey XOR, S-boxes, and then a linear transformation (*i.e.*, an SP round function). Our techniques are based largely on what they call the rebound attacks. As a result, our attacks work most effectively when the S-boxes have a "good" differential property (like the inverse function $x \mapsto x^{-1}$ in the finite field) and when the linear transformation has an "optimal" branch number (*i.e.*, a maximum distance separable matrix). We first describe known-key distinguishers on such Feistel block ciphers of up to 11 rounds, increasing significantly the number of rounds from previous work. We then apply our distinguishers to the Matyas-Meyer-Oseas and Miyaguchi-Preneel modes in which the Feistel ciphers are used, obtaining collision and half-collision attacks on these hash functions.

Keywords: known-key, block cipher, Feistel-SP, rebound attack, MDS, collision attack, hash function, MMO, Miyaguchi-Preneel.

1 Introduction

The security of block ciphers usually relies on the fact that the key value is kept secret in its encryption and decryption process. However, recently cryptographers have become interested in the security of block ciphers when the key value is known to the attackers. That is, in the traditional secret-key setting, it is the randomness and secrecy of a key that guarantees security of block ciphers. On the other hand, in the known-key setting, the value of the key is revealed to attackers; it becomes only the randomness of the key that retains (some) security of the cipher.

In practical applications, block ciphers are indeed used in such a way as their key values are known publicly. A typical example is the hash function constructed from block ciphers, including the Davies-Meyer (DM), Matyas-Meyer-Oseas (MMO) and Miyaguchi-Preneel hashing modes. In particular, the latter two modes make the known-key setting potentially relevant, because a known and essentially random (at least not under the attacker's full control) value is fed into the key input of the block cipher.

Recently, many papers have been published in the context of known-key attacks. In particular, there have been quite a few results on AES and Rijndael

A. Joux (Ed.): FSE 2011, LNCS 6733, pp. 397–415, 2011.

[13,19,17,12,24]. Among them is the work by Gilbert and Peyrin [12] of a known-key distinguisher on AES, which is one of the few attacks that can break 8 rounds of AES-128.[1]

On the other hand, there appear to be only a couple of results that analyzed the known-key security of the Feistel network [13,8]. Both pieces of work assume that a round function consists of a key XOR followed by a mixing function. In practice, the "mixing function" part is frequently realized by a combination of S-boxes (Substitution boxes) and a linear transformation.

In this paper, we show new attacks on the Feistel network. We consider the Feistel network which has SP round functions. That is, we assume that a round function consists of an XOR with a subkey, S-boxes, and then a linear matrix P. Our attacks work effectively up to 11 rounds, improving over the previous best 7 rounds.[2] We first present known-key distinguishers on such Feistel block ciphers and then translate them to collision and half-collision attacks on the MMO and Miyaguchi-Preneel hashing modes using these ciphers.

Our techniques are based on the so-called rebound attacks. Consequently, in our attacks we assume "attacker-unfriendly" properties of S-boxes and of P. Though not mandatory, these properties make our attacks most effective to work and simplest to describe. More specifically, we assume that the S-boxes have low maximum differential probabilities (like the multiplicative inverse function $x \mapsto x^{-1}$ in the finite field) and that the linear transformation P is an MDS (maximum distance separable) matrix. These are believed to be "good" choices of S-boxes and of P for designing a secure symmetric-key primitive.

Organization of the Paper. In Sect. 2 we review previous work, basic notions, and the rebound attacks. We present our new distinguishers on the Feistel network in Sect. 3 and apply them to the MMO and Miyaguchi-Preneel hashing modes in Sect. 4. In Sect. 5 we discuss the generality and applicability of our attack techniques. We conclude the paper in Sect. 6.

2   Preliminaries

2.1   Previous and Related Work

The concept of known-key distinguishers was introduced by Knudsen and Rijmen [13]. They showed known-key attacks on AES reduced to 7-rounds and on 7-round Feistel ciphers with a round function consisting of a round-key XOR followed by an arbitrary key-independent transformation. Their attack on AES(-128) leads to non-ideal behaviors of its MMO hashing mode. Their attack on Feistel ciphers has an even stronger impact, as a pair of blocks colliding in half of the output state can be found only with a complexity of two encryptions.

---

[1] Another attack on the 8-round AES-128 is [5], which is a "chosen-key" distinguisher.

[2] The 7-round attack [13] works for the Feistel network which has a round function consisting of an XOR with a subkey followed by any function f rather than SP.

Subsequently, Minier *et al.* [19] suggested a formalization of known-key attacks. They also presented known-key distinguishers on Rijndael having large blocks, which was further studied by [24].

As for AES, Mendel *et al.* [17] presented a new known-key distinguisher on 7-round AES-128 with a less attack complexity. Gilbert and Peyrin [12] attacked AES-128 and increased the number of rounds to eight.

The work [8] by Bouillaguet *et al.* presented new attacks on the Feistel network. This work is actually about analyzing certain hash functions (Lesamnta and SHAvite-3) but can be regarded as known-key attacks on two types of (generalized) Feistel ciphers. See also [11,20] for other attacks on SHAvite-3.

There is another type of attack in the open key scenario, namely, the "chosen-key" distinguishers. In the chosen-key setting, attackers can control, rather than know, the key value of block ciphers. For example, Biryukov *et al.* showed chosen-key distinguishers on the full-round AES-256 in the "related-key" scenario [4]. Biryukov and Nikolić showed chosen-key distinguishers on the 8-round AES-128 [5] and on several other block ciphers [6]. Lamberger *et al.* [14] presented a chosen-key distinguisher on the full Whirlpool compression function. See also [21] for other known-key and chosen-key distinguishers on several block ciphers.

2.2 Basic Notions

S-Boxes. An S-box

$$S : \{0,1\}^c \to \{0,1\}^c$$

is a substitution table, being most of the time a non-compressing fixed function. Popular choices are $c = 4, 8$. The main purpose of using S-boxes in block ciphers is to introduce non-linearity over the finite field $\mathbb{F}(2^c)$.

A typical example of an S-box is the multiplicative inverse function

$$S : x \mapsto x^{-1}$$

in the field $\mathbb{F}(2^c)$. This is a popular choice for an S-box in block ciphers, is believed to be a good choice for increasing the cipher's resistance to differential [2] and linear [16] cryptanalyses, and is indeed adopted by AES. With such an S-box the maximum differential and linear probabilities become 2^{-c+2}, being close to the best possible bounds [22,10].

Branch Numbers and MDS Matrices. Recall that the linear diffusion layer $P : \{0,1\}^n \to \{0,1\}^n$ can be written in the form of an $r \times r$ constant matrix over the field $\mathbb{F}(2^c)$, where c is the number of bits in a byte (*i.e.*, the size of an S-box) and r is the number of bytes (*i.e.*, the number of S-boxes) in the input to the round function, so that we have $n = r \cdot c$. Here we treat $\{0,1\}^n$ as $\mathbb{F}(2^c)^r$. For an input $X \in \{0,1\}^n = \mathbb{F}(2^c)^r$ we can write

$$P \cdot X = \begin{pmatrix} p_{11} & p_{12} & \cdots & p_{1r} \\ p_{21} & p_{22} & \cdots & p_{2r} \\ \vdots & \vdots & \ddots & \vdots \\ p_{r1} & p_{r2} & \cdots & p_{rr} \end{pmatrix} \begin{pmatrix} x_1 \\ x_2 \\ \vdots \\ x_r \end{pmatrix},$$

where each of p_{ij} and x_i is an element of the field $\mathbb{F}(2^c)$.

The *weight* $\mathtt{wt}(X)$ (over the field $\mathbb{F}(2^c)$) of a vector $X \in \mathbb{F}(2^c)^r$ is the number of non-zero components in X, so that we have $0 \leq \mathtt{wt}(X) \leq r$. The *branch number* $\mathtt{br}(P)$ of an $r \times r$ matrix P over the field $\mathbb{F}(2^c)$ is defined as

$$\mathtt{br}(P) := \min_X \{\mathtt{wt}(X) + \mathtt{wt}(P \cdot X)\},$$

where the minimum runs over all $X \in \mathbb{F}(2^c)^r$ such that $X \neq 0$. For block ciphers in the secret-key setting, it is desirable to use a linear transformation P having a high branch number in order for the cipher to be resistant to differential and linear cryptanalyses.

The Singleton bound [25] in coding theory implies that the highest branch number possible is $r + 1$. An $r \times r$ matrix P having such a branch number is called an *MDS (Maximum Distance Separable) matrix* and is frequently utilized in cryptography. For example, the block cipher AES [26] uses a 4×4 MDS matrix, whereas the hash function Whirlpool [1] uses an 8×8 MDS matrix.

Hashing Modes Using Block Ciphers. Matyas-Meyer-Oseas (MMO) and Miyaguchi-Preneel modes provide efficient ways to construct a compression function from a block cipher. They are among the 12 secure schemes [7] of PGV style [23].

Let E be a block cipher, and let E_K denote its encryption algorithm with a key K. The MMO compression function outputs H_i by computing

$$H_i = E_{H_{i-1}}(M_{i-1}) \oplus M_{i-1}$$

for a message block M_{i-1} and a previous chaining value H_{i-1}. Similarly, the Miyaguchi-Preneel mode computes H_i by

$$H_i = E_{H_{i-1}}(M_{i-1}) \oplus M_{i-1} \oplus H_{i-1},$$

given M_{i-1} and H_{i-1}.

In the above definitions, the Merkle-Damgård construction is implicitly assumed (by setting the initial chaining value to be a fixed constant IV and by letting the final output be the hash value). We keep it implicit, as all attacks presented in the current paper deal only with one-block input messages.

2.3   Rebound-Attack Technique

Here we briefly review a technique called rebound attacks, which was first introduced by Mendel *et al.* in analyzing AES-based hash functions [18]. Mendel *et al.* later applied it to a known-key attack on reduced-round AES [17], as already mentioned in Section 2.1. The attacks presented in the current paper mainly depend on this technique as well.

In rebound attacks, an attacker first builds a truncated differential path, fixing byte positions which have a difference. At this stage, the attacker is not interested

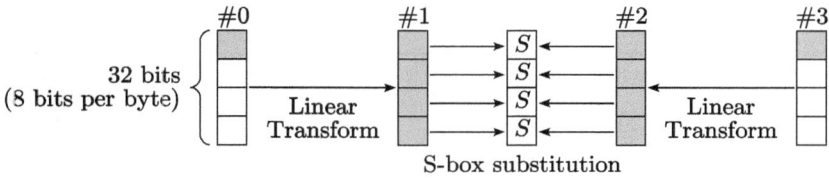

Fig. 1. Example of the inbound phase with 4-byte state (Gray bytes are active.)

in the exact differential value inside each byte. We call a byte position with a difference an *active byte*. Then the attacker tries to find efficiently a pair of values that follows the path.

The truncated differential path is divided into two phases: The *inbound phase* and the *outbound phase*. The inbound phase consists of a linear layer, a non-linear layer, and another linear layer in this order. The inbound phase corresponds to a low probability part of the truncated differential path. The attacker generates sufficiently many paired values that satisfy the truncated differential path of the inbound phase. This can be done with a small complexity on average. These paired values are called *starting points*. Then, the attacker computes the outbound phase with the generated starting points and checks whether or not any of the starting points satisfies the truncated differential path of the outbound phase. The attacker succeeds if he finds a starting point conforming to the truncated differential path of the outbound phase.

Let us explain the basic procedure of the inbound phase by using a 4-byte (1 byte = 8 bits) state with an 8-bit S-box as an example. In this example, the goal of an attacker is to find a pair of values (M, M') that satisfies the truncated differential path $1 \to 4 \to 1$ (one active byte diverging to four active bytes and then converging to one active byte again) which is illustrated in Fig. 1.

A naive method to obtain such a pair is to generate many pairs with a 1-byte difference at state #0 and then to compute the corresponding difference at state #3. After trying 2^{24} pairs, the attacker should be able to find a desired pair.

The rebound attack can generate such pairs more efficiently. It generates approximately 2^8 pairs satisfying the differential path with a complexity of approximately 2^8—in other words, each pair is generated with a complexity of 1 on average. The detailed attack procedure is as follows:

0. Prepare the differential distribution table (DDT) for the 8-bit S-box.
1. For all 2^8 (more precisely, $2^8 - 1$) possible differences of state #0, compute the corresponding 4-byte differences of state #1 and store them in a table T. Note that we can determine differential propagation irrespective of the byte values owing to the linearity of computation.
2. Choose a difference of state #3 and compute the corresponding 4-byte difference of state #2. For each 4-byte difference in T, check whether or not the computed 4-byte difference of state #2 can be output through the S-boxes by looking up the DDT. If the differences match, output such paired values.

Table 1. Our attack results for practical parameters of Feistel network

Block	Word	Byte	#Bytes	$2c \geq r$	Cipher	MMO Hash	
N	n	c	r	?	Known-key	Collision	Half-collision
128	64	8	8	✓	11R	9R	11R
		4	16		11R	9R	11R
64	32	8	4	✓	9R	7R	9R[†]
		4	8	✓	11R	9R	11R

[†] This 9R attack generates $(N-c)$-bit collisions rather than half-collisions.

In the case of the AES S-box (*i.e.*, the multiplicative inverse function in the finite field), for a pair of randomly determined input difference ΔS_{in} and output difference ΔS_{out}, an equation $S(x) \oplus S(x \oplus \Delta S_{\mathrm{in}}) = \Delta S_{\mathrm{out}}$ has approximately one solution with a probability of approximately 2^{-1}. If we find a solution x, then we will automatically obtain *two* paired values $(x, x \oplus \Delta S_{\mathrm{in}})$ and $(x \oplus \Delta S_{\mathrm{in}}, x)$ that satisfy the differential propagation through the S-box. In this example there are four S-boxes between state #1 and state #2. Therefore, if we have 2^4 pairs of $\Delta\#1$ and $\Delta\#2$, then one of these pairs can be expected to have a solution for each of the four S-boxes. Hence we obtain 2^4 paired values that satisfy the truncated differential path of the inbound phase.

Consequently, for a difference of state #3 in the above procedure, there are $2^8 \cdot 2^{-4} = 2^4$ differences in T that have approximately one solution for each of the four S-boxes. So we obtain $2^4 \cdot 2^4 = 2^8$ paired values that satisfy the truncated differential path (starting points). In other words, we can obtain one starting point with a complexity of 1 on average (Note that the average complexity to obtain one starting point is always 1 for any S-box).

We also need to count the number of starting points obtained. The above procedure can be iterated for all 2^8 differences of state #3 in Step 2. As a result, we obtain $2^8 \cdot 2^8 = 2^{16}$ starting points at maximum.

3   New Known-Key Attacks on Feistel Ciphers

Now we present known-key attacks on Feistel ciphers of the SP structure. Table 1 summarizes our attack results (R stands for rounds), where we use the following variables (which are fixed henceforth):

N: The block length of the cipher (in bits),
n: The word size in bits, equal to the size of the input and output of the round function, so that $n = N/2$,
c: The byte size in bits (In this paper the byte size is not fixed), equal to the size of an S-box,
r: The number of bytes in a word, so that $r = n/c$.

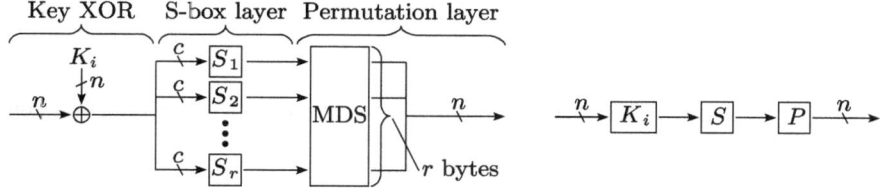

Fig. 2. Left: Detailed description of the SP round function **Right:** Simplified one

Recall that many of the block ciphers designed in practice are equipped with 128-bit or 64-bit blocks and use 8-bit or 4-bit S-boxes. Table 1 includes these practical parameters.

For the sake of simplicity, we assume $2c \geq r$ at the moment. Even with this restriction, our attacks cover most of the block ciphers that are used today: 128-bit block with 8-bit S-box, 64-bit with 8-bit, and 64-bit with 4-bit.

For the sake of completeness, in Sect. 5.3 we shall explain a simple extension of our attacks to treat the case $2c < r$. This case includes the remaining parameter of 128-bit block with 4-bit S-box, which is probably most unlikely to be used in practice among the 4 parameters listed in Table 1. This is because this case requires an MDS matrix acting on 16-byte state, where today such a large MDS matrix is still considered to be too costly to be implemented in systems.

Description of the SP Round Function. Before we proceed to describing our attacks, we specify the SP round function of the Feistel network to be analyzed. The round function is depicted in Fig. 2, consisting of the following three operations:

Key XOR: This layer computes the XOR of a round-function input and a round key K_i.

S-box layer: This layer substitutes each byte value by using one or several S-boxes; the S-boxes $S_1, S_2, \ldots, S_r$ may differ from each other. We assume that all S-boxes are designed to be resistant to differential and linear cryptanalyses, like the ones used in AES [9,26]. Hence, given a pair of randomly chosen input and output differences, there exist paired values following the given input/output differences with a probability of approximately 2^{-1}. If exist, then the number of such paired values is approximately two.

Permutation layer: This layer mixes values by multiplying the word value and an $r \times r$ matrix P over $\mathbb{F}(2^c)$ together. We make the assumption that P is an MDS matrix, so that the total number of active bytes in the input and output of P is always greater than or equal to $r + 1$, as long as there is at least one active byte.

The assumptions that we make about S and P are not quite mandatory. In Sect. 5 we discuss the feasibility of our attacks when S or P does not have these properties.

Table 2. Attack strategy

Input:[†]

- Subkeys generated from a random master key via a key schedule function
- S-boxes secure against differential and linear cryptanalyses
- A linear transformation with a branch number $r + 1$ (*i.e.*, an MDS)

Procedure:

1. Prepare DDTs for all S-boxes in use.
2. Find a pair of values that satisfies the truncated differential path of the inbound phase.
3. Verify that the pair found in Step 2 also follows the truncated differential path of the outbound phase.
4. Confirm that the above procedure requires a less complexity than finding such a pair for a random permutation.

[†] In Sect. 5 we discuss the case where S or P does not satisfy these conditions.

Overview of Our Attacks. Our attack procedure is summarized in Table 2. We shall explain the details of Steps 2 and 3 in the following sections; in Sect. 3.1 we first explain a basic version of our attacks which is effective up to 9 rounds, and then extend it to 11 rounds in Sect. 3.2.

Hereafter, we fix a system of notations as follows:

X^j: The j-th byte of a word X, where $1 \leq j \leq r$ and the size of X^j is c bits,
 0: A word where all bytes are non-active,
 1: A word where only one byte of the predetermined (j-th) position is active,
 F: A word where all bytes are active.

Our attacks amount to finding a pair of values whose input difference is of the form $(P(\mathbf{1}), \mathbf{F})$ and whose output difference is also $(P(\mathbf{1}), \mathbf{F})$. The point is that our attacks can find such a pair for the Feistel ciphers more efficiently than one could for a random permutation.

So let us mention in advance the time and memory complexities of the above procedure. The cost of Step 1 is $r \cdot 2^{2c}$ in time and $r \cdot 2^{2c}$ in memory for both 9R and 11R attacks. The cost of Step 2 is $r \cdot 2^c$ in time and $r \cdot 2^2$ in memory for 9R attacks and $r \cdot 2^{2c}$ in time and $r \cdot 2^{2c}$ in memory for 11R attacks. Step 3 can be done at a cost of 1. Overall, our attacks can find a pair of values following the truncated differential path with a complexity of $r \cdot 2^{2c}$ in time and $r \cdot 2^{2c}$ in memory.

This means that our known-key attacks work effectively, because for a random permutation such a pair cannot be found with that complexity. Namely, let us consider the complexity to find a pair of values that has the differential form of $(P(\mathbf{1}), \mathbf{F})$ for both of the input and output states in a random permutation. Attackers have an access to both encryption and decryption oracles.

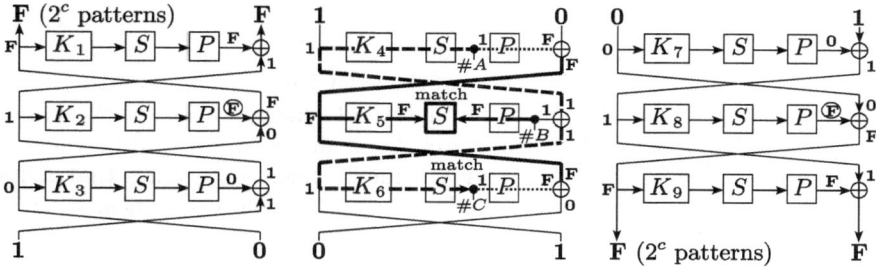

Fig. 3. Basic 9R attack

For such attackers, this problem is regarded as finding a 2^{n-c}-bit collision. Because enough freedom degrees are available to mount the birthday attack, this requires a complexity of $2^{(n-c)/2}$.

Note that we make use of the property of MDS and DDTs generated at Step 1 in performing Step 2 (the inbound phase) efficiently. Also, the outbound phase works for any S-box and any linear transformation.

3.1   Basic 9R Attack on the Feistel Ciphers

Entire Truncated Differential Path. Recall that in rebound attacks, an attacker needs to construct a differential path and divide it into inbound and outbound phases. Here the truncated differential path that we use is

$$(\mathbf{F}, \mathbf{F}) \xrightarrow{1^{\text{st}}\text{R}} (\mathbf{1}, \mathbf{F}) \xrightarrow{2^{\text{nd}}\text{R}} (\mathbf{0}, \mathbf{1}) \xrightarrow{3^{\text{rd}}\text{R}} (\mathbf{1}, \mathbf{0})$$

$$(\mathbf{1}, \mathbf{0}) \xrightarrow{4^{\text{th}}\text{R}} (\mathbf{F}, \mathbf{1}) \xrightarrow{5^{\text{th}}\text{R}} (\mathbf{1}, \mathbf{F}) \xrightarrow{6^{\text{th}}\text{R}} (\mathbf{0}, \mathbf{1})$$

$$(\mathbf{0}, \mathbf{1}) \xrightarrow{7^{\text{th}}\text{R}} (\mathbf{1}, \mathbf{0}) \xrightarrow{8^{\text{th}}\text{R}} (\mathbf{F}, \mathbf{1}) \xrightarrow{9^{\text{th}}\text{R}} (\mathbf{F}, \mathbf{F}),$$

which is shown in Fig. 3.

In the basic 9R attack, we use a 3-round differential path for the inbound phase. The difference propagates from $(\mathbf{1}, \mathbf{0})$ to $(\mathbf{0}, \mathbf{1})$ through the three rounds $(4^{\text{th}}\text{R} - 6^{\text{th}}\text{R})$.

The outbound phase consists of three rounds in backward direction $(3^{\text{rd}}\text{R} - 1^{\text{st}}\text{R})$ and three rounds in forward direction $(7^{\text{th}}\text{R} - 9^{\text{th}}\text{R})$, in total six rounds. In both directions, the differences propagate to $(\mathbf{F}, \mathbf{F})$. Here the patterns in $\mathbf{F}$ are limited in a certain way (Not all patterns are formed).

Given any paired value satisfying the truncated differential path of the inbound phase, the truncated differential path of the outbound phase is satisfied with a probability of 1. Hence, we need only one starting point from the inbound phase.

Three-Round Inbound Phase. In this phase, our goal is to find a pair of values whose difference will propagate as $(1,0) \xrightarrow{4^{\text{th}}\text{R}} (\mathbf{F},1) \xrightarrow{5^{\text{th}}\text{R}} (1,\mathbf{F}) \xrightarrow{6^{\text{th}}\text{R}} (0,1)$. We use the differential path depicted in the middle of Fig. 3.

We start by choosing differences of the form $\mathbf{1}$ in words $\#A$ and $\#B$ in Fig. 3. We propagate these differences through the linear operations with actual word values undetermined. We then search for a matched set of differences at the S-box operation in the 5th round and finally find word values that follow the desired differential path of the 3R inbound phase. More specifically, our attack procedure is as follows:

0. Prepare DDTs for all S-boxes. Choose a byte-position j $(1 \leq j \leq r)$ in a word to be activated in the differential form of $\mathbf{1}$.
1. For all 2^c possible differences in word $\#A$, compute the corresponding full-byte differences after applying the permutation layer and store the results in a table T. Note that the difference in word $\#C$ must be the same as that in $\#A$, which of course takes the differential form of $\mathbf{1}$.
2. For each of the 2^c possible differences in word $\#B$, compute the corresponding full-byte difference after applying the inverse permutation. For S-boxes in the 5th round, check whether or not we can match the full-byte difference evolved from $\Delta \#B$ with any of the 2^c differences stored in T. This can be done by looking up the DDTs.
3. Assuming that we find such a matched set of differences, choose one of them and fix word values in accordance with the chosen differences. Now the difference in word $\#A$ is fixed. Also, the word values on the bold lines drawn in Fig. 3 are all fixed.
4. For each of the 2^c possible values of $\#A^j$, compute the corresponding difference in $\#C^j$ and check whether or not the difference becomes equal to $\Delta \#A^j$. This computation is denoted by the broken lines in Fig 3. Namely, compute and check the following:

$$\Delta\left[S_j\left(S_j^{-1}(\#A^j) \oplus K_4^j \oplus \#B^j \oplus K_6^j\right)\right] \overset{?}{=} \Delta \#A^j. \tag{1}$$

5. A solution for (1) is the value we want for the active byte. Make an arbitrary choice for values of the remaining non-active bytes. Now all the values within the inbound phase are determined. In other words, we obtain a starting point.

Let us estimate the time and memory complexities necessary for each of the above steps. We also verify that the success probability of the inbound phase is sufficiently high.

- In Step 0 we need 2^{2c} computations and 2^{2c} memory to prepare the DDT for each c-bit S-box. When S-boxes are all different, Step 0 requires $r \cdot 2^{2c}$ computations and $r \cdot 2^{2c}$ memory.
- Step 1 requires 2^c computations and 2^c memory.
- In Step 2, with a complexity of 2^c, we can check the match of 2^{2c} pairs at maximum. Because each match succeeds with a probability of 2^{-r}, we can expect to find one or more matched pairs as long as $2c \geq r$.

- Step 3 requires negligible computations and memory.
- Step 4 requires 2^c 3-round computations and negligible memory. Note that this step needs to examine only one byte, and thus the actual complexity is lower than 2^c 3-round computations.
- In Step 5 we expect to find a solution for (1), because we can carry out a match check as many times as the number of possible patterns of $\Delta\#C^j$, which is 2^c.
- Therefore, the total complexity for Step 0 through Step 5 is $r \cdot 2^{2c}$ computations and $r \cdot 2^{2c}$ memory.

Outbound Phase. We explain the outbound phase for the last 3 rounds, which is shown in the right part of Fig. 3. As a result of the inbound phase, we obtain the difference $(\mathbf{0}, \mathbf{1})$ as an input to the 7th round. This propagates as $(\mathbf{0}, \mathbf{1}) \xrightarrow{7^{\text{th}}\text{R}} (\mathbf{1}, \mathbf{0}) \xrightarrow{8^{\text{th}}\text{R}} (\mathbf{F}, \mathbf{1}) \xrightarrow{9^{\text{th}}\text{R}} (\mathbf{F}, \mathbf{F})$ with a probability of 1. Because the input to the 8th round has a difference of the form $\mathbf{1}$, the corresponding output differences are of the form $P(\mathbf{1})$, which is emphasized by a circled $\mathbf{F}$ in Fig. 3. Hence, the number of possible differences in the output word is limited to 2^c. Therefore, although all bytes are active in ciphertexts (Recall that P is an MDS), only $2^c \cdot 2^n = 2^{c+n}$ differential patterns are formed in the ciphertexts.

The same applies to the differential propagation in backward direction for the first 3 rounds. The differential path reaches $(\mathbf{F}, \mathbf{F})$ after the 3 rounds with a probability of 1. Since differential patterns in the left half of the plaintext are limited to 2^c variations, only 2^{c+n} differential patterns appear in the plaintext pairs.

Comparison with a Random Permutation. The inbound phase finds a starting point with a time complexity of $r \cdot 2^{2c}$ using $r \cdot 2^{2c}$ memory. Given any solution of the inbound phase, the outbound phase, with a probability of 1, generates a pair of values that has a differential form of $(P(\mathbf{1}), \mathbf{F})$ for both plaintext and ciphertext.

By comparing the above complexity with the generic birthday bound, we can derive a condition of parameters with which our attacks work effectively. That is, an inequality $r \cdot 2^{2c} < 2^{(n-c)/2}$ gives us a condition

$$c < \frac{1}{5}(n - 2 \cdot \log_2 r), \tag{2}$$

which needs to be satisfied in order for the above attack to be successful.

Let us consider parameters in practical use with which the condition is satisfied. We see that the condition is met by 128-bit block ciphers having 4- or 8-bit S-boxes and by 64-bit block ciphers having 4-bit S-boxes. See Table 1. Unfortunately, the condition is not met by 64-bit block ciphers having 8-bit S-boxes. This case will be treated separately in Sect. 3.3.

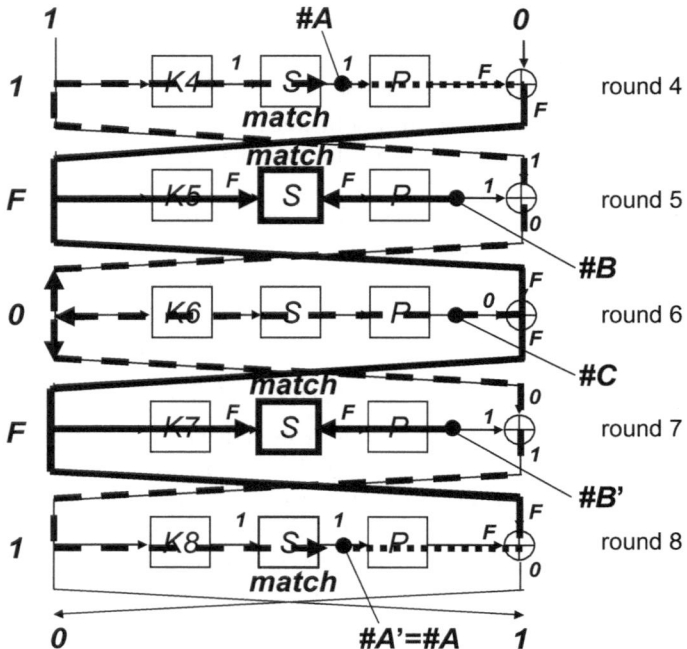

Fig. 4. 5-round inbound phase

3.2   Extended 11R Attack on the Feistel Ciphers

We extend the previous 9R attack to 11 rounds. We change the number of rounds in the inbound phase from three to five. The outbound phase is exactly the same as the basic 9R attack.

Five-Round Inbound Phase. The differential path of the new 5R inbound phase is

$$(1,0) \xrightarrow{4^{\text{th}}\text{R}} (\mathbf{F},1) \xrightarrow{5^{\text{th}}\text{R}} (0,\mathbf{F}) \xrightarrow{6^{\text{th}}\text{R}} (\mathbf{F},0) \xrightarrow{7^{\text{th}}\text{R}} (1,\mathbf{F}) \xrightarrow{8^{\text{th}}\text{R}} (0,1), \quad (3)$$

which is depicted in Fig. 4. We start with the same difference of the form **1** in words $\#A$ and $\#A'$ and then try to find matched sets of differences at the S-boxes in the 5th and 7th rounds, with the differences being expanded from words $\#B$ and $\#B'$, respectively. After finding matched sets, we determine the value of $\#C$ and compute values indicated by broken lines. We finally check the consistency of differences in words $\#A$ and that in $\#A'$. A detailed attack procedure is as follows:

0. Prepare DDTs for all S-boxes. Choose an active-byte position j for differential **1**.

1. For all 2^c differences of the active byte in word $\#A$, compute the corresponding full-byte differences after applying the (forward) permutation layer and store them in a table T. Set the difference in word $\#A'$ to be the same as that in $\#A$. This guarantees that the difference in word $\#C$ is $\mathbf{0}$.

2. For each of the 2^c differences in word $\#B$, compute the corresponding full-byte difference after applying the inverse permutation. For each difference stored in T, check whether we can match it with the above difference by looking up the DDTs. If a matched set of differences for $\#A$ and $\#B$ is found, we can instantly obtain a matched set for $\#A'$ and $\#B'$ by setting $\#B' = \#B$.

3. Now that a matched set of differences is found, we can fix word values and compute the value of word $\#C$. Here the values drawn in broken lines in Fig. 4 are fixed.

   (a) Check whether or not the computed differences in $\#A$ and $\#A'$ in Step 3 and the chosen difference in $\#A = \#A'$ at Step 1 are consistent. Namely, compute and check the following:

   $$\Delta\left[S_j\left(S_j^{-1}\left((P^{-1}(\#C))^j\right) \oplus K_6^j \oplus \#B^j \oplus K_4^j\right)\right] \stackrel{?}{=} \Delta\#A^j, \qquad (4)$$

   $$\Delta\left[S_j\left(S_j^{-1}\left((P^{-1}(\#C))^j\right) \oplus K_6^j \oplus \#B'^j \oplus K_8^j\right)\right] \stackrel{?}{=} \Delta\#A'^j. \qquad (5)$$

   (b) If we find a solution for the above two equations, then it means that we have found a starting point of the inbound phase.

Let us evaluate the time and memory complexities necessary for the above procedure:

- Step 0 requires $r \cdot 2^{2c}$ computations and $r \cdot 2^{2c}$ memory to prepare r-many DDTs.
- Step 1 requires 2^c computations and 2^c memory.
- In Step 2, we are expected to find 2^{2c-r} matches between $\#A$ and $\#B$ after trying 2^c differences in $\#B$. Since 2^r solutions are obtained from a match, we obtain 2^{2c} solutions for $\#A$ and $\#B$.
- Similarly, we obtain 2^{2c} solutions for $\#A'$ and $\#B'$.
- Step 3 requires just one computation for each solution and can be iterated $2^{2c} \cdot 2^{2c} = 2^{4c}$ times at maximum.
- In Step 3a, each match succeeds with a probability of 2^{-c}. Therefore, by iterating Step 3a 2^{2c} times, we will find a match.

To sum up, we can find a starting point for the 5R inbound phase with a complexity of $r \cdot 2^{2c}$ time and $r \cdot 2^{2c}$ memory.

Outbound Phase, Ideal Case, and Attackable Parameters. The outbound phase is exactly the same as the one used in the basic 9R attack. Any pair of values satisfying the differential path of the inbound phase, with a probability of 1, generates a pair of values that has a differential form of $(P(1), \mathbf{F})$ for both of the input and output states.

The attack complexity for the ideal case is also the same as before, which requires $2^{(n-c)/2}$. Therefore, (2) is also the condition that needs to be satisfied for our 11R attack to be applicable.

Since the condition is the same as the one for the basic 9R attack, the same parameters apply: $(N, c) = (128, 8)$, $(128, 4)$ and $(64, 4)$. See Table 1.

3.3 "Shrunken" 9R Attack: Case $(N, c) = (64, 8)$

Our attacks in previous sections cannot be applied to 64-bit block ciphers having 8-bit S-boxes. However, by reducing the number of outbound rounds of the 11R attack, we can attack up to 9 rounds of such ciphers. Namely, we let the differential path consist of

- 2-round backward outbound phase,
- 5-round inbound phase, and
- 2-round forward outbound phase.

Now the differences in plaintext and ciphertext are both $(1, P(1))$. The cost for finding a pair of values satisfying this differential form with a random permutation is equal to the one for finding a collision of $2(n - c)$ bits, which is $2^{2(n-c)/2} = 2^{n-c}$. Hence, the condition on the parameters becomes $r \cdot 2^{2c} < 2^{n-c}$, which is converted as

$$c < \frac{1}{3}(n - \log_2 r),$$

and thus, 64-bit block ciphers with 8-bit S-boxes can be attacked up to 9 rounds.

4   Application to MMO and Miyaguchi-Preneel Modes

We apply the known-key distinguishers to attacking the MMO and Miyaguchi-Preneel hashing modes using these Feistel ciphers. The attacks we present in this section can generate either full N-bit collisions or half n-bit collisions depending on the parameters.

We consider this aspect of our known-key attacks quite important, as it shows that the attacks endangers real-world security of these hash functions. Sometimes known-key attacks tend to become fairly complex. People may wonder what practical implications such results have, which is certainly not the case for our attacks.

Here we only describe our attacks on the MMO mode, but all the attacks can be trivially extended to the Miyaguchi-Preneel mode. This is because the key (randomly given constant) addition to the hash output state used by the Miyaguchi-Preneel mode does not make any impact upon the output value differences.

When we apply the distinguishers to collision attacks on the MMO mode, we iterate Step 2 of Table 2 in order to generate more paired values satisfying the inbound phase. We perform Step 3 of Table 2 and then check whether ciphertext differences cancel out plaintext differences. If they do, then we obtain a collision of the compression function (due to the feedforward used in the MMO mode).

We also present half-collision attacks. Generally, half-state collisions possibly become full collisions if more than half of the bits are truncated; systems in practice often truncate hash values to their desired length (*e.g.*, SHA-224 and SHA-384). Due to the nature of the Feistel network, in our half-collision attacks it is important which (left or right) half is chopped—our attacks work effectively on odd numbers (9 and 11) of rounds, whereas Feistel ciphers in practice usually have even number of rounds.

4.1   Eleven-Round Half-Collision Attack

We start with the parameters where the extended 11R attack applies. Recall that in our 11-round attack the differential forms of input and of output are both $(P(1), \mathbf{F})$. Therefore, a feed-forward operation (due to the MMO mode) makes the difference in the left half of the output state zero (cancellation) with a probability of 2^{-c}, which yields a half-state collision. Recall that our 11-round attack can generate up to x starting points with a complexity of $x \cdot r \cdot 2^{2c}$ and with $r \cdot 2^{2c}$ memory, where $x \leq 2^{2c}$. So by choosing $x := 2^c$, we can generate a half-state collision with a total complexity of $r \cdot 2^{3c}$. Note that it requires a complexity of $2^{N/4}$ to find a left-half-state collision of an N-bit ideal hash function due to the birthday attack. We can verify that for all the parameters in consideration the complexity $r \cdot 2^{3c}$ is faster than the birthday bound $2^{N/4}$.

4.2   Nine-Round Full-Collision Attack

Our attack can generate 9-round full collisions by using the outbound phase reduced to 2-rounds and using the 5-round inbound phase as described in Section 3.3 (but here we are treating the cases $(N, c) = (128, 8)$, $(128, 4)$ and $(64, 4)$, not the case $(N, c) = (64, 8)$). Now the differential forms of input and output are both $(1, (P(1)))$, and after a feed-forward operation, the xored values cancel each other with a probability of 2^{-2c}. Hence, setting $x := 2^{2c}$ allows us to generate a full collision. Of course to find a full collision of an N-bit ideal hash function should require a $2^{N/2}$ complexity, so our attack beats this birthday bound for all the parameters in consideration.

One may doubt that there is an adequate degree of freedom available. However, we can vary the degree of freedom as follows. In Step 0 of the attack procedure described in Section 3.2, we choose an active-byte position for 1. Because this can be chosen from r options, the degree of freedom becomes r times, and this makes the attack success probability almost 1.

4.3   Nine-Round Near-Collision Attack: Case $(N, c) = (64, 8)$

Now we discuss the remaining case of 64-bit blocks with 8-bit S-boxes. The shrunken 9-round attack can generate up to x starting points with a complexity of $x \cdot r \cdot 2^{2c}$ and with $r \cdot 2^{2c}$ memory, where $x \leq 2^{2c}$. The differential forms of input and output are both $\left(1, (P(1))\right)$. Hence, by choosing $x := 2^{2c}$ and by spending a running time of $r \cdot 2^{4c}$, a full collision could be generated. However, for this specific parameter, the cost of finding a collision for a random permutation is 2^{4c}, which is obviously faster than $r \cdot 2^{4c}$.

Therefore, instead of generating a full collision, we consider only cancelling the differences of the right half state. Such a cancellation occurs with a probability of 2^{-c}, and if this occurs, then we immediately obtain a collision of $(2r - 1) = 7$ bytes out of 8 bytes because the left half state only has the difference in one byte. Thus we can generate 7-byte collision with a complexity of $r \cdot 2^{3c}$, which is faster than the complexity for a random function, namely $2^{3.5c}$.

4.4   Seven-Round Full-Collision Attack: Case $(N, c) = (64, 8)$

If one wants to generate a full collision with the parameter $(N, c) = (64, 8)$, then one would need to use the previous 3R inbound phase and two sets of the 2R outbound phases. This can generate a full collision with a complexity of $r \cdot 2^{3c}$, which is faster than the birthday bound 2^{4c}.

5   Generality of Our Known-Key Distinguishers

So far our known-key distinguishers make the assumptions that a) S-boxes have balanced differential probabilities, b) the linear transformation has the branch number of $r + 1$, and c) the inequality $2c \geq r$ holds. In this section, we discuss possibilities of handling the situations where these conditions are not met.

5.1   When S-Boxes Are Biased

We have made the assumption that the S-boxes have the least maximum differential probabilities like the inverse function $x \mapsto x^{-1}$, so that we can estimate the matching probability roughly at $1/2$. This is not an essential requirement, and in fact S-boxes can be "suboptimal" in order for our attacks to work; one just needs more refined estimate of success probabilities in that case.

It is true that our attacks become infeasible when S-boxes have "very" biased differential distributions. However, block ciphers having such S-boxes are usually vulnerable to differential cryptanalysis, and the attacks should let us construct other types of distinguishers any way.

5.2   When P Is Not an MDS Matrix

The attacks described in Section 3 are based on the assumption that an MDS matrix is used in the P-layer. Again, this is not an absolute requirement. First note that the MDS property is used only in the inbound phase; it is completely irrelevant to the outbound phase. For example, in the inbound phase of the 9-round attack described in Section 3.1, we have used the fact that the branch number is $r + 1$ for satisfying the following three conditions:

1. Active-byte positions in $\Delta\#A$ and $\Delta\#B$ must be identical and should be minimized as much as possible,
2. Active-byte positions in $P(\Delta\#A)$ and $P^{-1}(\Delta\#B)$ must be identical, and
3. The degree of freedom for varying the differences in $\Delta\#A$ and in $\Delta\#B$ must be sufficient to find a match over the substitution (S-box) layer.

We have used the fact that the branch number is $r+1$ to simplify the description of our attacks and to minimize the attack complexity.

However, even if the branch number is smaller than $r + 1$, there is a fair chance that the attack would work. If P is not an MDS matrix, then we search for byte positions that satisfy the above three conditions. This can be done in the pre-computation stage, because properties of a linear transformation P can be analyzed independently of a key value (Recall that P is a constant matrix).

5.3 When $2c < r$

The attacks described in Section 3 do not work if $2c < r$, because the number of pairs is insufficient to find a match of differences over the S-box layer (with a probability about 1). This problem can be solved by increasing the number of active bytes in a word. The essence of this generalization is to activate $d \geq 1$ bytes in a word so that $2cd \geq r$. The minimum value of such a d is 1 for $(c, r) = (8, 8), (8, 4), (4, 8)$ and 2 for $(c, r) = (4, 16)$. Hence, by activating two bytes instead of using the differential form 1, we can attack, for example, Feistel ciphers having a 128-bit block size and 4-bit S-boxes.

With this generalization the total complexity becomes $r \cdot 2^{2cd}$ computations plus $r \cdot 2^{cd}$ memory for both 9-round and 11-round attacks. So for example, when $N = 128$, $c = 4$ and $d = 2$, these figures beat the birthday bound $2^{(n-cd)/2}$ for an ideal permutation.

Here we need to be careful about the fact that activating more than one byte may yield non-active bytes in the output of P or of P^{-1}. If that occurs, then most likely the match over S-boxes would fail (similarly to the case when P is not an MDS matrix). If we activate more than one byte, then such a case is inevitable even if P is an MDS matrix. However, the probability of output with non-active byte should not be so high, and we presume that a match can be found without a considerable increase in complexity even when we activate two bytes in a word.

6 Conclusion

We have presented new known-key distinguishers on block ciphers using the Feistel network whose round function consists of a key XOR, byte-oriented S-boxes, and an MDS matrix. We have considered most of the parameters used in practice and shown that "weak" parameters include 128-bit block with 8-bit S-box, where we can attack the cipher up to 11 rounds.

Our distinguishers can be extended to the MMO and Miyaguchi-Preneel hashing modes. With the weak parameters attackers can find full collisions up to 9 rounds and half-state collisions up to 11 rounds faster than the birthday bounds. To the best of our knowledge, as a generic attack on Feistel-SP ciphers, our attacks significantly increase the number of analyzable rounds from previous results.

References

1. Barreto, P.S.L.M., Rijmen, V.: The WHIRLPOOL hashing function. Submission to NESSIE (2003)
2. Biham, E., Shamir, A.: Differential cryptanalysis of DES-like cryptosystems. In: Menezes, A., Vanstone, S.A. (eds.) CRYPTO 1990. LNCS, vol. 537, pp. 2–21. Springer, Heidelberg (1991)
3. Biryukov, A., Khovratovich, D.: Related-key cryptanalysis of the full AES-192 and AES-256. In: Matsui, M. (ed.) ASIACRYPT 2009. LNCS, vol. 5912, pp. 1–18. Springer, Heidelberg (2009)
4. Biryukov, A., Khovratovich, D., Nikolić, I.: Distinguisher and related-key attack on the full AES-256. In: Halevi, S. (ed.) CRYPTO 2009. LNCS, vol. 5677, pp. 231–249. Springer, Heidelberg (2009)
5. Biryukov, A., Nikolić, I.: A New Security Analysis of AES-128. In: Halevi, S. (ed.) CRYPTO 2009. LNCS, vol. 5677. Springer, Heidelberg (2009), http://rump2009.cr.yp.to/
6. Biryukov, A., Nikolić, I.: Automatic search for related-key differential characteristics in byte-oriented block ciphers: Application to AES, camellia, khazad and others. In: Gilbert, H. (ed.) EUROCRYPT 2010. LNCS, vol. 6110, pp. 322–344. Springer, Heidelberg (2010)
7. Black, J.A., Rogaway, P., Shrimpton, T.: Black-Box Analysis of the Block-Cipher-Based Hash-Function Constructions from PGV. In: Yung, M. (ed.) CRYPTO 2002. LNCS, vol. 2442, pp. 320–335. Springer, Heidelberg (2002)
8. Bouillaguet, C., Dunkelman, O., Leurent, G., Fouque, P.A.: Attacks on hash functions based on generalized Feistel—application to reduced-round Lesamnta and SHAvite-3.512. In: Biryukov, A., Gong, G., Stinson, D. (eds.) SAC 2010. LNCS, vol. 6544, pp. 18–35. Springer, Heidelberg (2011)
9. Daemen, J., Rijmen, V.: AES Proposal: Rijndael. Submission to NIST (1998)
10. Dillon, J.F.: APN polynomials: An update. In: Fq9 Conference (2009)
11. Gauravaram, P., Leurent, G., Mendel, F., Naya-Plasencia, M., Peyrin, T., Rechberger, C., Schläffer, M.: Cryptanalysis of the 10-round hash and full compression function of SHAvite-3-512. In: Bernstein, D.J., Lange, T. (eds.) AFRICACRYPT 2010. LNCS, vol. 6055, pp. 419–436. Springer, Heidelberg (2010)
12. Gilbert, H., Peyrin, T.: Super-sbox cryptanalysis: Improved attacks for AES-like permutations. In: Hong, S., Iwata, T. (eds.) FSE 2010. LNCS, vol. 6147, pp. 365–383. Springer, Heidelberg (2010)
13. Knudsen, L.R., Rijmen, V.: Known-key distinguishers for some block ciphers. In: Kurosawa, K. (ed.) ASIACRYPT 2007. LNCS, vol. 4833, pp. 315–324. Springer, Heidelberg (2007)
14. Lamberger, M., Mendel, F., Rechberger, C., Rijmen, V., Schläffer, M.: Rebound distinguishers: Results on the full whirlpool compression function. In: Matsui, M. (ed.) ASIACRYPT 2009. LNCS, vol. 5912, pp. 126–143. Springer, Heidelberg (2009)
15. Lu, J., Dunkelman, O., Keller, N., Kim, J.-S.: New impossible differential attacks on AES. In: Chowdhury, D.R., Rijmen, V., Das, A. (eds.) INDOCRYPT 2008. LNCS, vol. 5365, pp. 279–293. Springer, Heidelberg (2008)

16. Matsui, M., Yamagishi, A.: A new method for known plaintext attack of FEAL cipher. In: Rueppel, R.A. (ed.) EUROCRYPT 1992. LNCS, vol. 658, pp. 81–91. Springer, Heidelberg (1993)
17. Mendel, F., Peyrin, T., Rechberger, C., Schläffer, M.: Improved cryptanalysis of the reduced Grøstl compression function, ECHO permutation and AES block cipher. In: Jacobson Jr., M.J., Rijmen, V., Safavi-Naini, R. (eds.) SAC 2009. LNCS, vol. 5867, pp. 16–35. Springer, Heidelberg (2009)
18. Mendel, F., Rechberger, C., Schläffer, M., Thomsen, S.S.: The rebound attack: Cryptanalysis of reduced whirlpool and grøstl. In: Dunkelman, O. (ed.) FSE 2009. LNCS, vol. 5665, pp. 260–276. Springer, Heidelberg (2009)
19. Minier, M., Phan, R.C.-W., Pousse, B.: Distinguishers for ciphers and known key attack against rijndael with large blocks size. In: Preneel, B. (ed.) AFRICACRYPT 2009. LNCS, vol. 5580, pp. 60–76. Springer, Heidelberg (2009)
20. Minier, M., Naya-Plasencia, M., Peyrin, T.: Analysis of reduced-SHAvite-3-256 v2. In: Joux, A. (ed.) FSE 2011 Preproceedings. LNCS, vol. 6733, pp. 68–87 (2011)
21. Nikolić, I., Pieprzyk, J., Sokołowski, P., Steinfeld, R.: Known and chosen key differential distinguishers for block ciphers. In: Rhee, K., Nyang, D. (eds.) Preproceedings of ICISC 2010, 1A-3 (2010)
22. Nyberg, K.: Differentially uniform mappings for cryptography. In: Helleseth, T. (ed.) EUROCRYPT 1993. LNCS, vol. 765, pp. 55–64. Springer, Heidelberg (1994)
23. Preneel, B., Govaerts, R., Vandewalle, J.: Hash functions based on block ciphers: A synthetic approach. In: Stinson, D.R. (ed.) CRYPTO 1993. LNCS, vol. 773, pp. 368–378. Springer, Heidelberg (1994)
24. Sasaki, Y.: Known-key attacks on Rijndael with large blocks and strengthening ShiftRow parameter. In: Echizen, I., Kunihiro, N., Sasaki, R. (eds.) IWSEC 2010. LNCS, vol. 6434, pp. 301–315. Springer, Heidelberg (2010)
25. Singleton, R.C.: Maximum distance q-nary codes. IEEE Trans. Inf. Theory 10, 116–118 (1964)
26. U.S. Department of Commerce, National Institute of Standards and Technology: Specification for the ADVANCED ENCRYPTION STANDARD (AES) (Federal Information Processing Standards Publication 197) (2001)

Author Index